LATE CENOZOIC HISTORY
OF THE PACIFIC NORTHWEST

CLARKIA FOSSIL SITE, LOCALITY P-33
Northern Idaho
Photographs courtesy of Maynard M. Miller

LATE CENOZOIC HISTORY OF THE PACIFIC NORTHWEST
Interdisciplinary Studies
on the Clarkia Fossil Beds of Northern Idaho

Editor
Charles J. Smiley
University of Idaho

Executive Editor
Alan E. Leviton
Pacific Division, AAAS and
California Academy of Sciences

Associate Editor
Margaret Berson
Pacific Division, AAAS and
California Academy of Sciences

San Francisco, California
1985

Library of Congress Catalog Card Number 85-72591
ISBN 0-934394-06-7

This volume has been typeset in *Press Roman* type on an IBM MagCard Composer in the Department of Herpetology, California Academy of Sciences.

Copyright © 1985 by the Pacific Division of the
American Association for the Advancement of Science
c/o California Academy of Sciences
Golden Gate Park
San Francisco, California 94118

Manufactured in the United States of America by the Allen Press, Lawrence, Kansas 66044

TABLE OF CONTENTS

	Page
DEDICATION	7

PREFACE
    Charles J. Smiley   9

Physical Setting of the Miocene Clarkia Fossil Beds, Northern Idaho
    Charles J. Smiley and W. C. Rember   11

Diatom Flora of the Miocene Lake Beds near Clarkia in Northern Idaho
    J. Platt Bradbury, K. V. Dieterich, and John L. Williams   33

Miocene Chrysophyta Cysts from a Lacustrine Deposit in Northern Idaho
    John L. Williams   61

Thecamoebian Scales from a Miocene Lacustrine Deposit in Northern Idaho
    John L. Williams   67

Miocene Freshwater Mollusks from the Clarkia Fossil Site, Idaho
    Dwight W. Taylor   73

Taxonomy of Fishes from Miocene Clarkia Lake Beds, Idaho
    G. R. Smith and R. R. Miller   75

Environmental Interpretation of Burial and Preservation of Clarkia Fishes
    G. R. Smith and R. L. Elder   85

Composition of the Miocene Clarkia Flora
    Charles J. Smiley and William C. Rember   95

Cuticular Analyses and Preliminary Comparisons of some Miocene Conifers from Clarkia, Idaho
    L. Maurice Huggins   113

Miocene Epiphyllous Fungi from Northern Idaho
    John L. Williams   139

Ultrastructural States of Preservation in Clarkia Angiosperm Leaf Tissues: Implications on Modes of Fossilization
    Karl J. Niklas, R. Malcolm Brown, Jr., and Richard Santos   143

The Paleobiochemistry of Fossil Angiosperm Floras. Part I. Chemosystematic Aspects
    David E. Giannasi and Karl J. Niklas   161

The Paleobiochemistry of Fossil Angiosperm Floras. Part II. Diagenesis of Organic Compounds with Particular Reference to Steroids
    Karl J. Niklas and David E. Giannasi   175

Interpretation of Co-occurring Megafossils and Pollen: A Comparative Study with Clarkia as an Example
    Jane Gray   185

Miocene Insects from the Clarkia Deposits of Northern Idaho
 Standley E. Lewis 245

Evolution of Freshwater Drainages and Molluscs in Western North America
 Dwight W. Taylor 265

Pollen Profiles of the Plio-Pleistocene Transition in the Snake River Plain, Idaho
 Estella B. Leopold and V. Crane Wright 323

Spicular Remains of Freshwater Sponges from a Miocene Lacustrine Deposit in Northern Idaho
 John L. Williams 349

Recent Climatic Variations, Their Causes and Neogene Perspectives
 Maynard M. Miller 357

Overview
 Charles J. Smiley 415

Respectfully Dedicated to the Memory of

E. Viet Howard

Executive Vice President and Chief Operating Officer
Sunshine Mining Company

for his encouragement and support

of the Study of the Clarkia Flora and of this Publication

PREFACE

The ideal study of an ecosystem represents a total-system evaluation of its component biological and physical interrelationships. In practice, as in the case of Miocene Clarkia Lake, such interdisciplinary research is restricted to a local area whose biota is bounded and governed by an arbitrarily designated set of physical parameters. The ecological study of modern organisms not only encompasses the interplay of observable physical and biological factors, but also must rely upon inferential foundations derived from historical perspectives based on a combination of known geological and paleontological data. For ecology, the "state of the art" remains far from precise, and therefore the field of paleoecology will continue to be a viable subject of intellectual pursuit that constantly changes in light of new factual data and inferential perspectives.

The Clarkia fossil area of northern Idaho is an exceptional window into the past. Seldom can we observe early Neogene evidence of (1) original physical setting, in combination with (2) essentially unaltered biological remains, in a condition that has (3) survived basically unchanged for millions of years to the present. The local area has remained topographically stable through the last few stages of the Cenozoic, although profound geologic changes occurred elsewhere. World climates differ today from those prevailing in earlier Cenozoic time. The Rocky Mountains in the middle of the North American continent, and the Sierra and Cascade ranges along its western margin, are now of considerably higher elevation than during the early Neogene. Later Neogene uplift of these north-south ranges has markedly affected the ecological setting of the western American interior, as they continued to develop into progressively more effective climatic barriers and more striking orographic influences on the biotic continuum.

Despite the apparent local geologic stability of northern Idaho, subsequent regional (orogenic) and global (climatic) changes in local physical conditions markedly influenced the biotic composition in the immediate vicinity of the Clarkia sedimentary basin during the early Neogene. This can readily be seen from the Miocene associations of animals and plants preserved in the Clarkia deposits, when compared to the quite different associations existing in the same area today. Modern societies of organisms comparable to those of the early Neogene in northern Idaho can now be found only in regions of the world where the climatic regime is markedly warmer and more humid than the present climate of the Clarkia area.

The Miocene Clarkia Lake represented an ecosystem conducive not only to the existence of aerobic organisms in the oxygenated surface waters of the lake, but also to the preservation of unaltered, or otherwise relatively undisturbed, organic remains in the anaerobic and toxic conditions of the lake bottom. Immediately overlooking the open water of the lake were the slopes of the surrounding hilly terrane, providing ready influx of allochthonous remains from a variety of terrestrial organisms. Such a combination of ecological, depositional and preservational factors has resulted in the uniqueness of the Clarkia fossil area for paleontological and paleoecological research: (1) the perpetuation of the early Neogene topography to the present day; and (2) the high taxonomic diversity of essentially unaltered and abundant fossil material of Miocene age.

Perhaps the most striking effect of this unaltered preservation is the vivid coloration of leaves (greens, reds, browns) when first exposed, occurring as abundant specimens in soft, finely laminated, lacustrine clays. It is equally striking to watch the leaf (and occasionally insect) colors rapidly vanish as the fossil begins to blacken immediately following exposure. Where chemical analyses have been made on leaves of specific Clarkia taxa, it can be demonstrated that some taxa have retained their own unique set of organic chemical constituents, including pigmentation components such as chlorophylls, xanthophylls, and carotinids. The cuticles of many Clarkia leaf taxa, commonly overlying crushed mesophyll tissue, show clear evidence of cellular features of the epidermis. Leaf compressions, furthermore, may show the intact morphology of epiphytic fungae, and leaf phytoliths have been observed in some of the compression fossils. Unaltered preservation is also exemplified in exceptionally well-preserved specimens, which display a variety of intracellular organelles under TEM examination, and some of the organelles appear to have been infested by Miocene bacteria. The potential for study of such unaltered remains from the early Neogene seems almost unlimited.

The Clarkia studies, begun in 1972 upon first discovery of the premier site P-33 by Francis Kienbaum of Clarkia, expanded during the early stage of collaborative investigations. By 1979 the studies had progressed enough that a symposium was organized for the Annual Meeting of the Pacific Division, American Association for the Advancement of Science, at the University of Idaho. This Clarkia symposium, entitled "Late Cenozoic History of the Pacific Northwest," is the central theme of the present volume.

Our purpose here is to provide, in a single volume, an insight into the extent of interdisciplinary research that is potential and necessary for a thorough understanding of this remarkable fossil area. Investigations to date involve its geologic setting, including the geomorphic evidence of the Miocene lacustrine basin and of the hilly terrane surrounding it; the paleontological evidence for inferences on paleoclimatic and other paleoecologic parameters that served as delimiting factors for the total Clarkia biotic association of animals and plants; and the sedimentologic/taphonomic evidence of the conditions of environment and preservation within the lacustrine basin of deposition. This has involved specific investigations on morphology/taxonomy/paleoecology both of (1) autochthonous remains representing those organisms that inhabited the lacustrine environment (sponges; mollusks; aquatic larvae of insects; fishes; diatoms and other limnitic microphytons), and (2) allochthonous remains derived from organisms that inhabited the adjacent land (foliage, reproductive structures and palynomorphs mainly of woody plants; fungal and other epiphytes preserved on leaves and fruits; forest insects; and insect mining and galling on leaves).

This compendium of research papers, focusing on the Clarkia fossil area, reflects the dedication of several individuals and organizations willing to participate in and to support an integrated total-system approach to complex scientific investigations. They range from the original discoverer, Francis Kienbaum; to the various scientists who were contacted for research in their different areas of expertise; to the Idaho Research Foundation for initial grant support during the developmental stages of the Clarkia research; and to the Pacific Division of AAAS in its recognition of the Clarkia study as a new discovery area of potentially high scientific importance. A significant source of encouragement has been a grant from Sunshine Mining Company of Kellogg, Idaho, supporting the publication of this symposium volume. Without the participation and efforts of all parties, the interest of the Pacific Division, AAAS, in publishing the book, and the financial assistance of Sunshine Mining Company to help defray publishing costs, this volume of papers reflecting the interdisciplinary potential of the Clarkia research would not have been possible. Collectively and separately, we are most appreciative of the support of all, and to all this volume is dedicated.

Charles J. Smiley
University of Idaho
Moscow, Idaho

May 1, 1984

PHYSICAL SETTING OF THE MIOCENE CLARKIA FOSSIL BEDS, NORTHERN IDAHO

CHARLES J. SMILEY AND W. C. REMBER
Department of Geology, University of Idaho, Moscow, ID 83843

The early Neogene fossil beds near Clarkia, Idaho were deposited as valley fill along the present drainage system of the St. Maries River and the lower courses of its major tributaries. This valley, occurring in an area of rugged hilly topography, was dammed probably by lava flows more than 30 km downstream from the present townsite of Clarkia. The result was an open body of water surrounded by densely forested hills. Anaerobic and toxic conditions developed in the lake bottom, facilitating the preservation of unaltered or undisturbed plant and animal remains from both limnic and terrestrial sources. The dominant fossils are plants representing taxonomic associations most correlative to modern forests of eastern Asia and of southeastern United States. This "exotic" character of the Clarkia flora, supplemented by modern distributions of relatives of Clarkia fish and insects, suggests that climates over this interior region of the Pacific Northwest were warm and humid, in contrast to the continental extremes of the present. Later Neogene orogeny in the Cascade Range, coupled with deteriorating world climates, has resulted in a more temperate, summer-drought regime that ultimately eliminated from the regional biota those taxa requiring warm and humid environments.

The St. Maries River in northern Idaho flows northwestward over a distance of about 50 km, from near the townsite of Clarkia to St. Maries, where it joins the St. Joe River just before entering Coeur d'Alene Lake. Coeur d'Alene Lake drains via the Spokane River to the Columbia River and ultimately the Pacific Ocean. The St. Maries River thus is now part of the Columbia River drainage system of the Pacific Northwest, and it seems to have been so connected since early Neogene time.

The bedrock in the St. Maries River area is Precambrian schist and quartzite that has been mapped as components of the Wallace Group of the Belt Supergroup (Hietanen 1963; Clark 1963). The mica schists locally contain star garnets of gem quality, and a thin surface mantle of garnet

Figure 1. Clarkia Valley of the St. Maries River in northern Idaho, from a composite of three photographs, showing Clarkia type-locality P-33 on the left and Clarkia townsite in the distance. Except for a thin mantle of Quaternary material, the flat valley represents the present surface of early Neogene valley fill. Surrounding hills represent the relict topography on Precambrian basement rocks, except for low hills immediately behind Clarkia, which represent a post-depositional local accumulation of Miocene basalt flows and near-vent pyroclastics.

Copyright ©1985, Pacific Division, AAAS.

Figure 2. Idealized sketch of the early Neogene Clarkia basin and surrounding topography, postulated from geomorphic and sedimentologic evidence. The designation "Miocene Clarkia Lake" would represent an extension down-valley of lacustrine conditions noted at locality P-33, assuming that damming occurred downstream from Mashburn Junction. "Embayments" on the southwest are named mainly from streams that still occupy the drainage channels: A=Clarkia; B=Emerald Creek; C=Carpenter Creek; D=Tyson Creek; E=Sheep Creek; F=Santa Creek. Locations of the five Clarkia fossil sites are indicated by appropriate numbers. (After Smiley and Rember 1979).

sands and gravels now covers the Miocene sediments in some of the valleys. Granites of the Idaho Batholith complex are exposed locally in the area, and one of these granite bodies crops out in the Hoodoo Mountains, where it was intruded into the Precambrian rocks during Cretaceous time. This topographic high now serves as the southwest watershed of the St. Maries River, and it probably was an important local topographic feature throughout the Cenozoic.

The upper 37 km of the river has a broad, in some places flat, valley that courses through the hilly topography of the western foothills of the Rocky Mountains (Figs. 1 and 2). Elevations of the valley floor decrease from 840 m near Clarkia to 790 m near Mashburn Junction, over a distance of 35 km. The northeast side of the valley follows a more or less straight line that trends northwesterly and probably is structurally controlled in conformance with other regional lineations. The land northeast of the valley is irregularly dissected with peak elevations of 1500-1650 m within a distance of 3-5 km. On the southwest the land rises within 6-8 km to a paralleling ridge (the Hoodoo Mountains) with peak elevations of 1400-1500 m. Several tributary streams flow down the northeast slope of this ridge to join the St. Maries River at regularly spaced intervals (see Fig. 2). The lower 2-3 km of each major tributary valley also are broad and flat and topographically represent lateral extensions of the main St. Maries River valley (Smiley and Rember 1979).

A break in slope between the flat valley floors and bordering hillsides is distinct in many places; this topographic feature has been plotted within the 2800-2960 ft (878-902 m) contour lines on available maps (Fig. 1). The area designated "Miocene Clarkia Lake" in Fig. 2 is considered to approximate the maximum extent of the Miocene lake, with the location of a dam downstream from Mashburn Junction. Outside the area of deposition, the patterns of drainages and intervening ridges reflect the present erosional surface on the metamorphic terrane of the region. Conformance between present stream patterns and Miocene valley deposition shows that the major patterns of early Neogene drainage and topography have changed little in the later Neogene history of the area.

CLARKIA FOSSIL LOCALITIES

Five fossil localities presently are known within the postulated boundaries of Miocene Clarkia Lake (Fig. 2). A second Miocene basin (the Oviatt Creek area) occurs 32 km due south of, and at the same elevation as, the Clarkia basin. These fossil sites are (UIMM=University of Idaho Mines Museum):

    UIMM P-33 (Clarkia basin): Sec. 12, T42N, R1E, Shoshone Co., Idaho
    UIMM P-34 (Clarkia basin): Sec. 32, T43N, R2E, Shoshone Co., Idaho
    UIMM P-37 (Clarkia basin): Sec. 33, T43N, R1E, Benewah Co., Idaho
    UIMM P-38 (Clarkia basin): Sec. 4, T42N, R2E, Shoshone Co., Idaho
    UIMM P-40 (Clarkia basin): Sec. 6, T42N, R1E, Latah Co., Idaho
    UIMM P-35 (Oviatt basin):   Sec. 12, T39N, R1E, Clearwater Co., Idaho

Localities P-33, P-34, and P-38 occur in the Clarkia valley, and their locations suggest that the sediments and terrestrial fossils are representative of different source areas. P-33 occurs at the mouth of a small tributary valley on the west side of the basin; this appears to have been a small bay of the Miocene lake, occupying the lower course of a valley now occupied by Bechtel Creek. Localities P-34 and P-38 probably received sediments and terrestrial organic remains from the land to the northeast of the lake. Localities P-37 and P-40 occur in the valley of Emerald Creek, where they are separated from the P-33 site by a ridge of metamorphic rocks now represented by Bechtel Butte, Cedar Butte and Clarkia Peak (Fig. 2). Miocene drainage in the Emerald Creek area appears to have been from the north-facing slope of this ridge and from the east-facing slope of the Hoodoo Mountains. The Oviatt Creek basin to the south is separated from the Clarkia basin by a ridge developed on the metamorphic basement of the area; it probably represented a different drainage in the Miocene as at present and probably connected with the Clearwater drainage system to the southwest. Because the Hoodoo Mountains do not extend as far south as the Oviatt Creek basin, no comparable upland existed just west of the Oviatt Creek site.

Locality P-33. This represents the initial discovery site in the Clarkia basin, and it is the most intensively studied to date. Locality P-33 has provided most of the fossil and sedimentologic data for this volume. The exposure occurs along the south flank of a small forested knoll that now forms the north end of a racetrack owned by Mr. Francis Kienbaum of Clarkia. The sedimentary section is exposed laterally for a distance of about 100 m and is known to be continuous for an additional 100 m toward the east. The section includes about 9 m of soft, highly fossiliferous, lacustrine clays with interbeds of barren volcanic ash (see Figs. 3, 4, 5). Below the exposed lacustrine section is a sequence of sands and clays probably representing valley alluvium (Unit 1, Fig. 4).

Figure 3. Photograph of a portion of the P-33 type-locality, showing a section approximately 9 m thick that is well exposed over a distance of about 100 m and is known to extend laterally for about another 100 m. Unit numbers at left correspond with those on Fig. 4.

Locality P-34. This fossil site is 1.3 km north of Clarkia townsite, where it appears to represent the drowned mouth of a small stream valley on the east side of Clarkia Lake. The locality is a roadcut in the bottom of a gully, where later Neogene erosion has cut through an overlying pile of basalt flows into the underlying Clarkia deposits. The exposed section is about 12 m thick and contains laminated silty clays with interbeds and lenses of sand lithologically resembling Unit 1 at locality P-33. The sediments here have been blackened in proximity to a dike that seems to have been the feeder dike to the overlying volcanic pile. Plants are the only fossils presently known from this site, but they are neither abundant, taxonomically diverse, nor well preserved. The small florule is represented by plants that occupied riparian and lower slope habitats.

Locality P-38. This locality, 0.6 km north of Clarkia townsite, is exposed in a bank of the forest access road leading to locality P-34. The site is separated from P-34 by an east-west ridge of metamorphic rocks, which was later capped by the local basalt pile. The exposed section underlies Miocene basalt of Priest Rapids type and contains about 4 m of fairly pure glass-shard sands and sandy clays. Finely fragmented wood chips and other disseminated plant debris, sometimes including isolated cones, seeds and fruits, occur in some beds near the base of the sandy unit. A cross-bed about 2 cm thick was noted in one of the sand layers. Non-fragmented leaves occur only on bedding surfaces, and they are commonly "stacked" one on top of another. The few plant taxa represented in this small florule are preserved as imprints, which are commonly iron-stained, and they are most indicative of riparian and lower slope habitats.

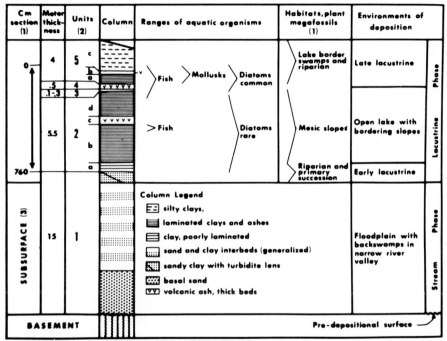

Figure 4. Clarkia P-33 section, showing stratigraphic occurrences of various fossils. Modified from Smiley et al. 1975 and Smiley and Rember 1979.

Locality P-37. This is one of two fossil localities in the Emerald Creek valley (Fig. 2). It is a roadcut exposure along the lower north flank of a small knoll of the Precambrian basement rocks that rises above the surrounding flat bottomland; this knoll is presumed to have been an "island" in the Emerald Creek area of the Clarkia basin. The exposed section, about 7 m thick, is composed of bedded silty clays and ashes lacking the fine laminations present in the P-33 section. Diatoms are present in some layers, and a single fish specimen represents one of the species common in the lacustrine section at P-33. Plant megafossils occur in abundance and taxonomic diversity. Fossil leaves are rarely abraded or fragmented, with no evidence of leaf "stacking." The coarser sediments and thick bedding are indicative of higher energy deposition than occurred at locality P-33. The combined evidence from sediments and fossils suggests a nearshore environment of deposition for the exposed P-37 section.

Locality P-40. This fossil site is an exposure along a national forest road near the postulated southern margin of the Emerald Creek embayment. A section about 5 m thick extends laterally for about 30 m. Sediments and plant fossils resemble those at locality P-37. At one end of the exposure the beds are fractured and displaced as much as 20 cm, and beds are inclined as much as 17°. Overlying these Clarkia deposits with an angular unconformity is a horizontal unit of poorly bedded sandy clays with a basal conglomerate 30 cm thick; no fossils have been found in this upper unit. The younger deposits exposed here represent a post-Clarkia stage of valley deposition, evidence of which can be seen in other marginal areas of the basin. The location of site P-40 suggests that the plant remains comprising the florule were derived from local bottomlands and from the lower north-facing slope of the ridge that separated it from the P-33 site.

Locality P-35. This fossil locality occurs in the Oviatt Creek basin 32 km south of the Clarkia basin. A measured section 5.3 m thick is composed of poorly bedded silty clays and discrete ash layers, with coarser-textured sediments in the upper part. A thin carbonaceous layer about 10 cm thick near the top of the section appears to represent a bog environment in the later stages of deposition. The exposure is capped by a basalt flow of Priest Rapids type, whose base is pillowed down into the uppermost part of the underlying sediments and in places has distorted the bog layer. The Oviatt Creek flora appears to be a typical "Latah" type, and many characteristic taxa of the Clarkia flora are not represented. Preliminary sampling at two other sites in the Oviatt Creek basin shows this to be a consistent difference between the Oviatt Creek and Clarkia floras, and suggests a younger age for the deposits in the Oviatt Creek basin.

Preliminary sampling at a new fossil plant locality 5.5 km downriver from Mashburn Junction, occurring outside the boundaries of the Clarkia basin, provides additional data bearing on the taxonomic and age differentials of the Oviatt Creek and Clarkia floras (see "Area Covered by Plateau Basalts" on Fig. 2). This locality (P-44) is an exposure along a logging railroad track, at the

Figure 5. Photograph showing the edge of sedimentary laminae from Unit 2 at locality P-33. Some of the white laminae represent volcanic ash; other laminae are more typical lacustrine clays. Each clay and ash lamina, regardless of thickness, is texturally graded (fining upward).

bottom of a narrow gorge where the St. Maries River has cut through a sequence of Miocene basalt flows. It is located near a site called Rover Siding on topographic maps. A section of sandy clays 6-7 m thick is "sandwiched" between flows near the base of the local basalt pile. Overlying basalts are of the Priest Rapids type and underlying basalts are of the Dodge type (Swanson, pers. commun.). The sedimentary interbed and contained flora thus seem correlative with the Middle Miocene Wanapum Formation of the Columbia Plateau basalt sequence. Plants collected from the Rover Siding interbed represent species characteristic of the Oviatt Creek rather than the Clarkia flora. Such correlation of a fossiliferous interbed by basalt stratigraphy seems to establish a Middle Miocene age for the Oviatt Creek type of vegetation in northern Idaho, and a minimum age for the older and floristically distinguishable Clarkia flora.

PRESERVATION OF CLARKIA FOSSILS

Plant Fossils

The soft, water-saturated and unoxidized Clarkia clays contain mostly plant megafossils, many of which have remained essentially unaltered since burial. They occur commonly as black compressions that can be removed intact and bleached to a translucent amber color. During sampling at the P-33 locality, it is common to uncover leaves that have retained what appear to be original green, brown or red colors; shortly after exposure, however, these colors disappear as the compressions turn black. Leaves are separated from one another on bedding surfaces, and they are not "stacked" one on top of another except at locality P-38 and at the top of the P-33 section. The megafossils show little or no evidence of pre-depositional deterioration either by abrasion from transport to the burial site or by biological decomposition. Teratological damage commonly is represented by insect galling or mining or by evidence of epiphytic fungi in leaf compressions.

A considerable diversity of plant organs is preserved in the Clarkia flora (see figured specimens throughout this volume). The great abundance of locally derived structures ensures that some of them will be preserved in more or less intact condition (see also Spicer 1981). For example, the Clarkia flora contains compound dicot leaves with leaflets attached (e.g. Juglandaceae, Leguminosae); terminal branches may have clusters of leaves attached (e.g. *Quercus*); leaves and reproductive structures may be attached to the same specimen (e.g. *Pseudofagus, Cercidiphyllum*); cones may be preserved with attached seeds (e.g. *Liriodendron, Magnolia*); most of the identified spermatophyte taxa are represented by leaves, and approximately half of them are represented by reproductive structures (cones, seeds, nuts, pods, calices, inflorescences, pollen).

In addition to the examples of dicots, Clarkia conifers are well represented by foliage and associated reproductive organs. For example, *Cunninghamia* and *Glyptostrobus* are preserved as shoots with cones attached; *Metasequoia* is preserved as distinctive shoots and seed-bearing cones; *Pinus* is preserved as fascicled needles sometimes attached to branches, and as pistillate and staminate cones, seeds and pollen; *Abies* is preserved as needled shoots, cone scales, seeds and pollen; *Chamaecyparis* and *Taxodium* are represented by distinctive shoots, cones and cone scales; *Amentotaxus* is preserved as compressed isolated needles showing the cellular anatomy of the epidermis.

Organ preservation of plant megafossils of less common occurrence, although some of them are taxonomically diversified, includes fronds of ferns, intact shoots of mosses, stems of horsetails with nodal leaves, grass stems with attached leaf blades, and clustered linear leaves of quillworts.

Preservation of Clarkia plant megafossils permits the removal from the soft clay matrix of intact leaves, seeds, cones and other structures for preservation between plates of glass or in bioplastic mounts. The total venation of leaves is naturally stained a dark brown color that persists through the bleaching process of the leaf compressions. Preserved also are the microscopic features of the leaf epidermis, including epidermal cells, stomatal complexes, trichomes and trichome bases, and intracellular and extracellular phytoliths (Smiley et al. 1975; Smiley and Huggins 1981; Huggins 1980, also this volume). For some leaf compressions, intracellular organelles are known to have been preserved (Niklas and Brown 1980; Niklas, Brown, and Santos, this volume). Additionally, some of the leaf compressions are known to have retained relatively undisturbed organic chemical constitutents (Giannasi and Niklas 1981, this volume). Compressed leaves and fruits commonly are preserved with intact fungal epiphytes (see Williams, this volume), and fungal spores and other microstructures are preserved in considerable taxonomic diversity (see Sherwood, *in* Gray, this volume). Insect-plant relationships similar to those of the present commonly are preserved as insect mining and galling on Clarkia foliage (see Lewis, this volume).

Another factor of preservation, in addition to quality of fossils and the diversity of taxa and organs, is the quantity of fossils in the Clarkia deposits. In many instances, Tertiary species of the Pacific Northwest are known only from a few poorly preserved, fragmental or abraded specimens that provide little evidence of morphological variation of foliar or reproductive organs within the species. For many of these taxa, the Clarkia record provides sufficiently large quantities of intact specimens to show such variations. A case in point is *Liriodendron hesperi* Berry, the Miocene tulip tree (Magnoliaceae), whose presence in other Tertiary floras of the Pacific Northwest is known only from rare isolated seeds. In the Clarkia flora, this species is represented by isolated seeds, by compressed cones with seeds attached, and by large numbers of leaves at certain levels. A study of this Tertiary species, involving seeds and cones, microscopic features of the leaf epidermis, and morphological variations of more than 100 well-preserved fossil leaves (supplemented by organic chemical studies by Giannasi and Niklas), has been completed at the University of Idaho by Nina Baghai (1983).

The abundance of plant megafossils available for study in the Clarkia flora (Fig. 6) can be illustrated by Rember's stratigraphic analysis through a 7.6-m high column of the P-33 exposure, which represents a fair estimate of the quantity of megafossils that can be found across the entire 100-m length of the exposure (see Smiley and Rember 1981). The column measured 30 x 45 cm across, which represented the surface area of individual beds (=1.5 sq ft). Megafossils were uncovered, identified and counted for each 0.5-1.0 cm interval, and were tabulated for each 30-cm vertical increment of the study column. The number of plant megafossils in 30-cm column increments (=1.5 cu ft) ranged from 138 to 1294, for a column total of 10,574. An analysis of blocks of similar size from the P-37 and P-40 localities showed the preservation of comparable numbers of plant megafossils in the Emerald Creek area of the Clarkia basin. For plant microfossils, a single 1cc of Clarkia clay contains innumerable quantities of palynomorphs and other structures of considerable taxonomic diversity (Gray, pers. commun., 1982).

Animal Fossils

Clarkia fossil fish are common at two levels of the P-33 locality (see Fig. 4), where they represent species of the families Cyprinidae and Centrarchidae (Smith and Miller, this volume). A single specimen of one of these species has been found also at the P-37 site in the Emerald Creek embayment. The family Salmonidae is represented by a single articulated specimen measuring 72 cm long at the P-33 locality. Most of the fish specimens are preserved as compacted and articulated skeletons on bedding surfaces, with a carbon film outlining the body and in some cases displaying the scale patterns. In most cases the mouths are "frozen" open at the time of burial as though the fish died of anoxia (Smith and Elder, this volume). Eyes are preserved as red-colored structures on the black carbon film of the head. Larger vertebral bones may be partially replaced by iron sulfides (crystalline pyrite or marcasite). Isolated coprolites, apparently of fish, are composed almost entirely of insect fragments.

Preservation of a variety of invertebrate fossils includes the remains of insects, mollusks, isolated spicules of freshwater sponges, and microscopic cysts of amoeboid organisms. Insects are

Figure 6. A bedding surface (about 30 x 45 cm) from near the base of Unit 5, locality P-33, showing physical separation and taxonomic diversity of leaves. Photograph is one-third natural size. A=*Alnus*; C=*Cocculus*; Cd=*Cercidiphyllum*; L=*Liquidambar*; Le=legume leaflet; Li=*Liriodendron*; M=*Metasequoia*; N=*Nyssa*; P=*Pseudofagus*; Q=*Quercus*; S=*Salix*; T=*Taxodium*.

the most common and taxonomically diverse megafossils, and are preserved as (1) impressions or compressions of the entire body, (2) beetle elytra that may show original bright colorations when first uncovered, (3) isolated wings or exoskeleton fragments, and (4) trace fossils such as mining or galling of leaves, leaf distortions from insect damage during ontogeny, and isolated cases of caddis fly larvae. Mollusks are rare and are found only in the uppermost part of the P-33 section (Fig. 4); these animals representing the limnic benthos are preserved as casts of a single pelecypod specimen and of two gastropod species, and by isolated opercula of one of the gastropods. Siliceous sponge spicules of varied forms are common microfossils in both the Oviatt Creek and Clarkia deposits, and several genera appear to be represented.

Compaction and Shrinkage Factors

The Clarkia sediments and contained fossils have been compacted to some degree following original deposition. Where fossils occur in coarser-textured sediments or within thicker layers that represent single episodes of rapid deposition, compaction ratios are not greater than about 2:1, based on the degree of flattening of stems, cones, nuts, and pollen grains. Where fossils were deposited on bedding surfaces, compaction ratios of 7:1 have been calculated.

Shrinkage factors that result from laboratory drying of the water-saturated clays and contained fossils are a potentially important consideration in determining original sizes of fossil specimens from measurements of dried laboratory material. For the lacustrine clays at locality P-33 (Unit 2), shrinkage resulting from dehydration has been calculated to range from 9-15% and to average about 10%. This was determined by measuring fossils *in situ* and by marking 10-cm lines on freshly exposed wet clays, then remeasuring the lines and fossils following laboratory drying that was slow enough to prevent noticeable dehydration cracks. We found that the shrinkage of fossils preserved as impressions coincides with that of the matrix, with no evidence of specimen distortion. To illustrate: the large salmonid fish found at the P-33 locality measures 62 cm in length after careful drying in the laboratory, yet measured 72 cm long *in situ* for a shrinkage factor of 14%. It has been noted further that the shrinkage of compressed plant tissue through drying exceeds that of the matrix. For example: spheroidal cones preserved as water-saturated tissue are smaller than their casts in the matrix following drying. Both shrinkage and compaction factors have been found to be considerably less for sediments with larger components of silts and sands in comparison to the high clay content of sediments in Unit 2 at locality P-33.

DEPOSITIONAL ENVIRONMENTS

The Clarkia fossil beds are restricted not only to the floor of the St. Maries River valley but also to the lower courses of its major tributaries on the southwest (Fig. 2). Such evidence shows that the major features of topography, and the patterns of drainage in the St. Maries River area, have changed little since early Neogene time. Evidence from hammer seismic surveys (R. Blum, pers. commun.) and from water well data indicates that more than 100 m of sediments are present in some parts of the basin, but little is known about the nature of the subsurface material. Our interpretations, therefore, rest on that portion of the Miocene section that is exposed at a few sites. For these considerations, the best and most studied exposure is locality P-33. Here the sedimentologic and paleontologic evidence indicates that a proto-St. Maries River valley was occupied by a lake during at least one stage of its depositional history. This early Neogene lake was named Clarkia Lake by Smiley and Rember (1979, 1981).

The lacustrine section at locality P-33 is composed of clays and silty clays characterized by fine laminations resembling varves. Clay layers ranging in thickness from about 0.5 cm to 10 cm are interspersed in the sequence of thinner bedding, as are volcanic ashes having thicknesses ranging from 1 or 2 mm to 50-60 cm (Figs. 3, 4, 5). The thicker clay units are texturally graded (fining upward) and the finer varve-like laminations examined by us and by Smith and Elder (this volume) show similar internal gradation. Laterally continuous and discrete ash deposits that can be traced across 100-200 m at this site also are graded; those less than about 10 cm thick show a single episode of upward fining, whereas ones more than 10 cm thick may show evidence of several episodes of gradation within a single ash unit. Single (graded) clay layers 5 mm or more thick rarely contain fossils except near their upper and lower surfaces; graded ash layers similarly are devoid of megafossils. These sedimentologic data have been interpreted (Smiley and Rember 1981) as most likely indicating rapid infilling of the lake resulting from frequent "muddying" during rain storms supplemented by rapid air-fall deposition of relatively thick layers of volcanic ash. The thicker ash layers may represent an episode of closely spaced volcanic eruptions, or they may represent a single eruption followed by rapid erosion of powdery ash that had fallen on the adjacent land surfaces. It has been estimated that deposition of the P-33 lacustrine section probably took less than 1000 years under the humid climatic conditions that prevailed (Smiley and Rember 1981; Smith and Elder, this volume).

In addition to the evidence of sedimentology, a lacustrine environment is indicated by paleontology and by the condition of preservation of organic remains. The fish, mollusks, sponges, and other aquatic organisms all indicate environments of sluggish streams or standing water of lakes. The abundance, taxonomic diversity, and unabraded condition of leaves, with high representation of slope taxa that commonly are rare or fragmental in other Miocene floras of the Pacific Northwest, can best be explained by the existence of an open lake in an area of rugged hilly terrane. With a depositional setting of this type, there would be little or no forest barrier to the rapid and short-distance dissemination of organs from the contributing plants of adjacent slopes to the open waters of the lake. As Spicer (1981) has shown, the abrasion and decomposition of leaves increase with the distance and with the duration of transport from the parent plant to the burial site. An additional factor is the separation of individual leaves on Clarkia bedding surfaces, which suggests an offshore depositional environment in relatively quiet water. The "stacking" of leaves one on top of another, which is prevalent only at the lake margin site of locality P-38 and at the top of the P-33 lacustrine section, occurs most commonly where leaves accumulate in close proximity to the contributing vegetation (e.g. riparian, swamp or forest floor sites).

A stagnant, anaerobic and probably toxic lake bottom site of burial is suggested by several factors. The fine laminations of organic-rich clays show little or no evidence of disturbance, indicating the virtual absence of a benthonic infauna. The numerous fish specimens and plant fossils show no evidence of benthonic scavenger activity. The remains of benthonic organisms such as mollusks are found only at the top of the P-33 section, representing a late stage in the history of the lake when the water was considerably shallower than before. The excellent preservation of leaves as intact cellular tissue retaining original chemical constituents seems explicable only in the absence of an oxidizing environment and decomposition by aerobic organisms. The surface waters of the lake, on the other hand, were oxygenated as evidenced by the abundance and common occurrence of fish remains and limnic (planktonic) micro-organisms. From their study of fish taphonomy relating to the P-33 locality, Smith and Elder (this volume) have estimated a water depth of not less than about 12 m to account for the presence and preservation of the fish remains in the lacustrine deposits.

Somewhat different depositional environments seem indicated for the four other Clarkia fossil sites. Locality P-38 was a depositional site of coarse sands near the lake border. Locality P-34 represents a narrow embayment on the northeast side of Clarkia Lake, apparently the drowned mouth of a small tributary stream. Localities P-37 and P-40 in the Emerald Creek embayment of Clarkia Lake appear to represent nearshore sites of limnic sedimentation.

VOLCANIC ACTIVITY

Syngenetic

Miocene Clarkia Lake was formed by damming of the proto-St. Maries River valley downstream from Mashburn Junction, presumably by a lava flow. In that area, however, a series of younger (Middle Miocene) basalt flows have covered and obscured the older rocks (see Fig. 2). If Clarkia lacustrine deposition was indeed initiated by lava-damming, neither the exact location nor the flow that was responsible are known at present. There are no instances where lava flows are known to be interbedded with or to underlie the fossiliferous Clarkia deposits.

Numerous layers of silicic volcanic ash occur throughout the exposed section at locality P-33 and at other sites. Their source is at present unknown. They occur as discrete and laterally continuous beds ranging in thickness from about 1 mm to more than 50 cm. The ashes may have originated from some eruptive source(s) outside the Clarkia basin; of interest here is the fact that 1-2 cm of air-fall ash accumulated in the Clarkia area during the 1980 eruption of Mt. St. Helens almost 500 km to the southwest. Periodic volcanic activity is known to have occurred in the Pacific Northwest region through most of Cenozoic time, and ash clouds from distant sources could well have extended over the Clarkia area during the early Neogene. An intensive study of these Clarkia ash deposits has yet to be accomplished and will be required in any attempt to resolve such problems.

Post-Clarkia

Just north of Clarkia townsite, in the area of localities P-34 and P-38, is a local pile of several basalt flows associated with near-vent pyroclastics, which contain pumice fragments, scoreaceous cinders and what appear to be bombs. Subsequent erosion has cut a gully through the basalt pile into the underlying Clarkia deposits at P-34, where a dike of the same basalt type (Priest Rapids) has cut through and blackened the older sediments. This younger "volcano" thus can be seen to post-date the episode of Clarkia deposition. Farther down the St. Maries River valley, and confined to its floor, are single sheets of Priest Rapids basalts that in places can be seen to rest on the older valley fill. Downstream from Mashburn Junction is a thick plateau-type sequence of basalt flows of the same type. A Dodge-type flow underlies the sequence of Priest Rapids basalts, from which it is separated by a sedimentary interbed (Locality P-44), in the area downriver from Mashburn Junction (see also Bishop 1974). The apparent basalt-sediment relations farther upvalley near Cedar Creek suggest that the extrusion of Dodge-type basalt in the St. Maries River area is post-Clarkia in age. Elsewhere on the Columbia Plateau, basalts of Priest Rapids type represent the upper member of the Wanapum Formation and basalt of the Dodge type represents a unit near the base of the same formation; the Wanapum has been designated a Middle Miocene series of flows, with radiometric dates falling within the range of about 13.6 and 14.5 my (see Rockwell International Report, 1979; Tolan and Beeson, 1984).

AGE CONSIDERATIONS

The assignment of an early Neogene age to the Clarkia fossiliferous deposits is based largely on the composition and regional comparisons of the Clarkia floral assemblage and its inferred climatic requirements. The floral composition is most characteristic of the Early to Middle Miocene for this part of the world (Smiley et al. 1975; Smiley and Rember, this volume; see also the Neogene floristics compiled by Wolfe, 1969). The Clarkia flora contains, moreover, a few taxa that elsewhere are more characteristic of, or have not been recorded in floras younger than, later Paleogene ages (e.g. *Amentotaxus, Cunninghamia, Metasequoia, Engelhardtia, Symplocos*, etc.). A mixture of "relict" Paleogene taxa in a flora that otherwise is indicative of a Miocene age suggests that the age of the Clarkia fossil beds most likely is Early Miocene.

The Clarkia flora is distinguishable floristically from Middle Miocene ("Latah" type) floras of the eastern Columbia Plateau region, such as those from Spokane, Coeur d'Alene, Grand Coulee, and Whitebird. The flora of the Oviatt Creek basin 32 km south of the Clarkia basin, and the small flora from the Rover Siding site about 5 km northwest of Mashburn Junction, are floristically of the Latah type. Whereas the Oviatt Creek and Rover Siding beds are overlain with apparent conformity by Middle Miocene basalts of Priest Rapids type, the Clarkia fossil beds have been intruded by Priest Rapids basalts. Physical stratigraphic relations of basalts that can be correlated with those of the Columbia Plateau sequence thus indicate an age for the Clarkia deposits that is older than the Oviatt Creek and equivalent fossil beds of the area, and an age for the Clarkia flora that is older than the widespread vegetation of "Latah" type.

Basalt correlations based on petrologic and chemical data seem more reliable and precise than those based on radiometric dating from K/A whole-rock methods. Two samples of Priest Rapids basalt were obtained from the same site (locality P-34) and sent to different laboratories, one responding with an age of ~22 my and the other, ~15.3 my. Another sample from the Oviatt Creek site (locality P-35) produced an age of ~12.8 my from one of these sources. On the other hand, various radiometric dates obtained by others for the entire Wanapum Formation (of which the Priest Rapids is the uppermost of three members) resulted in an estimate of ~13.6-14.5 my ago as the time span for this total sequence of Middle Miocene basalts (see Rockwell International Report, 1979). It is apparent, therefore, that the correlation of Miocene fossiliferous deposits in areas marginal to the Columbia Plateau may rest largely on biostratigraphic evidence, supplemented by the type of basalt that may be associated and its physical relations to the fossil beds. The age of the fossiliferous deposits in the Clarkia basin is older than the Middle Miocene basalt series of the Wanapum Formation, and regional floral comparisons seem most indicative of an Early Miocene age. Rember, however, considers an Early Miocene (pre-Latah) age to be too old on the basis of basalt-sediment relations in this region, and that the Clarkia flora represents a local facies of a widespread Middle Miocene flora.

PRESENT CLIMATE AND VEGETATION

Pacific Northwest Climates

Early Neogene floras of the Pacific Northwest are strikingly different from those existing in the same region today (for considerations of present plant geography of this region, see Daubenmire, 1969). Because regional (and also local) vegetation is well known to be dependent in large measure on prevailing climatic conditions for its existence and perpetuation, comparative factors of phyto-climatological relations are pertinent to our considerations of the Miocene Clarkia biota and of the physical parameters under which it lived. For the early Neogene, climates can be inferred from the association of taxa in a fossil assemblage. For the present, climatic factors can be documented from weather station data that have accumulated over several decades (Fig. 8, Table 1) and the extant biota can be observed directly.

Present climatic conditions over the Pacific Northwest are highly variable over relatively short distances, depending on factors that include (1) the proximity of sites to the moderating influence of the Pacific Ocean, (2) the location of the site relative to major topographic features such as the Cascade Mountains (that is, whether on the maritime or on the continental side of the range), (3) the elevations not only of the weather stations but also of the surrounding land surfaces, (4) local orographic factors, and (5) prevailing global climates. Present climatic factors in the Pacific Northwest can be illustrated by data from weather station sites extending from the maritime locations of Aberdeen, Olympia and Seattle eastward over 550 km through the intermontane locations of Ellensburg, Ritzville and St. Maries (Fig. 8, Table 1). Weather data from these stations are mainly from Kincer (1941).

For the maritime region, Aberdeen near the Washington coast has a mean annual precipitation exceeding 2100 mm, an average relative humidity of 80% throughout the year, and a mean length of growing season of more than six months. Toward the east near the southern end of Puget Sound, and partially shadowed by the Olympic Mountains on the northwest, Olympia has a similarly equable climate although precipitation is reduced to an annual mean of about 1330 mm. At Seattle on the Puget Sound due east of the Olympic Mountains, mean annual precipitation is reduced further to 890 mm. In the lowland maritime region west of the present Cascade Mountains, present climates reflect the moderating influence of the Pacific Ocean with local rainshadow effects of the Olympic Mountains: high relative humidity throughout the year, long growing seasons exceeding six months, and mean midwinter temperatures of 4°C.

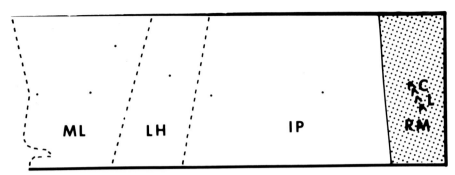

Figure 7. Part of the Pacific Northwest showing major physiographic provinces postulated for early Neogene time. Dots are present weather stations shown on Figure 8. ML=maritime lowlands, with some coastal embayments not shown; LH=area of low hills probably not exceeding about 700 m in elevation (Smiley 1963) other than local volcanoes; IP=interior province of predominantly plateau basalt extrusion; RM=western slopes of Miocene Rocky Mountains, with direction of drainage toward the interior province; CL=Clarkia Lake, with the Hoodoo Mountain highland on the west.

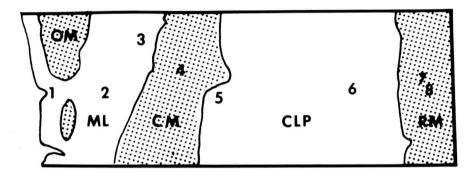

Figure 8. Part of the Pacific Northwest (see Figure 7 above) showing present major physiographic provinces and areas of greatest orographic influence on local and regional climates (stippled). Numbers depict sites of weather stations on Table 1. OM=Olympic Mountains; CM=Cascade Mountains; CLP=Columbia Lava Plateau; RM=Rocky Mountains.

For the interior region east of the Cascades, the rainshadow effect of the prevailing climatic barrier is pronounced. At Ellensburg 80 km east of the Cascade crest and at an elevation of 485 m, annual precipitation is 230 mm, midday relative humidity is about 70% in the winter and 30-40% in the summer, the growing season averages 142 days, and mean midwinter temperatures are -4°C. Another 190 km farther east at Ritzville near the center of the Columbia Plateau (elev. 555 m), mean annual precipitation is 255 mm, the growing season averages 136 days, midday relative humidity is within the range of 10-30% in the hot summer months, and mean midwinter temperatures are -2°C.

Weather conditions in the western foothill area of the Rocky Mountains near the eastern border of the Columbia Plateau can be illustrated by data at St. Maries (elev. 635 m) located 175 km east of Ritzville. Here the local orographic effect of the surrounding land that rises to elevations of 750-900 m results in an increase in mean annual precipitation to 635 mm. About 50 km southeast of St. Maries, a U.S. forest station in the Clarkia valley (about 840 m elevation) has recorded a mean annual precipitation of 965 mm, reflecting the higher elevations of surrounding land (up to 1525 m). About 80 km farther east near the Idaho-Montana border, several weather stations at elevations of 12-1500 m have recorded mean annual precipitation exceeding 1100 mm.

TABLE 1. MODERN CLIMATIC DATA, FOR PALEOCLIMATIC INTERPRETATIONS

Stations (Figs. 3-4)	Elev. (m)	Temperatures (°C)				Precipitation (mm)		Seasonal[5]	
		Jan.	July	R.[1]	GS[2]	M. Ann.[3]	Wet/Dry[4] Diff (%)	Min.	Max.
Pacific Northwest									
1. Aberdeen, WA	0	4	15	11	191	2133	361 (15)	S	W
2. Olympia, WA	0	4	17	14	191	1330	212 (16)	S	W
3. Seattle, WA	0	4	17	13	255	890	117 (15)	S	W
4. Snoqualmie Pass, WA	900	-4	14	18	115	2440	418 (17)	S	W
5. Ellensburg, WA	485	-4	20	24	142	230	32 (14)	S	W
6. Ritzville, WA	555	-2	22	24	136	255	37 (14)	S	W
7. St. Maries, ID	635	-2	20	22	137	635	76 (12)	S	W
8. Clarkia, ID	840	-3	20	23	136	965	115 (12)	S	W
Southeastern U. S.									
1. Charleston, SC	0	10	28	18	285	1020	111 (11)	F	S
2. Columbia, SC	90	8	27	19	248	1065	84 (8)	F	S
3. Spartanberg, SC	230	6	26	21	217	1265	64 (5)	F	S
4. Andrews, NC	550	6	23	18	177	1545	73 (5)	F	S
5. Tallahassee, FL	55	12	27	15	282	1395	114 (8)	Sp,F	S,W
6. Macon, GA	100	8	27	19	240	1120	67 (6)	F	S
7. Athens, GA	200	7	27	20	217	1265	67 (5)	Sp,F	S,W
8. Memphis, TN	36	6	27	22	238	1150	56 (5)	F	W,Sp
9. Nashville, TN	150	4	26	22	214	1140	60 (5)	F	W,Sp
10. Knoxville, TN	270	4	26	21	217	1280	76 (6)	F	W,Sp

1. Temperature range, midwinter to midsummer, mean monthly. 2. Growing seasons, days between killing frosts. 3. Mean annual precipitation. 4. Differential betweeen the wettest and driest months in mm (and percent of mean annual precipitation). 5. Sp=Spring; S=Summer; F=Fall; W=Winter.

In summary, the Cascadian climatic barrier now exerts considerable influence on weather systems that come into the interior of the Pacific Northwest from the Pacific Ocean. Farther east the rising land surfaces along the western slopes of the Rocky Mountains produce orographic effects that result in increased precipitation. West of the Cascades climates are maritime, with moderate to high mean annual precipitation, high relative humidity throughout the year, long growing seasons of more than six months, and mean midwinter temperatures appreciably above 0°C. East of the range climates are continental, with low mean annual precipitation, low relative humidity during the warmer months, a short growing season of about four months, and mean midwinter temperatures that are appreciably below 0°C. A tabulation of seasonal precipitation for the entire Pacific Northwest shows it to be a region of summer-dry climate, but west of the Cascade Range this is compensated for by a much higher summer relative humidity than occurs in the interior.

Forests of Northern Idaho

Present forests of northern Idaho are composed predominantly of conifers, with dicot shrubs and trees occupying more open sites. In the St. Maries River area the dominant conifers are grand fir (*Abies grandis*) and Douglas fir (*Pseudotsuga menziesii*) commonly associated with western white pine (*Pinus monticola*) and tamarack (*Larix occidentalis*). More locally in moist protected sites occur western red cedar (*Thuja plicata*), western hemlock (*Tsuga heterophylla*) and occasionally yew (*Taxus brevifolia*). Higher on the surrounding slopes and locally in frost pockets of the bottomlands are spruce (*Picea engelmannii*) and subalpine fir (*Abies lasiocarpa*), commonly admixed with tamarack. In drier habitats the conifer forests are dominated by Douglas fir and ponderosa pine (*Pinus ponderosa*). Toward the west and southwest, these conifer forests disappear near the Idaho-Washington boundary, where they are replaced by the open rolling grain fields of the Palouse country.

Larger dicot trees are black cottonwood (*Populus trichocarpa*), alder (*Alnus incana*) and quaking aspen (*Populus tremuloides*). Aspen and alder occur commonly as seral stands in disturbed areas. Smaller dicot trees and shrubs include willow (*Salix* spp.), blue elderberry (*Sambucus glauca*), cascara (*Rhamnus purshiana*), ninebark (*Physocarpus malvaceus*), dwarf maple (*Acer glabrum*), currants (*Ribes* spp.), roses (*Rosa* spp.), hawthorne (*Crataegus douglasii*), cherry (*Prunus demissa*), western syringa (*Philadelphus lewisii*), service berry (*Amelanchier alnifolia*), western blackcap (*Rubus leucodermis*), huckleberries or blueberries (*Vaccinium* spp.), and dogwood (*Cornus* spp.).

All conifer and dicot species are tolerant of long and severe winters and periods of summer drought, which is in striking contrast to the mixed mesophytic forests of the early Neogene. Only a few of the Clarkia genera are still represented in northern Idaho forests, and they commonly are represented in early Neogene fossil floras by rare shoots, seeds, cones, leaves or pollen. All of the extant conifer genera have survived in this region since the time of the Clarkia flora, but the characteristic early Neogene conifers of the family Taxodiaceae are no longer represented in modern forests of the Pacific Northwest. Of the approximately 75 ligneous dicot genera in the Clarkia flora, only a few have persisted to the present in this part of the world (e.g. *Populus, Alnus, Salix, Acer, Ribes, Rosa, Crataegus, Prunus, Philadelphus, Amelanchier, Cornus* and *Vaccinium*). All others have become locally extinct and most of them are now confined to forests of eastern North America or eastern Asia.

EARLY NEOGENE CLIMATE

Most of the Clarkia conifers and angiosperms are clearly referable to modern genera. Some of the genera are composed of numerous modern species that are indicative of different habitats, and an inherent problem is the recognition of which one (or more) of many extant species is the living correlative(s) of the fossil taxon. Other Clarkia genera are now monotypic or oligotypic, and identification to the level of genus usually is sufficient to show such modern species affinities. Where a significant proportion of taxa in a Tertiary flora is of the latter type, their coexistence in the fossil assemblage can serve as the basis for paleoclimatic and other paleoecologic inferences of the entire assemblage. For the interpretation of Tertiary paleoclimates, therefore, a synecological analysis of the requirements and tolerances of the various members of the fossil assemblage is required.

The problem to be resolved is what single set of a variety of climatic factors will best explain the fossil coexistence of numerous taxa that were derived from the local vegetation of the area. In instances where a fossil flora has high taxonomic diversity and is well preserved, as is the case with the Clarkia flora, such inferences on the general climatic requirements of the past biota can be made with a reasonable degree of confidence, by comparing the Tertiary assemblage of associated taxa with the requirements of modern forests that have similar taxonomic associations. Where the past topographic setting is known from geologic evidence, such floral comparisons should be made with comparable modern forests in areas of similar topography. Once it is established which modern forest(s) is the closest living counterpart of the fossil assemblage, the climatic parameters for this type of vegetation can be obtained from available data in weather station records (e.g. Kincer 1941). Major factors to be considered appear to be (1) the minimum length of growing season required by a significant number of the taxa, (2) the maximum incidence of freezing temperatures that can be tolerated by several of the taxa, and (3) the ability of taxa to withstand water-stress during periods of drought.

The Clarkia flora is remarkably similar, at the generic level, to living forests in and around the present *Metaseqouia* associations of eastern Asia (Wang 1961). For North America modern correlatives exist in the bottomland and slope forests of the southern Appalachian Mountain region (Smiley et al. 1975; Smiley and Rember 1981). Wolfe (1979) noted that climate in the Asian region of *Metasequoia* forests has a mean annual temperature in the range of 13-20°C, a cold month mean of not less than 1°C, and a mesic climate with precipitation fairly evenly distributed throughout the year (see also Chaney and Axelrod 1959; Wang 1961). Similar climatic conditions have been noted for the area in and around the modern *Taxodium* forests of southeastern United States (Smiley 1963; see also Fig. 9 and Table 1).

The early Neogene Clarkia flora contains both *Metasequoia* and *Taxodium*, occurring with other genera that are still associated with one or the other (or both) of these distinguishing conifers. These modern forest associations of eastern Asia and eastern North America occur in regions of similar climatic conditions, and it seems reasonable to assume that comparable plant associations of later Cenozoic time had approximately equivalent requirements. Northwest regional climates of the early Neogene, inferred from the Clarkia fossil associations, are considered to have been mesic, summer-wet, with long growing seasons possibly in excess of 8 or 9 months, and with mild winters having only occasional periods of light frost (i.e. a mean midwinter temperature greater than 0°C).

LATER NEOGENE EVENTS

Regional

Major later Neogene orogeny in the Pacific Northwest has involved mainly the rejuvenated uplift of the Rocky Mountains on the east and the Cascade orogeny on the west. Prior to and concomitant with this development of north-south trending ranges, fissure eruptions of basalt flows continued in the intervening region of the Columbia Plateau. The primary sources of these lava flows have been mapped near the eastern margin of the Plateau, and the direction of flow was toward the west (see Rockwell International Report, 1979). This indicates that the land

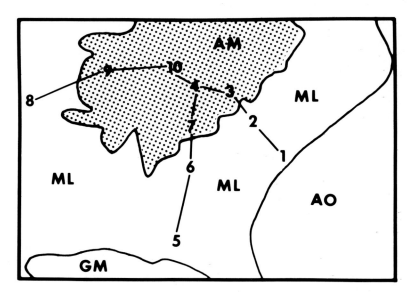

Figure 9. Part of the southeastern United States where modern forests are most like those of the Miocene Clarkia flora of northern Idaho, with locations of weather stations at various elevations (see Table 1). AM=Appalachian Mountains; ML=maritime lowlands, including *Taxodium* swamp forests; GM=Gulf of Mexico; AO=Atlantic Ocean.

surface sloped toward the west during the Miocene as at present, suggesting that the direction of flow of major drainage systems similarly was westward. The extension of some of the Miocene lava sheets through the Columbia River gorge where it crosses the Cascade Mountains suggests further that major drainage of surface water from the interior was via a route approximating that of the present Columbia River (Tabor et al. 1984; Tolan and Besson 1974). The tilting of lava sheets especially along the eastern flank of the Cascade Range indicates a subsequent uplift of the present montane area to elevations appreciably higher than those of the Miocene. From evidence of floral sequences and vegetational changes along the western margin of the Columbia Plateau, Smiley (1963) observed that the Neogene effect of this rising climatic barrier was not significant until the later Miocene time of Ellensburg (*sensu stricto*, Smith 1903) and Dalles (Chaney 1944) deposition some 10-12 million years ago. It was estimated that maximum elevations in the Cascadian uplands to the west probably did not exceed 700-800 m at that time.

For the life and environment of the interior region, the subsequent development of the Cascade Range as a barrier to the moist and moderating weather systems off the Pacific Ocean was a most significant event (cf. Figs. 7 and 8). This later Neogene uplift of the Cascade region post-dates the Clarkia fossil beds, and essentially maritime climates probably extended as far east as the Rocky Mountains during the time of the Clarkia flora. Furthermore, considerable evidence from many parts of the world indicate that global climates were appreciably warmer during the earlier part of the Neogene than were those that followed. The combined influence of the Cascadian orogeny and deteriorating world climates has resulted in progressively drier and colder conditions in the western interior during later Neogene time.

Post-Clarkia Sedimentation

In the area of eastern Washington and northern Idaho, Middle Miocene (post-Clarkia) lava flows dammed existing drainages creating numerous local basins of deposition. Fossil plants have been found in some of these deposits, depicting a "Latah" type of regional vegetation that was lower in taxonomic diversity and of somewhat cooler character than the Clarkia flora. The Oviatt Creek and Rover Siding floras in the vicinity of the Clarkia basin are of this type. In the valley of the St. Maries River are examples of valley deposits that rest on top of the Clarkia fossil beds. The contact in some exposures can be seen to be unconformable (e.g. localities P-33, P-34 and P-40). These younger deposits are poorly bedded, unfossiliferous sands and pebbly sands of distinctly different lithology than the silty clays and volcanic ashes of the underlying Clarkia fossil beds. Subsequent erosion has removed the younger sediments from the central areas of the basin, and those that remain occur only at the basin margins. One or two additional later Neogene episodes of sedimentation may also be represented locally. None of these unfossiliferous post-Clarkia deposits has been studied in detail.

Post-Clarkia Tectonic Activity

This region of northern Idaho and eastern Washington appears to have remained remarkably stable during later Neogene time. Evidence of geologic instability following Clarkia deposition seems to have been of local and minor significance. If regional uplift occurred during the Cascadian orogeny, it produced little noticeable effect on local topography or flow patterns of drainages. Evidence of minor faulting and associated local tilting of Clarkia clays has been observed at localities P-33 and P-40. The intrusion of a basalt dike at locality P-34 and the formation of a small "volcano" near Clarkia have been noted. The observed trends of post-Clarkia dikes and faults are northwesterly, and they parallel other lineations observed on topographic maps and aerial photographs of the region. If the northwest-trending St. Maries River valley is also structurally controlled, then this would indicate that the tectonic forces producing these structural trends were in operation both before and after the time of Clarkia deposition.

Early Neogene—Recent Comparisons

Comparisons between the present cool-temperate, conifer-dominated forests that now exist in northern Idaho, and the warm-temperate, dicot-dominated, mixed mesophytic forests that inhabited the same area in early Neogene time, show the degree to which vegetation and climate have changed during the course of the Neogene. From evidence largely of Tertiary floral changes that are well established for many parts of the world, global climates are known to have been much more equable early in the Neogene than near its end. In western North America, these climatic changes were influenced further by the later Neogene uplift of the Sierra and Cascade ranges, which created more continental (rainshadow) conditions in their lee (cf. Figs. 7 and 8). The trend toward longer and more severe winters and drier summers ultimately exceeded the limits of tolerance of taxa that characterized the mixed mesophytic forests of early Neogene time. Available data suggest that significant vegetational changes became noticeable in the Pacific Northwest interior in the Upper Miocene some 10-12 my ago (see Chaney and Axelrod 1959; Smiley 1963; Wolfe 1969). Although floral data are relatively sparse for this region in the later Neogene, Leopold (this volume) has documented the existence of steppe-type vegetation in southern Idaho during later Neogene time. For the Columbia Plateau region of the Pacific Northwest interior, therefore, climatic changes since the advent of the Clarkia vegetation have involved (1) significant decreases in mean annual precipitation, (2) significant increases in length and severity of winters, and (3) increasing drought conditions during the growing season.

CONCLUSIONS

Geomorphic evidence in the eastern Columbia Plateau region suggests that the topography, drainage patterns, and directions of stream flow have changed little since early Neogene (Clarkia) time. Extrusions of basaltic lavas have modified drainages only where they have dammed stream valleys. The resulting ephemeral basins were filled with sediments upvalley from the lava dams, and numerous examples exist in this area of northern Idaho showing broad and flat valley bottoms on the upstream side of basalt exposures. Because the present drainage of surface water is down the same valleys, such evidence indicates that the major drainage systems of the present were established prior to the time of the basalt extrusions. In the valley of the St. Maries River, for example, the valley-interfluve relations were the same at the time of the Clarkia lacustrine sedimentation as they are today. From such evidence it seems apparent that the eastern border of the Columbia Plateau has remained fairly stable geologically during later Cenozoic time.

Certainly some of the most striking changes in the later Cenozoic history of the Pacific Northwest interior involve the evolution of the biota in response to regional changes in physical conditions. These include: 1) a regional modification from a mesic warm-temperate vegetation to a summer-dry cool-temperate vegetation; 2) western North American extinctions of many plants now surviving only in Asia (e.g. *Cunninghamia, Glyptostrobus, Metasequoia, Cercidiphyllum, Pterocarya, Zelkova*, etc.); 3) regional extinctions of *Taxodium* forests closely resembling those now found only in southeastern United States; 4) regional extinctions of fish taxa that are now distributed over low latitudes of North America and Eurasia; and 5) continental extinctions of many larger mammals whose Tertiary remains are well documented in the John Day region of the Pacific Northwest (e.g. horses, rhinos, camels, proboscidians, etc.).

The early Neogene records of plants and animals reflect a more maritime influence on interior climates of the Pacific Northwest than now prevails. Later Neogene changes toward cooler world climates and a developing rainshadow east of the Cascades appear to be largely responsible for the local or regional extinctions of most members of the earlier "exotic" biota. Why many of these more mesic climate taxa did not invade and survive in the western coastal areas is unclear, as apparently suitable habitats have persisted there to the present time. Neogene uplift in the

Cascade area may have barred the coastward migration of some of the exotic taxa from the interior, or they may have been incapable of competing successfully with elements of an established biota that were occupying appropriate ecological niches. Had they been able to establish themselves in the western maritime regions, their survival to the present probably would have been assured by latitudinal shifts in distribution commensurate with the climatic fluctuations of later Neogene time.

LITERATURE CITED

Baghai, N. L. 1983. The Miocene Clarkia flora: *Liriodendron*. M.S. Thesis. University of Idaho, Moscow, Id.

Bishop, D. T. 1974. Petrology and geochemistry of the Purcell Sills, Boundary County, Idaho and adjacent areas. Ph.D. Thesis. University of Idaho, Moscow, Id.

Chaney, R. W. 1944. The Dalles flora. Carnegie Inst. Wash. Pub. 553. (11):285-322.

Chaney, R. W., and D. I. Axelrod. 1959. Miocene floras of the Columbia Plateau. Carnegie Inst. Wash. Pub. 617. 237 pp.

Clark, A. L. 1963. Geology of the Clarkia area, Idaho. M.S. Thesis. University of Idaho, Moscow, Id.

Cole, M. R., and J. M. Armentrout. 1979. Neogene paleogeography of the western United States. Pages 292-323 *in* J. M. Armentrout, M. R. Cole, and H. Terbest, eds. Cenozoic Paleogeography of the Western United States. Pacific Division, Society of Economic Paleontologists and Mineralogists, Los Angeles, Calif.

Daubenmire, R. 1969. Ecological plant geography of the Pacific Northwest. Madroño 20:83-110.

Giannasi, D. E., and K. J. Niklas. 1981. Paleobiochemistry of some fossil and extant Fagaceae. Amer. J. Bot. 68(6):762-770.

Giannasi, D. E., and K. J. Niklas. 1985. The paleobiochemistry of fossil angiosperm floras. Part I. Chemosystematic aspects. Pages 161-174 *in* C. J. Smiley, ed. Late Cenozoic History of the Pacific Northwest. Pacific Division, Amer. Assoc. Adv. Sci., San Francisco, Calif.

Gray, J. 1985. Interpretation of co-occurring megafossils and pollen: A comparative study with Clarkia as an example. Pages 185-244 *in* C. J. Smiley, ed. Late Cenozoic History of the Pacific Northwest. Pacific Division, Amer. Assoc. Adv. Sci., San Francisco, Calif.

Hietanen, A. 1963. Metamorphism of the Belt Series in the Elk River-Clarkia area of Idaho. U. S. Geol. Surv. Prof. Pap. 344C. 49 pp.

Huggins, L. M. 1980. Laboratory methods and analysis of fossil leaves and their application to educational curricula. Ph.D. Thesis. University of Idaho, Moscow, Id.

Huggins, L. M. 1985. Cuticular analyses and preliminary comparisons of some Miocene conifers from Clarkia, Idaho. Pages 113-138 *in* C. J. Smiley, ed. Late Cenozoic History of the Pacific Northwest. Pacific Division, Amer. Assoc. Adv. Sci., San Francisco, Calif.

Kincer, J. B. 1941. Climate and weather data for the United States. Pages 685-1201 *in* G. Hambridge, ed. Climate and Man. Yearbook of Agriculture for 1941. U.S. Dep. Agric., Washington, D.C.

Leopold, E. B., and V. Crane Wright. 1985. Pollen profiles of the Plio-Pleistocene transition in the Snake River plain, Idaho. Pages 323-348 *in* C. J. Smiley, ed. Late Cenozoic History of the Pacific Northwest. Pacific Division, Amer. Assoc. Adv. Sci., San Francisco, Calif.

Lewis, S. E. 1985. Miocene insects from the Clarkia deposits of northern Idaho. Pages 245-264 *in* C. J. Smiley, ed. Late Cenozoic History of the Pacific Northwest. Pacific Division, Amer. Assoc. Adv. Sci., San Francisco, Calif.

Niklas, K. J., and R. M. Brown, Jr. 1980. Ultrastructure and paleobiochemical correlations of fossil leaf tissue from St. Maries (Clarkia) area, Northern Idaho. Amer. J. Bot. 68(3):332-341.

Niklas, K. J., R. M. Brown, and R. Santos. 1985. Ultrastructural states of preservation in Clarkia angiosperm leaf tissues: Implications on modes of fossilization. Pages 143-159 *in* C. J. Smiley, ed. Late Cenozoic History of the Pacific Northwest. Pacific Division, Amer. Assoc. Adv. Sci., San Francisco, Calif.

Rockwell International. 1979. Geologic studies of the Columbia Plateau. Status Rep. RHO-BWI-ST-4.
Smiley, C. J. 1963. The Ellensburg flora of Washington. Univ. Calif. Pub. Geol. Sci. 35:159-276.
Smiley, C. J., J. Gray, and L. M. Huggins. 1975. Preservation of Miocene fossils in unoxidized lake deposits, Clarkia, Idaho. J. Paleont. 49(5):833-844.
Smiley, C. J., and L. M. Huggins. 1981. *Pseudofagus idahoensis* n. gen. et sp. (Fagaceae) from the Miocene Clarkia flora of Idaho. Amer. J. Bot. 68(6):741-761.
Smiley, C. J., and W. C. Rember. 1979. Guidebook and roadlog to the St. Maries River (Clarkia) area of northern Idaho. Idaho Bur. Mines Geol. Inf. Circ. 33. 27 pp.
Smiley, C. J., and W. C. Rember. 1981. Paleoecology of the Miocene Clarkia Lake (northern Idaho) and its environs. Pages 551-590 (Chapter 17) *in* J. Gray, A. J. Boucot, and W. B. N. Berry, eds. Communities of the Past. Hutchinson and Ross Publishing Co., Stroudsburg, Pa.
Smiley, C. J., and W. C. Rember. 1985. Composition of the Miocene Clarkia flora. Pages 95-112 *in* C. J. Smiley, ed. Late Cenozoic History of the Pacific Northwest. Pacific Division, Amer. Assoc. Adv. Sci., San Francisco, Calif.
Smith, G. O. 1903. Description of the Ellensburg (quadrangle), Washington. U. S. Geol. Surv. Geol. Atlas Ellensburg Folio 86. 7 pp.
Smith, G. R., and R. L. Elder. 1985. Environmental interpretation of burial and preservation of Clarkia fishes. Pages 85-93 *in* C. J. Smiley, ed. Late Cenozoic History of the Pacific Northwest. Pacific Division, Amer. Assoc. Adv. Sci., San Francisco, Calif.
Smith, G. R., and R. R. Miller. 1985. Taxonomy of fishes from Miocene Clarkia lake beds, Idaho. Pages 75-83 *in* C. J. Smiley, ed. Late Cenozoic History of the Pacific Northwest. Pacific Division, Amer. Assoc. Adv. Sci., San Francisco, Calif.
Spicer, R. A. 1981. The sorting and deposition of allochthonous plant material in a modern environment at Silwood Lake, Silwood Park, Berkshire, England. U. S. Geol. Surv. Prof. Pap. 1143. 77 pp.
Tabor, R. W., V. A. Frizzell, Jr., J. A. Vance, and C. W. Naeser. 1984. Ages and stratigraphy of lower and middle Tertiary sedimentary and volcanic rocks of the central Cascades, Washington: Application to the tectonic history of the Straight Creek fault. Geol. Soc. Amer. Bull. 95(1): 26-44.
Tolan, T. L., and M. H. Beeson. 1984. Intracanyon flows of the Columbia River Basalt Group in the lower Columbia River Gorge and their relationships to the Troutdale Formation. Geol. Soc. Amer. Bull. 95(4):463-477.
Wang, C-W. 1961. The forests of China. Maria Moors Cabot Found. (Harvard Univ.) Pub. 5. 313 pp.
Williams, J. L. 1985. Miocene epiphyllous fungi from northern Idaho. Pages 139-142 *in* C. J. Smiley, ed. Late Cenozoic History of the Pacific Northwest. Pacific Division, Amer. Assoc. Adv. Sci., San Francisco, Calif.
Wolfe, J. A. 1969. Neogene floristic and vegetational history of the Pacific Northwest. Madroño 20(3):83-110.
Wolfe, J. A. 1979. Temperature parameters of humid to mesic forests of eastern Asia and relation to forests of other regions of the Northern Hemisphere and Australasia. U. S. Geol. Surv. Prof. Pap. 1106. 37 pp.

DIATOM FLORA OF THE MIOCENE LAKE BEDS NEAR CLARKIA IN NORTHERN IDAHO

J. PLATT BRADBURY[1], K. V. DIETERICH[1], AND JOHN L. WILLIAMS[2]

[1] U.S. Geological Survey, Denver, CO 80225
[2] Cities Service Oil & Gas Corp., 1600 Broadway, Suite 900, Denver, CO 80202

The Miocene Clarkia and Oviatt Creek lacustrine deposits contain freshwater diatoms that provide paleolimnological insights into Miocene climates, hydrography, and tectonics in northern Idaho. In addition, the diatom floras in these and related deposits in the western and northwestern United States contribute to an understanding of Neogene diatom evolution and distribution. A hypothetical model that accounts for the distinctive features of Neogene diatom distribution and biostratigraphy is constructed on the basis of the potential distribution of lacustrine environments in time and space.

The dominant diatoms in the Clarkia and Oviatt Creek deposits are *Melosira* sp. aff. *M. distans* and *Actinocyclus* sp. aff. *A. normanii* f. *subsalsa*. Subdominant diatoms include several species of *Eunotia, Tabellaria, Fragilaria,* and *Tetracyclus*. The majority of the diatom species appear very close to their modern counterparts, and the paleolimnological interpretation is based on the assumption of a similar ecology for the fossil and modern forms.

Except for *Actinocyclus* sp., the fossil diatoms are common inhabitants of inland lakes of circumneutral or slightly acidic pH, low conductivity and productivity, and often with a comparatively high influx of allochthonous organic material. Such lakes are common in north temperate North America, especially on granitic terranes, but the diatom flora can also be found in slow-moving rivers and shallow lakes of warm temperate, subtropical and even tropical regions where appropriate water chemistry exists. The modern distribution of *Actinocyclus* sp. in freshwater systems is considerably more restricted and appears to be related to lacustrine systems of similar but perhaps somewhat more eutrophic character that drain to the sea by low-gradient rivers. These limnological analogs suggest that the Clarkia and Oviatt Creek basins existed under equitable climates of increased effective moisture and that they were integrated and drained to the Pacific Ocean by the ancestral Columbia River drainage. The climatic and hydrographic conditions inferred suggest low relief in northern Idaho and the absence of a significant mountain range to the west.

A comparison of the "modern-looking" Clarkia and Oviatt Creek diatom floras with the distinctive extinct but partly contemporaneous diatom floras of the western Snake River Basin suggests that Neogene diatoms have evolved along two different pathways: (1) the maintenance of a relatively ancient, conservative group of generalist diatoms that occupied short-lived lacustrine environments similar chemically and physically to modern counterparts, and (2) the establishment of distinctive extinct diatom populations in large, stable lacustrine systems, which in the past had much wider distributions than their modern endemic analogs do today.

The Miocene lake beds near the town of Clarkia, Idaho (Fig. 1) have been known to paleontologists for a number of years for their great diversity of excellently preserved fossil plants and

Copyright ©1985, Pacific Division, AAAS.

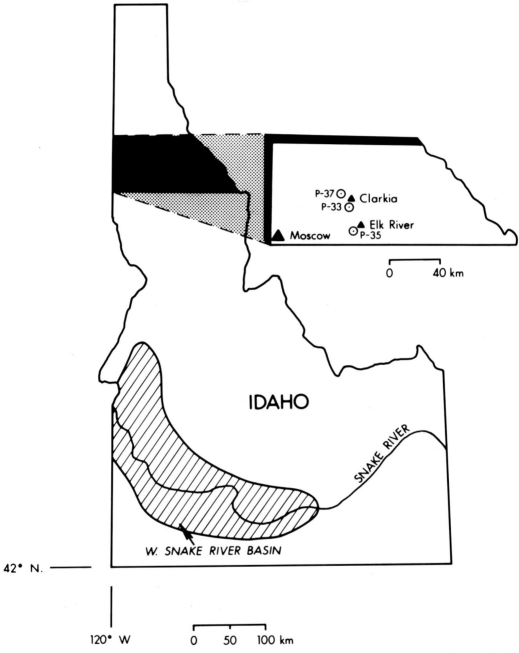

Figure 1. Index map showing the location of the Clarkia and Oviatt Creek lake beds. The Oviatt Creek lake beds are southwest of Elk River, Idaho.

animals (Smiley and Rember 1979; Smiley, Gray, and Huggins 1975). In some units of the Clarkia deposits, freshwater diatoms are also well preserved, and their study and documentation contribute to a fuller understanding of the paleolimnology of this ancient lacustrine deposit. At the same time, documentation of the fossil diatom flora provides relevant information about the age and correlation of nearby disjunct diatomaceous deposits. Lastly, detailed morphological studies of some of the characteristic diatoms of these lake beds allow useful comparisons with diatoms from deposits of similar age in other parts of western North America and the Northern Hemisphere, and thus permit speculations about the nature of freshwater diatom evolution.

Three localities of Miocene diatomaceous sediments are considered in this paper. Two are near the town of Clarkia and are known respectively as locality P-33 near Clarkia and locality P-37 in the Emerald Creek valley (Smiley and Rember 1979). A third, floristically similar locality is called the Oviatt Creek locality P-35, which is about 13.6 km southwest of the town of Elk River, Idaho in the NE 1/4 section 12, T. 39 N., R. 1 E. (Fig. 1).

LITHOLOGY OF THE CLARKIA AND OVIATT LACUSTRINE DEPOSITS

Thirty-eight samples of the lacustrine sediments at the principal locality of the lake beds near Clarkia (P-33) were examined for diatoms. At this locality the sediments are predominantly clayey silts with variable amounts of micaceous (muscovite?) minerals. The total exposed thickness of the deposit is approximately 10.3 m, of which the upper 3.8 m consist of oxidized, reddish-brown and buff-colored sediments, whereas the lower part of the section is composed of unoxidized grey and olive-colored deposits (Smiley and Rember 1979). Several volcanic ashes are present in the section, and for the most part the sediments are finely laminated and highly fissile. The large number of plant macrofossils (leaves) and generally silty nature of the sediments reflect a nearshore depositional environment for this locality, probably not far from an incoming stream. Logically these sediments are poor in diatoms and many of the diatom frustules are broken, abraded, or otherwise corroded. Nevertheless, well-preserved diatoms occur in several units of this section, particularly in the upper, oxidized portion of the deposit. The specific location of this outcrop (Smiley and Rember 1979) indicates that it is near the margin of the hypothesized Clarkia basin, and in fact near the modern course of Bechtel Creek, which may also have functioned as an incoming stream in Miocene time.

The Emerald Creek locality (P-37, Smiley and Rember 1979) is situated more in the limnetic region of the ancient lake, although it is close to what apparently was an island in the lake, according to Smiley and Rember (1979). Only selected samples from this locality were available for study. They are highly diatomaceous, although fine micaceous silt is still an important constituent. The greater diatom content of the Emerald Creek samples probably reflects their greater distance from the shore of the Clarkia lake, in environments dominated by biogenic, fine-grained sediments.

The Oviatt Creek locality is about 30 km south of the Clarkia localities. The deposits are at approximately the same location (between 850 and 900 m) as the outcrops of lacustrine sediment in the St. Maries River valley (the Clarkia deposits), although the modern topography of the region does not suggest that the two lacustrine systems were connected. The faithfulness with which the Miocene lacustrine deposits and basalt flows follow the modern valley systems in this part of northern Idaho has led Smiley and Rember (1979) to conclude that the present topography was established by early Neogene times.

The diatomaceous deposit at Oviatt Creek is at least 5.3 m thick and is composed of a greyish organic diatomite with minor amounts of silt and clay. Leaf fossils are locally abundant in the sediment.

AGE OF THE DIATOMACEOUS DEPOSITS IN THE CLARKIA REGION

Two potassium-argon radiometric dates are available for the Clarkia lake deposits: a date of 22 million years (my) (Geochron Laboratories, Cambridge, Mass.) and a date of 15.3 ± 0.3 my by the U.S. Geological Survey (William Greenwood, pers. commun., 1981). The age of the Oviatt Creek deposit is placed at about 12.8 ± 0.9 my (Smiley and Rember 1979).

Preliminary geologic information from the Clarkia localities suggests that the Priest Rapids Member of the Wanapum Basalt (part of the Columbia River Basalt Group) overlies the lacustrine deposits. Swanson et al. (1979) suggest that the Priest Rapids Member is between 12 and 14-16.5 my in age, providing a minimum age for the Clarkia deposits if the basalt correlations are correct. The Priest Rapids Basalt Member also overlies the deposits at Oviatt Creek (William Rember, pers. commun., 1981). The younger radiometric dates associated with these deposits are all from basalts that postdate the lacustrine deposits in question and provide no information about the maximum age of the lake sediments.

DIATOM FLORA OF THE CLARKIA AND OVIATT CREEK DEPOSITS

No attempt is made in this paper to list all the diatom species encountered in the lacustrine deposits of the Clarkia region, nor can the identification of the dominant and characteristic species of these localities be regarded as final. Considerable uncertainty exists regarding the significance of minor (and sometimes major) morphological variation for taxonomic subdivision, and fossil freshwater diatoms are often distressingly similar to their modern counterparts. Until such problems are resolved by monographic study of important taxa and culture studies of modern diatoms in different environmental settings, it seems best to present only tentative identifications that stress affinity or comparability with described taxa. The scanning electron micrographs (Plates 1-12) provide photographic documentation of some of the more abundant taxa in these deposits, and will allow detailed morphological comparisons with both modern and fossil diatom floras.

Table 1 provides a more complete list of the floral assemblages of the Clarkia and Oviatt Creek localities. The rationale for including the Oviatt Creek locality in a discussion of the Clarkia material is found in the close floristic similarity with the Clarkia deposits, evident in Table 1.

The Clarkia and Oviatt Creek deposits are both dominated by species of *Melosira*, apparently closely related to *Melosira distans*. In modern samples this species is characterized by considerable morphologic variation, as are many members of the freshwater *Melosira* flora, and consequently it is difficult to assign specific and varietal epithets with certainty. The problem is aggravated in Miocene samples because of the temptation to consider morphological variants as distinct (and often extinct) species or varieties. Comparison of SEM micrographs, for example Plate 4, Figures 1 and 2, and Plate 6, Figures 1-7 with modern examples of *Melosira distans* (Plate 13) suggests that the differences between the Miocene and modern forms are so slight that specific identity seems likely.

Many of the other diatoms in these deposits also show striking similarities to modern forms, and although the tentative taxonomic listings (Table 1) are intended to be conservative, in most cases the diatoms could be assigned to modern taxa without undue difficulty. Perhaps the apparent similarities will dissolve when modern type specimens are critically examined with the scanning electron microscope, and until this is done some caution is advisable in accepting a taxonomic identification without question.

The most distinctive diatom of the Clarkia and Oviatt Creek deposits is a species of *Actinocyclus* (Plates 1, 2, 3). This species is difficult to identify with the light microscope, because the presence of the pseudonodulus, a definitive characteristic of the genus *Actinocyclus*, cannot be clearly detected by this means. For this reason, diatoms of this general form were often identified as species of *Coscinodiscus* (e.g. VanLandingham 1964). SEM micrographs, however, clearly show the presence of this feature (Plates 1, 2, 3).

The *Actinocyclus* species in the Clarkia and Oviatt Creek deposits is closely related to *A. normanii* f. *subsalsa* (Juhl.-Dannf.) Hust. SEM micrographs of *A. normanii* f. *subsalsa* (Hasle

TABLE 1. FLORISTIC LIST OF DIATOMS FROM THE CLARKIA
AND OVIATT CREEK LACUSTRINE DEPOSITS[1]

	Clarkia P-33				Emerald Creek P-37	Oviatt Creek
	4	6	9	13		
Actinocyclus sp.	++	+	+	+	++	++
Melosira teres Brun. ?	+	+	+	+	+	+
Melosira praeislandica Jouse				+		++
*Melosira sp. aff. *M. distans* (Ehr.) Kutz	+	++	++	++	++	+
*Melosira distans v. *alpigena* Grun.	++	++	++	++	+	++
Melosira sp. cf. *M. ambigua* (Grun.) O.M.	+					+
Tetracyclus lacustris Ralfs	+	+	+	+	+	+
Tetracyclus ellipticus (Ehr.) Grun.	+	+	+	+	+	+
*Tabellaria sp. cf. *T. flocculosa* (Lyngb.) Kutz.			+			+
Tabellaria sp.	+	+	+	+	+	+
Fragilaria construens (Ehr.) Grun.			+			
Fragilaria construens var. *pumila* Grun.			+			
Fragilaria construens var. *subsalina* Hust.	+				+	+
Fragilaria construens var. *venter* (Ehr.) Grun.	+	+		+		
Fragilaria construens var. *binodis* (Ehr.) Grun.						+
Fragilaria elliptica Schu.				+		
*Fragilaria constricta Ehr.	+		+		+	+
Fragilaria sp. cf. *F. bicapitata* A. Mayer	+	+				
*Fragilaria sp. cf. *F. virescens* Ralfs	+					+
Fragilaria leptostauron var. *dubia* Grun.						+
Meridion circulare Agardh					+	+
*Eunotia incisa Wm.Sm. *ex* Greg.	+	+	+	+	+	+
Eunotia pectinalis (Kutz.) Rabh.	+	+	+			+
*Eunotia vanheurckii Patr. var.	+	+			+	+
Achnanthes sp. cf. *A. lapidosa* var. *lanceolata* Hust.	+		+			+
Achnanthes sp. cf. *A. subsalsa* Petersen			+			
Anomoeoneis sp. cf. *A. styriaca* (Grun.) Hust.				+		
*Anomoeoneis sp. cf. *A. serians* (Breb.) Cleve		+				
*Neidium iridis (Ehr.) Cleve ?			+			
Stauroneis anceps fo. *fossilis* Cleve ?			+			
Stauroneis anceps var. *obtusa* Grun.					+	
Stauroneis phoenicentron Ehr.						+
Pinnularia sp. cf. *P. appendiculata* (Ag.) Cleve		+			+	
Pinnularia sp. cf. *P. makariensis* Okuno		+				
Pinnularia sp. cf. *P. cardinalculus* Cleve			+			
Amphipleura sp.					+	
*Frustrulia sp. aff. *F. rhomboides* (Ehr.) DeT.					+	
Navicula sp. cf. *N. maeandrina* Cleve					+	
Navicula bacillum Ehr.?		+			+	
Navicula sp. cf. *N. aggesta* Hust.		+				
Navicula minuscula Grun.?				+		
Navicula pelliculosa (Breb) Hilse ?				+		
Cymbella sp. cf. *C. minuta* Hilse *ex* Rabh.			+			
Cymbella sp. cf. *C. hauckii* V.H.	+					
Gomphonema sp. aff. *G. subtile* var. *sagitta* (Schum.) Cl.	+	+	+			
Gomphonema gracile Ehr. emend. V.H. ?	+	+				
Gomphonema dichotomum Kutz. ?			+		+	
Gomphonema sp. cf. *G. grovei* M. Schmidt				+		+

[1] Identifications are tentative. + = present, ++ = very abundant or dominant.
* Species commonly recorded in low conductivity, oligotrophic lakes of northern Minnesota and Adirondack Mountains.

1977) indicate that minor differences exist, however, and it seems likely that the Miocene species from Clarkia and Oviatt Creek may be a variety or form of this species.

It is unusual to find *Actinocyclus* species in freshwater diatom associations, and such occurrences give Neogene diatom deposits throughout the world a distinctive character that may have some biostratigraphic value.

The remaining diatoms listed from the Clarkia and Oviatt Creek deposits are comparatively rare. Of these, the species of *Eunotia, Tetracyclus, Tabellaria*, and *Fragilaria* are the most frequent, while species of the other genera are scarce.

PALEOECOLOGY OF THE CLARKIA AND OVIATT CREEK DEPOSITS

Paleoecological interpretations from fossil diatoms, as with many other fossil organisms, depend most heavily on knowledge of ecological and physiological requirements of modern species analogs. To a lesser extent, growth habit or generalized morphological features provide some information about ancient habitats occupied by fossil diatom species. In the Clarkia and Oviatt Creek deposits, many of the diatoms show striking similarities to modern forms, a fact that gives some confidence in inferring paleoecological and paleolimnological conditions similar to those of modern species analogs.

Modern diatom floras in which *Melosira distans* dominates or is very common are found today in temperate lakes and ponds that are characterized by low alkalinity, low amounts of total dissolved solids, neutral or slightly acidic pH and low nutrient levels. These aquatic environments are characteristic of granitic terranes, such as the Precambrian Laurentian Shield, which extends into the U.S. in the Lake Superior region and in the Adirondack Mountains of upper New York state. The chemical characteristics of modern lakes with a *Melosira distans*-dominated flora (Table 2) may represent the surface-water chemistry of the Clarkia and Oviatt Creek environments. This assumption is strengthened by the fact that several other diatom species present in the Clarkia and Oviatt Creek deposits are also found in the low-conductivity, oligotrophic lakes of northeastern Minnesota and the Adirondack Mountains. These species are marked with an asterisk in Table 1.

Melosira distans var. *alpigena* is often very abundant in north temperate lakes in Finland and the U.S.S.R. where it often dominates the phytoplankton production (Simola 1977; Davidova 1976). Ecological work in varved Finnish lake sediments suggests that this species reaches its maximum population in the fall (Simola 1977), presumably in response to increased nutrient levels associated with fall overturn. Despite the common presence of *Melosira distans* var. *alpigena* in north temperate acidic lakes and ponds, the species is not unknown from tropical and subtropical regions. It is common in the Rio Nanay, an acidic tributary of the Amazon River near Iquitos, Peru (Hohn 1966), and Patrick et al. (1967) have reported it consistently in the Savannah River, Georgia. It also occurs commonly in ponds and streams in Kentucky (K. E. Camburn, pers. commun., 1981). The chemical composition of these slow riverine habitats is similar to that reported for the Minnesota lakes in which the species is reported (Table 2) although it is not excluded from more eutrophic aquatic environments. Evidently, the species is not faithful to a particular climatic regime.

Like many species of *Melosira, M. distans* and its varieties may be found in planktonic or benthonic habitats, depending partly on the availability of limnetic nutrients for growth and sufficient turbulence to support it in the photic zone in open water. The scarcity of obligate benthonic diatoms in the Clarkia and Oviatt Creek samples suggests that the lake was comparatively deep and that *Melosira distans* var. *alpigena* occupied a planktonic niche in these lacustrine systems.

Some paleoecological and paleolimnological comments can be made about the second most common diatom in the Clarkia and Oviatt Creek deposits, *Actinocyclus* sp. As noted previously, this species shows close resemblances to *Actinocyclus normanii* f. *subsalsa*. The species *A. normanii* is a marine or brackish-water diatom that commonly occurs in estuaries, and Hustedt (1957)

TABLE 2. CHEMICAL CHARACTERISTICS OF NORTHEASTERN MINNESOTA LAKES IN WHICH *MELOSIRA DISTANS* VAR. *ALPIGENA* IS A DOMINANT DIATOM[1]

	Conductivity μmhos/cm	pH	Chlorophyll a μg/l	Total phosphorus μg/l	NO_3 μg/l	SiO_2 mg/l	Alkalinity mg $CaCO_3$/l	Ca mg/l	Mg mg/l	Na mg/l	K mg/l	SO_4 mg/l	Cl mg/l	Transparency (m)	*Melosira distans* (%)
Nickel Lake	53	6.1	0.6	10	90	5.0	10	9.4	0.2	3.2	0.4	11.8	0	0.9	16.6
Gabbro Lake	79	7.3	0.2	15	148	14.5	26	10.3	6.7	3.2	0.8	6.8	0	1.5	27.6
Bald Eagle Lake	82	7.4	0.2	11	148	8.5	25	10.9	2.0	1.8	0.5	5.6	0	1.8	35.5
Bassett Lake	87	7.4	0.4	10	80	16.0	28	9.5	2.6	2.4	0.6	2.6	0	2.0	20.6
Savannah River, Ga.															
maximum	–	7.0	–	–	7	14.6	34	5.8	3.3	–	–	11.0	7.0	–	–
minimum	–	6.5	–	–	1	7.2	15	2.0	0.3	–	–	2.2	0.6	–	–
Rio Nanay, Peru	–	5.2	–	–	8	6.5	4.2	0.4	0.5	–	–	20.2	1.6	–	[2]

[1] Data from Richard B. Brugam, Southern Illinois University, Edwardsville (unpublished). These data are compared with the water chemistry of the Savannah River (Patrick et al. 1967) and with the Rio Nanay, Peru (Patrick 1966); subtropical and tropical localities where *M. distans* var. *alpigena* has been recorded.

[2] Common.

considered the form *A. normanii* f. *subsalsa* to be simply an ecological variant of the species that inhabits low-salinity environments. It is found sparingly in the upper reaches of the Weser River estuary, northern Germany, where presumably it is restricted by high salinities. However, it is common in Zwischenahner Lake (32 mg/l Cl⁻) in the same region (Hustedt 1957). *A. normanii* f. *subsalsa* is also becoming common in the Laurentian Great Lakes, especially in nearshore localities where pollution is responsible for increased nutrients and slightly elevated salinities (Stoermer et al. 1974). In Lake Ontario it reaches its maximum populations in the eastern end of the lake in the late summer, and Stoermer et al. (1974) concluded that this diatom apparently requires relatively high temperatures for maximum growth.

Despite the ability of *A. normanii* f. *subsalsa* to establish itself in freshwater environments, it is apparently not known from lakes that do not have drainage to estuarine or marine environments, perhaps because it cannot withstand the rigors of subaerial transportation. Nevertheless, marine and brackish water diatoms can be transported far inland in low-gradient rivers such as the Amazon (Gessner and Simonsen 1967), and this mechanism may be responsible for the introduction of species tolerant of low salinity into freshwater environments. Even though Miocene sea levels were high, and the general relief of Idaho, Oregon, and Washington was low during this time (Cole and Armentrout 1979), it seems doubtful that direct tidal introduction of *A. normanii* and its low-salinity variant, f. *subsalsa*, occurred in the Clarkia region, which is presently more than 600 km from the sea. However, if the modern analogs of the Clarkia and Oviatt Creek diatom floras can be taken at face value, they suggest that the Miocene lakes in this region drained to the Pacific through relatively slow-moving, low-gradient rivers. The extensive distribution of *Actinocyclus* species in other mid-Miocene lacustrine deposits of this area implies that similar conditions existed at various times in the Miocene, and/or that widely separated lake basins were interconnected by an extensive drainage system.

STRATIGRAPHIC PALEOLIMNOLOGY OF THE CLARKIA DEPOSIT: P-33

Because of poor preservation and sparse representation of diatoms at locality P-33, a detailed paleolimnological study of the Miocene Clarkia lake at this site is not feasible. The abundance of allochthonous fossils (plant megafossils, insects, etc.) and the generally coarse nature of the sediments indicate comparatively near shore deposition, although the lake may have been at least 10 m deep at this site (Smiley and Rember 1979). In such circumstances the diatoms are often abraded and corroded, and diluted by influx of clastic sediments. The best preservation of diatoms is in the upper 3 m of oxidized sediments. In samples studied from this section, the dominant form is *Melosira distans* v. *alpigena*. Species of *Tetracyclus, Fragilaria, Eunotia, Tabellaria* and *Actinocyclus* are subdominants. Although the abundance ranking of the subdominants varies throughout this section, no substantial limnological changes are indicated. The generally better preservation of diatoms in the upper oxidized sediments and their greater abundance here as compared to the lower unoxidized sediments suggest that the lake was more productive and probably shallower during the later phase of its history when the diatom-rich sediments were being deposited. The greater organic content of the lower sediments presumably reflects large influxes of allochthonous organic material rather than autochthonous organic production in the lake.

CORRELATION OF THE CLARKIA AND OVIATT CREEK DEPOSITS WITH OTHER LOCALITIES

Although it is premature to propose precise correlations between the diatomaceous deposits of the Clarkia region and other lacustrine sequences in the western United States, it is nevertheless important to point out that closely similar species of *Actinocyclus* exist at other localities. A diatomite, the Squaw Creek Member of the Ellensburg Formation in Yakima Canyon, south central Washington contains species of *Actinocyclus* very near *A. normanii* f. *subsalsa*. This diatomite is dated at between 12 and 14 my (Swanson et al. 1979) and is apparently older than the Priest Rapids Member, which also overlies the Clarkia and Oviatt Creek deposits.

A diatomite about 1 km south of the city limit of Spokane, Washington, also contains a centric diatom that in the light microscope appears identical to the form identified as *Actinocyclus* sp. aff. *A. normanii* f. *subsalsa* in the Clarkia and Oviatt Creek deposits. In addition, the diatom flora of this deposit contains many other species that characterize the Clarkia and Oviatt Creek sediments, such as *Melosira distans* v. *alpigena, M. distans, M. praeislandica*, and species of *Tabellaria, Tetracyclus*, and *Eunotia*. This suggests that these aquatic environments were of similar ecology. Although precise geologic correlations have not been made, the diatomaceous sediments south of Spokane apparently belong to the Latah Formation, which has a date of 14.5 my (Evernden and James 1964), and thus falls generally in the range of ages suggested for the lacustrine deposits in the Clarkia region.

The lake beds near Jungle Point, about 10 km north of Lowell, Idaho, also contain an *Actinocyclus* sp. that is related to but not identical to the forms from Clarkia, Oviatt Creek, and the other localities mentioned. The beds near Jungle Point are considered to be about the same age as the Latah flora, based on a 15.5 my radiometric date obtained from Geochron Laboratories, Cambridge, Mass. (Smiley and Rember 1979).

Most of the fossil forms of *Actinocyclus* sp. aff. *A. normanii* f. *subsalsa* appear to fit within the known range of morphological variation of extant populations of *A. normanii* and *A. normanii* f. *subsalsa*. Consequently, these diatoms are assumed to have a long geological range. They do not, therefore, necessarily demonstrate contemporaneity between the deposits discussed above, because environments suitable for this species probably existed at different times throughout the Miocene lacustrine history of the northwestern United States. Nevertheless, the radiometric dates associated with deposits containing *Actinocyclus* sp. aff. *A. normanii* f. *subsalsa* confirm that such conditions were common in this area during the early Neogene, and apparently this species can be considered a paleoenvironmental indicator that has some biochronologic utility on a regional basis.

EVOLUTION AND DISTRIBUTION OF NEOGENE FRESHWATER DIATOMS

The diatomaceous deposits at Clarkia, Oviatt Creek and other previously mentioned localities in the northwestern United States provide important information that bears on our understanding of the nature of evolution of freshwater diatoms and the mechanisms by which they become distributed to distant areas. A complete understanding of this problem will require extensive documentation and dating of freshwater diatomaceous deposits, and will not likely be forthcoming in the near future. However, some general aspects of the question can be outlined as a framework for future research. The general lack of rigorous data concerning the age and morphology of both fossil and modern diatom collections is a significant impediment to a responsible discussion of this topic, but hopefully this theoretical approach will be of some value.

In a study of the diatomaceous deposits of the western Snake River Basin, southwestern Idaho, Bradbury and Krebs (1982) documented a number of extinct species of centric diatoms, particularly of the genera *Actinocyclus, Cyclotella* and *Stephanodiscus*. The comparison of these diatoms to similar or identical species in other areas of the western United States and other areas of the world led to the development of the following model of the evolution and distribution of lacustrine diatoms.

Consideration of the problem of diatom evolution and biogeographic distribution must incorporate the following observations, considered to be generally true if not without exception.

1. Paleogene and Neogene lacustrine diatoms are morphologically modern. This statement refers to the apparent fact that the earliest (Eocene to Oligocene) lacustrine diatoms have structures, such as raphes, labiate processes, areolae and so forth, that are identical to or closely resemble those of modern species. Their placement with respect to each other on the frustule may vary. Consequently, the form can often be described as an extinct diatom, but it has not been possible to convincingly demonstrate the existence of a primitive feature that undergoes progressive change over a period of millions of years to arrive at a modern homologous structure. It appears that the genetic disposition to produce these structures has existed for a long time, and that throughout much of the Tertiary, diatom evolution has occupied itself largely with the rearrangement, reduction or multiplication of these structures, presumably to fit specific environmental needs.

   Phylogenetic interpretations of diatom structures are most convincingly derived from the study of modern forms. Ancestral forms are either hypothetical and usually without a fossil record, or are represented by modern "more generalized" species.

2. Many diatom species appear to have long stratigraphic records. It is not uncommon that a significant proportion of a fossil diatom flora will consist of forms that can be approximately classified as modern species using standard descriptions from European or North American diatom floras. Although this may result from non-rigorous description of type materia, or from non-rigorous examination of fossil forms, the most critical taxonomist must admit that the difference between presumably extinct taxa and their modern counterparts is often slight. This circumstance is a reflection of the first observation: that Tertiary lacustrine diatoms are morphologically modern. The first appearance of new (modern) genera in Tertiary lacustrine sediments may represent a more important migrational or biostratigraphic event than the first appearance of a modern species.

3. The biogeographic distribution of modern diatoms is very nearly cosmopolitan. This statement obviously becomes more correct with loose specific descriptions than with

rigid ones. The recent trend to synonymize a number of taxa described from different parts of the world by diatomists of the early and middle 20th century (e.g. Lange-Bertalot and Simonson 1978) suggests that the geographic location of a sample may have played too great a role in a diatom's taxonomy at that time.
4. In general, endemic diatom species are rare and often confined to large, stable and ancient lacustrine systems. This partly follows from the previous statement, and if that statement is true, its corollary would be that unstable, temporary, small, and young aquatic environments would be the least likely to contain endemic species.

If these four statements are acceptable in a general way, they provide insights into the mechanisms that determine the stratigraphic and geographic distribution of lacustrine diatoms. Lacustrine, or inland aquatic environments, can be considered in terms of their temporal duration (long-lived, stable systems vs. short-lived, temporary and variable systems) and in terms of their geographic distribution (widespread over many parts of the earth vs. restricted local occurrence). A simple matrix involving combinations of these two possibilities can be constructed (Fig. 2), and the evolutionary result of colonization of the possible aquatic habitats by a community of generalist diatoms can be hypothesized.

It is assumed that generally successful colonizing diatoms will come from a community of eurytopic species that are tolerant of wide and rapid fluctuations in limnological parameters. They will be replaced in stable limnologic systems by species clones, or ecotypes that become adapted to specific limnologic conditions that persist, perhaps with predictable fluctuations in response to seasonal changes, over long periods of time. The distribution of these distinctive specialized diatoms will depend upon the size and distribution of the special habitat available to them. If such habitats are widespread, as appears to have been the case in southern Idaho during Miocene and early Pliocene time, the distinctive diatoms will have widespread geographic distribution (case 1 in Figure 2), and consequently the diatoms can become useful biostratigraphic and biochronologic markers (Bradbury and Krebs 1982).

If stable, long-lived lacustrine habitats are comparatively rare, as they are in the world today (examples are Lake Baikal, the African Rift Valley lakes, and Lake Ohrid), then the specialized diatoms that develop in them are known as local endemics (case 2 in Figure 2). Such local endemics coexist with a widely distributed cosmopolitan community of generalist diatoms that are ecologically sorted out among common aquatic habitats such as acid bogs and alkaline ponds, which exist in many parts of the world for short periods of time (case 3 in Figure 2). Case 4 of Figure 2, a theoretical extreme of case 3, would in general not favor the prolonged development of any type of aquatic life. Only the most eurytopic species, perhaps also adapted to survival in terrestrial environments, would be expected in this situation.

This model of diatom evolution adaptation relates in some respects to the concept of K vs. r selection (specialists vs. opportunists) (McArthur and Wilson 1967) and seems to fit comfortably with the concepts of evolution via punctuated equilibria (Gould and Eldredge 1977). The diverse reproductive strategies available to diatoms (Drebes 1977) insure that distinctive forms, responding to specific environmental and interspecific selective pressures, could quickly proliferate in stable lacustrine systems. Coupling such adaptive strategies to temporal and geographic distributions of suitable environments appears to account for the observed features of fossil diatom distributions. The fact that temporal and spatial distribution of environments promoting specialist diatoms (K strategists) is closely linked to paleoclimatic and climatic factors makes it logical that some Miocene diatomaceous deposits, which formed under comparatively equitable climates with increased effective moisture, should be more easily correlated on the basis of distinctive and widespread diatoms than Pleistocene deposits forming under circumstances of rapidly changing environments.

The diatom sequences of the western Snake River Basin represent an example of specialized

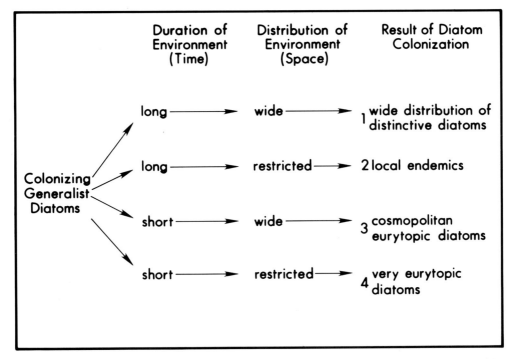

Figure 2. A matrix classification of aquatic environments on a temporal and spatial basis, and the evolutionary result of colonization by generalist diatoms (from Bradbury and Krebs 1982).

diatoms inhabiting large, stable lacustrine systems in mid-Miocene to Pliocene time (Bradbury and Krebs 1982). The Clarkia and Oviatt Creek deposits, however, appear to represent examples of case 3 (Fig. 2) in the scheme of evolutionary options of Neogene freshwater diatoms. These aquatic environments, although they existed under similar stable and equitable climatic conditions in the mid-Miocene, were apparently too short-lived to acquire a distinctive flora of diatom species uniquely adapted to then existing limnological conditions. The diatoms are morphologically similar or identical to modern forms that exist in similar environments at present, even though they are as old as or older than the distinctive floras of the western Snake River Basin.

Additional radiometric control and morphological documentation and comparison of Neogene freshwater diatoms will be required to confirm or modify these speculations. Ultimately, detailed morphological and taxonomic studies of dated diatomaceous sediments of Miocene age in the western United States will provide valuable insights into the regional tectonic and paleoclimatic history of this area.

LITERATURE CITED

Bradbury, J. P., and W. N. Krebs. 1982. Neogene lacustrine diatoms of the western Snake River Basin, Idaho-Oregon, U. S. A. Acta Geologica Academiae Scientarium Hungaricae 25(12): 97-122.

Cole, M. R., and J. M. Armentrout. 1979. Neogene paleogeography of the western United States. Pages 297-323 in J. M. Armentrout, M. R. Cole, and H. Terbest, Jr., eds. Cenozoic Paleogeography of the Western United States: Pacific Coast Paleogeography Symposium 3, Pacific Section, Soc. Econ. Paleont. Mineral.

Drebes, G. 1977. Sexuality. Chapter 9 in D. Werner, ed. The Biology of Diatoms: Botanical

Monographs, Vol. 13. University of California Press, Los Angeles, Calif. 498 pp.
Evernden, J. F., and G. T. James. 1964. Potassium-argon dates and the Tertiary floras of North America. Amer. J. Sci. 62:945-974.
Gessner, F., and R. Simonsen. 1967. Marine diatoms in the Amazon. Limnol. Oceanog. 12:709-711.
Gould, S. J., and N. Eldredge. 1977. Punctuated equilibria: The tempo and mode of evolution reconsidered. Paleobiology 3(2):115-151.
Hasle, G. R. 1977. Morphology and taxonomy of *Actinocyclus normanii* f. *subsalsa* (Bacillariophyceae). Phycologia 16(3):321-328.
Hohn, M. H. 1966. The Catherwood Foundation Peruvian-Amazon Expedition, XVII-Bacillariophyta. Monogr. Acad. Nat. Sci. Phila. 14:459-495.
Hustedt, F. 1957. Die Diatomeenflora des Fluss-systems der Weser im Gebiet des Hansestadt Bremen. Abdhl. naturw. Ver. Bremen 34(3):181-440.
Lange-Bertalot, H., and R. Simonson. 1978. A taxonomic revision of the Nitzschiae Lanceolatae Grunow. Bacillaria 1:11-111.
MacArthur, R. H., and E. O. Wilson. 1967. The theory of island biogeography. Princeton University Press, Princeton, N. J. 203 pp.
Patrick, R. 1966. The Catherwood Foundation Peruvian-Amazon Expedition. Introduction. I-Limnological observations and discussion of results. Monogr. Acad. Nat. Sci. Phila. 14:1-40.
Patrick, R., J. Cairns, Jr., and S. S. Roback. 1967. An ecosystematic study of the fauna and flora of the Savannah River. Proc. Acad. Nat. Sci. Phila. 118(5):109-407.
Simola, H. 1977. Diatom succession in the formation of annually laminated sediment in Lovojarvi, a small, eutrophicated lake. Ann. Bot. Fenn. 14:143-148.
Smiley, C. J., J. Gray, and L. M. Huggins. 1975. Preservation of Miocene fossils in unoxidized lake deposits, Clarkia, Idaho. J. Paleont. 49(5):833-844.
Smiley, C. J., and W. C. Rember. 1979. Guidebook and road log to the St. Maries River (Clarkia) fossil area of northern Idaho. Idaho Bur. Mines Geol. Inf. Circ. 33. 27 pp.
Stoermer, E. F., M. M. Bowman, J. C. Kingston, and A. L. Schaedel. 1974. Phytoplankton composition and abundance in Lake Ontario during IFYGL, Final Report, U. S. Environmental Protection Agency Research Grant R-800605: Special Report No. 53, Great Lakes Research Division, University of Michigan. 373 pp.
Swanson, D. A., T. L. Wright, P. R. Hooper, and R. D. Bentley. 1979. Revisions in the stratigraphic nomenclature of the Columbia River Basalt Group. U. S. Geol. Surv. Bull. 1457-G. 59 pp.
VanLandingham, S. L. 1964. Miocene non-marine diatoms from the Yakima region in south-central Washington. Beihefte zur Nova Hedwigia, vol. 14, 78 pp.

PLATES

PLATE EXPLANATIONS:

Scale bars on all figures equal 5 micrometers unless otherwise noted.
Authorship is provided for taxa not listed in Table 1.

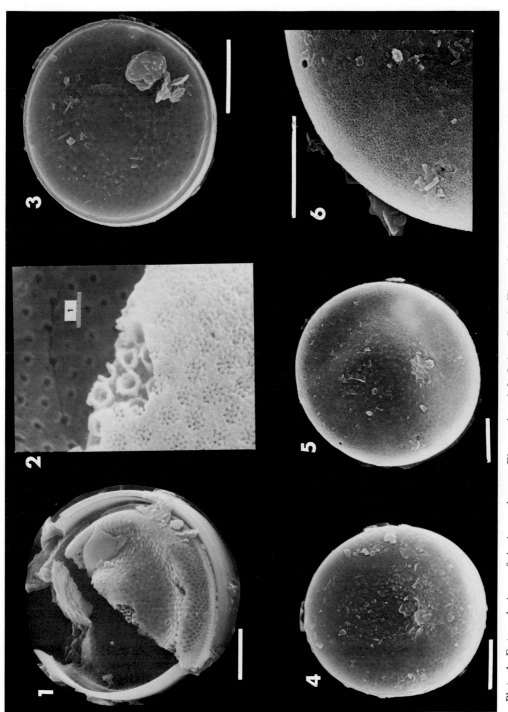

Plate 1. External views of *Actinocyclus* sp. Figures 1 and 2, Oviatt Creek. Figure 1 shows internal view with labiate process. Figure 2, detail of areolae. Figures 3-6, Clarkia, Emerald Creek locality. Note presence of pseudonodulus in Figures 5 and 6.

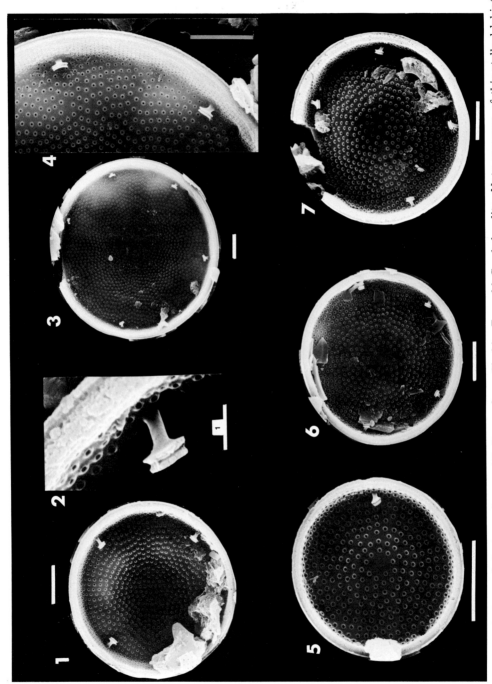

Plate 2. Internal views of *Actinocyclus* sp. All figures from Clarkia, Emerald Creek locality. Note presence of thin-stalked labiate processes and the internal opening of the pseudonodulus in Figures 3-6.

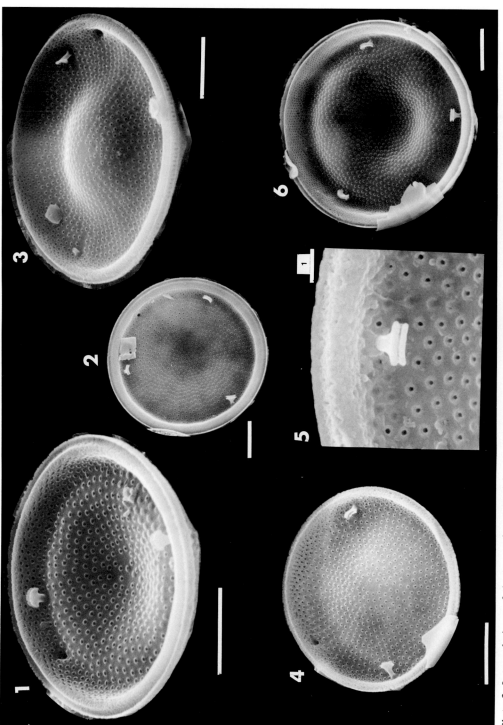

Plate 3. Internal views of *Actinocyclus* sp. All figures from Oviatt Creek. Note presence of pseudonodulus in Figures 1-4 and 6.

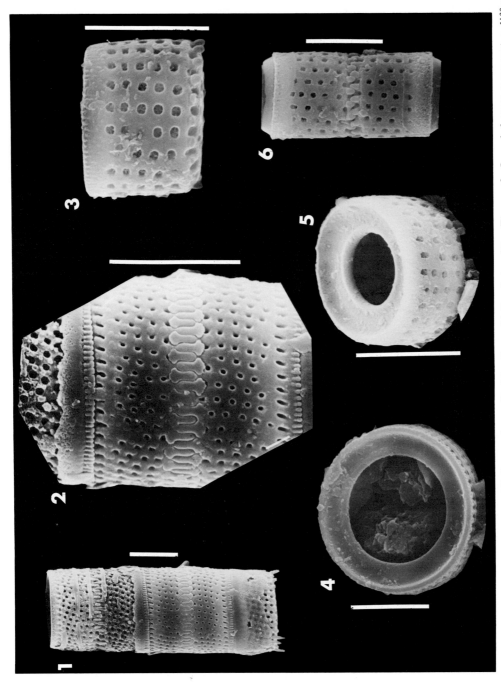

Plate 4. *Melosira* sp. aff. *M. distans*. All figures from Clarkia, Emerald Creek locality. Figures 3, 5, and 6 may represent a different species, similar in several respects to specimens identified as *Melosira praeislandica* Jouse, but not corresponding precisely to published descriptions of that taxon.

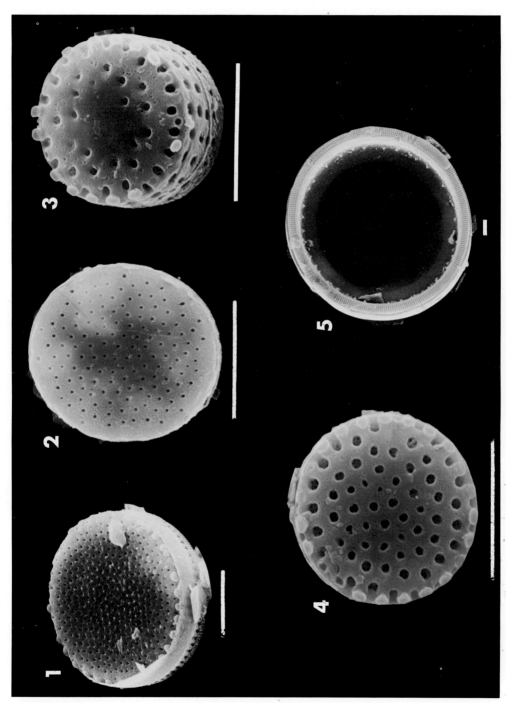

Plate 5. Figures 1 and 2, *Melosira* sp. aff. *M. distans*. Figures 3 and 4, *Melosira* sp. cf. *M. praeislandica*. Figure 5, *Melosira teres*. All figures from Clarkia, Emerald Creek locality.

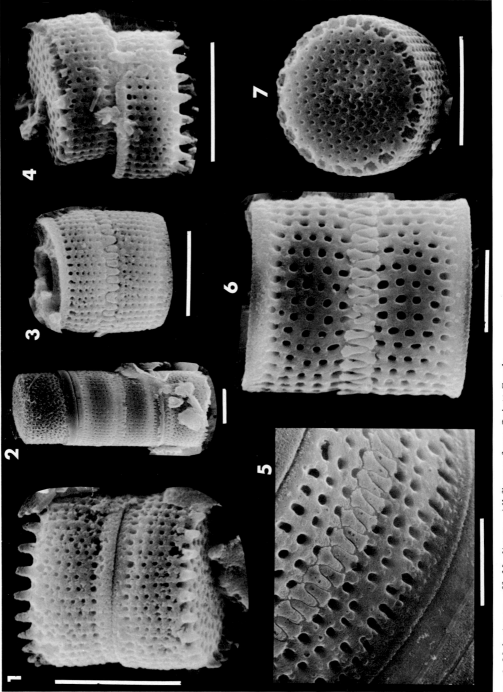

Plate 6. *Melosira* sp. aff. *M. distans*. All figures from Oviatt Creek.

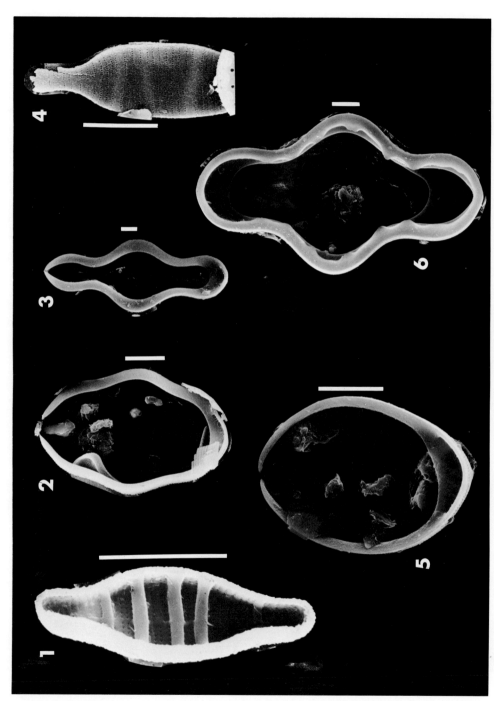

Plate 7. Figures 1 and 4, *Diatoma* sp. aff. *D. anceps* (Ehr.) Grun. or *D. capitata* Lauby. Figures 2, 3, 5, 6, *Tetracyclus* septae. Figures 2, 3 and 6, *T. lacustris*. Figure 5, *T. ellipticus*. All specimens from Clarkia, Emerald Creek locality.

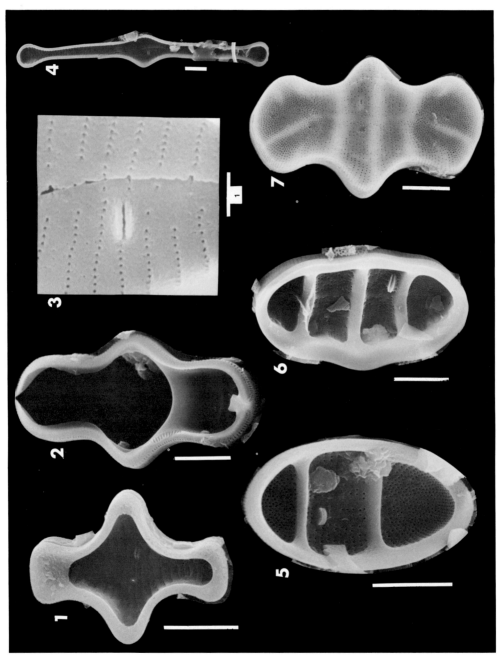

Plate 8. Figure 1, *Tetracyclus lacustris*, septum and internal view of valve. Figure 2, septum of *T. lacustris*. Figures 3 and 4, internal view of *Tabellaria* sp. cf. *T. fenestrata* showing central labiate process. Figure 5, internal view of *T. ellipticus*. Figure 6, internal view of *T. lacustris*. Figure 7, external view of *T. lacustris*. All specimens from Oviatt Creek.

Plate 9. Figures 1, 2, 5, 6, *Fragilaria* sp. aff. *F. construens* var. *subsalina*. Figure 6 shows two valves with interlocking marginal spines. Figure 3, *Fragilaria* sp. aff. *F. pinnata* Ehr. Figure 4, *Fragilaria* sp. aff. *F. constricta*. All specimens from Oviatt Creek.

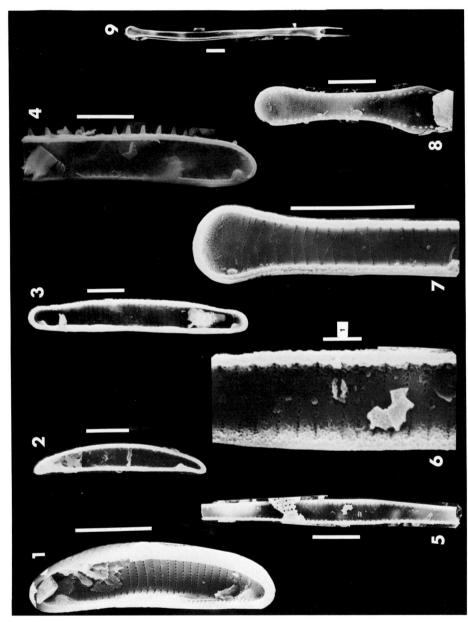

Plate 10. Figure 1, internal view, *Eunotia* sp. cf. *E. vanheurckii*. Figure 2, internal view, *Eunotia* sp. cf. *E. incisa*. Figure 3, internal view, *Eunotia* sp. cf. *E. pectinalis*. Figure 4, internal view, *Eunotia* sp. aff. *E. denticulata* (Breb.) Rabh. Figures 5 and 6, *Tabellaria* sp., internal view showing labiate process. Figure 7, internal view, *Tabellaria* sp. cf. *T. fenestrata*. Figure 8, external view, *Tabellaria* sp. cf. *T. fenestrata*. Figure 9, septum, *Tabellaria* sp. All specimens from Clarkia, Emerald Creek locality.

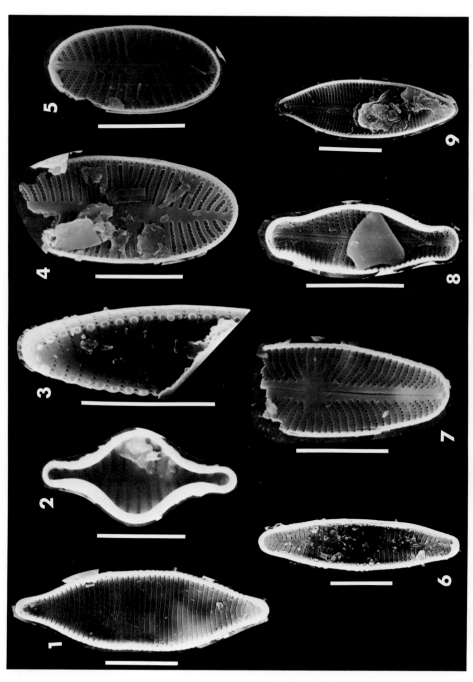

Plate 11. Figure 1, *Fragilaria* sp. aff. *F. constricta*, internal view. Figure 2, *Fragilaria* sp. cf. *F. construens*, internal view. Figure 3, *Fragilaria* sp. cf. *F. construens* var. *subsalina*, external view. Figures 4 and 5, *Achnanthes* sp. cf. *A. kryophila* Petersen (assuming raphe and pseudoraphe valves are from same taxon). Figure 6, *Achnanthes* ? sp. Figures 7-9, *Achnanthes* ? spp. Figures 4-9 are all internal views. All specimens from Clarkia, Emerald Creek locality.

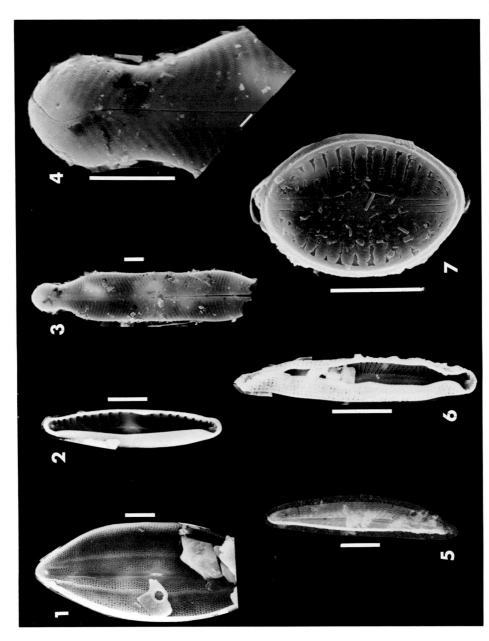

Plate 12. Figure 1, *Neidium* sp. aff. *N. dubium* (Ehr.) Cleve, internal view. Figure 2, *Gomphonema* sp. cf. *G. dichotomum*, internal view. Figures 3 and 4, *Pinnularia* sp. cf. *P. mesolepta* (Ehr.) Wm. Sm. Figure 5, *Cymbella* ? sp., external? view. Figure 6, *Navicula* sp. cf. *N. subtilissima* Cl., internal view. Figure 7, *Diploneis*? sp., external view. All specimens from Clarkia, Emerald Creek locality.

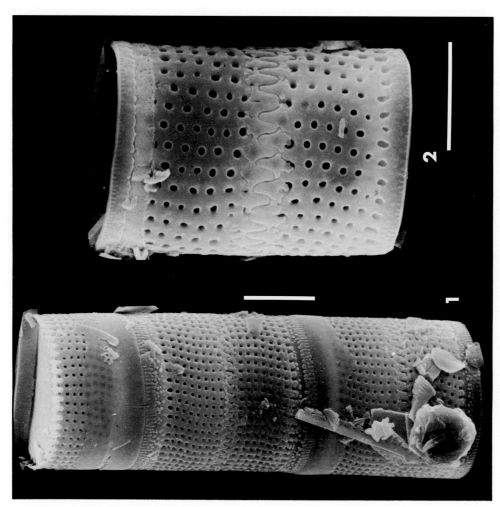

Plate 13. Figures 1 and 2. External views of *Melosira distans*. Holocene, Heart Lake, Adirondack Mountains, New York. Specimens and photos courtesy of Susan Reed, University of Indiana, Bloomington.

MIOCENE CHRYSOPHYTA CYSTS FROM A LACUSTRINE DEPOSIT IN NORTHERN IDAHO

JOHN L. WILLIAMS
Cities Service Oil & Gas Corp., 1600 Broadway, Suite 900, Denver, CO 80202

The Clarkia, Idaho fossil locality (P-33) yielded an abundance of Chrysophyta cysts. Fifteen distinct types were noted from this well-preserved assemblage. Many of these forms are similar to previously described cysts, whereas others appear significantly different.

A common function in many unicellular organisms is that of encystment. This function serves as a survival mechanism during adverse environmental conditions and is particularly common among the Chrysophyceae. Cornell (1969) stated that of the fifteen orders of living, flagellated, yellow-brown algae, ten produce siliceous cysts.

These cysts are frequently preserved in the fossil record. The freshwater forms are assigned to the cyst-family Chrysostomataceae, based on the geologic occurrence of the cysts and on their similarity to cysts produced by living taxa (Cornell 1969). Bourrelly (1963) noted an abundance of these freshwater forms in diatomaceous lake deposits. The Clarkia, Idaho locality is another diatomaceous lacustrine deposit in which such cysts are both abundant and diverse.

VanLandingham (1964) noted that the identification of these cysts is very difficult due to problems of variation in cyst types. This is expressed by the fact that one chrysophycean species may produce very different cysts, while two other species may produce identical or nearly identical cysts.

The cysts described here were categorized by their general morphological features. To avoid taxonomic confusion, these morphological groupings have been referred to as "cyst type" 1, 2, etc. Affinities of the cysts with the vegetative phase of the algae are not dealt with in this study.

METHODS

Several small samples were obtained from the upper part (100-175 cm below the top of the exposure) of the Clarkia (P-33) locality (NW¼NE¼ Sec. 13, T. 42 N., R. 1 E.). These samples were combined into one composite sample and chemically treated using the "dry sample" technique described by Schrader (1973). The siliceous residue was mounted in Hyrax, a highly refractive mounting medium.

The preparation of this material for scanning electron microscopy was very similar to that employed for light microscopy. Following the chemical process, the residue was diluted with distilled water and a few drops of this solution were then placed on a cover glass and allowed to dry. When completely dry, double-sided tape was used to mount the cover glass to an aluminum SEM stub. The last step involved coating the cover glass mount with a thin layer of gold-palladium. A JEOL-35 scanning electron microscope was used to examine this material, using an accelerating voltage of 25 kV.

Copyright ©1985, Pacific Division, AAAS.

SYSTEMATIC PALEONTOLOGY

Incertae sedis

CYST TYPE 1 (Figure 1)

Description. This ovate cyst generally has a smooth surface with few shallow depressions of various sizes. The widest end is typically flattened. Although numerous cysts were observed, their orientation was such that the pore was not visible.

Discussion. Kaczmarska (1976, Pl. 4, fig. 7) illustrated a form comparable to cyst type 1. Another similar form was described by VanLandingham (1964) as cyst type 30. His specimen displays a pore, which is not evident in the present study, possibly due to improper orientation. The size of VanLandingham's cyst type 30 is typically larger than cyst type 1. However, VanLandingham (1964) notes that smaller forms (possibly within the range of cyst type 1) of his cyst type 30 do exist.

CYST TYPE 2 (Figure 5)

Discussion. This cyst is quite similar to cyst type 1, except that the ends are more tapered.

CYST TYPE 3 (Figure 2)

Description. The surface ornamentation of this cyst consists of a fine reticulation with the lumina arranged in linear, longitudinal rows. A narrow, cylindrical collar occurs at the abruptly-tapering apical end. The specimen in figure 2 was photographed at low focus and does not show the ornamentation.

CYST TYPE 4 (Figures 3, 4)

Description. The surface of this globose cyst is finely reticulate, with no linear pattern of the lumina. The collar is stout and cylindrical.

CYST TYPE 5 (Figures 6, 7)

Description. This cyst is globose and coarsely reticulate with irregularly shaped lumina. The apical end is ornamented with short, slightly curved spines of equal length.

CYST TYPE 6 (Figure 8)
cf. *Carnegia armata* Frenguelli, 1932

Description. Short, rather slender, straight to curved spines sparsely cover this ovoid cyst. Longer spines of this same nature are concentrated at the apical end.

Discussion. This cyst is similar to *Carnegia armata* (see Gritten, 1977, Pl. 6, fig. 4) in size, shape, and spinal arrangement, but lacks the stoutness and denticulate flange terminations that encircle the pore of *C. armata*.

CYST TYPE 7 (Figure 9)

Description. The surface of this ovoid cyst is sparsely ornamented with stout, blunt spines of nearly equal length.

CYST TYPE 8 (Figure 10)
cf. *Carnegia johannis* Andrieu, 1937

Description. This ovoid cyst is wider towards its antapical end and has coarse reticulation in which the muri form polygonal lumina. The apical end is characterized by a collar, which is higher near the site of insertion of a short aliform hook.

Discussion. This cyst is nearly identical to *C. johannis* (see Gritten, 1977, Pl. 7, fig. 6). The only difference is the width of the muri, which are slightly wider in cyst type 8 than they are in *C. johannis*. This minor difference may be insignificant, and cyst type 8 may be conspecific with *C. johannis*.

Gritten (1977) distinguished *C. johannis* from *C. willingtoniensis* Conrad, 1940 by its more delicate ribbing and shorter, more acute hook. This hook is short and extends out away from the cyst in cyst type 8, whereas in *C. willingtoniensis* (Gritten, 1977, Pl. 7, fig. 4), it is long, curves over the pore, and extends down the side of the cyst in a sinuous fashion.

CYST TYPE 9 (Figures 11-13)

Description. The surface of this spherical cyst is sub-reticulate to reticulate, with broadly based acuminate spines at the intersections of the muri. The cylindrical collar is short and broad, thus giving rise to a very conspicuous pore aperture.

Discussion. This cyst compares closely with VanLandingham's (1964) cyst type 5.

CYST TYPE 10 (Figure 14)

Description. Numerous short, pointed, broad-based spines are irregularly distributed over the surface of this cyst. The majority of these spines are isolated, but some are connected at their bases by small ridges. The cylindrical collar is very broad and short.

Discussion. Cyst type 10 differs from cyst type 9 in having less conspicuous interconnecting ridges between spines. This may be the result of corrosion, and cyst type 10 may be merely an etched form of cyst type 9. On the other hand, these cyst types may represent variations within a species.

CYST TYPE 11 (Figures 15, 16)

Description. The surface of this subspherical cyst is covered with short, dome-shaped to acuminate protuberances, which are incompletely connected at their bases by short ridges.

CYST TYPE 12 (Figures 17, 18)

Description. This cyst is subspherical and typically flattened at the apical end. It is ornamented with numerous short spines which are broad-based and sharply acuminate. The apical end bears a conspicuous obconical collar.

CYST TYPE 13 (Figure 19)
cf. *Cysta subsphaerica* Nygaard, 1956

Description. This form is characteristically smooth, elliptical to ovate, and has a very conspicuous pore near the middle of its longest axis.

Discussion. This cyst is comparable with *C. subsphaerica* (see Leventhal, 1970, fig. X-11).

CYST TYPE 14 (Figure 20)
cf. *Cysta decollata* Playfair, 1915

Description. This cyst is ovate and entirely smooth. The pore is easily recognized and located at one end.

Discussion. Cyst type 14 is comparable to *C. decollata* as illustrated by Gritten (1977, Pl. 2, fig. 5). Cyst type 280 of Adam and Mahood (1980, Pl. D, fig. 3) also closely resembles this form.

CYST TYPE 15 (Figures 21, 22)
cf. *Cysta subsphaerica* Nygaard, 1956

Description. The finely verrucate surface of this globose cyst is interrupted by the presence of a pore that is inset in the inner wall.

Discussion. This cyst type closely resembles *C. sphaerica* as described and illustrated by Gritten (1977, Pl. 2, fig. 1) and Nygaard (1956, Pl. 10, figs. 1-7).

ACKNOWLEDGMENTS

I am grateful to David P. Adam, U.S.G.S. (Menlo Park), and Dr. Frederick H. Wingate, Cities Service Oil & Gas Corporation, for reading the manuscript and giving helpful advice. The scanning electron micrographs were taken by Miss Kathryn V. Dieterich, U.S.G.S. (Denver).

EXPLANATION OF FIGURES

Figures 13, 14, and 22 are scanning electron micrographs. All other figures are light micrographs using bright field illumination. The measurements correspond to the largest diameter of the specimen. Specimen locations are in parentheses; letters correspond to the slide and numbers to the ring on that slide.

Fig. 1. Cyst type 1, (D-7), 23 μm.
    2. Cyst type 3, (C-2), 25 μm.
    3. Cyst type 4, (F-15), 15 μm, high focus.
    4. Cyst type 4, (F-15), 15 μm, low focus.
    5. Cyst type 2, (F-6), 23 μm.
    6. Cyst type 5, (B-4), 13 μm, high focus.
    7. Cyst type 5, (B-4), 13 μm, low focus.
    8. Cyst type 6 cf. *Carnegia armata*, (B-9), 17 μm.
    9. Cyst type 7, (B-9), 15 μm.
    10. Cyst type 8 cf. *Carnegia johannis*, (A-14), 13 μm.
    11. Cyst type 9, (B-4), 11 μm.
    12. Cyst type 9, (B-4), 10 μm.
    13. Cyst type 9, 12 μm.
    14. Cyst type 10, 11 μm.
    15. Cyst type 11, (D-5), 10 μm, high focus.
    16. Cyst type 11, (D-5), 10 μm, low focus.
    17. Cyst type 12, (B-2), 11 μm, high focus.
    18. Cyst type 12, (B-2), 11 μm, low focus.
    19. Cyst type 13 cf. *Cysta subsphaerica*, (B-13), 16 μm.
    20. Cyst type 14 cf. *Cysta decollata*, (C-7), 10 μm.
    21. Cyst type 15 cf. *Cysta subsphaerica*, (C-8), 8 μm.
    22. Cyst type 15 cf. *Cysta subsphaerica*, 9 μm.

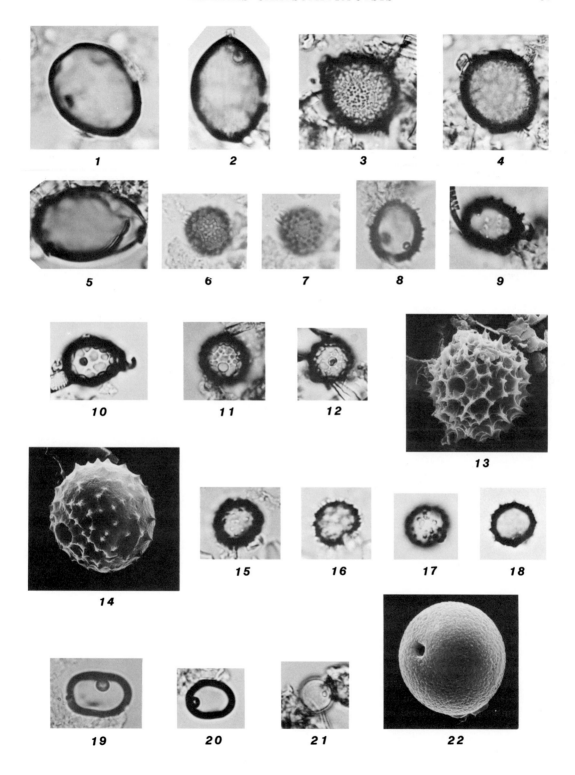

LITERATURE CITED

Adam, D. P., and A. D. Mahood. 1980. Modern chrysomonad cysts from Alta Morris Lake, El Dorado County, California. U.S. Geol. Surv. Open-File Rep. No. 80-822. 13 pp.

Bourrelly, P. 1963. Loricae and cysts in the Chrysophyceae. Ann. N. Y. Acad. Sci. 108(2):421-429.

Cornell, W. C. 1969. The chrysomonad cyst-families Chrysostomataceae and Archaeomonadaceae: Their status in paleontology. Pages 958-965 *in* E. L. Yochelson, ed., Proc. N. Amer. Paleont. Conv. Vol. 2, pt. G.

Gritten, M. M. 1977. On the fine structure of some Chrysophycean cysts. Hydrobiologia 53(3):239-252.

Kaczmarska, I. 1976. Diatom analysis of Eemian profile in fresh-water deposits at Imbramowice near Wroclaw. Acta Palaeobot. 17(2):3-34.

Leventhal, E. A. 1970. The Chrysomonadina. *In* G. E. Hutchinson, ed. Ianula: An Account of the History and Development of the Lago di Monterosi, Latium, Italy. Trans. Amer. Phil. Soc. 60:123-142.

Nygaard, G. 1956. Ancient and Recent flora of diatoms and Chrysophyceae in Lake Gribsø. *In* K. Berg and I. E. Peterson, eds. Studies on the Humic Acid Lake Gribsø. Folia Limnol. Scand. 8:32-94.

Schrader, H. J. 1973. Proposal for a standardized method of cleaning diatom-bearing deep-sea and land-exposed marine sediments. *In* R. Simonsen, ed. Second Symposium on Recent and fossil marine diatoms. Beihefte zur Nova Hedwigia 45:403-409.

VanLandingham, S. L. 1964. Chrysophyta cysts from the Yakima Basalt (Miocene) in south-central Washington. J. Paleont. 38(4):729-739.

THECAMOEBIAN SCALES FROM A MIOCENE LACUSTRINE DEPOSIT IN NORTHERN IDAHO

JOHN L. WILLIAMS
Cities Service Oil & Gas Corp., 1600 Broadway (Suite 900), Denver, CO 80202

Thecamoebian scales are both abundant and diverse in the upper portion of the Clarkia (P-33) locality. These siliceous scales strongly resemble those from modern forms in the family Euglyphidae; the majority appear to represent species of *Euglypha*. The suggested paleoenvironment of these thecamoebians, based on the studies of modern forms, appears to be that of a lentic habitat in which submerged vegetation, moss, and sphagnum are common.

While examining diatoms from the Clarkia (P-33) locality, an abundance of ovoid, siliceous "plates" was noted. Detailed investigation of these "plates" suggests they are comparable with the siliceous scales that cover the test of extant amoeboid organisms belonging to the Euglyphidae, a family of Protozoa (Fig. 1).

These animals have been referred to by paleontologists as thecamoebians. They are common in modern environments and have consequently been studied by numerous protozoologists. Fossil thecamoebian scales, on the other hand, have been studied to a much lesser degree. Scales from various Quaternary sediments have been found to be identical with those of modern forms (Takahashi, pers. commun., 1980).[1] The majority of the scales from the Miocene Clarkia locality closely resemble those of modern species of the genus *Euglypha* Dujardin.

METHODS

Several small samples were obtained from the upper part (100-175 cm below the top of the exposure) of the Clarkia (P-33) locality (NW¼ NE¼ Sec. 13, T. 42 N., R. 1 E.). These samples were combined into one composite sample and chemically treated using the "dry sample" technique described by Schrader (1973). The siliceous residue was mounted in Hyrax, a highly refractive mounting medium.

The preparation of this material for scanning electron microscopy was very similar to that employed for light microscopy. Following the chemical process, the residue was diluted with

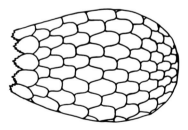

Figure 1. Generalized drawing showing lateral view of a thecamoebian test. This ovate shell is composed of symmetrical scales made of silica. The relatively small pseudostome is encircled by a row of denticulate scales; the rest of the lorica is ornamented with overlapping, oval body scales. The scales used in this reconstruction resemble those of *Euglypha* cf. *E. crenulata* from this study.

[1] Dr. Eiji Takahashi, Department of Biology, Kobe University, Nada, Kobe 657, Japan.

Copyright ©1985, Pacific Division, AAAS.

distilled water and a few drops of this solution were then placed on a cover glass and allowed to dry. When completely dry, double-sided tape was used to mount the cover glass to an aluminum SEM stub. The last step involved coating the cover glass with a thin layer of gold-palladium. A JEOL-35 scanning electron microscope was used to examine this material, using an accelerating voltage of 25 kV.

DISCUSSION

The Euglyphidae appears to be the dominant thecamoebian family at the Miocene Clarkia deposit (Plate I, Figs. 2-24; Plate II, Figs. 25-43). The scales recovered from this locality are assigned to the family Euglyphidae, primarily as species of *Euglypha*, based on individual scale morphology. Three forms are identified as *E. brachiata* Leidy, *E. crenulata* Wailes, and *E. nana* Decloitre. More tentative assignments have been made to the species *E. crenulata* var. *minor* Wailes, *E. scutigera* Penard, and *E. loevis* (Ehrenberg) Perty.

The Euglyphidae are amoeboid organisms that live within tests composed of siliceous scales. These tests have single apertures through which pseudopodia protrude. According to Loeblich and Tappan (1964), these cytoplasmic projections function in locomotion and food-capturing, with the primary food supply being diatoms, green algae, flagellates, and ciliates. This food supply is typically abundant in the general thecamoebian habitat proposed by Leidy (1879). He found these freshwater forms to be quite abundant in and adjacent to clear, partially shaded lakes and ponds.

The general habitat that these Euglyphidae occupied during the Miocene is inferred from ecological studies that have been done on the extant family. Decloitre (1962), Leidy (1879), Penard (1902), and Wailes (1911-1916), have all noted the general habitats of numerous species of *Euglypha*. Based on these studies of modern forms, the species from the Clarkia locality suggest an aquatic habitat in which the thecamoebians were associated with submerged vegetation, moss, and sphagnum of lakes, ponds, or swamps.

According to Loeblich and Tappan (1964), the known geologic range of the Euglyphidae, and in particular the *Euglypha*, is from Middle Eocene to Recent. Most of these occurrences have been from lake and other freshwater deposits. Continued studies of fossil thecamoebians may lead to refinement of the ranges for individual species and increased utility of these fossils as biostratigraphic tools. These fossil forms may also prove to be valuable paleoecological indicators as additional studies become available regarding the ecology of modern freshwater species. Also, continued studies of fossil Euglyphidae should lead to a better understanding of phylogenetic relationships within the group. Until such continued studies are forthcoming, our knowledge of the geologic history of the thecamoebians will remain sketchy and incomplete.

ACKNOWLEDGMENTS

I am indebted to Dr. Eiji Takahashi, Kobe University, and Dr. Frederick H. Wingate, Cities Service Oil & Gas Corporation, for reading the manuscript and for helpful discussions during the preparation of this paper. The technical assistance of Miss Kathryn V. Dieterich, USGS (Denver), in the preparation of most of the scanning electron micrographs is also gratefully acknowledged.

LITERATURE CITED

Decloitre, L. 1962. Le genre *Euglypha* Dujardin. Arch. Protistenk. 106:51-100.

Leidy, J. 1879. Fresh-water rhizopods from North America. U. S. Geol. Surv. Terr. Rep. 12:1-324.

Loeblich, A. R., and H. Tappan. 1964. Sarcodina-chiefly "thecamoebians" and Foraminiferida. Pages 16-54 *in* R. C. Moore, ed. Treatise on Invertebrate Paleontology. Protista 2, pt. C, vol. 1. University of Kansas Press, Lawrence, Kan.

Penard, E. 1902. Faune Rhizopodique du Bassin du Léman:. Kümdig, Genève. 714 pp.

Schrader, H. J. 1973. Proposal for a standardized method of cleaning diatom-bearing deep-sea and land-exposed marine sediments. *In* R. Simonsen, ed. Second Symposium on Recent and Fossil Marine Diatomes. Beihefte zur Nova Hedwigia 45:403-409.

Wailes, G. H. 1911-1916. Freshwater Rhizopoda and Heliozoa from the states of New York, New Jersey, and Georgia, U.S.A.; with a supplemental note on Seychelles species. Linn. Soc. J., Zool. 32:121-161.

PLATES I & II

The figures in these plates were photographed using either differential interference contrast (Nomarski) or scanning electron microscopy. All measurements represent the greatest length of the specimen. Specimen locations are in parentheses; letters correspond to the slide and numbers to the ring on that slide. (b.s.=body scale; p.s.=pseudostome scale; d.p.s.=denticulate pseudostome scale.)

PLATE I

Fig. 2. Euglyphidae, b.s., (F-19), 8 μm.
Fig. 3. Euglyphidae, b.s., (C-10), 7.5 μm.
Fig. 4. Euglyphidae, b.s., 10 μm.
Fig. 5. Euglyphidae, b.s., 8 μm.
Fig. 6. Euglyphidae, b.s., (C-8), 15 μm.
Fig. 7. Euglyphidae, b.s., (B-28), 6 μm.
Fig. 8. ?*Euglypha* sp., b.s., (F-10), 13 μm.
Fig. 9. *Euglypha* ?*scutigera*, b.s., (F-18), 10 μm.
Fig. 10. *Euglypha* cf. *E. nana*, p.s., (B-29), 10 μm.
Fig. 11. *Euglypha* cf. *E. brachiata*, b.s., 10 μm.
Fig. 12. *Euglypha brachiata*, b.s., (A-7), 15 μm.
Fig. 13. *Euglypha* cf. *E. brachiata*, b.s., (E-5), 10 μm.
Fig. 14. *Euglypha* ?*crenulata* var. *minor*, b.s., (F-22), 14 μm.
Fig. 15. *Euglypha* ?*crenulata* var. *minor*, b.s., (F-17), 13 μm.
Fig. 16. ?*Euglypha* sp., b.s., (E-9), 15 μm.
Fig. 17. ?*Euglypha* sp., b.s., (E-10), 15 μm.
Fig. 18. Euglyphidae, scale indeterminate, (C-5), 12.5 μm.
Fig. 19. Euglyphidae, scale indeterminate, (F-1), 15 μm.
Fig. 20. ?*Euglypha* sp., b.s., (F-21), 12 μm.
Fig. 21. ?*Euglypha* sp., b.s., (E-9), 13 μm.
Fig. 22. ?*Euglypha* sp., b.s., (B-7), 16 μm.
Fig. 23. ?*Euglypha* sp., b.s., 8 μm.
Fig. 24. *Euglypha* ?*brachiata*, b.s., 14 μm.

PLATE II

Fig. 25. *Euglypha* ?*brachiata*, b.s., (C-4), 12.5 μm.
Fig. 26. ?*Euglypha* sp., b.s., (F-13), 15 μm.
Fig. 27. ?*Euglypha* sp., b.s., (B-17), 12.5 μm.
Fig. 28. *Euglypha* ?*brachiata*, b.s., (B-23), 14 μm.
Fig. 29. ?*Euglypha* sp., b.s., (B-16), 26 μm.
Fig. 30. ?*Euglypha loevis*, p.s., (D-2), 28 μm.
Fig. 31. ?*Euglypha* sp., b.s., (F-2), 18 μm.
Fig. 32. ?*Euglypha* sp., b.s., 13 μm.
Fig. 33. ?*Euglypha* sp., b.s., (B-28), 14 μm.
Fig. 34. ?*Euglypha* sp., b.s., (C-8), 15 μm.
Fig. 35. *Euglypha* cf. *E. crenulata*, d.p.s., (A-15), 12.5 μm.
Fig. 36. *Euglypha* cf. *E. crenulata*, d.p.s., (F-21), 11 μm.
Fig. 37. *Euglypha* cf. *E. crenulata*, d.p.s., 11 μm.
Fig. 38. *Euglypha* cf. *E. crenulata*, denticulate pseudostome scale on a frustule of the diatom *Actinocyclus*, 10 μm.
Fig. 39. *Euglypha* cf. *E. brachiata*, d.p.s., (F-8), 13 μm.
Fig. 40. *Euglypha* cf. *E. brachiata*, d.p.s., (A-2), 12.5 μm.
Fig. 41. *Euglypha* cf. *E. brachiata*, d.p.s., (B-24), 14 μm.
Fig. 42. *Euglypha* cf. *E. brachiata*, d.p.s., (E-13), 20 μm.
Fig. 43. Bristle, ?possibly from a thecamoeba or chrysophyte (?*Mallomonas lychenensis*), 55 μm.

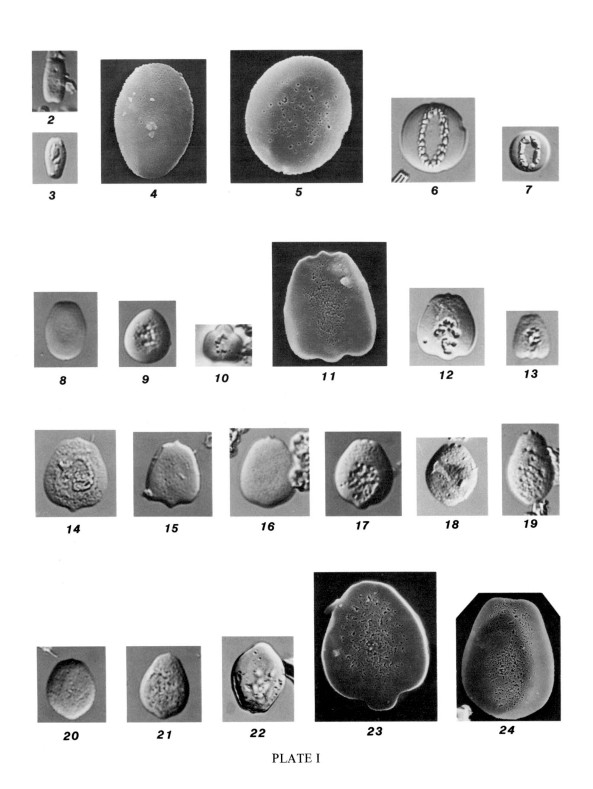

PLATE I

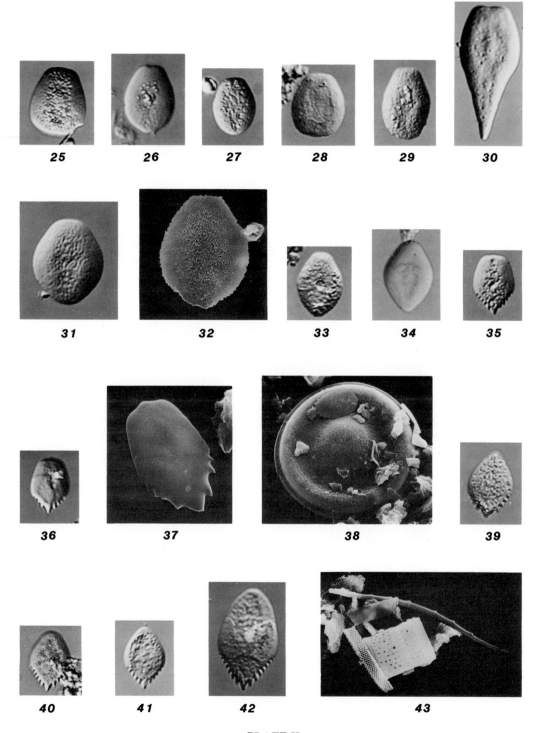

PLATE II

MIOCENE FRESHWATER MOLLUSKS FROM THE CLARKIA FOSSIL SITE, IDAHO[1]

DWIGHT W. TAYLOR
Tiburon Center for Environmental Studies, Tiburon, CA 94920[2]

Five specimens representing three species of mollusks have been collected at the Clarkia fossil site by Charles J. Smiley and colleagues. As is usually the case with fossil shells found with well-preserved leaves, the original lime of the shell has been dissolved and the specimens are compressions of the external periostracum with little of the detail ideally suited for study. The three species are a clam, *Anodonta*, and a gastropod, *Juga*, referred to modern species living in western Oregon and Washington, and a gastropod, *Bellamya*, that is found living in Old World tropical to temperate regions but is not native in North America. Even though the species are so few, they provide significant evidence of environmental change that is in agreement with that derived from the flora.

ACCOUNTS OF SPECIES

Anodonta oregonensis Lea. One specimen, probably half-grown, cited as a Unionid clam by Smiley and Rember (1979:10). The outline of the shell is preserved, showing the arcuate ventral margin of the shell characteristic of this species. The species lives in lakes and slow streams from northernmost California to southernmost Alaska. In Washington and Oregon most recorded localities are in the western halves of the states, but there is a record from Latah Creek, Spokane, about 70 miles NW of the fossil site. This is the nearest modern occurrence.

Bellamya species. Two compressed shells, not sufficiently well preserved to show what surface sculpture there might have been, and one operculum. Recorded by Smiley and Rember (1979:10) as Viviparidae.

The shells are so lacking in detail that they are identifiable in themselves only as generalized Viviparidae. The known late Tertiary species in the western United States are all classified as *Bellamya*, hence the Clarkia fossils are presumed to represent that group. The operculum measures 15 mm long, 12 wide, in outline like living species of the genus. None of these specimens could be identified confidently as distinct from, or conspecific with, the Miocene *Bellamya turneri* (Hannibal), found widely in southwestern Idaho, southeastern Oregon, and Nevada (Taylor 1963).

Juga ?*silicula* (Gould). One specimen, a compression with mold of exterior sculpture. This specimen demonstrates that solution of the living shell was relatively slow, so that compression of the sediments impressed the exterior sculpture on the fine compact mud before the lime had been entirely dissolved.

The specimen represents most of the spire, showing about six remaining whorls. Axial sculpture consists of opisthocyrt ribs, rounded in cross-section, nearly straight on the anterior two-thirds of the whorl, curved abruptly toward the posterior suture but sharply terminated before it. Spiral sculpture is incompletely preserved, consisting of about six spiral cords passing over the ribs and not preserved in the interspaces. The strongest spiral cord lies next to the anterior suture.

---

[1] Research supported (in part) by National Science Foundation grant DEB-7822584.

[2] Present address: Department of Geology, Oregon State University, Corvallis, OR 97331

Copyright ©1985, Pacific Division, AAAS.

These features are most like *Juga silicula* among all Pleuroceridae of the Pacific Northwest. The species is abundant in creeks and rivers from northernmost California to southern Puget Sound, Washington, on the west slope of the Cascade Range. The quality of the material makes a confident identification of species impossible, but the fossil is at least closely related to the modern form. Nearest modern occurrences are about 200 miles to the west.

HABITAT

The gastropod *Juga silicula* is a characteristic stream species. It is found occasionally in small lakes where wave action produces oxygenated water throughout the year. This species could not have lived in the quiet water and soft substratum of the Clarkia lake, but was evidently washed in. Both *Bellamya* and *Anodonta oregonensis* are commonly found in lakes, but evidence of toxic or anaerobic conditions in the Clarkia lake (Smiley and Rember 1979) indicates that they too were washed in. These two combine relatively thin shells with large body size, and it is plausible that the gases from decomposing bodies floated the shells until they were carried to the site of burial.

Like the plants, the mollusks are an association of genera that are not living together in Idaho, or even in the western United States. *Juga silicula*, now living only on the humid western slopes of the Cascades, is evidence of a Miocene climate more humid than at present, especially in summer. *Bellamya* is not found in regions of severe winters, and agrees with the botanical evidence for a warmer climate with milder winters.

ACKNOWLEDGMENTS

Charles J. Smiley provided the specimens for study and invited me to participate in the symposium, "Late Cenozoic History of the Pacific Northwest," held at the University of Idaho, June 3-7, 1979. At that time I had a chance to visit the fossil site and talk with others studying the biota from Clarkia Lake. All five specimens cited herein are from the type-section at locality P-33, NW¼ NE¼ T. 42 N., R. 1 E., 2 miles south of Clarkia, Clearwater County, Idaho. Smiley and Rember (1979) have published a summary of geological data.

LITERATURE CITED

Smiley, C. J., and W. C. Rember. 1979. Guidebook and road log to the St. Maries River (Clarkia) fossil area of northern Idaho. Idaho Bur. Mines Geol. Inf. Circ. 33, 27 pp.

Taylor, D.W. 1963. Mollusks of the Black Butte local fauna. Pages 35-41 *in* J. A. Shotwell, ed. The Juntura Basin: Studies in earth history and paleoecology. Trans. Amer. Philos. Soc. 53.

TAXONOMY OF FISHES FROM MIOCENE CLARKIA LAKE BEDS, IDAHO

G. R. SMITH AND R. R. MILLER
Museum of Zoology and Museum of Paleontology
University of Michigan, Ann Arbor, MI 48109

Clarkia fish fossils include three or four species from the Salmonidae (trouts and salmon), Cyprinidae (minnows), and Centrarchidae (sunfishes). The trout belongs to a different genus than recent North American salmonids. It has large jaws and teeth, coarse scales, and 12 dorsal rays. Preliminary studies indicate that it belongs to the Eurasian genus *Hucho*. The minnows, probably *Gila turneri* (Lucas), have 8 dorsal and 8 anal rays; dorsal origin over or behind pelvics, depth 15-20% standard length; 42 vertebrae; large mouth; pharyngeal teeth in two rows, with a grinding surface and a conical point; head about 0.27 in standard length (SL=44-100 mm). These characters are generalized among North American minnows. The sunfish have 9-11 dorsal spines and 11-14 rays; 6-7 anal spines and 11-13 rays; 30-32 vertebrae (18-19 caudal); 4 predorsal bones; serrate preopercle and lacrimal; long pelvic spine; notched opercle (SL=63-200 mm). They are a species of *Archoplites*, known previously from the Miocene of Idaho and the Pliocene of the Northwest, as well as the Recent fauna of the Great Valley of California.

Miocene freshwater fishes from western North America are poorly known. Available samples indicate rather low diversity representing nine families. Among the most widely distributed are the Salmonidae (trouts and salmon), Cyprinidae (minnows), and Centrarchidae (sunfishes). Fishes collected from the lacustrine deposits in the St. Maries River valley near Clarkia, in Latah and Shoshone Counties, Idaho, belong to one species of Salmonidae, one species of Centrarchidae, and one, perhaps two, species of Cyprinidae. Several dozen specimens collected by C. J. Smiley and colleagues were made available to us for study. The specimens show remarkably detailed skeletal preservation, allowing inferences about the early evolution of the western American fish fauna.

Most of the fossils are from the transitional brown, ashy silts and silty clays (level 376-397 cm) below the ash in the type section (P-33) (Smiley and Rember 1979; Smiley et al. 1975), although they have been collected at other horizons (120-150 and 106-236 cm) at locality P-33 and the Emerald Creek locality (P-37). In this paper we describe the fishes and discuss their relationships to other western American forms. Smith and Elder (this volume) discuss taphonomy.

Order SALMONIFORMES
Family Salmonidae
Genus *Hucho* Günther
Figure 1

A single large trout, nearly complete excepting some details of the skull, was collected at locality P-33 on June 12, 1980. The specimen is estimated to be 668 mm in total length and 588 mm to the end of the hypural bones (standard length). It was buried without disturbance—even the lateral line is discernable.

Characteristics of the trout are, in combination, unlike any other genus of North American salmonid. The jaws and teeth are large (Fig. 1), like trout and salmon, but unlike grayling. There are between 55 and 58 vertebrae and 22 or 23 rows of scales on the caudal peduncle posterior to

Copyright © 1985, Pacific Division, AAAS.

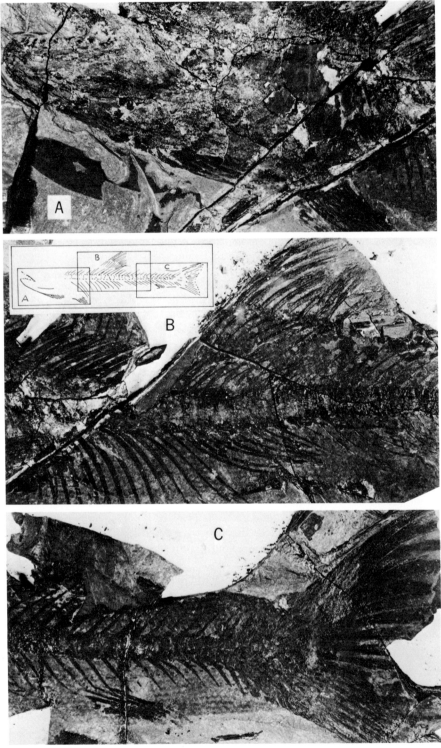

Figure 1. *Hucho* sp. (locality P-33). Head (A), trunk (B), caudal (C) of specimen 668 mm long. Note lateral line on caudal peduncle scales below vertebrae. Photography by C. J. Smiley.

the insertion of the anal fin. The scales are larger than in other trout. The pelvic fin has more than seven rays; its origin is below the center of the dorsal fin, which has 12 rays. A large, isolated salmonid pelvic girdle from the same locality has 10-10 pelvic rays. The anal and pectoral fins may be incomplete, but the anal had at least 9 rays and the pectoral at least 10. There were 14 predorsal bones and at least 11 branchiostegal bones in the right series. The caudal fin has 10/9 principal rays.

This is the earliest known North American trout after the grayling-like ancestral trout, *Eosalmo driftwoodensis* (Wilson 1977:15) from the Eocene of British Columbia. It differs from that form especially in the large jaws and teeth. The Clarkia fossils are distinguished from Pacific salmon (*Oncorhynchus*) by the small number of vertebrae, scales, and anal fin rays. It differs from *Salvelinus* in the large vertebrae, large scales, and 12 dorsal rays. The large vertebrae and scales also distinguish the Clarkia fossil from North American *Salmo*. In all of these features it is similar to the Eurasian genus *Hucho*. Circumstantial evidence for relationship to *Hucho* comes from the discovery of bones of this genus in the late Miocene sediments of Lake Idaho on the Snake River Plain (Smith 1975:18, Fig. 6B; Kimmel 1975:71, Figs. 1, 2A). Additional material of the form from the Snake River Plain shows a transverse rather than longitudinal row of teeth on the prevomer, a large patch of basibranchial teeth, large vertebrae, and large scales—all characteristic of *Hucho*, not *Salmo, Oncorhynchus,* or *Salvelinus*.

There is no indication that the lineage represented by the Clarkia trout and its Miocene relative from southern Idaho was a descendant of *Eosalmo* or an ancestor of any living North American salmonids. Its relationships seem to be with Eurasian forms.

Order CYPRINIFORMES
Family Cyprinidae
Genus *Gila* Baird and Girard
Figure 2 A-C (see also Smith et al. 1975: Pl. 1, Fig. 1; Smiley and Rember 1979: Pl. 4, Fig. 4; Smith and Elder 1985: Fig. 4 A-C)

The osteological diagnosis of this group (Uyeno 1960, unpublished; Smith 1975) is not applicable to the material at hand. The name is also currently applied to Miocene and Pliocene specimens known only from lateral aspects of skeletons and body outlines on lacustrine slabs. These share the following (generally plesiomorphic) characters: minnows usually 10-30 cm in length, with terminal mouth, rather long jaws, slender body, usually 8 to 10 dorsal and anal rays, 36-42 post-Weberian vertebrae, forked caudal fin, and conical to slightly hooked pharyngeal teeth in one or two rows. (Several recent genera in western North America may be characterized similarly; several specialized species of *Gila* in the Colorado River drainage do not fit all of these characters.)

Gila milleri Smith from the Pliocene of Glenns Ferry Formation in southwestern Idaho is known from detailed osteology and is related to the Recent *Gila caerulea* (Girard) of the Klamath drainage. *Gila turneri* (Lucas), *G. esmeralda* La Rivers, and *G. traini* Lugaski, from late Miocene and Pliocene lake slabs of Nevada, are not necessarily in the genus *Gila* and not obviously different species; they and the forms from Clarkia fit the above diagnosis.

Because of the taxonomic uncertainty and the lack of suitable type material, the Clarkia cyprinids are referred to as follows:

Gila sp.—Small (adults 12-18 cm), slender minnows with a large terminal mouth (reaching to below eye); 8 dorsal rays (8 in 11 specimens, possibly 9 in one); 8 or 9 anal rays (8 in seven specimens, 9 in one, possibly 7 in one); caudal with 19 rays; 12 or 13 pectoral rays; 9 or 10 pelvic rays (usually 9); 18-21 precaudal vertebrae; 21 caudal vertebrae; 42 post-Weberian vertebrae (in two); pharyngeal teeth (Fig. 2B), conical, hooked, in two rows; caudal fin deeply forked, with equal lobes; caudal peduncle slender; eye large, 0.27-0.33 of head length.

The dorsal origin is over the pelvin origin in most, including the smallest specimen, 55 mm in total length (Fig. 2A), but behind the pelvic origin in two larger specimens (Fig. 2C). If the difference is a reflection of the morphology in life, it is an indication that two species of minnows were present. Although this character is frequently used, it is unreliable in fossils because of the possible shift of the abdominal wall and pelvic girdle during preservation.

Nineteen specimens have been studied; 17 are from adults 9.5-18 cm, and two are small, 55-65 mm in length. The larger specimens lack well-preserved heads, but are proportionally and meristically similar to *Gila turneri* (Lucas) from the Miocene Esmeralda Group in Esmeralda Co., Nevada (see Lucas 1900:Fig. 1). The observations on teeth are based on four specimens (Fig. 2A, B).

Similar but not necessarily identical late Miocene or Pliocene fossils from the following areas are being studied by Ted Cavender, R. R. Miller, and G. R. Smith: Madison Valley, Gold Creek, and Drummond, Montana; Sentinel Butte, North Dakota; Bear Valley, California; Cache Valley, Utah; Stewart Valley, Cedar Mountain, Black Valley, Jersey Valley, and Big Smokey Valley, Nevada. Related forms were described from the Bidahochi Formation, in eastern Arizona, by Uyeno and Miller (1965).

In summary, the *Gila* from the Clarkia beds are members of a widespread group. It is not a particularly primitive cyprinid, notwithstanding its early place in the history of North American minnows. It is a generalized, midwater fish with body form and dentition remarkably similar to its widespread relatives now living in lakes and streams throughout the Basin and Range Province south into Mexico. Compared with the Pleistocene and Recent distribution, the Miocene distribution of the group was broader to the north and east.

Order PERCIFORMES
Family Centrarchidae
Genus *Archoplites* Gill
Figures 2D, 3; Smith and Elder 1985, Figs. 3 A, B, C, 4D

Diagnosis. Miocene to Recent sunfish with the combination of teeth on the endopterygoid, ectopterygoid, and posterior basibranchial; vomer with small teeth; premaxilla with short ascending process; dentary truncate with tooth patch expanded anteriorly and teeth small; opercle weakly notched; preopercle angular, normally with 6 distinct pores, a deep adductor fossa, strong serrae ventrally, and weak serrae posteriorly; lacrimal serrate but rounded posteriorly; 3 or 4 predorsal bones; long pelvic spine; and 5 to 8 anal spines (Smith 1975).

Relatives of this genus are known from several Miocene and Pliocene localities in western North America. Cope (1883) described *Plioplarchus sexspinosus* and *whitei* from Sentinel Butte, North Dakota, and in 1889 described *Plioplarchus septemspinosus* from the John Day Basin, Oregon. Schlaikjer (1937) described *Boreocentrarchus smithi* from Alaska (its status is uncertain according to Uyeno and Miller 1963:17-18). Bailey (1938, unpublished) described specimens from Trout Creek, Oregon, and recognized (as did Schlaikjer) that the Oregon specimens represented a genus different from *Plioplarchus* (see Table 1).

Plioplarchus whitei has a short pelvic spine, 5 anal spines, and 9 dorsal spines. Other nominal forms of this genus, plus *Boreocentrarchus*, have longer pelvic spines and more spines in the dorsal and anal fins. *Archoplites* is similar to the latter group, but has stronger serrations on the preopercle and lacrimal. Specimens from Trout Creek, Oregon, are intermediate. Relatives of this group from the Humboldt Formation, Nevada, and from Bear Valley, California, are recognized on the basis of the opercle shape, strongly serrate preopercles, long pelvic spines, and meristic characters. On the basis of our continuing studies, it would appear that one or two genera and three or four species are represented in the diversity of fossil sunfishes outlined above. The species from Clarkia is sufficiently distinct and well represented to be described.

Figure 2. *Gila* sp. from Clarkia lake beds. (A) Small form 55 mm SL with dorsal origin over pelvic fins; (B) Enlarged (x10) view of head of (A) showing pharyngeal teeth scattered between preopercle and pectoral fin; (C) Large form (x.8) with dorsal origin behind pelvic origin; (D) Holotype of *Archoplites clarki*, natural size.

TABLE 1. SUMMARY OF CHARACTERISTICS OF SEVERAL FOSSIL CENTRARCHIDS FOR WHICH MERISTIC DATA ARE AVAILABLE (Modal counts in boldface type)

| | *Archoplites* | | *Boreocentrarchus* | *septemspinosus* | *Plioplarchus* | | |
	clarki	*interruptus*	*smithi*	John Day	*septemspinosus* Trout Cr.	*whitei*	*sexspinosus*
Dorsal spines	9, **10**, 11	12, **13**, 14	11	10, 11	9, 10	9	10, 11
Dorsal rays	11, **12**, 13, 14	10, **11**, 12	12, ?13	12	11, 12	11, 12	12, 11
Anal spines	**6**, 7	6, **7**, 8	7	7, **8**	5, 6	5	6
Anal rays	11, 12, **13**, 14	9, **10**, 11, 12	12, 13	?	11-14	14	11-13
Preopercular serrae	+	+	-?	-?	3	?	?
Precaudal vertebrae	13, 14	13	?	?	13, 14	?	?
Caudal vertebrae	18, 19	18	?	?	18	?	?
Pelvic spine	long	long	?	?	long	short	long
Predorsal bones	4	3	?	?	4	?	?

Figure 3. *Archoplites clarki* (A) Small adult (x.75) showing four predorsal bones, dorsal spines and rays, and some scale pattern (above anal fin); (B) serrate lacrimal; (C) serrate lower limb of preopercle.

Archoplites clarki, new species

Holotype UMMP V 74202 (Fig. 2D), an imprint of a sunfish 123 mm in standard length, 155 mm in total length, and 50 mm in body depth, 42 mm in head length, and 17 mm in caudal peduncle depth. The eye diameter is 9.5 mm, maxilla 16 mm, lower jaw 22 mm, pelvic spine 17 mm, longest anal spine (6th of 6) 17.2 mm, and longest dorsal spine (7-10 of 10) 18.2 mm. The specimen has 13 dorsal rays, 13 anal rays, 17 caudal rays, 19 ± 1 caudal vertebrae, 15 ± precaudal vertebrae (with 13 bearing primary ribs), and 4 predorsal bones. The specimen has an estimated 44-48 scales in the lateral line.

The following description is summarized from 21 specimens plus the holotype. Sizes range from 63-200 mm in standard length; dorsal spines 9(2), 10(12), 11(2); dorsal rays 11(1), 12(4), 13(4), 14(1); anal spines 6(9), 7(3), anal rays 11+(4), 12(1), 13(5), 14(1); precaudal vertebrae

14(3), 15(9), caudal vertebrae 18(2), 19(6); principal caudal rays 9/8(7); pectoral rays 13-15(7); pelvic fin with long spine (equal to the 4th D spine) and 5 rays (7); predorsal bones 3(1), 4(11); supramaxilla large; lacrimal strongly serrate; preopercle strongly serrate; opercular margin weakly notched dorsal to longitudinal strut; branchiostegals at least 6.

The four largest specimens (ca. 200 mm SL) have scales 4-5 mm in diameter. Single isolated scales, 6.0 and 7.9 mm long, are referrable to the same species; they appear to have six and nine growth rings respectively. The largest scale is the size of those belonging to a Trout Creek specimen ca. 300 mm SL.

This species is named for Captain William Clark of the Lewis and Clark Expedition.

In summary, *Archoplites clarki* is the most abundant of the three species in the Clarkia beds. It was a rather large sunfish whose ecology probably included predatory habits like those of bass (*Micropterus*) in the Recent fauna of eastern North America. Its lineage represented the more primitive and most western of known sunfish groups in the Miocene, just as it does today.

DISCUSSION

The lake in which the Clarkia beds were deposited was inhabited by a fish fauna of low diversity by modern standards. Two of the three (possibly four) species—*Gila* sp. and *Archoplites clarki*—belonged to groups widely distributed in the Miocene of western North America. They occurred in warm-water faunas of low species diversity in other western localities, as well.

The centrarchids occupied ranges far to the north (probably to Alaska) of their present distribution in the Miocene. Miocene Cyprinidae extended from southern Nevada to northern Idaho and will probably be found in the Miocene of Alaska as well, since they probably reached North America from Asia across the Bering Straits in the early Miocene or late Oligocene.

The Clarkia salmonid, *Hucho* sp., belongs to a Eurasian genus and is especially similar to *Hucho perryi* of Japan. *Hucho* was probably also present in fresh waters of Alaska during the Miocene.

A warm-temperate climate analogous to that of the southern Appalachians is inferred from the Clarkia flora (Smiley and Rember 1979). Fish evidence is similar. The Miocene distribution of *Archoplites* indicates warm winters in northwestern North America, but the salmonid indicates moderately cool summers at Clarkia. Salmonids are not present in most Recent lakes dominated by centrarchids except at the latitude of the Great Lakes. However, *Hucho* is a rather southern salmonid, being found in southern Europe and Japan. Later in the Miocene, three genera of salmonid fishes, including *Hucho*, were sympatric with *Archoplites* in southern Idaho. By late Pliocene, following a cooling trend, *Archoplites* was restricted in the western U.S., *Hucho* was extinct in Lake Idaho, and other salmonids were abundant and widespread (Smith 1975).

LITERATURE CITED

Bailey, R. M. 1938. A systematic revision of the centrarchid fishes, with a discussion of their distribution, variations, and probable interrelationships. Ph.D. Thesis. University of Michigan, Ann Arbor, Mich.

Cope, E. D. 1883. On a new genus and species of Percidae from Dakota Territory. Amer. J. Sci. 25(3):414-416.

Cope, E. D. 1889. On a species of *Plioplarchus* from Oregon. Amer. Natur. 23:625-626.

Kimmel, P. G. 1975. Fishes of the Miocene-Pliocene Deer Butte Formation, southeast Oregon. Univ. Mich. Mus. Paleont. Pap. Paleont. 14:69-87.

LaRivers, I. 1966. A new cyprinid fish from the Esmeralda (Pliocene) of southeastern Nevada (Cypriniformes, Cyprini, Cyprinoidei, Cyprinidae). Biol. Soc. Nev. Occ. Pap. 11:1-4.

Lucas, F. A. 1900. A new fossil cyprinoid, *Leuciscus turneri*, from the Miocene of Nevada. Proc. U.S. Nat'l. Mus. 23:333-334.

Lugaski, T. 1979. *Gila traini*, a new Pliocene cyprinid fish from Jersey Valley, Nevada. J. Paleont. 53(5):1160-1164.

Schlaijker, E. M. 1937. New fishes from the continental Tertiary of Alaska. Bull. Amer. Mus. Nat. Hist. 74:1-23.

Smiley, C. J., J. Gray, and L. M. Huggins. 1975. Preservation of Miocene fossils in unoxidized lake deposits, Clarkia, Idaho. J. Paleont. 49(5):833-844.

Smiley, C. J., and W. C. Rember. 1979. Guidebook and road log to the St. Maries River (Clarkia) fossil area of northern Idaho. Idaho Bur. Mines Geol. Inf. Circ. 33:1-27.

Smith, G. R. 1975. Fishes of the Pliocene Glenns Ferry Formation, southwest Idaho. Univ. Mich. Mus. Paleont. Pap. Paleont. 14:1-68.

Smith, G. R., and R. L. Elder. 1985. Environmental interpretation of burial and preservation of Clarkia fishes (Miocene, Idaho). Pages 85-93 *in* C. J. Smiley, ed. Late Cenozoic History of the Pacific Northwest. Pacific Division, Amer. Assoc. Adv. Sci., San Francisco, Calif.

Uyeno, T. 1960. Osteology and phylogeny of the American cyprinid fishes allied to the genus *Gila*. Ph.D. Thesis. University of Michigan, Ann Arbor, Mich.

Uyeno, T., and R. R. Miller. 1963. Summary of late Cenozoic freshwater fish records for North America. Occ. Pap. Mus. Zool. Univ. Mich. 631:1-34.

Uyeno, T., and R. R. Miller. 1965. Middle Pliocene cyprinid fishes from the Bidahochi Formation, Arizona. Copeia 1965(1):28-41.

Wilson, M. V. H. 1977. Middle Eocene freshwater fish from British Columbia. Life Sci. Contrib. Roy. Ontario Mus. 11B:1-61.

ENVIRONMENTAL INTERPRETATION OF BURIAL AND PRESERVATION OF CLARKIA FISHES

G. R. SMITH AND R. L. ELDER
Museum of Paleontology, University of Michigan, Ann Arbor, Michigan 48109

Most Clarkia fish fossils are buried in thin, usually fining-upward, alternately light (quartz-rich) and dark (organic-rich) silts. Contacts between organic-rich laminae and overlying quartzose laminae are usually sharp, while quartz-rich layers frequently grade upward into darker laminae, indicating cyclic deposition of couplets. Stable, cold conditions in the hypolimnion are indicated by fish fossils whose preservation is sufficiently perfect to indicate that the specimen rested permanently on the bottom after death, rather than floating as occurs when water temperatures are above 15° C. Decay and burial were slow, but bones are usually undisturbed by scavengers or currents, which indicates anoxic conditions. The surrounding silts infrequently exhibit ripple cross-lamination. Some fish died of poisoning or anoxia, as indicated by tetany of fins and mouths. Lack of distortion of most fins indicates flat substrate and minimal soft-sediment deformation. Compression of silts between fins and hard parts of known thickness was on the order of 7:1. Sedimentation rates of certain fish-rich units are judged to have averaged 2-10 mm per year (compacted), based on estimated fish decay and burial times.

In this analysis we use observations on fossil fish skeletons to infer characteristics of death and burial, then combine that information with sedimentary data to interpret aspects of limnology and the environment of deposition in Clarkia Lake of Miocene age, Shoshone Co., Idaho. Consider the possibility of using a dead fish body—that is, an organized assembly of soft and bony tissues surrounding a float structure sensitive to temperature—as a recording device dropped into a lake, to be recovered later with imprinted information about temperature, depth, energy, scavengers, and oxygen concentration. We will use data from experimental and field observations to show that decay time and floating or sinking response of fish are directly dependent on temperature and pressure, that presence or absence of scavengers is an indication of oxygen concentration at the lake bottom, and that the preserved positions of various skeletal elements reliably record the effects of temperature, current energy, and scavengers. The positions of the jaws and fins often provide clues about the cause of death. Finally, the pattern of sedimentary sequences and structure allows interpretations regarding substrate, seasonality, rate of deposition, and current energy in the environment.

BASIN SETTING

According to Smiley and Rember (1979), the Clarkia beds were deposited in a moderately steep-walled valley, cut into Pre-Cambrian schist, and now drained by the St. Maries River. The present rolling topography was probably established by the Miocene, and Miocene beds now form much of the flat valley floor. A lava dam probably created the lake, and the thickness of the sedimentary sequence at locality P-33 suggests that initial lake depths were not less than 10 m at this site. Fossils occur in finely laminated organic-rich silts of a 760-cm lacustrine sequence, which is underlain by thick-bedded alluvial sandstones and silts and overlaid by poorly bedded floodplain silts. Numerous thin beds of fine ash occur throughout the sequence. Fish are most

Copyright ©1985, Pacific Division, AAAS.

common in three layers in the type section (Fig. 1, locality P-33). The highest of these units (121-150 cm level) is near the base of an oxidized zone of yellow ashy silt (base of Unit 5; Smiley et al. 1975; Smiley and Rember 1979). The middle fish-bearing stratum (206-236 cm level) is in Unit 3 and the top of Unit 2, in a thinly bedded, brown ashy silt immediately underlying the 56-cm thick Unit 4 ash bed. The lower fish horizon is in the upper part of Unit 2 (367-397 cm level), in unoxidized blue-gray silty clays immediately below a 70-cm thick devitrified ash layer. The unoxidized zone has remained water-saturated since deposition, approximately 20 million years ago, as evidenced by intact cellular tissue and original organic chemistry of leaves (Smiley and Rember 1979:9). Approximately two-thirds of the sedimentary layers in the column at P-33 are varved couplets that average about 1 mm thick, and about one-third of the layers occur as silty clays or ashes more than 4-5 mm thick. According to Smiley and Rember (this volume), the excellent preservation of fossils, the fining-upward texture observed in most layers, and the occurrences

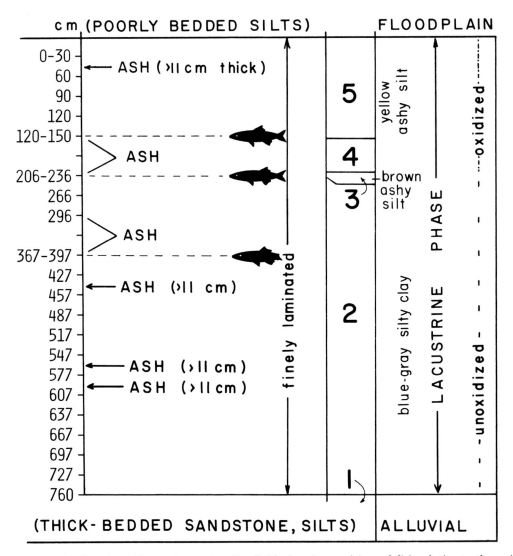

Figure 1. Stratigraphic section at locality P-33 showing position of fish relative to the units of Smiley et al. (1975), and Smiley and Rember (1979). Data provided by C. J. Smiley.

of single leaves with connected halves preserved on stratigraphically separated laminae suggest rapid, storm-generated input of sediments in a geologically brief period. A total of at least 170 cm of the column is represented by graded ash layers that fine upwards, owing to differential settling by particle size during episodes of air-fall deposition into the lake water. Fish and leaves are associated in the same couplets and the excellent preservation of each is probably related to the same factors.

FISH TAPHONOMY

The nature of skeletal disarticulation and the position of parts of the skeleton provide the most important evidence relating to fish taphonomy. Clarkia fishes belong to one species of salmonid, one (possibly two) species of minnows, and one species of sunfish (Smith and Miller, this volume) that were actively swimming, quasi-neutrally buoyant, with large, gas-filled swimbladders. The bodies were scaled and the flexible fin rays were plated by small, bony lepidotrichia held together by dermal connective tissue. Upon death, these fish would sink to the bottom and begin to decay. The immediate effect of the decay process on gas formation and arrangement of the skeleton is temperature-dependent. This relationship is sufficiently well-defined to allow formulation of hypotheses about decay time, temperature, and limnology, based on the form of the fish skeletons preserved in the lake sediments.

Temperature

The most critical process in fish taphonomy is the eventual floating of the carcass by decay gasses, plus the swimbladder, if the temperature is above $15°$ C (Schäfer 1972). Our observations show that once a carcass floats, only an immediate temperature drop can cause it to sink again, and after some hours of decay at temperatures over $15°$ C, gasses forming in the tissues are sufficient to cause buoyancy even if the swimbladder is deflated. Under these circumstances, the carcass will float until disarticulation (the heavy jaw bones disarticulate first), unless it floats onto a beach and/or is preyed upon by scavengers. (In a large lake, the carcass can float long enough for the tissues to decay and drop the skeleton—minus the jaws—and the semi-empty skin to the bottom.) In each of the above cases, however, the eventual fate of the floating specimen is the scattering of its bones.

In order to determine the threshold of the floating response to temperature, we have observed the decay of specimens in families represented at Clarkia under a variety of controlled temperatures and pressures. The preliminary results are shown in Fig. 2. Specimens decaying above $10\text{-}15°$ C eventually float (Fig. 2A). The effect of pressure (e.g. with depth) is to delay the time to buoyancy (Fig. 2B). A specimen near neutral buoyancy can be sent rapidly up or down with pressure changes of just a few centimeters of water, or a temperature change of a few degrees. At temperatures warmer than about $20°$ C, fishes float within about one-half day, unless they become caught in vegetation (which cases would be apparent by orientation of the fossils). Exceptions are catfish or other dense benthic fishes and exceptionally large or small specimens, which have a slower, less predictable buoyancy response to decay. At around $15°$C fish may take over a week to float; below $10°$ C fish remain on the bottom for over 60 days—these decay without becoming buoyant. In summary, immediately after death a fish will not float unless the water is warm and shallow. After several hours or days, however, low temperature or high pressure is required to keep it down. These results are consistent with the observations of fisheries biologists who have studied the censusing of fish kills (Parker 1970; Rupp and De Roche 1965).

The Clarkia specimens we have examined show few examples of scattering and disarticulation of bones. The majority of specimens are not disorganized or disarticulated except at small, superficial attachment points (see below) and must be interpreted as carcasses that never floated.

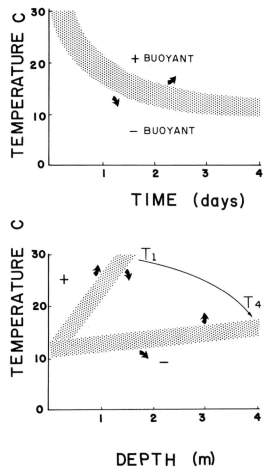

Figure 2. Graphic summary of observed buoyancy responses to temperature, pressure, and time. (A) Carcasses float quickly at high temperatures (data from shallow water). (B) Increased pressure increases the temperature necessary to float a carcass, but decay time decreases the temperature at which buoyancy occurs (i.e. high pressures are required to keep the carcass on the bottom if the temperature is above 15° C).

The specimens, with few exceptions (Fig. 3A), are flat on their sides with no sign of entrapment by vegetation (Fig. 3B, C; Smith and Miller, this volume, Figs. 1-3). We interpret this as unequivocal evidence of decay at temperatures lower than 15° C.

The concentration of fish at several 30-cm zones is evidence for a relatively stable hypolimnion with temperatures of approximately 15° C or below during those episodes. Relative scarcity of fishes in other zones can only be interpreted as lack of evidence at this stage of our knowledge. Preliminary data suggest that the few specimens in the zones of sparse fossils are less articulated. If continued studies verify this, it can be taken as evidence for higher temperatures in the hypolimnion.

Figure 3. Sunfish (*Archoplites*) showing (A) partially upright position and penetration of several laminae (arrow shows laminae that extended over head but under tail), (B) tetany of jaws and branchial apparatus, and (C) tetany of dorsal and anal fins.

Oxygen

Fish skeletons that have been cleaned by scavengers show disturbance proportional to scavenger size. Our observations on skeletons cleaned in the laboratory by bacteria, snails, amphipods, small fish, and crayfish indicate that the strength of the dissociated connective tissue, as well as the bone size and distance moved, correspond directly to scavenger body size and feeding mechanisms. Macroscopic scavengers cause considerable scattering of elements, especially of the head and thorax. Microscopic decay organisms cause separation of connective tissue in such a way that observed movements of bones are clearly attributable to gravity and current energy. Most of the Clarkia skeletons show no macroscopic scavenger disturbance whatever, indicating that macroscopic scavengers were rarely present between the time of death and burial (Smith and Miller show one of the few exceptions [this volume, Fig. 1B]). Either burial was extremely rapid or the environment did not usually permit macroscopic scavengers to live; evidence against rapid burial of the fish is presented below. The lack of scavengers is evidence of toxic or anaerobic conditions (Figs. 3, 4; Smith and Miller, this volume, Figs. 1-3). This interpretation is consistent with other evidence for anoxic or toxic conditions (unoxidized sediments and leaf tissue; Smiley and Rember 1979). Although examples of two kinds of caddis-fly larvae have been identified from the sediments, most of the insect fossils are of terrestrial origin. Molluscs are rare (Taylor, this volume).

The evidence of cold temperatures and lack of oxygen at the bottom of the lake suggests density stratification with only partial or infrequent mixing. The presence of fish requires suitable habitat in shallow and surface waters. Brown, humus-rich waters with low calcium carbonate and few nutrients (dystrophic) are indicated by the abundance of leaves and wood and the upper organic layer of each varve (compare Wilson 1977).

Specimens frequently show tetany of the jaws, branchial apparatus and fins (Fig. 3B, C). This response is seen in laboratory specimens killed by anoxia or rotenone. This evidence suggests possible causes of death. Fishes that ventured into the hypolimnion could suffocate there, but this should have been rare because there was little food to attract them, and they would have possessed a surfacing response to anoxia.

In summary, undisturbed articulation and jaw tetany of the fish fossils argue for an anaerobic environment in Clarkia Lake. This evidence allows us to infer a minimum depth for the lake. Water deeper than 8-12 meters is required for the development of a cold, anoxic hypolimnion in a warm-temperate climate such as that hypothesized on the basis of the floral community. However, more experimental evidence is needed to describe the effects of tannins on decay time, especially in interaction with anaerobic bacteria, temperature, and pressure.

Energy and Rate of Deposition

Clarkia fossils occur in rhythmites or varves (Fig. 4E). Burial of fishes was usually by light-colored, quartz-rich, fining-upwards layers that grade upward into thin, dark organic-rich layers. Coarse grains tend to lie on a sharp contact. The varves are not disturbed by burrowing invertebrates at the type locality (P-33), but sediments at Emerald Creek (P-37) are burrowed. Laminae compressed to about 1 mm were originally as thick as 7 mm, based on inference from layers that apparently filled spaces of known thickness such as mouths and spaces between fins (Fig. 4C). The ratio of about 7:1 is observed in few organic-rich examples and may be a maximum estimate for varved couplets. Only one specimen penetrates several layers. The sunfish (Fig. 3A) was buried partly upright, with the head slightly down, penetrating three layers of what must have been soft flocculent sediment.

Compression of sedimentary layers does not seem to have been uniform throughout the column at locality P-33 (Smiley, pers. comm., 1980). Single leaves occupying two separated layers (whole leaves folded to exist in two sequential sedimentary layers) could be used to set

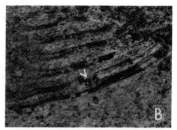
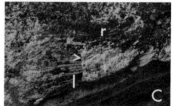

Figure 4. (A) Cyprinid fish (*Gila* sp.) showing burial position and lack of scavenger disturbance, including intact contents of digestive system (arrows). (B) Displacement of lepidotrichs (arrow) from anal fin of *Gila* sp. These fins would have been about 7 mm apart during burial. (D) Scattering of caudal lepidotrichs (arrow) of *Archoplites*. (E) Thin section showing fining upward of light, quartz-rich laminae, grading into thin, dark laminae of humus.

lower limits on the compaction estimates mentioned above. Some layers show no compaction whatever. Whole, split-level leaves are interpreted by Smiley and Rember (1979) as evidence of rapid sedimentation rates; they suggest an average rate of up to 15 cm per year (but possibly as slow as approximately 1 cm per year). The rapid-sedimentation hypothesis requires that a part of the leaf be vertical for days or a few weeks before being flattened by a subsequent depositional cycle. The alternative requires an exposed vertical edge for months.

The rapid-sedimentation hypothesis is supported by the evidence presented by Smiley and Rember (1979, this volume) but is contradicted by occasional evidence from the fish skeletons. Minute, scale-like lepidotrichs are commonly seen displaced a few millimeters from their fin ray, but in the same plane of preservation (Fig. 4B, D). Decay and detachment of these bony elements should take 8 weeks or more at low temperature. Transport was probably by slow currents; faint ripple marks are occasionally seen in the sediments. If transport occurred after initial deposition of sediments, the lepidotrichs would be unlikely to come to rest in the original plane. This evidence suggests a minimum of 2 months of decay before burial by the next sedimentation episode. The humus was probably allochthonous and entered the system with the inorganic input, during storms.

If the above interpretation is correct, the estimated interval for deposition of the 760-cm section would be at least 760 years, near the maximum time suggested by Smiley and Rember. Two months might be a rather short time for accumulation of a 0.5-5 ($\bar{x}=2$) mm varve couplet (up to 35 mm estimated original, wet thickness). These estimates are consistent with the sequence of hypotheses discussed above. However, because they are heavily interrelated, some additional, independent tests would be useful. In any case these estimates apply to the lower two fish-rich zones; different conditions probably prevailed higher in the section (Smiley, pers. commun., 1980).

DISCUSSION

Although the lake went through a complete lacustrine cycle (Smiley and Rember 1980), the fish evidence from the middle brown ashy silt horizon (206-236 cm level) and the lower blue-gray silty clays (367-397 cm level) indicate similar conditions in the hypolimnion at these two stages. These fish-rich layers each occur immediately below thick ash layers. (The upper fish-rich level [121-150 cm] occurs near the base of a yellow ashy silt.) It is tempting to attribute the deaths of these preserved fishes to volcanic effects, but the evidence for slow burial indicates that the fish lived several years before the eruptions. That is, 30 cm of sediment represents at least 30 years of deposition, according to our estimates; each of the ash layers was probably deposited within one or two years of its eruption, several decades after the lower part of the fish layer was buried. The alternative (ash-kill) hypothesis requires that the fish-rich layers represent hourly or daily episodes after eruption; in this case, however, they should show more ash content. If this slight displacement of small surface bones could be attributed to soft-sediment deformation rather than gentle surface currents, as we have argued, the rates of deposition could have been more than 10-100 times as rapid as our estimates, and the ash-kill hypothesis would be more acceptable.

The difficulty of inferring sedimentation rates from taphonomic evidence is related to the problem of inferring conditions in the hypolimnion from dead fish that must have lived in the epilimnion. A fundamental taphonomic dilemma confronts us here: dead fish are the only evidence for abundance of living fish and the major evidence for bottom conditions at the time of death, therefore population abundance and condition at the time of death are not completely separable problems. Alternative hypotheses remain viable. Either fish abundance varied greatly and the abundances are reflected in the sediments, or the abundance of fossils in the sediments is independent of living population density and reflects the stability of a cold hypolimnion. No matter how abundant the living fish, burial of perfectly preserved fossils cannot occur in warm or well-oxygenated bottom sediments unless burial occurs within several hours of death. The present evaluation of the alternative hypotheses is based on experimental fish taphonomy that is just beginning. Preliminary results contradict current ideas in this field. Experimental work coupled with continuing analysis of the lacustrine sequence at Clarkia should clarify the present contradiction.

CONCLUSIONS

The Clarkia beds contain moderately abundant fossil fishes buried in varved sediments with extremely abundant leaves. A near-shore environment and geologically rapid burial is indicated by the abundance and excellent preservation of the leaves.

Relatively deep, cold, anoxic waters are required for the preservation of the fishes, which show evidence of slow, undisturbed decay by bacteria followed by displacement of minute external bones by gentle currents. Evidence that the fishes did not float after death requires a hypolimnion cooler than about 15°C and lake depths of at least 8-12 m. Infrequent mixing, maintaining cool bottom waters in the warm-temperate climate (inferred from the morphology and taxonomic diversity of the plant community), is consistent with an equitable environment with cool summers and warm winters. During the deposition of the fish-rich horizons, sedimentation apparently occurred in several storm-generated episodes per year, each usually resulting in a couplet consisting of a 0.5-5 mm quartz-rich layer fining and grading upward into a thin, argillaceous, humus-rich layer. Deposition of the 760-cm cycle, beginning on alluvium and ending in flood plain, might have taken as little as 760 years, according to estimates based heavily on interpretation of the fish taphonomy.

ACKNOWLEDGMENTS

The fossil fish were collected and prepared for study by C. J. Smiley and his colleagues, who also facilitated work at the field site and helpful discussion of these conclusions at the Clarkia Symposium and Field Conference in 1979. C. J. Smiley, B. H. Wilkinson, and A. McCune made helpful comments on the manuscript. J. C. Schneider and P. Yant contributed helpful discussion. K. Steelquist prepared the photographs. B. H. Wilkinson prepared the thin sections.

LITERATURE CITED

Parker, R. O., Jr. 1970. Surfacing of dead fish following application of rotenone. Trans. Amer. Fish Soc. 99(4):805-807.

Rupp, R. S., and S. E. De Roche. 1965. Standing crops of fishes in three small lakes compared with C^{14} estimates of net primary productivity. Trans. Amer. Fish. Soc. 94(1):9-25.

Shäfer, W. 1972. Ecology and palaeoecology of the marine environment. Oliver and Boyd, Edinburgh. 568 pp.

Smiley, C. J., J. Gray, and L. M. Huggins. 1975. Preservation of Miocene fossils in unoxidized lake deposits, Clarkia, Idaho. J. Paleont. 49(5):833-844.

Smiley, C. J., and W. C. Rember. 1979. Guidebook and road log to St. Maries River (Clarkia) fossil area of northern Idaho. Idaho Bur. Mines Geol. Info. Circ. 33:1-27.

Smiley, C. J., and W. C. Rember. 1985. Physical setting of the Miocene Clarkia fossil beds, Northern Idaho. Pages 11-31 *in* C. J. Smiley, ed. Late Cenozoic History of the Pacific Northwest. Pacific Division, Amer. Assoc. Adv. Sci., San Francisco, Calif.

Smith, G. R., and R. R. Miller. 1985. Taxonomy of fishes from Miocene Clarkia lake beds, Idaho. Pages 75-83 *in* C. J. Smiley, ed. Late Cenozoic History of the Pacific Northwest. Pacific Division, Amer. Assoc. Adv. Sci., San Francisco, Calif.

Taylor, D. W. 1985. Miocene freshwater mollusks from the Clarkia fossil site, Idaho. Pages 73-74 *in* C. J. Smiley, ed. Late Cenozoic History of the Pacific Northwest. Pacific Division, Amer. Assoc. Adv. Sci., San Francisco, Calif.

Wilson, M. V. H. 1977. Paleoecology of Eocene lacustrine varves at Horsefly, British Columbia. Can. J. Earth Sci. 14:953-962.

COMPOSITION OF THE MIOCENE CLARKIA FLORA

CHARLES J. SMILEY AND WILLIAM C. REMBER
Department of Geology, University of Idaho, Moscow, ID 83843

> The Miocene Clarkia flora of northern Idaho is preserved in lacustrine clays that accumulated in a narrow valley surrounded by rugged hilly terrane. Leaves and other megascopic organs show only minor (if any) abrasion due to transport, and they appear to have been carried to the depositional site either directly by wind dissemination from overlooking slopes or over short distances by small tributary streams or by slope wash. Preservation occurs as compression or impression fossils. Spermatophytes of diverse taxonomic affinities (at least 100 genera in 50 families) dominated the forests that surrounded Clarkia lake. Other plants include a diversity of fungi, mosses, ferns, horsetails, club mosses and quillworts. A microstratigraphic study of megafossils through the exposed lacustrine section at locality P-33 has been interpreted to reflect the development and expansion of a lake-border swamp vegetation as the Miocene lake silted in, whereas both megafossil and microfossil evidence suggests that the forests of adjacent slopes remained in a stable and essentially climax condition.

The topographic setting of Miocene Clarkia Lake—as a narrow body of open water surrounded by rugged hilly terrane—has resulted in short-distance surface transport, or in direct wind transport, of an abundance and diversity of plant taxa and organs derived from a variety of habitats in the immediate vicinity (see Fig. 2, Smiley and Rember, this volume). The contributing vegetation was contained within a restricted topographic basin that is still bounded within 3-5 km by ridge crests that rise 800-900 m above the surface of the Miocene valley fill. Miocene topography, which has remained essentially unchanged to the present, shows that forested slopes would have descended directly to the lake margins and would have facilitated wind-dispersal of plant organs from slope to open water. In addition, surface (water) transport would have been restricted to the dimensions of the topographic basin, where headwaters of tributary streams would have been no more than 1-3 km from the lake that occupied their lower courses (see Smiley et al. 1975; Smiley and Rember 1979, 1981, this volume). This topographic setting thus enhanced the transport and accumulation of allochthonous plant material that was little decomposed or abraded when buried in the lake sediments.

Much of the Clarkia compression material has remained essentially unaltered since original burial in the anaerobic bottom sediments of the Miocene lake (see examples in Plates 1-7). Many of the leaf compressions show an original green, red, or brown pigmentation when first exposed, and a variety of retained organic chemical constitutents in leaves has been identified by Giannasi and Niklas (1981, this volume). In exceptional cases, electron microscope (TEM) studies show the preservation of intracellular organelles in a variety of taxa (Niklas and Brown 1980, this volume), and both intracellular and extracellular phytoliths also are identified in leaf compressions (Huggins 1980, this volume; see also Smiley and Huggins 1981).

CLARKIA FLORISTICS

The majority of Clarkia plant megafossils appear referable to Tertiary species that were described originally in other early Neogene (Miocene) floras of the Columbia Plateau region of the

Copyright © 1985, Pacific Division, AAAS.

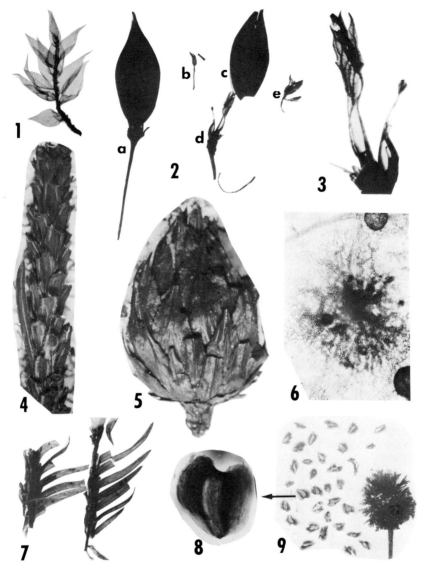

Plate 1. Clarkia Compression Fossils. Fig. 1. Moss shoot (Brachytheciaceae), transmitted light (X 6). Loc. P-33. Figs. 2, 3. *Pseudofagus idahoensis* Smiley and Huggins. 2(a) = nut in cupule, Loc. P-40 (X1). 2(b-e) = parts of single fruit (X1), separated during processing, loc. P-33. 3 = cupule of Fig. 2(d), enlarged (X3.5) to show details of appendages, transmitted light. Figs. 4, 5, 7. *Cunninghamia chaneyi* Lakhanpal. 4 = shoot (X3) showing bands of stomata as white lines on leaf and leaf bases, reflected light with main source at bottom. 5 = cone (X1.3), reflected light with main source at bottom. 7 = shoot fragments (X1), mounted in bioplastic. Loc. P-33. Fig. 6. Unidentified fungus, common on compression fossils of various taxa (X20). From cleared legume pod, transmitted light. Loc. P-33. Figs. 8, 9. *Metasequoia occidentalis* (Newb.) Chaney. 9 = cone partially decomposed during processing, with release of 32 seeds, mounted in bioplastic (X1). 8 = one of the seeds (see arrow) enlarged (X7). Loc. P-33.

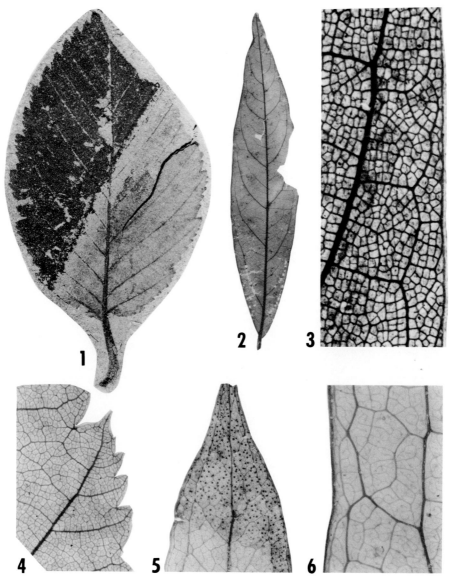

Plate 2. Clarkia Compression Fossils (except Fig. 1). Figs. 1, 4. *Betula* cf *B. vera* Brown. 1 = leaf partially coated with carbon residue (X1). 4 = portion of cleared leaf compression showing details of secondary and finer veins and unabraded margin (X15), transmitted light. Loc. P-33. Figs. 2, 3. Leaf of Lauraceae (?*Lindera*). 2 = leaf removed from bedding surface and cleared (X1). 3 = portion of leaf margin enlarged (X15) to show naturally stained secondary and finer veins, transmitted light. Loc. P-33. Figs. 5, 6. *Smilax* sp., cleared leaf compressions, transmitted light. 5 = leaf apex showing fungal infestation as dots (X1). 6 = leaf margin enlarged (X5) to show details of venation. Loc. P-33.

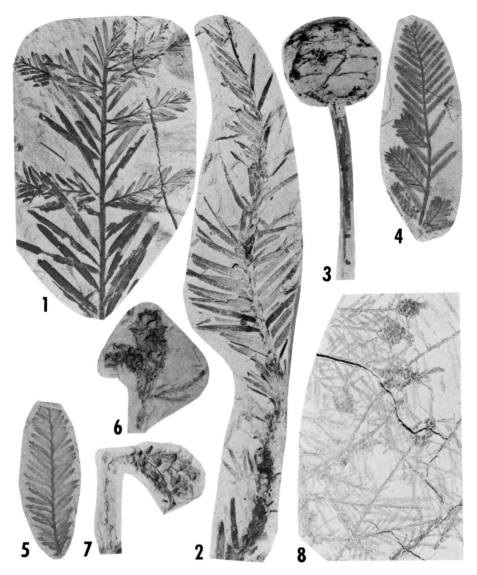

Plate 3. Clarkia Impression Fossils (Taxodiaceae). All figures natural size. Figs. 1, 8, *Taxodium dubium* (Sternb.) Heer. 1 = shoot with young side branches. 8 = mature branch (see lower left) with attached annual shoots; cone-like structures are dipterid galls resembling ones found on living Taxodium. Loc. P-37. Fig. 2. *Cunninghamia chaneyi*. Branch showing 4-5 seasons' growth. Loc. P-37. Figs. 3, 4. *Metasequoia occidentalis*. 3 = cone. 4 = shoot with young side branches. Loc. P-33. Fig. 5. Shoot, cf. *Sequoia affinis* Lesq. Loc. P-33. Figs. 6, 7. *Glyptostrobus oregonensis* Brown. Leafy shoots with cones. Loc. P-37.

Pacific Northwest, especially in the Mascall and "Latah" floras (Knowlton 1902, 1926; Berry 1929, 1934; Ashlee, 1932; Brown 1937; Chaney and Axelrod 1959). Others seem referable to species described in late Paleogene (Oligocene) floras such as the Weaverville of northern California (MacGinitie 1937), the Rujada of northwest Oregon (Lakhanpal 1958), and the Bridge Creek of central Oregon (Chaney 1927). A few Clarkia species were described originally in the later Miocene Ellensburg flora of central Washington (Smiley 1963). Others represent undescribed species or perhaps new combinations, and one Clarkia taxon has been classified as an extinct genus (*Pseudofagus* Smiley and Huggins 1981).

The new information present in preserved Clarkia material indicates that taxonomic revisions and additions will be required particularly at the level of species, and that affinities with modern species need to be re-evaluated. Because of this potential for species modifications, the composition of the Clarkia flora (Table 1) is exemplified for the present by the approximately 100 genera and 50 families of the predominating spermatophytes (conifers and angiosperms). Some of the megafossils are identified only provisionally, either because diagnostic organs have not yet been found or because a more intensive study is required for both the fossil taxon and its putative living relatives. Furthermore, pollen of certain taxa can be identified only to the level of family or to a complex of families or genera (see Gray, this volume).

The organ representation of Clarkia spermatophytes listed in Table 1 is indicated as leaves or shoots (L), fruits or other reproductive megaossils (F), or pollen (P). Reproductive megafossils and/or pollen have been found for more than one-half the taxa listed. In addition to spermatophytes, the Clarkia flora is known to contain a diverse assemblage of other plants including fungi, mosses, ferns, horsetails, club mosses, quillworts and Gnetaceae (*Ephedra*). Fungal epiphytes are common on foliage and fruits of angiosperms and conifers (Plate 1, Fig. 6; see also Williams, this volume), and fungal spores and other microstructures presently are being investigated by Sherwood at the University of Oregon (preliminary report, *in* Gray, this volume). Several genera of mosses are preserved as intact foliar shoots, some of which were extracted in HF solutions (Miller and Anderegg, in progress). Ferns occur as rare fragmental fronds and spores representing 7-8 species including *Osmunda* and *Polypodium*. Quillworts are represented by megafossils of *Isoetes* at localities P-37 and P-40 and by microfossils of the genus at P-33 (Gray, this volume). Pollen of *Ephedra* has also been identified in the P-33 microflora.

In addition to the allochthonous fossils of the surrounding forests, numerous autochthonous microfossils of the limnitic realm also are present. These include various diatoms (Bradbury et al., this volume) and algal cysts (Williams, this volume). Other limnitic microfossils include spicules of sponges and scales of amoeboid organisms (Williams, this volume), and ?Dinoflagellates (Gray, this volume).

The mixed mesophytic forests surrounding Clarkia Lake contained a diversity of conifers dominated by the family Taxodiaceae (see Plates 1 and 3), with lesser representation of the Pinaceae and Cupressaceae. The family Taxaceae (cf *Taxus*) is represented by rare shoots and possibly pollen ("TCT"). Isolated compressed leaves of *Amentotaxus* (Amentotaxaceae) are common in some of the Clarkia beds. As presently identified from foliage, cones, seeds and pollen, Clarkia conifers are now known to represent not less than 5 families, 17 genera and 19-20 species.

Among the angiosperms, the monocotyledons are dominated both numerically and taxonomically by the family Smilacaceae, which is represented by two species of *Smilax* and one that appears referable to the Asian *Heterosmilax*. Other monocotyledons listed in Table 1 occur as rare fossils in the Clarkia flora. Based on both mega- and microfossil data, the flora contains 5 families, 5 genera and 6-7 species of monocotyledons.

The dicotyledons are the taxonomic dominants of the Clarkia mesophytic forests and are the numerical dominants among the plant megafossils. The families most commonly represented by fossils at Clarkia localities are the Magnoliaceae, Lauraceae, Menispermaceae, Hamamelidaceae,

TABLE 1. FAMILIES AND GENERA OF CLARKIA CONIFERS AND ANGIOSPERMS

Families	Genera (and organs represented)
CONIFERS	
Amentotaxaceae	*Amentotaxus* (L).
Taxaceae (P*)	cf. *Taxus* (L, F).
Pinaceae	*Abies* (L, F, P), *Cedrus* (P), cf. *Keteleeria* (F), *Picea* (P), *Pinus* (L, F, P), *Tsuga* (P), ?*Pseudotsuga* (P).
Taxodiaceae (P*)	*Cunninghamia* (L, F), *Glyptostrobus* (L, F), *Metasequoia* (L, F), *Sequoia* (L, ?F), *Taxodium* (L, F).
Cupressaceae (P*)	*Calocedrus* (*Librocedrus*) (L, F), *Chamaecyparis* (L, F), *Thuja* (L).
(P* = TCT)	
MONOCOTYLEDONS	
Alismataceae	*Caldesia* (L).
Smilacaceae	cf. *Heterosmilax* (L), *Smilax* (L).
Gramineae (P)	*Graminites* (L).
Cyperaceae (P)	------
Typhaceae	*Typha* (L, ?P).
DICOTYLEDONS	
Magnoliaceae	*Liriodendron* (L, F, P), *Magnolia* (L, F, P).
?Anonaceae	cf. *Asimina* (L).
Lauraceae	*Lindera* (L), *Persea* (L), *Sassafras* (L).
Nymphaeaceae	*Nuphar* (L).
Menispermaceae	*Cocculus* (L).
Cercidiphyllaceae	*Cercidiphyllum* (L, F).
Hamamelidaceae	*Exbucklandia* (L), *Hamamelis* (L), *Liquidambar* (L, F, P), cf. *Parrotia* (L).
Platanaceae	*Platanus* (L, F, P).
Ulmaceae	*Celtis* (L, P), *Ulmus* (L, ?P), *Zelkova* (L, ?P).
Fagaceae	*Castanea* (L, ?F, ?P), cf. *Castanopsis* (L), *Fagus* (L, F, P), cf. *Lithocarpus* (L, F), *Quercus* (L, F. P), *Pseudofagus* (L, F).
Betulaceae	*Alnus* (L, F, P), *Betula* (L, F, P), *Corylus* (L, F, P), *Ostrya* (L, F, ?P).
Comptoniaceae	*Comptonia* (L).
Myricaceae	*Myrica* (P).
Juglandaceae	*Carya* (L, ?F, P), *Engelhardtia* (F, ?P), *Juglans* (L, P), *Pterocarya* (L, F, P).
Chenopodiaceae (P)	----------
?Theaceae	cf. *Gordonia* (L).
Salicaceae	*Populus* (L), *Salix* (L).
Ericaceae (P)	cf. *Arbutus* (L), cf. *Pieris* (L, F), cf. *Rhododendron* (L), cf. *Vaccinium* (L)
Styraceae	*Halesia* (L, F).
?Symplocaceae	?*Symplocos* (L, ?P).
Ebenaceae	*Diospyros* (L, F).
Tiliaceae	*Tilia* (L, F, P).
Malvaceae	?*Malva* (F).
Saxifragaceae	*Philadelphus* (L).
Rosaceae (P)	*Amelanchier* (L), *Crataegus* (L), cf. *Malus* (L), *Prunus* (L), *Rosa* (L), cf. *Spryea* (L).
Leguminosae	cf. *Amorpha* (L), cf. *Bauhinia* (L), *Cercis* (L), cf. *Derris* (F), *Gleditsia* (L, F), *Gymnocladus* (L, F), *Robinia* (L), cf. *Zenia* (F).
Anacardiaceae	*Rhus* (L, P).

TABLE 1. (Continued)

Families	Genera (and organs represented)
DICOTYLEDONS (continued)	
Meliaceae	*Cedrela* (L, F).
Aceraceae	*Acer* (L, F, P).
Hippocastanaceae	*Aesculus* (L, P).
Cornaceae	*Cornus* (L).
Nyssaceae	*Nyssa* (L, F, P).
Aquifoliaceae	*Ilex* (L, P).
?Celastraceae	cf. *Perrottetia* (L).
Rhamnaceae (P)	*Berchemia* (L), *Paliurus* (L, F).
Vitaceae (P)	*Ampelopsis* (L), cf. *Parthenocissus* (L, P), *Vitis* (L).
Oleaceae	*Fraxinus* (L, F, P).
Elaeagnaceae	*Shepherdia/Elaeagnus* (P).
Caprifoliaceae	*Viburnum* (L).
Scrophulariaceae	*Paulownia* (L).

Platanaceae, Fagaceae, Betulaceae, Styracaceae, Symplocaceae, Leguminosae, Aceraceae, Nyssaceae, and Vitaceae. For the Clarkia flora as a whole, based on fossils obtained from 5 collecting sites in the Clarkia basin, the dicotyledons are presently known to comprise not less than 40 families and 80 genera, and appear to be represented by more than 100 species.

Megafossil representation of taxa may vary from locality to locality or from level to level within an exposed section. The conifers *Amentotaxus* and *Cunninghamia*, for example, are common fossils in some beds but are rare or absent at other levels even of the same exposure. *Cercidiphyllum* generally is rare in Clarkia florules, but it is common and is one of the dominant megafossils in one 50-60 cm zone of the P-40 site. At locality P-33, the Platanaceae, Betulaceae, Aceraceae, Nyssaceae and Taxodiaceae (especially *Taxodium*) are well represented by megafossils at the bottom and top of the exposed section, but they are much less common in the middle part of the section. The Fagaceae and Leguminosae are common and taxonomically varied in the middle part of the P-33 section, but these families are represented by uncommon fossils of low taxonomic diversity near the bottom and top. The genera *Acer* and *Liriodendron* are represented by numerous leaves at certain levels of 3 localities (P-33, P-37 and P-40), but leaves of these genera are rare or absent at other levels or localities.

It would appear, therefore, that our concept of the taxonomic composition of the vegetation surrounding Miocene Clarkia Lake—that is, the empirical fossil evidence of plant taxa that lived within the confined limitations of the topographic basin—needs to be distinguished from theoretical considerations of the temporal (seral) changes that seem to have been taking place in the more local lake-border vegetation. The taxonomic composition of the Clarkia forests is based on the totality of taxa from all Clarkia localities, whether they are of common and widespread occurrence (e.g. *Metasequoia, Taxodium, Cocculus, Persea, Pterocarya, Betula, Quercus, Pseudofagus*, etc.) or are rare and from a single locality (e.g. *Caldesia* as two leaves from P-33 and *Engelhardtia* as an isolated fruit at P-38). Seral changes of vegetation that occupied lake-border habitats near the P-33 site, on the other hand, are inferred from documented vertical trends of dominant megafossil organs that show little if any abrasion due to transport.

INFERRED CLARKIA FOREST ASSOCIATIONS

On the basis of preferred habitats of known or putative modern relatives, a variety of ecological niches appears to have been present in the vicinity of Miocene Clarkia Lake (Smiley and

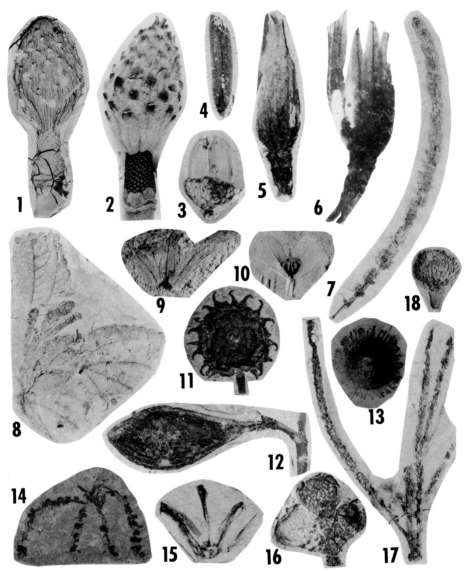

Plate 4. Selected Clarkia Fruits and Aments (except Fig. 2). All figures natural size. Fig. 1. *Magnolia* sp. Immature cone. Loc. P-33. Fig. 2. Immature cone of living *Magnolia* sp. cultivated in San Jose, CA, for comparison with Fig. 1. Fig. 3. *Quercus* sp. Acorn in cup. Loc. P-33. Figs. 4-6. *Liriodendron hesperia* Berry. 4 = seed. 5 = cone with appressed seeds. 6 = compressed cone and seeds, mounted in bioplastic. Loc. P-33. Figs. 7, 17. Erect staminate aments of ?*Lithocarpus* sp. (cf. extant *L. densiflora* Rehd.). Loc. P-37. Fig. 8. *Cercidiphyllum crenatum* (Unger) Brown. Parts of three leaves and fruit cluster. Loc. P-40. Fig. 9. *Engelhardtia* sp. Tri-winged seed. Loc. P-38. Fig. 10. *Pterocarya* sp. Bi-winged seed. Found with leaves of *P. mixta* (Kn.) Brown. Loc. P-33. Fig. 11. *Liquidambar* cf. *L. pachyphyllum* Knowlton. Fruit "head." Loc. P-37. Fig. 12. *Gleditsia* sp. Seed pod (cf. extant *G. aquatica* Marsh). Loc. P-37. Fig. 13. ?*Lithocarpus* sp. Cupule with acicular appendages (cf. extant *L. densiflora*). Loc. P-37. Fig. 14. Pendant staminate aments of *Quercus* sp. Loc. P-33. Fig. 15. *Libocedrus* sp. Opened cone (cf. extant *L. decurrens* Torrey). Loc. P-37. Figs. 16, 18. ?*Lithocarpus* sp. Cupules with small scale-like appendages. Loc. P-33.

Rember, 1981, Table 17-3). The three general associations that have been inferred are: (1) a swamp association, (2) a floodplain-slope association, and (3) a drier habitat association. The swamp association probably inhabited poorly drained swampy sites of floodplains or lake borders, including swamp margins and swamp hummocks. The floodplain-slope association would have occupied well-drained mesic sites of streamsides, or deep-soil, relatively steep slopes of adjacent hillsides. The more zeric association, which is a relative category for the mesic Clarkia vegetation, includes those taxa which would have inhabited rocky or sandy sites with thin soil cover, or which may have been primarily species on new landslide scars or on fire-denuded slopes. A few conifer taxa are represented in the Clarkia flora by only winged seeds or pollen grains, and they are considered most likely to have represented the forests of higher elevations such as nearby ridge crests or peaks. Subsocieties of Clarkia vegetation undoubtedly were present, but their identification is much less tenable when based on the ecological tolerances of putative living descendants. Clarkia taxa that are here considered to be most characteristic of, although not necessarily restricted to, each of the inferred associations are listed in Table 2.

INFERRED SERAL CHANGE OF LAKE-BORDER VEGETATION

The outward progression of marginal land that occurs during the sedimentary infilling of modern lakes, and that results in an expanding fringe of swamp or bog habitats, is applicable also to Miocene Clarkia Lake. The abrupt establishment of a Tertiary lake where none had existed before, and its subsequent infilling by rapidly deposited sediments and volcanic ashes, is well exemplified by the exposed section at locality P-33 (Smiley and Rember, 1979, 1981, this volume). This site of accumulation seems to have been a small bay of fairly deep and quiet water on the western edge of Clarkia Lake, representing the drowned mouth of a minor tributary valley now occupied by Bechtel Creek. As sedimentation continued in this bay, new marginal ground would have resulted and the shoreline would have progressed outward toward the present site of locality P-33. Judging from responses of modern plants to similar conditions (see Muller-Dombois and Ellenberg, 1974) a seral change of lake-border vegetation would occur. An application of this general observation to Clarkia vegetational interpretations would imply that the Tertiary lake-border forests necessarily experienced notable changes as lacustrine sedimentation progressed. That such a change occurred seems evident from the stratigraphically documented trends of allochthonous plant megafossils through the exposed lacustrine section at locality P-33 (Rember 1979; see also Smiley and Rember, 1981).

The stratigraphic analysis conducted by Rember at P-33 extended through 7.6 m (~25 ft) of the exposed lacustrine section. Blocks that were cut from sequential levels of the outcrop, and transported for microstratigraphic analysis in the laboratory, provided bedding surfaces approximately 30 cm by 45 cm across (=~1.5 sq ft). Megafossils that were preserved in the soft finely laminated clays were uncovered layer-by-layer with a knife and were identified and counted for each 0.5-1.0 cm interval (see Smiley and Rember, 1981, Fig. 17-6). Because the quantity of megafossils was considered to be not significant on a statistical basis (unlike pollen: see Gray, this volume), the numerical and taxonomic data were tabulated for 30-cm increments (~1.5 cu ft) and for 21 successive increments of the study column. Although this method will tend to obscure any short-term trends that may have occurred within the temporal span of a 30-cm column increment (see, for example, Smiley and Rember, 1981, Fig. 17-6, showing the decline of *Castanea*-type leaves within the 120.5-148.9 cm increment), longer-term trends of commonly represented taxa could be plotted through the entire column on the basis of significant numbers. Actual numbers of megafossils observed in the 21-column increments (excluding 2 thick ash beds that were devoid of fossils) ranged from 138 to 1294, for a mean to 504 and a total of 10,574 (Fig. 1A). The taxonomic diversity within 30-cm column increments ranged from 22 species to 50 species for a mean of 35 species (Fig. 1B).

TABLE 2. INFERRED CLARKIA FOREST ASSOCIATIONS

SWAMP ASSOCIATION
Bryophytes (mosses)
Equisetum (horsetail)
Isoetes (quillwort)
Osmunda (royal fern)
Polypodium (polypody fern)
Taxodium (bald cypress)
Chamaecyparis (white cedar)
Caldesia (Asian monocot)
Typha (cattail)
Populus (poplar)
Salix (willow)
Quercus (live oak type)
Nuphar (water lily)
Magnolia (sweetbay type)
Lindera (benzoin)
Persea (avocado)
Platanus (sycamore)
Gleditsia (water locust type)
Acer (red maple type)
cf. *Ampelopsis* (grape family)
Vitis (muscadine grape type)
cf. *Gordonia* (loblolly bay)
Nyssa (water tupelo type)
Cornus (dogwood)
Viburnum (arrow-wood type)

DRY-SLOPE ASSOCIATION
Ephedra (Gnetaceae)
Pinus (pines)
Quercus (white oak type)
Celtis (hackberry)
Comptonia (sweet-fern)
Amelanchier (service berry)
Crataegus (hawthorn)
Philadelphus (mock orange)
Rosa (rose)
cf. *Amorpha* (false indigo)
Rhus (sumac)
cf. *Parthenocissus* (creeper)
Shepherdia (buffalo berry)

FLOODPLAIN-SLOPE ASSOCIATION
Various mosses
Lycopodium (club moss)
Various ferns
Taxus (yew)
Abies (fir)
Amentotaxus (Asian conifer)
Cunninghamia (Asian conifer)
Glyptrostrobus (Asian conifer)
Metasequoia (Asian conifer)
Sequoia (coast redwood)
Thuja (red cedar)
Calocedrus (incense cedar)
Smilax (greenbriar)
Carya (hickory)
Engelhardtia (walnut family)
Juglans (walnut)
Pterocarya (Asian wingnut)
Alnus (alder)
Betula (birch)
Corylus (hazel nut)
Ostrya (hop hornbeam)
Castanea (chestnut)
Fagus (beech)
Quercus (red oak type)
Pseudofagus (extinct Fagaceae)
Ulmus (elm)
Zelkova (Asian "elm")
Cercidiphyllum (Asian katsura)
Cocculus (moonseed)
Liriodendron (tulip tree)
Magnolia (cucumber tree type)
Sassafras (sassafras)
Hydrangea (hydrangea)
Hamamelis (witch hazel)
Liquidambar (sweet gum)
cf. *Malus* (apple)
Prunus (cherry)
Leguminosae (various genera)
Cedrela (cedrela)
Ilex (holly)
Acer (sugar maple type)
cf. *Asimina* (pawpaw)
Aesculus (horse chestnut)
Berchemia (buckthorn family)
Paliurus (Asian, buckthorn family)
Vitis (fox grape type)
Nyssa (black tupelo type)
Cornus (dogwood)
Ericaceae (various genera)
Diospyros (persimmon)
Symplocos (sweet leaf)
Halesia (silver-bell tree)
Fraxinus (ash)
Paulownia (Asian empress tree)

INFERRED UPLAND CONIFERS

Cedrus (Cedar)
Keleleeria (Asian conifer)
Picea (spruce)

Pseudotsuga (Douglas fir)
Tsuga (hemlock)

Numerical trends of taxa represented by 5% or more of the fossils in any one column increment were plotted through the 7.6 m of the study column, producing trend diagrams similar to ones commonly used in palynological studies (Fig. 2). Gray (this volume) has conducted a similar study of the pollen representation through much of the section that provided the megafossil data, showing both similarities and differences in proportional representation and in trends. This most likely reflects the complex interplay of variables that influence the transport, sorting and eventual accumulation of megascopic as compared to microscopic remains. As Gray has observed, empirical studies relating to the transport and accumulation of pollen (and seeds) have progressed considerably further than similar studies on leaves and larger reproductive structures. Despite such limitation of specific information for uniformitarian considerations, however, it seems evident that the more readily transportable pollen (and winged seeds) would be more reflective of the total vegetation that inhabited the confined Clarkia basin, whereas the remains of leaves and other megascopic

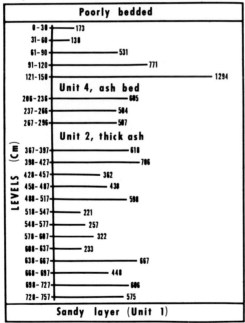
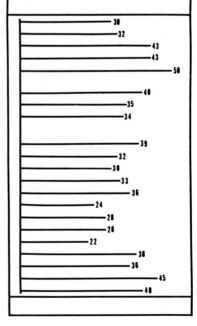

Figure 1. Numerical data obtained from 21 column increments through the exposed section at locality P-33. Each 30-cm increment of the study column measured approximately 30 cm X 45 cm across (= 1.5 cu ft). A = the number of megafossil specimens counted in each column increment. B = the number of species identified in each column increment. The lower numbers of specimens in the 518-637 cm levels reflect the common occurrence of thick (upward-fining) clay layers that were devoid of megafossils. The lower numbers in the 0-60 cm level at the top of the column reflect the accumulation of fewer megafossils at the site, in sediments that are siltier in texture and less laminated than in the column below.

Plate 5. Clarkia Fossils. All figures natural size. Fig. 1. *Quercus* sp. Cleared leaf compression preserved in glass mount, transmitted light. Loc. P-33. Fig. 2. *Caldesia* sp. (Alismataceae). Leaf (cf. extant *C. reniformis* Makino of eastern Asia). Loc. P-33. Fig. 3. Small leaf of Vitaceae (*Vitis* or *Ampelopsis*), coated with carbon residue. Loc. P-33.

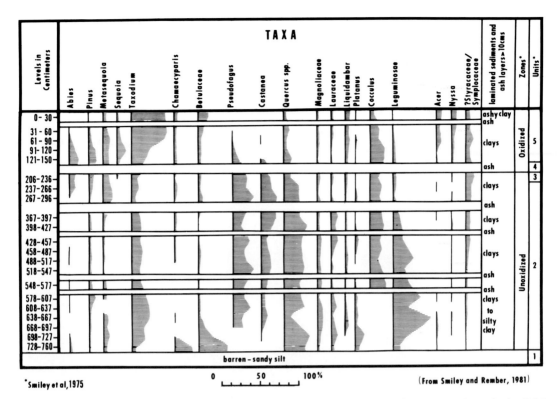

Figure 2. Numerical trends of dominantly preserved plant megafossil taxa through the P-33 study column. Included are those taxa that represent 5% or more of the specimens in any one column increment. See comparative pollen trends of Gray (this volume).

structures would be more indicative of the composition of vegetation that inhabited the nearby shorelines and immediately overlooking slopes (see also Spicer 1981).

In the case of the Clarkia condition, where the time span represented by the P-33 lacustrine section was too short for significant climatic changes to have notably modified the taxonomic composition of the surrounding slope forests, the establishment and maturation of a lake-border swamp forest would have served as an increasingly effective filter to the transport of megascopic plant structures from slope to lake. This would be reflected in the trends of accumulating megafossils during lacustrine sedimentation, even though the slope vegetation experienced little appreciable change in taxonomic content. Microscopic structures such as pollen, on the other hand, could continue to be transported to the depositional site in significant quantities, which would account for some of the differences that have been observed between the megafossil and microfossil trends. For specific taxa, however, explanations of trends are not readily apparent (e.g. the abrupt appearances of megafossils of *Cocculus* and *Pseudofagus*), and these and other questions that remain are likely to be unresolved until a study is conducted on a modern analogue of the Tertiary Clarkia basin in an area of similar climate and vegetation.

The vertical changes in numerical representation of common Clarkia megafossils (Fig. 2) reflect the ability of the various plant organs to reach the site of deposition under the prevailing conditions. The number, variety and condition of detached organs that were carried to the depositional site by wind or water are at least partly dependent upon the distance and duration of transport. Abrasion and ultimate destruction of plant organs is known to be accelerated during surface transport in running water, especially when the water is turbulent and sand-laden. Therefore, the

Plate 6. Terminal leaf cluster of ?*Quercus* (cf. *Q. simulata* Knowlton) (slightly reduced). Loc. P-33.

closer the proximity of plants to a depositional site, the greater the opportunity of their organs to survive surface transport (see Chaney 1959); the shorter the time of transport, the less degradation occurs (see Spicer 1981); the fewer obstacles that are present to organ dispersal, the less the "filtering" effect. The numerical representation of a species in a fossil assemblage undoubtedly is affected by a complex of other factors that may not be clearly evident from studies of modern conditions nor from analyses of fossil assemblages. Some of the major concerns in the interpretation of past vegetation, where necessarily based on detached organs that have already been transported and deposited, involve factors that are not preserved as observable, identifiable or quantifiable data. Such concerns will always be the subjects of interpretations (and conjecture) by both megafossil paleobotanists and palynologists: (1) For any given fossil species, how precisely can we infer habits, habitats, or resistance of organs to decomposition on the basis of known or putative living descendants? (2) How many individuals of a species occurred in the local vegetation? (3) Was a species widely distributed through the local vegetation, or was it confined to localized special habitats? (4) What size did an individual of a species attain, and how many organs of different kinds did it produce? (5) Were individuals of a species large canopy trees, smaller understory trees, shrubs, climbing vines, members of the ground cover? (6) What was the resistance of organs of a fossil species to physical, chemical or biological decomposition—that is, how soon after detachment must transport and deposition occur in order to produce a fossil that was essentially unaltered and showed no evidence of degradation? As any one of these intangibles can affect the quantitative representation of a fossil species at a site of deposition, depending upon prevailing conditions, statistical data—regardless of how impressive it may be—will provide only one of the several bodies of information that must be considered.

In the case of the Clarkia flora, the Tertiary conditions were conducive to high numerical representation of a diversity of taxa and organs, and to the preservation of megafossils in a "fresh" appearing state at time of burial. The study of a variety of Clarkia leaves by Niklas and Brown (1980, this volume), on degradation of intracellular organelles, suggests that many of the Clarkia leaves apparently were detached (during storms?) from the contributing plants and were deposited essentially unaltered in the preserving medium of Clarkia Lake. The taxonomic and organ diversity of the Clarkia flora, its excellent state of preservation, the incorporation of a large variety of forest insects in the lacustrine deposits (Lewis, this volume), and the upward fining of individual lacustrine laminations, all can be best explained by frequent rain storms and attendant gusty winds, commonly with wind-dispersal of organs directly from parent plant to depositional site. The "muddying" of the lake water during such storms further would enhance the preservation of essentially unaltered remains, as plant organs were blown out onto the lake surface, sank to the anoxic/toxic realm of the lake bottom, and were buried in rapidly accumulating fine sediments.

The plant megafossil sequence, and the vertical trends of individual taxa, seem to reflect vegetational changes that were taking place in lake-border habitats during the sedimentary infilling of Clarkia Lake. Plant megafossils at the bottom of the P-33 study column are dominated by taxa that are most indicative of riparian and floodplain conditions; ones in the middle are dominantly slope taxa; swamp and riparian taxa are the statistical dominants near the top. It seems apparent that a seral change occurred in the lake-border vegetation near the site of P-33 (see Fig. 2). Originally it was a riparian/floodplain type of vegetation characterized by birches, alders, sycamores, white cedar and bald cypress. After the dammed valley had filled with water, and the shoreline became established higher on the valley walls, the dominant taxa to be represented in the lacustrine deposits were those that characterized the slope forests (see Table 2). Finally, as lacustrine infilling progressed to a termination of lake conditions, a *Taxodium-Nyssa* type of swamp community developed, expanded and matured, providing both a major source of plant remains in the accumulating sediments and a barrier to the dispersal of organs from forests that occupied the slopes beyond. A relatively short time has been inferred for the existence of Clarkia Lake as

Plate 7. Terminal leaf cluster of *Quercus* sp. (cf. *Q. payettensis* Knowlton). Loc. P-33.

represented by the P-33 column, with a life-span that has been estimated at less than 1000 years (Smiley and Rember 1981; see also Smith and Elder, this volume). Despite the rapid change postulated for the lake-border vegetation, the forests of adjacent slopes and uplands probably remained in a stable condition during the short interval of time involved (see also the palynological analysis of Gray, this volume).

SUMMARY

The taxonomic and organ diversity and the excellent preservation of the Clarkia flora have resulted apparently from close proximity of a variety of habitats to an open lake of sufficient depth to have an anaerobic and toxic bottom. Supplementing such optimum conditions of topographic setting and depositional environment was a warm and humid climate with frequent storms, and periodic air-fall deposition of volcanic ash, that rapidly filled the lake basin with fine sediments. The unabraded and essentially unaltered condition of many of the plant megafossils seems to have resulted from direct wind dispersal of organs from slopes immediately overlooking the lake, or from short-distance surface transport within the confined Clarkia basin. Essentially unaltered cellular and chemical preservation further were enhanced by rapid burial in the anoxic/toxic realm of the lake bottom, and by the retention of a water-saturated and unoxidized state of the clays and fossils to the present day.

The importance and uniqueness of the Clarkia flora in the Tertiary history of North America lie in its potential for clarifying the relationships between Tertiary and modern species, for many of the plant taxa that were widely distributed in Tertiary floras of the Pacific Northwest region. The new fossil information that can be derived from study of Clarkia material includes: (1) retained organic chemistry, (2) microscopic features of the leaf epidermis, (3) intracellular organelles, (4) leaf phytoliths, (5) architectural details of whole leaves that can be removed intact from the soft clays, (6) the availability of numerous leaves of many of the Clarkia species, showing range of morphological variation, and (7) compressions or impressions of reproductive structures (sometimes attached) that are available for approximately half the taxa in the Clarkia flora. The excellence of preservation, and examples of organ diversity, are illustrated in Plates 1-7 and in other reports contained in this volume.

LITERATURE CITED

Ashlee, T. R. 1932. A contribution to the Latah flora of Idaho. Northwest Sci. 6(2):69-82.

Berry, E. W. 1929. A revision of the flora of the Latah formation. U. S. Geol. Surv. Prof. Pap. 154-H:225-265.

Berry, E. W. 1934. Miocene plants from Idaho. U. S. Geol. Surv. Prof. Pap. 185-E:97-125.

Bradbury, J. P., K. V. Dieterich, and J. L. Williams. 1985. Diatom flora of the Miocene lake beds near Clarkia in Northern Idaho. Pages 33-59 in C. J. Smiley, ed. Late Cenozoic History of the Pacific Northwest. Pacific Division, Amer. Assoc. Adv. Sci., San Francisco, Calif.

Brown, R. W. 1937. Additions to some fossil floras of the western United States. U. S. Geol. Surv. Prof. Pap. 186-J:163-206.

Chaney, R. W. 1927. Geology and palaeontology of the Crooked River basin, with special reference to the Bridge Creek flora. Carnegie Inst. Wash. Pub. 346:45-138.

Chaney, R. W., and D. I. Axelrod. 1959. Miocene floras of the Columbia Plateau. Carnegie Inst. Wash. Pub. 617:1-237.

Giannasi, D. E., and K. J. Niklas. 1981. Paleobiochemistry of some fossil and extant Fagaceae. Amer. J. Bot. 68(6):762-770.

Giannasi, D. E., and K. J. Niklas. 1985. The paleobiochemistry of fossil angiosperm floras. I. Chemosystematic aspects. Pages 161-174 in C. J. Smiley, ed. Late Cenozoic History of the Pacific Northwest. Pacific Division, Amer. Assoc. Adv. Sci., San Francisco, Calif.

Gray, J. 1985. Interpretation of co-occurring megafossils and pollen: A comparative study with Clarkia as an example. Pages 185-244 in C. J. Smiley, ed. Late Cenozoic History of the Pacific Northwest. Pacific Division, Amer. Assoc. Adv. Sci., San Francisco, Calif.

Huggins, L. M. 1980. Laboratory methods and analyses of fossil leaves and their application to educational curricula. Ph.D. Thesis. University of Idaho, Moscow, Ida.

Knowlton, F. H. 1902. Fossil flora of the John Day Basin, Oregon. U. S. Geol. Surv. Bull. 204: 1-153.

Knowlton, F. H. 1926. Flora of the Latah Formation of Spokane, Washington and Coeur d'Alene Idaho. U. S. Geol. Surv. Prof. Pap. 140:17-81.

Lakhanpal, R. N. 1958. The Rujada flora of west central Oregon. Univ. Calif. Pub. Geol. Sci. 35(1):1-66.

Lewis, S. E. 1985. Miocene insects from the Clarkia deposits of Northern Idaho. Pages 245-264 in

C. J. Smiley, ed. Late Cenozoic History of the Pacific Northwest. Pacific Division, Amer. Assoc. Adv. Sci., San Francisco, Calif.

MacGinitie, H. D. 1937. The flora of the Weaverville beds of Trinity County, California. Carnegie Inst. Wash. Pub. 465(3):83-152.

Muller-Dombois, D., and H. Ellenburg. 1974. Aims and methods of vegetation ecology. John Wiley and Sons, Inc., New York, N.Y. 547 pp.

Niklas, K. J., and R. M. Brown. 1980. Ultrastructural and paleobiochemical correlations of fossil leaf tissue from St. Maries River (Clarkia) area, northern Idaho. Amer. J. Bot. 68(3):332-341.

Niklas, K. J., R. M. Brown, and R. Santos. 1985. Ultrastructural states of preservation in Clarkia angiosperm leaf tissues: Implications on modes of fossilization. Pages 143-159 *in* C. J. Smiley, ed. Late Cenozoic History of the Pacific Northwest. Pacific Division, Amer. Assoc. Adv. Sci., San Francisco, Calif.

Rember, W. C., and C. J. Smiley. 1979. Dominant plants and seral analyses of the Clarkia fossil beds. Pacific Division, AAAS Sixtieth Ann. Mtg., Abstracts:23.

Smiley, C. J. 1963. The Ellensburg flora of Washington. Univ. Calif. Pub. Geol. Sci. 35(3):159-276.

Smiley, C. J., J. Gray, and L. M. Huggins. 1975. Preservation of Miocene fossils in unoxidized lake deposits, Clarkia, Idaho. J. Paleont. 49(5):833-844.

Smiley, C. J., and L. M. Huggins. 1981. *Pseudofagus idahoensis* n. gen. et sp. (Fagaceae) from the Miocene Clarkia flora of Idaho. Amer. J. Bot. 68(6):741-761.

Smiley, C. J., and W. C. Rember. 1979. Guidebook and road log to the St. Maries River (Clarkia) area of northern Idaho. Idaho Bur. Mines Geol. Inf. Circ. 33:1-27.

Smiley, C. J., and W. C. Rember. 1981. Paleoecology of the Miocene Clarkia Lake (northern Idaho) and its environs. Pages 551-590 (Chapter 17) *in* J. Gray, A. J. Boucot, and W. B. N. Berry, eds. Communities of the Past. Hutchinson and Ross Publishing Co., Stroudsburg, Pa.

Smiley, C. J., and W. C. Rember. 1985. Physical setting of the Miocene Clarkia fossil beds, Northern Idaho. Pages 11-31 *in* C. J. Smiley, ed. Late Cenozoic History of the Pacific Northwest. Pacific Division, Amer. Assoc. Adv. Sci., San Francisco, Calif.

Smith, G. R., and R. L. Elder. 1985. Environmental interpretation of burial and preservation of Clarkia fishes. Pages 85-93 *in* C. J. Smiley, ed. Late Cenozoic History of the Pacific Northwest. Pacific Division, Amer. Assoc. Adv. Sci., San Francisco, Calif.

Spicer, R. A. 1981. The sorting and deposition of allochthonous plant material in a modern environment at Silwood Lake, Silwood Park, Berkshire, England. U. S. Geol. Surv. Prof. Pap. 1143:1-77.

Williams, J. L. 1985. Miocene Chrysophyta cysts from a lacustrine deposit in Northern Idaho. Pages 61-66 *in* C. J. Smiley, ed. Late Cenozoic History of the Pacific Northwest. Pacific Division, Amer. Assoc. Adv. Sci., San Francisco, Calif.

Williams, J. L. 1985. Thecamoebian scales from a Miocene lacustrine deposit in Northern Idaho. Pages 67-71 *in* C. J. Smiley, ed. Late Cenozoic History of the Pacific Northwest. Pacific Division, Amer. Assoc. Adv. Sci., San Francisco, Calif.

Williams, J. L. 1985. Miocene epiphyllous fungi from Northern Idaho. Pages 139-142 *in* C. J. Smiley, ed. Late Cenozoic History of the Pacific Northwest. Pacific Division, Amer. Assoc. Adv. Sci., San Francisco, Calif.

Williams, J. L. 1985. Spicular remains of freshwater sponges from a Miocene lacustrine deposit in Northern Idaho. Pages 349-355 *in* C. J. Smiley, ed. Late Cenozoic History of the Pacific Northwest. Pacific Division, Amer. Assoc. Adv. Sci., San Francisco, Calif.

CUTICULAR ANALYSES AND PRELIMINARY COMPARISONS OF SOME MIOCENE CONIFERS FROM CLARKIA, IDAHO

L. MAURICE HUGGINS
Science Program Coordinator, Federal Way School District, Federal Way, WA 98003

Compressed and impressed conifer shoots, cones, cone scales, seeds and isolated needles are well represented in the Miocene Clarkia flora of northern Idaho. Many of these conifer taxa are represented in other Tertiary floras of western North America where they are preserved as imprints with little evidence of cuticular features. The Clarkia flora, therefore, provides a unique opportunity to examine the microscopic characters of conifers (and dicots) as expressed in preserved cuticles, and to compare each cellular feature with ones produced in living counterparts. This paper includes an analysis and a preliminary comparison of two Miocene species of *Pinus*, three species of Taxodiaceae representing the genera *Cunninghamia, Metasequoia* and *Taxodium*, and a species of the Cupressaceae representing the genus *Chamaecyparis*. Identifications of the foliage are determined from foliar morphology supported by distinctive reproductive structures. Preliminary cuticular comparisons of Miocene and Recent conifer species of the same genus suggest trends towards: (1) increase in cell size, (2) increase in numbers of stomatal rows per stomatal band, and (3) increase in numbers of and crowding of stomata in stomatal rows. Such apparent trends have been noted for all conifers examined to date. These trends are particularly significant in the case of the Taxodiaceae, where genera are presently monotypic (or perhaps oligotypic) and probably were represented by a single species per genus in Miocene floras of western North America.

Richly fossiliferous, partly unoxidized, Miocene lake deposits near Clarkia, Idaho, contain compression fossils of a variety of conifers and dicots. Six of the more commonly represented conifer taxa are described here in reference to their preserved cuticular features (see also Huggins 1980). Single leaves or leafy shoots and branches are preserved as compressions with both epidermal and mesophyll cellular details retained. This is a rare first opportunity to study such microscopic features of Miocene conifers and to compare them to selected extant species of the same genus having similar morphology.

The purpose of this report is to document the microscopic features of foliar organs of Miocene species preserved in the Clarkia deposits, which represent taxa that occur as common fossils in other Tertiary floras of the circum-North Pacific region. The processing methods required for Clarkia compression fossils, developed over several years of experimentation, also are recorded (see also Smiley and Huggins 1981). Detailed descriptions of the cuticular features of Miocene and Recent species are included, as well as tabulations of observed distinctions. These data are most significant for the representatives of the Taxodiaceae (*Cunninghamia, Metaseqouia, Taxodium*), presently monotypic (or perhaps oligotypic), which probably were represented by a single species in Miocene floras of western North America. For the other conifer taxa considered (*Pinus, Chamaecyparis*), comparisons are preliminary, since only one of many extant species has been examined to date.

Documentary specimens are retained in the collections of the University of Idaho Mines Museum (UIMM).

Copyright ©1985, Pacific Division, AAAS.

MATERIALS AND METHODS

The conifers described here are preserved as compressed leafy shoots or fascicled needles. All are represented by common fossils in the soft, unoxidized Clarkia clays, and they apparently have remained water-saturated since time of burial (Smiley, Gray and Huggins 1975; Smiley and Rember 1979, 1981). Identifications of fossils are based on similarities of foliar morphology, supplemented by distinctive reproductive structures. Representing *Pinus* are fascicled needles of two kinds: (1) long 3-needled, and (2) short 3-needled; pine cones, seeds and pollen also occur in the Clarkia flora. Of the three taxodiaceous forms considered here, *Cunninghamia* is represented by numerous shoots, some of which have cones attached; *Metasequoia* by leafy shoots and isolated cones, some of which are known to have retained seeds; and *Taxodium* by leafy shoots and isolated cones and cone scales, with some of the shoots bearing cone-like dipterid (insect) galls resembling those found on modern trees. Several genera of the Cupressaceae are now known to be represented in the Clarkia flora, but only *Chamaecyparis* is considered here; it is abundantly represented by compressed branches and isolated cones. Fossil material thus is adequate for positive generic identifications and for microscopic analyses of intact foliar structures.

The conifer material is represented by intact foliar units, which may be very fragile and brittle as preserved. As a consequence, new methods of extraction and processing were developed by Huggins (1980) for study of the Miocene cuticles. Because Clarkia leaf compressions seem not to have experienced dehydration since time of burial, drying commonly reduces intact specimens to microfragments, which become disseminated in the processing solutions. For this reason, specimens are collected in plastic bags and are processed within 2-3 days in their original state of water saturation. The conifer leaves are lifted physically from the bedding surface using a razor blade, and are transferred to a petri dish containing water. While immersed in water, the leaf is cut along the margin, using a glass slide as a cutting surface, and the apex and base are removed to facilitate separation of the upper and lower epidermis.

A slow bleaching process is required for Clarkia material, and sodium hypochlorite (5.25%, 8 drops per 10 cc of water) applied over a period of 5-15 minutes obtained good results. During the bleaching process, the crushed mesophyll material adhering to the epidermis is physically removed. Two tools were designed: a loop scraper and a two-pronged fork, both made from human hair. The scraper, constructed as a loop 2 mm in diameter glued to the end of a toothpick, is used to scrape off the adhering mesophyll tissue. The fork, fashioned by cutting a loop to produce tines about 5 mm long, is used to separate the two epidermal surface and to otherwise handle the delicate tissue during processing. Constant washing of the tools in water is required to prevent the bleach from dissolving the hair. After bleaching, the specimen is thoroughly washed in water; if the bleach is allowed to react with the fossil tissue for more than 15 minutes, the specimen will be reduced to an amorphous mush.

After experimentation with several kinds of dyes (see Gurr 1965), I found that 100% saturated aqueous Saffarin "O," or Defields Hematoxin 5% alcohol, produced the best results for staining Clarkia material. Again, care is taken not to destroy the fragile material during the staining process as a result of turbulence that may result from application of the staining solution or from chemical reactions. (If the bleach has not been thoroughly removed by washing, a dark scum or granular precipitate will form during the staining process.) A slide is then made, using 50% glycerine and 50% water as mounting medium.

For measuring the various microscopic features, a stage micrometer (0.01 mm scale) is photographed over a range of magnifications from 16X to 63X, to serve as a standard of reference. Measurement of the same cuticular feature at different magnifications was used to check accuracy. Variations in size and shape (range, average) for each type of cell or structure were determined by measuring (or comparing) 50 examples of each for both fossil and modern species. Exceptions where 50 examples are not available are so indicated. Experimentation has shown little difference in statistical values derived from measurements of 100-200 examples (e.g. Smiley and Huggins 1981) and measurements of the 50 examples used in this report. Measurements are made of length and width of a microscopic feature (in microns), the ratio of length to width (L/W), and the cell "area" (in square microns) obtained from multiplying the length by the width as expressed on the leaf cuticle.

SYSTEMATIC CONSIDERATIONS

Family Pinaceae
Genus *Pinus* L.
Pinus cf. *P. harneyana* Chaney & Axelrod
(Plate 1, figs. 1, 2; Plate 2, fig. 1)

Pinus harneyana Chaney & Axelrod. Carnegie Inst Wash. Pub. 617, pp. 141-142, Pl. 13, figs. 1, 2. 1959.

Cuticular Description

Lower epidermis. Cells linear with long axis parallel to midrib; length 330-650 (avg. 471) microns; width 11-21 (avg. 18) microns; L/W ratio 25.6; cell area 8666 square microns; end walls oblique, corners angled; anticlinal walls straight, with fine zig-zag undulations; phytoliths lacking. Stomatal bands composed of 4 rows of stomata, one band on each side of midrib, stomatal rows separated by 8-10 normal cell widths, bands 30 cell widths from both midrib and leaf margin; stomata widely and uniformly spaced along rows, long axis parallel to midrib; polar cells 4, lacking phytoliths; subsidiary cell length 40-50 (avg. 46) microns, width 8-18 (avg. 13) microns, L/W ratio 3.6, cell area 590 square microns; anticlinal walls straight, lacking ornamentation; phytoliths absent; guard cell length 32-40 (avg. 36) microns, width 12-19 (avg. 15) microns, L/W ratio 2.5 surface smooth, outer ledge thick and conspicuous, poles marked by a wide wedge of cutin.

Pinus ponderosa Lawson
Cuticular Description

Lower epidermis. Cells linear with long axis parallel to midrib; length 85-175 (avg. 147) microns; width 10-20 (avg. 16) microns; L/W ratio 16; cell area 869 square microns; end walls squared, corners angled; anticlinal walls straight, finely undulate; phytoliths lacking. Stomatal bands composed of 13 rows of stomata, one band on each side of midrib, stomatal rows separated by 6-8 (rarely 4) normal cell widths, specialized cells between stomatal rows forming lines of tooth-like topography; stomata oriented with long axis parallel to midrib; one polar cell at each end of guard cells, lacking phytoliths; subsidiary cells usually 6 (rarely 7-8), length 25-70 (avg. 42) microns, width 16-27 (avg. 21) microns, L/W ratio 2.0, cell area 869 square microns; guard cell length 26-39 (avg. 30) microns, width 10-14 (avg. 12) microns, L/W ratio 2.5, surface smooth, plain outer ledge thick and conspicuous, poles marked by wide wedge of cutin. Tooth-like structures of stomatal bands composed of 3-7 (avg. 4) modified epidermal cells of rectangular shape and rounded walls, tooth width 22 microns, length 95 microns, apically pointing.

Cuticular comparison of P. *cf.* P. harneyana *and* P. ponderosa. Upper epidermal tissue was not available on fossil material and comparisons thus are not made. The lower epidermis of the two species (Tables 1, 2) differ in several regards. The Miocene species has a band consisting of 4 rows of stomata on each side of the midrib, whereas *P. ponderosa* has bands consisting of as many as 13 rows of stomata. Epidermal cells are longer and broader in *P.* cf. *P. harneyana* than in the modern species, and the L/W ratio and cell area also are greater in the fossil species. Cells of *P. ponderosa* have anticlinal walls showing a fine rounded undulation and square end walls,

TABLE 1. OBSERVED DISTINCTIONS OF LOWER
EPIDERMAL CELLS, LONG 3-NEEDLED *PINUS*[1]

Characters	*P.* cf. *harneyana*	*P. ponderosa*
Cell length	330-650 (471)	85-175 (147)
Cell width	11-21 (18)	10-20 (16)
L/W ratio	(25.6)	(9.0)
Cell area	(8666)	(2396)
End walls	Oblique	Square
Anticlinal walls	Straight	Finely undulate
Ornamentation	Fine, zig-zag	None

TABLE 2. OBSERVED DISTINCTIONS OF STOMATAL
COMPLEX, LONG 3-NEEDLED *PINUS*[1]

Characters	*P.* cf. *harneyana*	*P. ponderosa*
Subsidiary cells (no.)	4	6-8 (6+)
Subsidiary cell length	40-50 (46)	25-70 (42)
Subsidiary cell, L/W ratio	(3.6)	(2.0)
Subsidiary cell area	(591)	(869)
Polar cells (no.)	4	2
Guard cell length	32-40 (36)	26-39 (30)
Guard cell width	12-19 (15)	10-14 (12)

[1] Measured in microns, averages in parentheses.

whereas the Miocene *P.* cf. *P. harneyana* has straight (rarely zig-zag) anticlinal walls and oblique end walls. Cell size also is reduced in the modern species (Table 1). The numer of subsidiary cells in the stomatal complex is 4 in the Miocene species and 6 in the modern species; polar cells are 4 in the fossil and 2 in the extant species; the guard cells appear to be more sunken in *P. ponderosa*, and cuticular thickening is less apparent. A major and striking difference is the presence of rows of tooth-like topographic features in the stomatal bands of *P. ponderosa*, not observed in the Miocene fossil material.

Pinus sp.
(Plate 3, figs. 1, 2; Plate 4, figs. 1, 2)

Cuticular Descriptions

Lower epidermis. Cells linear with long axis parallel to leaf midrib; length 170-250 (avg. 206) microns; width 8-13 (avg. 9) microns; L/W ratio 23; cell area 1874 square microns; end walls oblique corners angled; anticlinal walls straight, with fine undulations, ornamentation a fine zigzag pattern; phytoliths lacking. Stomatal bands one on each side of midrib, composed of a single row of stomata; stomata uniformly and widely spaced, with long axis parallel to midrib; stomatal complex lacking polar cells and phytoliths; subsidiary cells 4, length 22-45 (avg. 29) microns, width 4-17 (avg. 7) microns, L/W ratio 4.3, cell area 197 square microns; guard cell length consistently 40 microns, width 8-10 (avg. 9) microns, L/W ratio 4.3, surface smooth, outer ledge plain with a thin ring of cutin.

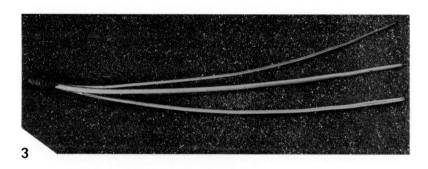

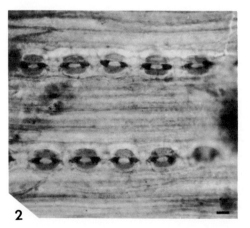

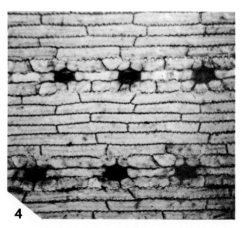

Plate 1. Figure 1. *Pinus* cf. *P. harneyana* Chaney and Axelrod. Long 3-needled pine. Clarkia flora (0.75X). Figure 2. *P.* cf. *P. harneyana*. Lower epidermis, showing long linear cells, and stomata with thick wedge of cutin at polar ends. Clarkia flora (190X). Figure 3. *Pinus ponderosa* Douglas. Extant species of long 3-needled pine, for comparison (0.75X). Figure 4. *P. ponderosa*. Lower epidermal tissue, showing linear cells with fine undulations and squared end walls (190X).

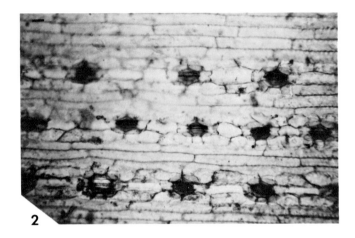

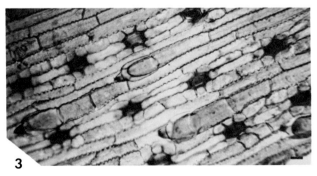

Plate 2. Figure 1. *P.* cf. *P. harneyana*. Three rows of stomata, showing heavy cutin at polar ends, and two polar cells at ends of guard cells. Clarkia flora (190X). Figure 2. *P. ponderosa*. Three rows of stomata, showing sunken guard cells, and one polar cell at ends of guard cells (190X). Figure 3. *P. ponderosa*. Several rows of "teeth" between rows of stomata. "Teeth" composed of 3-4 modified epidermal cells, with "teeth" in lower left pointing toward leaf apex (190X).

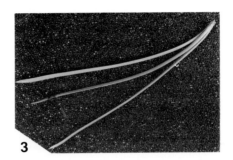

Plate 3. Figure 1. *Pinus* sp. Short 3-needled pine. Clarkia flora (0.75X). Figure 2. *Pinus* sp. Lower epidermis, showing long linear cells, with fine zig-zag undulation of cell wall, and long axis of cells parallel to axis of leaf. Clarkia flora (190X). Figure 3. *Pinus taeda* L. Extant species of short 3-needled pine, for comparison (0.75X). Figure 4. *P. taeda*. Lower epidermis near midrib, showing linear cells with square end walls, and characteristic lack of wall undulations (or rarely a few minute rounded undulations) (190X).

Pinus taeda L.
(Plate 3, figs. 3, 4; Plate 4, figs. 3, 4)

Cuticular Descriptions

Lower epidermis. Cells linear with long axis parallel to leaf midrib; length 120280 (avg. 134) microns; width 12-21 (avg. 16) microns; L/W ratio 8; cell area 2202 square microns; end walls squared, rarely oblique, corners angled; anticlinal walls straight with finely rounded undulations rare; phytoliths lacking. Stomatal bands one on each side of midrib, composed of 4 rows of stomata; stomata uniformly and widely spaced in rows, long axis parallel to leaf midrib; stomatal rows separated by 16-22 (rarely 6) cell widths, specialized cells between rows forming lines of tooth-like topography; polar cells present at each end of guard cells, lacking phytoliths; subsidiary cells 4, length 23-52 avg. 44) microns, width 14-23 (avg. 20) microns, L/W ratio 2.2, cell area 891 square microns; guard cell length 52-60 (avg. 53) microns, width 10-14 (avg. 12) microns, L/W ratio 4.8, surface smooth, outer ledge plain with a thin ring of cutin. Tooth-like structures of stomatal bands composed of 3-4 (usually 3) modified epidermal cells of rectangular or squared shape, tooth width 25 microns, width 168 microns, apically pointing.

Cuticular comparisons of Pinus *sp. and* P. taeda. Upper epidermal tissue was not available on fossil material and comparisons thus were not made. The lower epidermis of the two species (Tables 3 and 4) differ in several regards. The Miocene *Pinus* sp. has a stomatal band on each side of the midrib composed of a single row of stomata, whereas the extant *P. taeda* has bands composed of 4 rows of stomata. Epidermal cells of *P. taeda* have a much smaller L/W ratio and a larger cell area than in the Miocene species, and cell walls of the extant species have fine rounded undulations whereas those of the fossils are sharply zig-zag. In the stomatal complex, the number of subsidiary cells is 4 in the fossil *Pinus* sp. and 6 in *P. taeda*, accompanied by an increase in cell size. *Pinus* sp. lacks polar cells in the stomatal complex, and *P. taeda* has one polar cell at each end of the guard cells. Size of guard cells is greater in the extant than in the fossil species. The presence of cells modified into rows of tooth-like structures in *P. taeda* is another distinctive difference (see also comparisons between the Miocene *P.* cf. *P. harneyana* and the living *P. ponderosa*) (Table 4).

TABLE 3. OBSERVED DISTINCTIONS OF LOWER EPIDERMAL CELLS, SHORT 3-NEEDLED *PINUS*[1]

Characters	*Pinus* sp.	*P. taeda*
Cell length	170-250 (206)	120-280 (134)
Cell width	8-13 (9)	12-21 (17)
L/W ratio	(22.6)	(8.1)
Cell area	(1875)	(2203)
End walls	Oblique	Square or oblique
Wall thickness	Thin	Medium to thin
Ornamentation	Zig-zag walls	Fine, rounded, undulate

[1] Measured in microns, averages in parentheses.

TABLE 4. OBSERVED DISTINCTIONS OF STOMATAL COMPLEX, SHORT 3-NEEDLED *PINUS*[1]

Characters	*Pinus* sp.	*P. taeda*
Subsidiary cells (no.)	4	6
Subsidiary cell length	22-45 (29)	23-52 (44)
Subsidiary cell width	4-11 (7)	14-23 (20)
Subsidiary cell, L/W ratio	(4.3)	(2.2)
Subsidiary cell area	(198)	(891)
Ornamentation	Finely undulate	none
Polar cells (no.)	none	1
Guard cell length	40 (consistent)	52-60 (56)
Guard cell width	8-10 (9)	10-14 (12)
Guard cell, L/W ratio	(4.3)	(4.8)

[1] Measured in microns, averages in parentheses.

Family Taxodiaceae
Genus *Cunninghamia* R. Braun
Cunninghamia chaneyi Lakhanpal
(Plate 5, figs. 1-10)

Cunninghamia chaneyi Lakhanpal. Univ. Calif. Pub. Geol. Sci. 35(1), pp. 22, 23; Pl. 1, fig. 9; Pl. 2, figs. 1-3. 1958.

Cuticular Descriptions

Upper epidermis. Cells linear to rectangular, long axis parallel to axis of leaf; length near leaf margin 44-85 (avg. 62) microns, length near center of leaf 50-130 (avg. 93) microns; width near margin 11-27 (avg. 20) microns, width near center of leaf 10-23 (avg. 17) microns; anticlinal walls straight, of medium thickness, ornamentation occurring as alternating dark and light bands; end walls oblique or square. Stomata sometimes present on upper epidermis as a single incomplete row of stomata in apical part of leaf, stomatal characters similar to those of the lower epidermis.

Lower epidermis. Cells of similar shape and orientation to those of the upper epidermis; length near leaf margin 73-143 (avg. 97) microns, length near center of leaf 60-160 (avg. 107) microns; width near leaf margin 9-19 (avg. 13) microns, width near leaf center 9-20 (avg. 14) microns; phytoliths 11-26 per cell near leaf margin, fewer near leaf center, discoid shape, 1.5-6 (avg. 4) microns in diameter. Two stomatal bands, one on each side of midrib, composed of modified tissue 16-25 (avg. 21) cells wide; cells irregular, random orientation, length 22-55 (avg. 39) microns, width 12-26 (avg. 18) microns; bands, extending from near leaf base to near leaf apex, containing 7-9 rows of stomata; subsidiary cells irregular and randomly oriented, polycytic, 5 or more in number, length 19-55 (avg. 37) microns, width 12-25 (avg. 18) microns; polar cells shorter and broader than other subsidiary cells; guard cells sunken, length 50-65 (avg. 56) microns, width across aperture 24-37 (avg. 30) microns, surface smooth, outer stomatal ledge plain and conspicuous, polar cutin forming T's. Marginal teeth composed of tissue from both upper and lower epidermis, consisting of 3-26 (avg. 15) irregular cells, tooth length 170-340 (avg. 230) microns, width at base 21-70 (avg. 45) microns; tooth apex composed of a single cell with a length and width of 23-40 microns, tapering to a fine point; teeth apically trending.

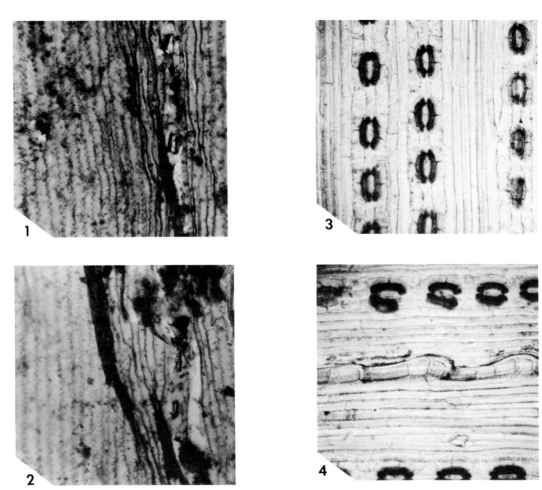

Plate 4. Fig. 1. *Pinus* sp. Lower epidermis, showing 2 stomata w/cutin on outer ledge of guard cells. Clarkia flora. Fig. 2. *Pinus* sp. Lower epidermis, showing 2 stomata w/long axis parallel to axis of normal epidermal cells. Clarkia flora. Fig. 3. *P. taeda*. Lower epidermis, showing 3 rows of stomata, w/guard cells and short, wide, modified subsidiary cells. Fig. 4. *P. taeda*. Lower epidermis, showing 3 "teeth" composed of 3-4 cells end-to-end, with "teeth" at right pointing toward apex of leaf. All 305X.

Plate 5. (opposing page). Figs. 1-10: *Cunninghamia chaneyi* Lakhanpal. Clarkia flora. Fig. 1. Area where tissue was removed. (0.75X). Fig. 2. Upper epidermis near margin, showing cells w/straight walls that are banded. (190X). Fig. 3. Upper epidermis near center of leaf, showing rectangular cells, oblique to square end walls. (190X). Fig. 4. Lower epidermis near leaf margin, showing straight anticlinal walls of medium thickness, lacking undulations. (305X). Fig. 5. Lower epidermis, showing 3 rows of stomata. (190X). Fig. 6. Lower epidermis near center of leaf, showing squared end walls. (190X). Fig. 7. Lower epidermis, showing bands of modified tissue containing rows of stomata. (190X). Fig. 8. Lower epidermis, showing modified tissue composing subsidiary and polar cells of stomatal complex. (190X). Fig. 9. Lower epidermis, showing sunken guard cells. Cutin at poles of guard cells forms a Y. (305X). Fig. 10. Marginal tooth formed from modified tissue of lower and upper epidermis. (190X). Figs. 11-20: *Cunninghamia lanceolata* (Lamb.) Hook. Extant species, for comparison. Fig. 11. Area where tissue was removed (0.75X). Fig. 12. Upper epidermis near leaf margin, showing highly undulate cell walls, and enclosing end walls (305X). Fig. 13. Upper epidermis, near margin, showing highly undulate cell walls, and enclosing end walls (305X). Fig. 14. Upper epidermis, showing modified tissue area containing stomatal rows (190X). Fig. 15. Lower epidermis, showing bands of modified tissue containing stomata. Fig. 18. Lower epidermis, showing stomata w/basically hexacytic subsidiary cells (190X). Fig. 19. Lower epidermis, showing guard cells that are broad, and lacking cuticular structures at polar ends (305X). Fig. 20. Marginal tooth formed from modified tissue of lower and upper epidermis (190X).

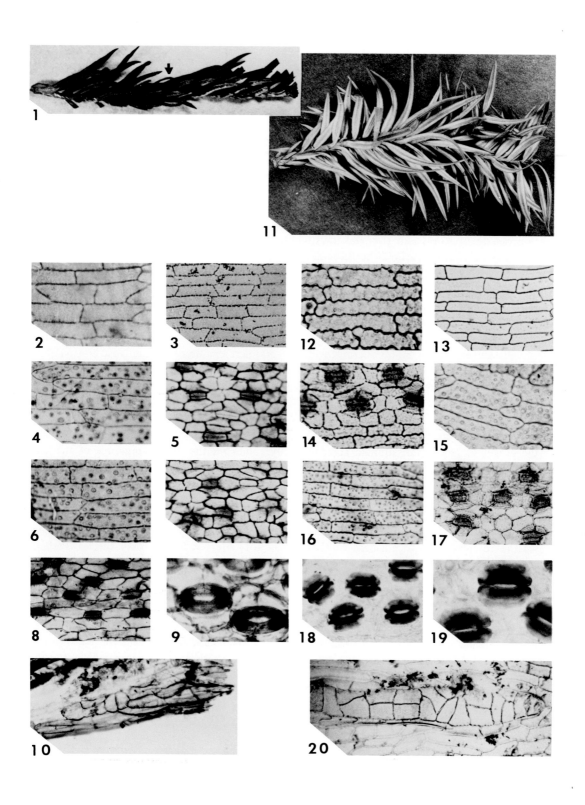

Cunninghamia lanceolata (Lamb.) Hook
(Plate 5, figs. 11-20)

Cuticular Descriptions

Upper epidermis. Cells linear to rectangular, long axis parallel to axis of leaf; length near leaf margin 31-112 (avg. 71) microns, length near center of leaf 50-135 (avg. 91) microns; width near laef margin 10-30 (avg. 17) microns, width near leaf center 11-27 (avg. 19) microns; anticlinal walls highly undulate near leaf margin, finely undulate near leaf center, of medium thickness, lacking ornamentation; end walls square or oblique, sometimes enclosing near leaf margin; phytoliths 5-9 per cell, circular in shape, 1-4 microns diameter. Stomata commonly present on upper epidermis, occurring as rows within bands of specialized tissue, one band on each side of midrib; bands composed of cells in random orientation, more or less rectangular, length 20-70 (avg. 43) microns, width 10-30 (avg. 22) microns, in part comprising stomatal subsidiary cells; stomatal rows 1-2 at base of band, usually 7-9 near middle of leaf, rows variably terminating in apical portion, length of bands and stomatal rows highly variable and commonly absent on upper epidermis; stomatal characters similar to those of the lower epidermis.

Lower epidermis. Cells of similar shape and orientation to those of the upper epidermis; bands similar to upper epidermis but broader, and contain 17-20 stomatal rows, with rows and stomatal axes parallel with leaf midrib; subsidiary cells of irregular shapes and random orientation, 4-7 in number, length 29-80 (avg. 50) microns, width 13-29 (avg. 21) microns, polar cells shorter and broader than others, walls usually straight without surface ornamentation, phytoliths lacking; guard cells sunken, length 50-60 microns, width 35-50 microns; outer stomatal ledge conspicuous, plain, smooth; no thickening of cutin at poles of guard cells. Marginal teeth composed of tissue from both lower and upper epidermis, consisting of 2-26 (avg. 23) irregular cells, tooth length 230-470 (avg. 340) microns; tooth apex composed of a single cell with a length and width of 30-40 microns, tapering to a fine point; teeth apically trending.

Cuticular comparisons of Cunninghamia chaneyi *and* C. lanceolata. (See also Table 5). In the extant *C. lanceolata*, stomata are commonly found on the upper epidermis of the leaf, where they may occur in 7-9 rows within specialized bands of modified tissue. In contrast, the Tertiary *C. chaneyi* in the Clarkia flora has rare stomata on the upper epidermis, and where these are present, they occur as a single row of few stomata in the apical part of the leaf. Upper epidermal cells of the extant species have highly undulate walls, whereas cell walls of the Miocene species are straight. In the extant *C. lanceolata* the cell ends have a "dog-bone" shape that encloses the rounded ends of adjacent cells, which is not observed in the fossil species. Perhaps the most striking difference between the Miocene and Recent species is the number of stomatal rows that determine the width of stomatal bands: *C. lanceolata* has 17-20 rows per band, whereas *C. chaneyi* has only 7-9 rows per band. In the stomatal complex of the modern species, the poles of guard cells are not marked by a thickening of cutin (Pl. 5, figs. 18, 19), compared to the fossil species, which has poles distinctly marked by a "T" configuration (Pl. 5, fig. 9). Although the size of guard cells is similar in both species, those of *C. lanceolata* appear distinctly wider due to the presence of "flanges" or broadened "wings" on the outer margins of the guard cells, not found in the fossil species (Pl. 5, figs. 9, 18, 19). The number of teeth along the leaf margin is greater in the extant than in the fossil species, and the increase in tooth size seems to reflect an increase in cell size rather than in cell number. These comparisons do not differ significantly from those noted by Sveshnikova (1963, pp. 226-227, Pl. XVII, figs. 1-15) for extant and Oligocene species of *Cunninghamia* from Eurasia (see also Cross 1942).

TABLE 5. OBSERVED DISTINCTIONS OF UPPER
EPIDERMAL CELLS OF *CUNNINGHAMIA*[1]

Characters	*C. chaneyi*	*C. lanceolata*
Cell length	44-85 (62)	31-112 (71)
Cell width	11-27 (20)	10-30 (17)
L/W ratio	(3.1)	(4.2)
Cell area	(1240)	(1258)
End walls	Square or oblique, not undulate	Square or enclosing, sinusoid
Anticlinal walls	Straight	Highly undulate
Cell corners	Square, angled	Rounded, enclosing
Ornamentation	Bands	None
Phytoliths	None	5-9 per cell

[1] Measured in microns, averages in parentheses.

Genus *Metasequoia* Miki

Metasequoia occidentalis (Newberry) Chaney

(Plate 6, figs. 1, 2; Plate 7, figs. 1, 2)

Metasequoia occidentalis (Newberry) Chaney. Amer. Phil. Soc. Trans. 40, p. 225, Pl. 12, fig. 3; Pl. 13, figs. 1-3. 1951.

Cuticular Descriptions

Upper epidermis. Cells linear to rectangular, long axis parallel to leaf axis, length 65-160 (avg. 70) microns, width 19-32 (avg. 25) microns, L/W ratio 2.8; cell area 1723 square microns; cell walls straight, medium in thickness, lacking ornamentation; end walls usually square, sometimes oblique, enclosing.

Lower epidermis. Cells highly irregular in shape, long axis parallel to leaf axis, length 75-152 (avg. 99) microns, width 21-37 (avg. 28) microns, L/W ratio 3.5, cell area 2797 square microns, inclusions numerous; anticlinal walls straight to rounded, medium in thickness, lacking surface ornamenation; end walls square, sometimes oblique, with angled corners. Stomata occur in bands, one on each side of midrib, 5-7 rows of stomata per band; stomata uniformly spaced in rows, parallel to axis of leaf; subsidiary cells 5 in number, length 15-40 (avg. 27) microns, width 12-20 (avg. 17) microns, L/W ratio 1.7, cell area 452 square microns, one polar cell at each end of guard cells; guard cell length 22-32 (avg. 29) microns, width 6-10 (avg. 9) microns, L/W ratio 3.3, surface smooth, outer ledge plain, thin rim of cutin on outer margins.

Metasequoia glyptostroboides Hu et Cheng

(Plate 6, figs. 3, 4; Plate 7, figs. 3, 4)

Cuticular Description

Upper epidermis. Cells rectangular, walls highly undulate, long axis parallel to leaf axis, length 60-100 (avg. 78) microns, width 30-45 (avg. 38) microns, L/W ratio 2.0, cell area 2913 square microns; anticlinal walls undulating, medium to thick, lacking ornamentation; end walls square to rounded, enclosing.

Lower epidermis. Cells similar to those of upper epidermis. Stomata occur in bands, one on

126   LATE CENOZOIC HISTORY

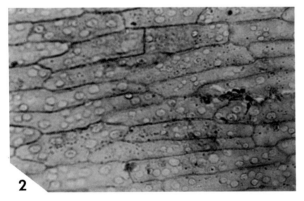

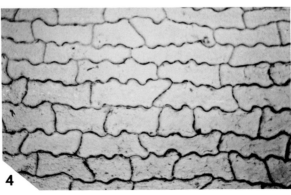

each side of midrib, bands containing a single row of stomata; stomata uniformly spaced along rows, with long axis parallel to leaf axis; subsidiary cells five in number, length 33-62 (avg. 46) microns, width 20-32 (avg. 25) microns, L/W ratio 1.8, cell area 1146 square microns, walls straight to rounded and lacking surface ornamentation, containing a moderate number of inclusions, polar cells one at each end of guard cells; guard cell length 33-41 (avg. 36) microns, width 8-14 (avg. 12) microns, L/W ratio 3.0, surface smooth, outer ledge plain, thin rim of cutin on outer margins.

Cuticular comparisons of Metasequoia occidentalis *and* M. glyptostroboides. Stomatal bands in the Miocene *M. occidentalis* from Clarkia are composed of 5-7 rows of stomata, whereas bands in the extant *M. glyptostroboides* are composed of 10-12 stomatal rows. The living species has thick anticlinal walls that are highly undulate, whereas the Tertiary species has thinner cell walls that are straight. The cells are larger, and the end walls are more enclosing, on the living *M. glyptostroboides*; the size of guard cells is markedly increased in the modern species (Tables 6-8). These comparisons do not differ significantly from those noted by Sveshnikova (1963, p. 214, Pl. VI, figs. 1-10; Pl. VII, 1-13) for extant and Cenozoic species of *Metasequoia* from Eurasia (see also Sterling 1949).

TABLE 6. OBSERVED DISTINCTIONS OF UPPER
EPIDERMAL CELLS OF *METASEQUOIA*[1]

Characters	*M. occidentalis*	*M. glyptostroboides*
Cell length	65-160 (70)	60-100 (78)
Cell width	19-32 (25)	30-45 (38)
L/W ratio	(2.8)	(2.0)
Cell area	(1723)	(2914)
End walls	Square to oblique, enclosing	Rounded to enclosing
Cell corners	Angled	Squared to rounded, enclosing
Anticlinal walls	Straight	Undulate
Phytoliths	Common	Few

[1] Measured in microns, averages in parentheses.

Plate 6. Fig. 1. *Metasequoia occidentalis* (Newb.) Cheny. Compressed specimen. Clarkia flora (0.75X). Fig. 2. *M. occidentalis*. Upper epidermis near midrib, showing enclosing end walls, and intracellular inclusions. Clarkia flora (190X). Fig. 3. *Metasequoia glyptostroboides* Hu et Cheng. Extant species, for comparison (0.75X). Fig. 4. *M. glyptostroboides*. Upper epidermis near midrib, showing short rectangular cells with undulate walls, and square to oblique end walls (190X).

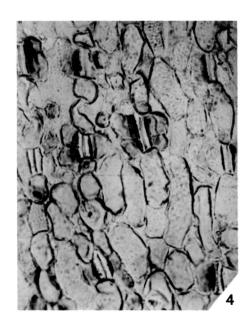

TABLE 7. OBSERVED DISTINCTION OF LOWER
EPIDERMAL CELLS OF *METASEQUOIA*[1]

Characters	*M. occidentalis*	*M. glyptostroboides*
Cell length	75-152 (99)	77-130 (97)
Cell width	21-37 (38)	20-35 (30)
L/W ratio	(3.5)	(3.2)
Cell area	(2797)	(2923)
Cell shapes	Irregular	Rectangular
End walls	Square to oblique	Square to rounded, some enclosing
Cell corners	Angled, enclosing	Rounded
Anticlinal walls	Straight, rounded	Highly undulate
Phytoliths	Numerous	Few

TABLE 8. OBSERVED DISTINCTIONS OF STOMATAL
COMPLEX OF *METASEQUOIA*[1]

Characters	*M. occidentalis*	*M. glyptostroboides*
Subsidiary cell length	15-40 (27)	33-62 (46)
Subsidiary cell width	12-20 (17)	20-32 (25)
Subsidiary cell, L/W ratio	(1.7)	(1.8)
Subsidiary cell area	(452)	(1147)
Phytoliths	Numerous	Common
Guard cell length	22-32 (29)	33-41 (36)
Guard cell width	6-10 (8)	8-14 (12)
Guard cell, L/W ratio	(3.3)	(3.0)

[1] Measured in microns, averages in parentheses.

Plate 7. Fig. 1. *M. occidentalis*. Lower epidermis, showing three rows of stomata parallel to leaf axis. Clarkia flora (190X). Fig. 2. *M. occidentalis*. Lower epidermis, showing four rows of stomata, and rounded subsidiary cells. Clarkia flora (305X). Fig. 3. *M. glyptostroboides*. Lower epidermis, showing two rows of stomata near midrib (at bottom) (305X). Fig. 4. *M. glyptostroboides*. Lower epidermis, showing four rows of stomata, with sunken guard cells and polar cells (305X).

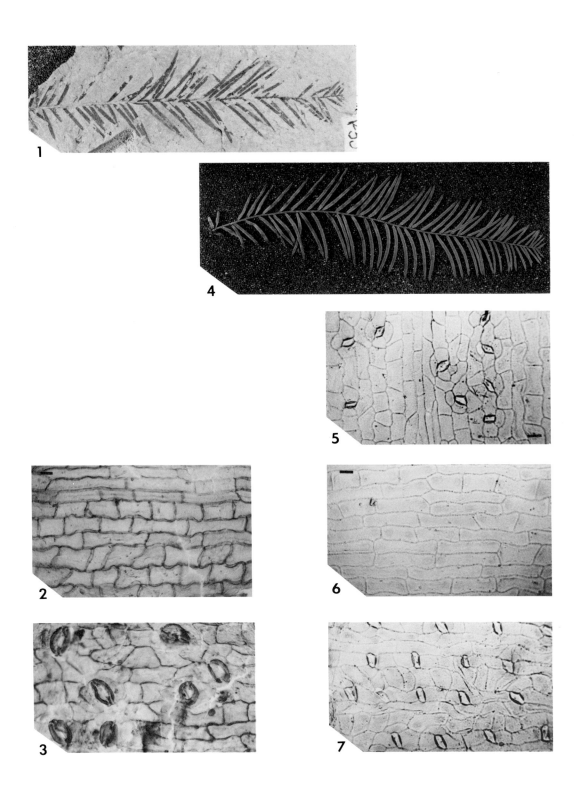

Genus *Taxodium* Richard
Taxodium dubium (Sternb.) Heer
(Plate 8, figs. 1-3)

Taxodium dubium (Sternb.) Heer. Flora Tert. Helv. I, p. 49, Pl. 14, figs. 4-17. 1855.

Cuticular Descriptions

Upper epidermis. Cells short, rectangular, long axis parallel to leaf axis, length 75-245 (avg. 143) microns, width 19-26 (avg. 22) microns, L/W ratio 6.6, cell area 3093 square microns; anticlinal walls straight to slightly undulating, thin, without surface ornamentation; phytoliths lacking.

Lower epidermis. Cells of similar shape and orientation to upper epidermis, length 52-100 (avg. 81) microns, width 18-30 (avg. 23) microns, L/W ratio 3.6, cell area 1818 square microns; anticlinal walls slightly undulate, medium in thickness, lacking surface ornamentation; end walls oblique, enclosing, with angled corners. Stomatal bands one on each side of midrib, bands containing 5 rows of stomata; stomata randomly spaced in rows, with stomatal axis variably oriented; subsidiary cells 4, length 30-65 (avg. 47) microns, width 17-33 (avg. 30) microns, L/W ratio 1.6, cell area 1408 sq. microns, polar cells lacking; guard cells sunken, length 43-53 (avg. 49) microns, width 12-19 (avg. 16) microns, L/W ratio 3.0, surface smooth, outer ledge plain and thick with cutin, cuticular thickening at poles forming "T" configurations; cellular phytoliths present but rare.

Taxodium distichum (L.) Richard
(Plate 8, figs. 4-7)

Cuticular Descriptions

Upper epidermis. Cells rectangular, long axis parallel to leaf axis, length 55-135 (avg. 125) microns, width 19-35 (avg. 25) microns, L/W ratio 3.2, cell area 2049 square microns; anticlinal walls straight to broadly undulate, medium in thickness, lacking surface ornamentation; cellular phytoliths present but few. Stomata present on upper epidermis, occurring as a single row of stomata on each side of midrib; stomatal characters resemble those of the lower epidermis.

Lower epidermis. Cells similar in shape and orientation to upper epidermis, length 42-72 (avg. 57) microns, width 17-27 (avg. 22) microns, L/W ratio 2.6, cell area 1210 sq. microns; anticlinal walls slightly undulate; of medium thickness, without surface ornamentation; end walls oblique or squared, enclosing, with angled corners; cellular phytoliths present but rare. Stomatal bands, 1 on each side of midrib, composed of 7 rows of stomata, stomatal axes variously oriented; subsidiary cells 5, polar cells not differentiated, length 22-32 (avg. 26) microns, width 12-21 (avg. 18) microns, L/W ratio 1.5, cell area 465 sq. microns; guard cells sunken, length 30-40 (avg. 34) microns, width 10-18 (avg. 15) microns, L/W ratio 2.3, surface smooth, outer ledge plain and thick with cutin, cuticular thickening at poles forming "T" configurations; cellular phytoliths present but rare.

Plate 8. Fig. 1. *Taxodium dubium* (Sternb.) Heer. Shoot showing leaves arranged alternately. Clarkia flora (0.75X). Fig. 2. *T. dubium.* Upper epidermis, showing short wide cells, with slightly undulate anticlinal walls. Clarkia flora (190X). Fig. 3. *T. dubium.* Lower epidermis, showing stomata randomly oriented, with most guard cells oriented at right angles to leaf axis. Clarkia flora (190X). Fig. 4. *Taxodium distichum* (L.) Richard. Extant species, for comparison. Shoot showing leaves arranged alternately (0.75X). Fig. 5. *T. distichum.* Upper epidermis, showing short wide cells, and several rows of stomata near midrib of leaf (190X). Fig. 6. *T. distichum.* Upper epidermis, showing areas of long and rectangular cells with square to oblique end walls (190X). Fig. 7. *T. distichum.* Lower epidermis, showing stomata with long axis at right angles to axis of leaf (190X).

Cuticular comparisons of Taxodium dubium *and* T. distichum. In most regards, the epidermal features of the fossil and modern species of *Taxodium* are similar (see also Cross 1940). A major difference is the presence of stomata on the upper epidermis of *T. distichum* whereas none has been observed on the upper epidermis of the Tertiary species. On the lower epidermis, there is an increase from 5 rows of stomata per band in the Clarkia *T. dubium* to 7 rows per band in the extant *T. distichum* (see Tables 9 and 10). On the modern species, the stomata are more numerous and crowded along a row. Also, there is an apparent increase in number of subsidiary cells, from 4 in *T. dubium* to 5 in *T. distichum*, associated with an increase in size of subsidiary cells in the extant species (see also cuticular comparisons in Sveshnikova 1963).

Family Cupressaceae
Genus *Chamaecyparis* Spach
Chamaecyparis cf. *C. linguaefolia* (Lesquereux) MacGinitie
(Plate 9, figs. 1-11)

Chamaecyparis linguaefolia (Lesquereux) MacGinitie. Carnegie Inst. Wash. Pub. 599, p. 89, Pl. 21, figs. 1-6. 1953.

Cuticular Descriptions

Upper epidermis. Cells rectangular, sometimes broadened at ends, long axis parallel to axis of leaf, length 48-110 (avg. 79) microns, width 11-27 (avg. 18) microns; anticlinal walls straight, thin, surface ornamentation lacking; end walls oblique, rarely squared, enclosing; large intracellular phytoliths present, 4-15 (avg. 9) per cell and apparently attached to cell wall, hexagonal or tetragonal in shape, some apparently cracked and refused, length 2-10 (avg. 7) microns, width 1-18 (avg. 4) microns. Cells at leaf base irregular, of variable shapes, long axis commonly at right angle to leaf axis, length 25-60 (avg. 46) microns, width 16-26 (avg. 20) microns.

Lower epidermis. Cell shape, size and orientation similar to those of the upper epidermis. Stomata on lower epidermis only, located mainly along basal margin and central area of leaf; stomatal areas composed of 8 rows of stomata, with 6-8 stomata per row, stomatal axes parallel to axis of leaf; each stoma surrounded by 4-6 subsidiary cells, which may be shared by as many as 3 adjacent stoma, shapes variable, length 19-55 (avg. 33) microns, width 16-31 (avg. 21) microns; guard cells shunken, size undetermined, surface smooth, outer ledge plain. Large intracellular phytoliths occurring as crystals in rectangular cells and in subsidiary cells of the stomatal complex, generally lacking near base and apex of leaf, 1-11 (avg. 5) per cell, length 4-10 (avg. 7) microns, width 2-5 (avg. 4) microns.

Chamaecyparis lawsoniana Parlatore
(Plate 9, figs. 12-22)

Cuticular Descriptions

Upper epidermis. Cells rectangular, sometimes broadened at ends, long axis parallel to axis of leaf, length 48-90 (avg. 67) microns, width 16-22 (avg. 20) microns; anticlinal walls straight, medium in thickness, surface ornamentation lacking; end walls oblique and enclosing; large intracellular phytoliths present, 1-11 (avg. 5) per cell, length 4-10 (avg. 7) microns, width 2-5 (avg. 4) microns, in rectangular cells and near stomata, rare or absent in cells near leaf base and apex. Cells at leaf base irregular, of variable shapes and orientation, length 25-60 (avg. 46) microns, width 16-26 (avg. 20) microns. Stomata present on upper epidermis, occurring near base of leaf, similar to those of lower epidermis.

Lower epidermis. Cell size, shape and orientation similar to those of upper epidermis. Stomata located along basal margins and central area of leaf; stomatal areas composed of 8 or more stomatal rows, with 8-10 stomata per row, stomatal axes parallel to axis of leaf except near base where they are parallel to margin; each stoma surrounded by 4-6 subsidiary cells, which may be shared by as many as 3 adjacent stomata, shaped dominantly pentagonal or tetragonal, length 23-52 (avg. 35) microns, width 13-28 (avg. 23) microns; anticlinal walls of subsidiary cells forming a ring 4-5 microns thick having a "jointed" appearance; guard cells sunken, length 25-38 microns, surface smooth, outer ledge plain. Large intracellular phytoliths present, having size, shape and distribution similar to those of upper epidermis.

Cuticular comparisons of C. *cf.* C. linguaefolia *and* C. lawsoniana. Epidermal cells of the Clarkia species are longer, narrower and are of larger size than in the extant *C. lawsoniana*. Cell shapes are also more variable in the fossil species where they are predominantly pentagonal, whereas the shape is predominantly tetragonal in the living species. Cell walls appear to be thinner in the fossil species. Both the fossil and extant species have intracellular phytoliths of the same size, shape and number. Also, the stomatal complex of both species is similar, although the average number of subsidiary cells is larger in the fossil (6) than in the modern (5) species (Tables 12-14).

I emphasize here that this comparison does not imply a single evolutionary lineage between the fossil and extant species of *Chamaecyparis*. Numerous extant species of Cupressaceae have foliage of generally similar morphology. Although the Clarkia foliage is most typical of that

TABLE 9. OBSERVED DISTINCTIONS OF UPPER
EPIDERMAL CELLS OF *TAXODIUM*[1]

Characters	T. dubium	T. distichum
Cell length	75-145 (143)	55-135 (81)
Cell width	19-26 (22)	19-35 (25)
L/W ratio	(6.6)	(3.2)
Cell area	(3093)	(2049)
Cell shapes	Linear, rectangular	Rectangular
Cell corners	Angular	Rounded
Anticlinal walls	Straight	Broadly undulate
Wall thickness	Thin	Medium
Phytoliths	None	Few

TABLE 10. OBSERVED DISTINCTIONS OF LOWER
EPIDERMAL CELLS OF *TAXODIUM*[1]

Characters	T. dubium	T. distichum
Cell length	52-100 (81)	42-72 (56)
Cell width	18-30 (23)	17-27 (22)
L/W ratio	(3.6)	(2.6)
Cell area	(1818)	(1211)
End walls	Oblique to square, enclosing	Oblique
Cell corners	Angled	Angled or rounded

[1] Measured in microns, averages in parentheses.

TABLE 11. OBSERVED DISTINCTIONS OF
STOMATAL COMPLEX OF TAXODIUM[1]

Characters	T. dubium	T. distichum
Subsidiary cell length	30-65 (47)	22-32 (26)
Subsidiary cell width	17-33 (30)	12-21 (18)
Subsidiary cell, L/W ratio	(1.6)	(1.5)
Subsidiary cell area	(1409)	(465)
Anticlinal walls	Irregular	Rounded
Guard cell length	43-53 (49)	30-40 (34)
Guard cell width	12-19 (16)	10-18 (15)
Guard cell, L/W ratio	(3.0)	(2.3)
Cutin	Thick ledge, small polar T	Thin strip on outer ledge

[1] Measured in microns, average values in parentheses.

Plate 9. Fig. 1. *Chamaecyparis* cf. *C. linguaefolia* (Lesq.) Macginitie. Shows area where tissue was removed. Clarkia flora (0.75X). Fig. 2. *C.* cf. *C. linguaefolia*. Upper epidermis, leaf base at lower left. Clarkia flora (190X). Fig. 3. *C.* cf. *C. linguaefolia*. Upper epidermis, showing irregular cells, usually oblique (or square) end walls, and straight anticlinal walls. Clarkia flora (190X). Fig. 4. *C.* cf. *C. linguaefolia*. Upper epidermis, showing elongate cells, with straight anticlinal walls of medium thickness, and intracellular inclusions. Clarkia flora (305X). Fig. 5. *C.* cf. *C. linguaefolia*. Lower epidermis, showing irregular cells and stomata near leaf base. Clarkia flora (190X). Fig. 6. *C.* cf. *linguaefolia*. Tissue near leaf margin. Clarkia flora (190X). Fig. 7. *C.* cf. *C. linguaefolia*. Lower epidermis, showing large, thin-walled, irregular cells near leaf base. Clarkia flora (305X). Fig. 9. *C.* cf. *C. linguaefolia*. Lower epidermis, showing cells with enclosing end walls, and intracellular phytoliths. Clarkia flora (190X). Fig. 10. *C.* cf. *C. linguaefolia*. Lower epidermis, showing stoma with 7 subsidiary cells forming a thick rampart. Clarkia flora (475X). Fig. 11. *C* cf. *C. linguaefolia*. Lower epidermis near leaf base, showing irregular cells and stomata. Clarkia flora (305X). Fig. 12. *Chamaecyparis lawsoniana* Parlatore. Extant species, for comparison. Shows area where tissue was removed (0.75X). Fig. 13. *C. lawsoniana*. Upper epidermis, leaf base at lower left. (190X). Fig. 14. *C. lawsoniana*. Upper epidermis, showing irregular cells with straight anticlinal walls, and oblique to enclosing (sometimes squared) end walls. (305X). Fig. 15. *C. lawsoniana*. Upper epidermis, showing elongate cells with straight anticlinal walls of medium thickness, with intracellular inclusions. (305X). Fig. 16. *C. lawsoniana*. Lower epidermis, showing irregular cells and stomata near leaf base. (190X). Fig. 17. *C. lawsoniana*. Lower epidermal tissue, near leaf margin. (190X). Fig. 18. *C. lawsoniana*. Lower epidermis, showing cells with phytoliths. (305X). Fig. 19. *C. lawsoniana*. Lower epidermis, showing large irregular cells with thin walls, near leaf base (190X). Fig. 20. *C. lawsoniana*. Lower epidermis, showing cells with enclosing end walls, and intracellular phytoliths. (190X). Fig. 21. *C. lawsoniana*. Lower epidermis, showing stoma with 5 subsidiary cells forming a thick rampart. (475X). Fig. 22. *C. lawsoniana*. Lower epidermis, showing irregular cells and stomata near leaf base (305X).

HUGGINS: CONIFERS

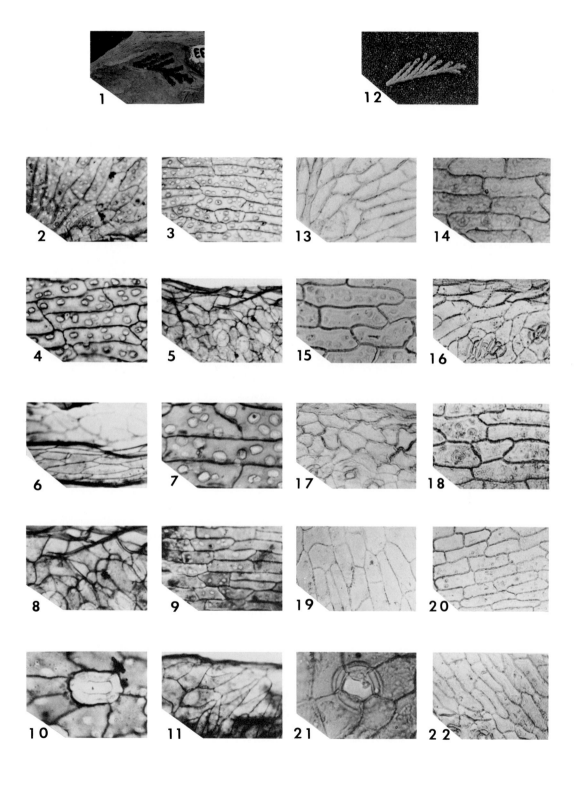

TABLE 12. OBSERVED DISTINCTIONS OF
UPPER EPIDERMAL CELLS OF *CHAMAECYPARIS*[1]

Characters	*C.* cf. *linguaefolia*	*C. lawsoniana*
Cell length	48-110 (79)	48-90 (67)
Cell width	11-27 (18)	16-22 (20)
L/W ratio	(4.4)	(3.4)
Cell area	(1422)	(1341)
Cell sides (no.)	4	3
Cell shapes	Mainly pentagonal and hexagonal	Tetragonal, pentagonal, hexagonal
Cell corners	Angled	Square to rounded
Wall thickness	Thin	Medium
Phytoliths	4-15 (9)/cell	3-11 (0)/cell

TABLE 13. OBSERVED DISTINCTIONS OF LOWER
EPIDERMAL CELLS OF *CHAMAECYPARIS*[1]

Characters	*C.* cf. *linguaefolia*	*C. lawsoniana*
Cell length	57-105 (72)	37-73 (64)
Cell width	9-30 (16)	12-22 (17)
L/W ratio	(4.5)	(3.7)
Cell area	(1152)	(1107)
Cell shapes (no.) and dominant shapes	4. Pentagonal and hexagonal	4. Pentagonal and tetragonal
End walls	Straight	Oblique, enclosing
Cell corners	Angled	Angled or rounded
Wall thickness	Thin	Medium

TABLE 14. OBSERVED DISTINCTIONS OF
STOMATAL COMPLEX OF *CHAMAECYPARIS*[1]

Characters	*C.* cf. *linguaefolia*	*C. lawsoniana*
Subsidiary cell length	19-55 (33)	23-52 (35)
Subsidiary cell width	10-23 (18)	13-28 (23)
Subsidiary cell, L/W ratio	(1.8)	(1.5)
Subsidiary cell area	(594)	(805)
Phytoliths	Many	Few

[1] Measured in microns, averages in parentheses.

produced by *Chamaecyparis*, and isolated cones in the same deposits can definitely be referred to this genus. Additional cuticular comparisons with the various extant species of the Cupressaceae are required. This comparison, as with the two Miocene species of *Pinus*, is included to demonstrate the quality of preservation and the variety of cuticular comparisons that can be obtained from Clarkia material.

CONCLUSION

The comparative investigations exemplified in this paper have revealed many differences in cell structure and tissue organization among conifers. Although the foliar structures (and also associated reproductive organs) of the fossil conifers examined here show only minor morphological variations from the compared living species, significant differences are noted in some of the epidermal features. For *Pinus* and *Chamaecyparis*, such comparisons are preliminary in that only one of many extant species has been presently examined for each fossil taxon. In the case of the Taxodiaceae (*Cunninghamia, Metasequoia, Taxodium*), these comparative studies are much more significant because the genera are now monotypic (or perhaps oligotypic) and each seems to have been represented also by a single species in later Tertiary floras of western North America. Evolutionary trends for the taxodiaceous conifers thus would seem implied, whereas specific lineages are yet to be determined for the pines and white cedars.

For all the conifer genera examined here, apparent general trends have been noted between the early Neogene fossils from Clarkia and the living examples used for comparison. These are particularly evident in the stomatal bands. If they can, in fact, be inferred to represent Neogene evolutionary changes, then the Miocene to Recent trends would appear to involve: (1) a change from long linear cells to shorter and more rectangular cells in the modified tissue of stomatal bands; (2) a change from cells in stomatal bands having predominantly straight walls and angular corners to more undulate walls and rounded corners; (3) an increase in the number of cells per unit area of bands; (4) a change in orientation of cell axes in bands, from more random in fossil material to a distinct alignment of long axes parallel with the axis of the leaf (except for *Taxodium*, in which the orientation has remained random); (5) a change from little specialization of stomatal subsidiary cells to more modified cells of consistent orientation relative to guard cells, including specialization into polar cells; (6) an increase in amount of cutin in the stomatal complex, as expressed by a thickening of the outer stomatal ledge, the rampart, and the development of polar markers (e.g. polar T's); (7) an increase in the number of stomatal rows per band; and (8) an increase in the number of stomata, and a greater crowding of stomata, per stomatal row.

More specifically, both of the Miocene species of pines have fewer stomatal rows per band than their morphologically similar counterparts (1 vs 4, and 4 vs 13); and the two fossil species lack the rows of specialized cells that form tooth-like topographic features within bands, which are present in both living species. For *Chamaecyparis*, the Clarkia species has epidermal cells that are larger and of more variable shapes than in the compared living species; and stomata are confined to the lower epidermis of the fossils but occur on both leaf surfaces of the modern species. For the pines and white cedar, these comparisons are for documentation purposes only, and are not necessarily to be inferred as representing specific Neogene lineages.

Concerning the Taxodiaceae, the observed Miocene to Recent differences in number of stomatal rows per band would seem to be genetically significant, and to show that the Tertiary and extant forms do not represent the same species. The number of rows per band in modern *Cunninghamia* is virtually double that of the fossils (7-9 vs 17-20), as is the case also with *Metasequoia* (5-7 vs 10-12). There is a 40% increase in number of rows per band for *Taxodium* (5 vs. 7), and leaves of the extant bald cypress have stomata on both surfaces whereas those of the fossil species have stomata on the lower epidermis only.

Further study is required for each fossil species and its putative living descendant(s), in order to better understand the affinities between Tertiary and extant plants and the apparent evolutionary trends that may be indicated. It has been demonstrated here, particularly regarding the taxodiaceous conifers, that detailed examination and comparisons of cuticular features can provide significant new data for this purpose. The cuticular investigations, centering on the compression fossils that are abundantly represented in the Clarkia flora, are continued in the present study of selected conifer taxa. Cuticular studies of other Clarkia conifers, and of some of the angiosperms, similarly have been undertaken (Smiley et al. 1975; Huggins 1980; Smiley and Huggins 1981). Ultimately the Clarkia and other Miocene floras of northern Idaho should provide important new fossil data for better understanding of the relationships between Tertiary and modern plant species, and for better insights pertaining to the rates and directions of evolution in many of the plant taxa represented in Tertiary floras of western North America.

ACKNOWLEDGMENTS

I wish to thank C. J. Smiley for professional advice and critical review of the manuscript, and W. C. Rember for extensive field and technical assistance. My appreciation is also extended to Sandra McGilchrist for manuscript typing.

LITERATURE CITED

Chaney, R. W. 1951. A revision of fossil *Sequoia* and *Taxodium* in western North America based on the recent discovery of *Metasequoia*. Amer. Phil. Soc. Trans. 40(3):171-263.

Chaney, R. W., and D. I. Axelrod. 1959. Miocene floras of the Columbia Plateau. Carnegie Inst. Wash. Pub. 617:1-237.

Cross, G. L. 1940. Development of the foliage leaves of *Taxodium distichum*. Amer. J. Bot. 27: 471-482.

Cross, G. L. 1942. Structure of the apical meristem and development of the foliage leaves of *Cunninghamia lanceolata*. Amer. J. Bot. 29:288-301.

Gurr, E. 1965. The rational use of dyes in biology. Williams and Wilkins Co., Baltimore, Md. 422 pp.

Heer, O. 1855. Flora Tertiaria Helvatiae, Tertiaer Flora der Schweiz. Winterthur J. Wurter and Co.

Huggins, L. M. 1980. Laboratory methods and analysis of fossil leaves and their application to educational curricula. Ph.D. Thesis. University of Idaho, Moscow, Ida.

Lakhanpal, R. N. 1958. The Rujada flora of west central Oregon. Univ. Calif. Pub. Geol. Sci. 35(1):1-66.

MacGinitie, H. D. 1953. Fossil plants of the Florissant beds, Colorado. Carnegie Inst. Wash. Pub. 559:1-198.

Smiley, C. J., J. Gray and L. M. Huggins. 1975. Preservation of Miocene fossils in unoxidized lake deposits, Clarkia, Idaho. J. Paleont. 49:833-844.

Smiley, C. J., and L. M. Huggins. 1981. *Pseudofagus idahoensis* n. gen. et sp. (Fagaceae) from the Miocene Clarkia flora of Idaho. Amer. J. Bot. 68:741-761.

Smiley, C. J., and W. C. Rember. 1979. Guidebook and road log to the St. Maries River (Clarkia) fossil area of northern Idaho. Idaho Bur. Mines Geol. Inf. Circ. 33:1-27.

Smiley, C. J., and W. C. Rember. 1981. Paleoecology of the Miocene Clarkia Lake (northern Idaho) and its environs. Pages 551-590 (Chapter 17) *in* J. Gray, A. J. Boucot, and W. B. N. Berry, eds., Communities of the Past. Hutchinson and Ross Publishing Co., Stroudsburg, Pa.

Sterling, C. Some features in the morphology of *Metasequoia*. Amer. J. Bot. 36:461-471.

Sveshnikova, I. N. 1963. Atlas and key for the identification of the living and fossil Sciadopityaceae and Taxodiaceae based on the structure of the leaf periderm. Acad. Sci. U.S.S.R., Paleobotanica IV:207-237.

MIOCENE EPIPHYLLOUS FUNGI FROM NORTHERN IDAHO

JOHN L. WILLIAMS
Cities Service Oil & Gas Corp., 1600 Broadway, Suite 900, Denver, CO 80202

A well-preserved fossil *Smilax* leaf from the Clarkia (P-33) fossil locality in northern Idaho yielded numerous epiphyllous fungi consisting of fruiting bodies of *Phragmothyrites* sp. and associated microthyriaceous germlings (possibly belonging to this same genus). Based on ecological studies of modern fungi, the paleoclimate of this Miocene locality was probably quite moist, possibly having more than 1000 mm of annual rainfall. Maceration and preparation techniques were employed, allowing the fungal material to be effectively examined with light and scanning electron microscopy.

Alvin (1971) stated that "the leaves of vascular plants have in the past, as today, provided suitable substrata or hosts for a wide range of different fungi and, moreover, that these fungi may, under certain conditions, fossilize sufficiently well to lend themselves to worthwhile study." These fossilizing conditions prevailed at the Clarkia, Idaho (P-33) locality during the Miocene. The exceptional preservation led to this study of the epiphyllous fungi found *in situ* on *Smilax*.

This study of fossil epiphyllous fungi arose incidentally to the examination of fossil *Smilax* leaf cuticles. Most fossil fungi seem to be discovered in this fortuitous manner, as well as during the examination of fossil pollen and spores from palynological residues.

METHODS

A complete *Smilax* leaf was carefully excavated from the Clarkia (P-33) locality (NW¼ NE¼ Sec. 13, T. 42 N., R. 1E.). Hydrofluoric acid was used to separate the leaf from its matrix without breakage. The leaf and adhering sediment were immersed in HF and after approximately five minutes the intact leaf floated free and was quickly removed and submerged in water. At this point separate steps were employed to prepare for light and scanning electron microscopical examination of the material.

To view this epiphyllous fungi with transmitted light, it was necessary to macerate the leaf. Small pieces of leaf (½ in.^{2}) were placed in a petri dish containing Schulze solution ($KC1O_3$ and HNO_3). After 2-3 minutes they were transferred to another petri dish containing a low concentration of KOH. Again after 2-3 minutes, the leaf fragments were carefully removed and placed back in the Schulze solution. This cycle was repeated until the leaf cuticles separated from the mesophyll. When the leaf cuticles were completely free, they were submerged in water to await slide preparation.

Placing properly oriented cuticles on cover glasses was very tedious, as these small, delicate tissues were less than cooperative. This task was accomplished by immersing a piece of cuticle in a few drops of alcohol on a cover glass. A sharp probe was then used to tease the tissue into the desired orientation. Once this was attained, the alcohol was allowed to evaporate to a point where the cuticle still remained moist. At this time, approximately 4 drops of polyvinyl alcohol were uniformly placed on the cover glass and permitted to dry. When dry, the cover glass was placed on a glass slide containing 3-5 drops of elvacite. The slide was cured for 24 hours on a slide warmer before microscopical examination.

A much easier method was used for preparing the leaf for examination with the scanning

140 LATE CENOZOIC HISTORY

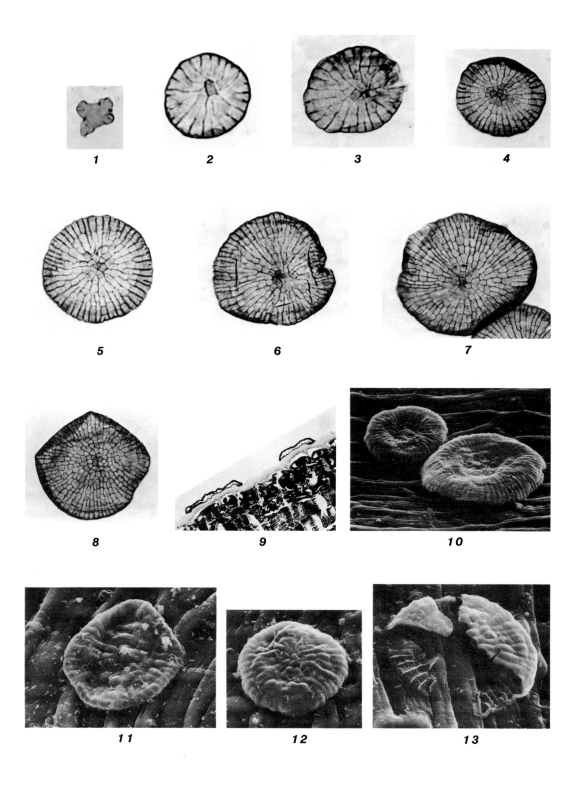

electron microscope (SEM). First, the original leaf material was cut up into numerous pieces (1 in.^{2}) and placed in a fixative (3% glutaraldehyde in 0.1M cacodylate/HCL buffer, pH of 7) for 2 hours. Next, the material was washed with the same buffer for 10 minutes. Post-fixation followed with a 1½ hour treatment in an osmium tetroxide solution (2% osmium tetroxide in 0.1M cacodylate/HCL buffer, pH of 7). After this, the specimens were again washed for 10 minutes in the same buffer. Dehydrating the leaf material with an ethanol series was the final step in this fixation procedure. This involved immersing for 10 minutes in each step of the series (40, 50, 70, 80, 90, 95 and 100% ethanol).

Lastly, the specimens were critical-point dried using CO_2 as the transition fluid. The dried leaf pieces were then mounted, with their upper surfaces up, on aluminum SEM stubs and coated with gold-palladium. The leaf surfaces were scanned for epiphyllous fungi using an ISI-60 scanning electron microscope on an accelerating voltage of 30 kV.

SYSTEMATIC PALEONTOLOGY

Class ASCOMYCETES
Order MICROTHYRIALES
Family Microthyriaceae
Genus *Phragmothyrites* Edwards, 1922

Phragmothyrites sp.
Figures 4-13

Description. These fruiting bodies are flattened, circular to subcircular, and range from 60-135 μm in diameter. They are uniformly dark or darker in the periphery, have entire or slightly crenate margins, and non-ostiolate and unmodified centers. Radiating hyphae are interconnected, forming pseudoparenchymatous tissue; the subtending layers having thick radial cell walls (Fig. 13). All cells are aporate, with the central ones typically small, dark and squarish to subcircular, and the outer cells larger and more elongated. The fructifications were usually grouped on or near the veins of the *Smilax* leaf.

Discussion. Figures 4 through 8 suggest a sequence of increasing maturity. Figures 4 and 5 appear to be "early" mature forms (round thyriothecia having small number of rather large cells), while Figures 6-8 appear to be "late" mature forms (larger, more irregularly shaped thyriothecia having a very large number of much smaller cells).

EXPLANATION OF FIGURES

Figures 1-9 are light micrographs. Figures 10-13 are scanning electron micrographs. All measurements represent the greatest diameter of the specimen.

Fig. 1. Microthyriaceous germling, 20 μm.
2. Microthyriaceous germling, 30 μm.
3. Microthyriaceous germling, 55 μm.
4. *Phragmothyrites* sp., 80 μm.
5. *Phragmothyrites* sp., 85 μm.
6. *Phragmothyrites* sp., 100 μm.
7. *Phragmothyrites* sp., 105 μm.
8. *Phragmothyrites* sp., 135 μm.
9. *Phragmothrites* sp., 90 μm, 58 μm. CROSS SECTION. The smaller thyriothecium may be an immature form or possibly a microthyriaceous germling.
10. *Phragmothyrites* sp., 67 μm, 82 μm.
11. *Phragmothyrites* sp., 77 μm. Note the hypha running over the top of the fructification and continuing onto the leaf surface.
12. *Phragmothyrites* sp., 72 μm.
13. *Phragmothyrites* sp., 69 μm. This thyriothecium was intentionally broken for internal viewing.

Young forms (germlings) of microthyriaceous fungi Dilcher, 1965
Figures 1-3

Discussion. Different stages of development are shown by these figures. There is an increase in maturity from Figure 1 to Figure 3. Because of their immaturity, these germlings lack the cell structure that is essential in the identification of a specific genus, and are therefore placed under this general heading of microthyriaceous germlings. Even though these young forms were only found associated with the thyriothecia of *Phragmothyrites*, they could possibly be germlings of various species or genera of other microthyriaceous fungi. It is my opinion, however, that Figures 1-8 represent a partial ontogenetic series for *Phragmothyrites* sp.

DISCUSSION

Dilcher (1965) reported the geologic range of microthyriaceous fungi to be from Early Cretaceous to Recent. He further notes that epiphyllous forms are not common until the Late Cretaceous (based on sparse records), but were abundant worldwide by the Eocene. Tiffney and Barghoorn (1974) stated that most ascomycetes reported from the Tertiary are epiphyllous forms associated with leaf cuticles. The fruiting bodies of *Phragmothyrites* found in the Clarkia fossil locality have a geologic range of Albian to Early Pleistocene (Elsik 1978). The microthyriaceous germlings from this locality have a known range of Lower-Middle Eocene to the present.

The presence of epiphyllous fungi suggests environmental conditions of high moisture (Edwards 1922). Smith (1980) states that fruiting bodies of epiphyllous fungi from Tertiary fossil deposits indicates subtropical to tropical climatic conditions. Elsik (1978) and Edwards (1922), however, both pointed out that while most extant microthyriaceous fungi are tropical, several forms occur at higher latitudes and that their distribution is controlled more by precipitation than by temperature. Elsik (1978) correlates this distribution with those areas of the world having not less than 1000 mm of annual precipitation.

ACKNOWLEDGMENTS

The author thanks Dr. F. H. Wingate, Cities Service Oil & Gas Corporation, and Mr. E. T. Peterson, Amoco Production Company, for their useful comments and criticisms on various portions of this manuscript.

LITERATURE CITED

Alvin, K. L. 1971. The study of fossil epiphyllous fungi by scanning electron microscopy. Pages 297-306 *in* V. H. Heywood, ed. Scanning Electron Microscopy. The Systematics Association, Spec. Vol. 4.

Dilcher, D. L. 1965. Epiphyllous fungi from Eocene deposits in Western Tennessee, U.S.A. Palaeontographica 116(B):2-4):1-54.

Edwards, W. N. 1922. An Eocene microthyriaceous fungus from Mull, Scotland. Trans. Brit. Mycol. Soc. 8:66-72.

Elsik, W. C. 1978. Classification and geologic history of the microthyriaceous fungi. Proc. IV Int'l. Palynol. Conf., Lucknow (1976-1977). 1:331-342.

Smith, P. H. 1980. Trichothyriaceous fungi from the Early Tertiary of Southern England. Palaeontology 23(1):205-212.

Tiffney, B. H., and E. S. Barghoorn. 1974. The fossil record of the fungi. Occ. Pap. Farlow Herb Cryptogam. Bot. 7:1-42.

ULTRASTRUCTURAL STATES OF PRESERVATION IN CLARKIA ANGIOSPERM LEAF TISSUES: IMPLICATIONS ON MODES OF FOSSILIZATION

KARL J. NIKLAS[1], R. MALCOLM BROWN, JR.[2], AND RICHARD SANTOS[2]

[1] Div. Biological Sciences, Cornell University, Ithaca, NY 14853
[2] Dept. Botany, University of Texas, Austin, TX 78712

> Ultrastructural analyses of Miocene leaf tissues from the Clarkia flora indicate the preservation of various protoplasmic features such as chloroplasts, mitochondria, and nuclei. Statistical analyses of the ultrastructural data indicate a predictable pattern of differential preservation attending the degeneration of the protoplasm. Of the 2,300 randomly sampled cells from the type locality (P-33), 26% contain mitochondria, 90.1% contain chloroplasts, and 4.3% contain nuclei. Comparable analyses of tissues from site P-40 indicate that 34.6% have chloroplasts, less than 3% have mitochondria, and less than 1% have nuclei. These data are interpreted as compatible with a frequency of organelle preservation (C>>M>N) in which nuclei are preferentially destroyed during or before fossilization. Simulated fossilization involving the compression of living leaf tissues and studies of senescent and dehydrated tissues are used to evaluate the probable effects of these factors on the sequence of events leading to the observed ultrastructure of the Clarkia fossils. From these studies we conclude that 63% of the sampled tissues appear to have entered the depositional environment as the result of leaf abscission and 15% were detached from their branches as a result of trauma; 89% of the tissues examined show evidence of compression wall failure, while 17% show concertina cell wall distortion similar to that seen in dehydration of living tissues.

The opportunity to study fossil cell contents with the electron microscope and thereby draw inferences as to the frequency and extent of protoplasmic preservation in the fossil record has been rare. In part, this is due to the vagaries in preservation and sampling of fossil floras that preclude the consistent application of electron microscopy and the gathering of adequate sample sizes for requisite statistical analyses. The abundant angiosperm leaf compressions recently found in the St. Maries River area of northern Idaho (cf. Smiley, Gray and Huggins 1975; Smiley and Rember 1979; Huggins, this volume) provide an opportunity to study fossil protoplasm in a detailed and systematic manner hitherto unparalleled. This paper presents a survey of the ultrastructural states of preservation found in a variety of fossil plant tissues, collected in the Miocene strata near the town of Clarkia, Idaho, as well as a preliminary investigation of the physical and biochemical processes that appear to have affected protoplasmic survival in different environments of geologic preservation. Representative angiosperm leaf fossils were collected from two sites, and their ultrastructural states of preservation were determined by means of transmission electron microscopy. Intra- and intertaxonomic variability of preservation was determined within and between these two sites on the basis of the statistical examination of randomly selected cells from which the frequency of organellar preservation was determined. To determine the extent to which various physical factors affected the integrity of protoplasmic features, modern taxa referable to the fossils studied (*Betula, Hydrangea, Platanus,* and *Quercus*) were subjected to dehydration and compression and then examined ultrastructurally. In addition, leaves of the modern forms that had undergone abscission or senescence were studied. Data derived from both the fossil and modern

Copyright ©1985, Pacific Division, AAAS.

plants studied will be placed within the context of protoplasmic degeneration seen to attend fossilization, necrosis, and senescence.

MATERIALS AND METHODS

Fossil tissues of *Betula, Hydrangea, Platanus*, and *Quercus* were collected from freshly exposed strata at sites P-33 and P-40 of the Clarkia flora. Site P-33 is interpreted by Smiley and Rember (1979) to be derived from quiet water, offshore lacustrine sedimentation, while Site P-40 in the Emerald Creek embayment is interpreted to be due to nearshore sedimentation (cf. Smiley and Rember, this volume). The vertical sequence of sediments exposed at P-33 shows a temporal change from the base of the outcrop (alluvial sands) to the top (probable flood plain deposits) indicative of a progressive infilling of the Miocene Clarkia Lake. Thus the precise environment of deposition associated with any particular horizontal suite of fossils must be largely inferred from their relative position within the column. The Clarkia fossils are considered to be Lower Miocene in age, on the basis of paleofloristic and potassium/argon dating (cf. Smiley and Rember, 1979, this volume).

Fossils were handled at a minimum, as described by Niklas (1979, 1981), and reference samples were taken for cuticular and chemical analyses, and for light and transmission electron microscopy (TEM). Tissues were placed either directly in a fixative (2%, vol/vol glutaraldehyde; 1% wt/vol tannic acid; 50 mM cacodylate buffer at pH 7.2, cf. Niklas et al. 1978; Niklas and Brown 1981) or stored in nitrogen for subsequent fixation and TEM analysis. Referable modern tissue samples were fixed with glutaraldehyde and with tannic acid to determine fixation-induced variability in ultrastructure.

For purposes of statistical analyses, 100 cells from random areas of random TEM sections taken from every specimen were photographed. The number of specimens for each taxon collected is given in Table 1. Tracheids and vessel members were excluded from the random selection of cells, since these cell types lack protoplasmic structure and maturity and would skew the data. Transmission electron micrographs were printed at standard magnifications (10,000 and 16,000X), and their identifiable contents tabulated.

The percentage of cells having an ultrastructural feature (e.g. chloroplasts, nuclei, etc.) for any taxon from a specific locality is given by x/100k x 100%, where x is the number of cells having the feature, and k is the number of fossil specimens examined. The total number of sampled cells for a taxon is given by 100k or N. The variance, σ, or standard deviation of cells having a feature for any specific taxon is given by:

$$\sigma = \frac{[(p_1 - \pi)^2 + (p_2 - \pi)^2 + ...(p_k - \pi)^2]^{1/2}}{k}, \text{ Equation 1}$$

where p is the proportion of cells having the feature for a fossil specimen and π is the proportion of cells having the feature for the taxon. Equation 1 must be used to compute the intrataxonomic variance in ultrastructural preservation instead of a more generally used standard deviation formula, $s = (\xi x^2/N)^{1/2}$, since the data are in the form of a "presence vs. absence" tabulation (cf. Croxton 1953). The Kolmoganov-Smirnov one-sample test was used to compute the cumulative distribution of cells possessing various permutations of organelles within the between fossil localities. Subcellular features such as a nucleus, chloroplast, or mitochondrion, which were present in >50 of the random cells, were entered as a "typical" protoplasmic feature of the taxon—noted by a "+" in Table 1. If the feature was seen in <50, then a "−" was tabulated. Some of the raw data upon which the + and − entries were made are given in Table 2.

Observed sequences in the differential preservation of various organelles were quantified by means of an index of fidelity to the presumed living state. Each organelle and cell walls/cuticles,

TABLE 1. DIFFERENTIAL PRESERVATION OF ULTRASTRUCTURAL FEATURES IN CLARKIA FOSSIL LEAF TISSUES

	Betula (9)[1]		Hydrangea (2)		Platanus (6)		Quercus (6)	
Site P-33								
ER, golgi	0[2]	-	0	-	0	-	0	-
Nuclei	5 ± 0.8	-	1 ± 0.6	-	5 ± 0.9	-	2 ± 0.5	-
Mitochondria	10 ± 1.9	-	52 ± 6.6	+	42 ± 6.2	-	26 ± 6.5	-
Chloroplasts	83 ± 2.6	+	85 ± 3.3	+	96 ± 9.2	+	98 ± 9.5	+
Starch grains	76 ± 5.2	+	82 ± 9.2	+	59 ± 1.3	+	73 ± 6.2	+
Cell wall	100	+	100	+	100	+	100	+
Cuticle	100	+	100	+	100	+	100	+
Biotic effects	10 ± 0.7	-	5 ± 2.3	-	8 ± 5.5	-	10 ± 5.2	+
FI-value		10		15		10		9
	Betula (0)		Hydrangea (2)		Platanus (4)		Quercus (8)	
Site P-40								
ER, golgi	0	-	0	-	0	-	0	-
Nuclei	0	-	0	-	0	-	1 ± 0.3	-
Mitochondria	3 ± 1.2	-	0	-	1 ± 3.6	-	3 ± 0.3	-
Chloroplasts	42 ±20	-	39 ± 9.3	-	39 ±20	-	23 ± 6.9	-
Starch grains	0	-	0	-	0	-	0	-
Cell wall	69 ± 9.3	+	53 ± 7.2	+	59 ±5.2	+	72 ±6.2	+
Cuticle	100	+	100	+	100	+	100	+
Biotic effects	50 ± 7.2	+	48 ± 5.2	-	40 ± 3.1	-	36 ± 5.2	-
FI-value		2		3		3		3

[1] Number of specimens.

[2] Percent of cells possessing the feature.

typical for a living cell, were assigned an arbitrary value in inverse order of the frequency of their preservation. ER and golgi = 7, nucleus = 6, mitochondria = 5, chloroplasts = 4, starch grains = 3, cell wall with cellulose microfibrils = 2, cuticle = 1. The fidelity index, FI, of a tissue was computed as the total of the values for each component, C_i, which was present in 50% of the 100 random cells, i.e. $FI = \sum_{1}^{7} C_i$. In addition, biotic degradation due to fungi and/bacteria was denoted by FI-1. The fidelity index provides a rough quantitative expression of ultrastructural preservation, as well as assigning a tissue to a generalized numerical sequence indicative of its protoplasmic deterioration.

RESULTS

Ultrastructural States of Preservation

The macroscopic and ultrastructural appearance of *Betula, Hydrangea, Platanus*, and *Quercus*, as well as some other genera from P-33 and P-40, have been reported elsewhere (Niklas 1981, 1982, 1983; Niklas and Brown 1981). Details of *Betula* (Fig. 1), *Quercus* (Fig. 2), *Platanus* (Fig. 3), and *Hydrangea* (Fig. 4) are reviewed here for purposes of comparison with their modern counterparts. Quantitative expressions of organelle frequencies are given in Table 1.

Betula. Gultaraldehyde fixed tissues of *Betula* from site P-33 lack golgi bodies and endoplasmic reticula, occasionally possess nuclei and mitochondria, and show chloroplasts with excellent infrastructure revealing the grana fretwork membrane system (Fig. 1). The cell walls of most

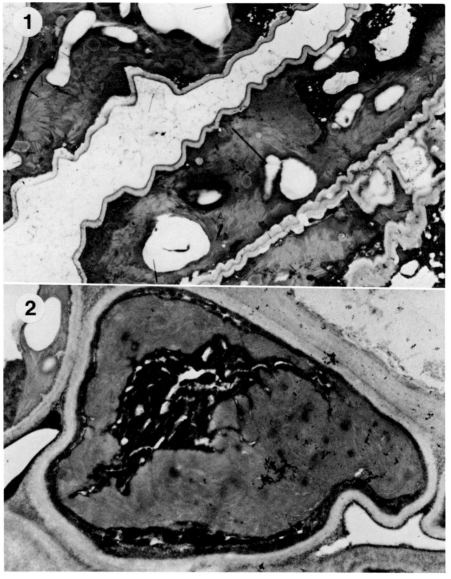

Figure 1. Transmission electron micrograph (TEM) of fossil *Betula* palisade cells showing concertina folding of cell walls and protoplasmic remnants. X 7,600.

Figure 2. TEM of fossil *Quercus* spongy mesophyll cells revealing occluded cell lumin. Portions of chloroplasts are seen to be contiguous with amorphous, electron dense regions. X 11,000.

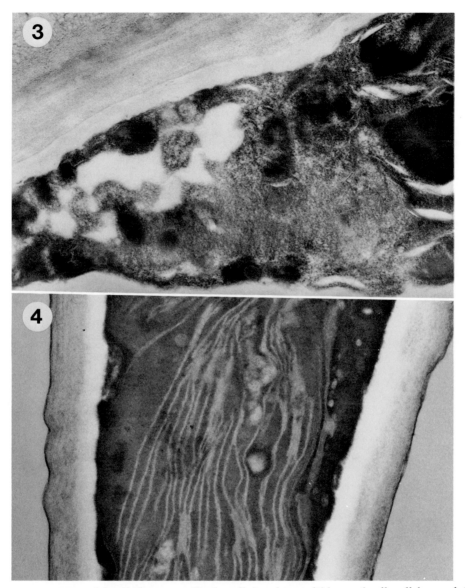

Figure 3. TEM of fossil *Platanus* palisade cell showing multi-layered cell wall (upper left) and heterogenously preserved protoplasm. X 32,000.

Figure 4. TEM of fossil *Hydrangea* floral bract showing "negative staining" property of chloroplast infrastructure and slightly crinulate cell wall. X 25,000.

mesophyll tissues are well preserved. *Betula* tissues from Site P-40 lack nuclei, in addition to Golgi apparati, and rarely possess demonstrable mitochondria. Chloroplasts are present in some tissue samples but are not abundant.

Quercus. Tissues from P-33 fixed in both glutaraldehyde and tannic acid show chloroplast infrastructure and occasionally possess nuclei and mitochondria. Chloroplasts are present in the majority of the cells examined (Fig. 2). *Quercus* tissues from P-40 are poorly preserved; the most common organelle is the chloroplast.

Platanus. Tissues from leaves identified as *Platanus* fixed in tannic acid show moderate to good preservation from Site P-33. Evidence for cell wall compression is common, with lacunae occluded with electron-dense cytoplasmic remnants (Fig. 3). Some cell lacunae show nuclei and mitochondria, and a majority have chloroplasts. Tissues from P-40 lack nuclei and possess infrequent mitochondria; chloroplasts are present in about a third of the cells examined.

Hydrangea. Glutaraldehyde-fixed tissues of *Hydrangea* bracts from P-33 possess plastids, which appear as if in negative contrast (Fig. 4), as well as mitochondria and occasional nuclei. The only organelle detected in P-40 tissues is the chloroplast.

Statistical Analyses of Ultrastructural Preservation

The percentages of randomly sampled cells (100 cells per specimen) containing various organelles, of cell walls with microfibril structure, and of specimens with *bona fide* cuticles are given in Table 1 for each of the four taxa studied from Sites P-33 and P-40. Some of the actual numbers of cells having various permutations of nuclei, chloroplasts, and mitochondria are given in Table 2. Variation of the percentage of cells possessing any of the protoplasmic or ergastic features examined is given in the form of standard deviation values. A chi-square, 2x2 table for each taxon is presented to test for the significance of differences between the number of cells with and without nuclei in regard to the presence of chloroplasts and mitochondria (cf. Table 2). In all but one case (*Hydrangea*), there is a significant difference between groups of cells with and without nuclei with regard to the presence of chloroplasts ($p<.25$ to $<.001$). Only two cells from P-33 showed evidence of nuclei and lacked chloroplasts, while 1,994 cells out of the 2,300 sampled had chloroplasts and lacked nuclei (213 cells lacked both nuclei and chloroplasts). Two cells had mitochondria and no chloroplasts, 600 cells had both, while 1,481 had chloroplasts and no evidence of mitochondria (Table 2). Statistical examination of Site P-40 cells (data not shown) indicates an even more pronounced trend for a preferential preservation of chloroplasts: 8 of all the cells examined (2,300) have nuclei, while all cells showing some ultrastructural features have chloroplasts (cf. Niklas 1982, 1983).

To test the significance of the data with regard to the relative number of chloroplasts per living cell, fossil cells for each taxon were stereologically examined and the chloroplast-to-nucleus ratio for the fossil specimen (C/N) computed. An overestimate of the C/N ratio in modern angiosperm palisade parenchyma was used ($C/N \cong 250$). On the basis of chi-square tests (Table 3), the probability of finding the fossil C/N ratios observed by chance is less than 0.15 to 0.001. A similar computation for mitochondria-to-nucleus (M/N) ratios was not performed owing to the lack of reliable M/N ratios in modern angiosperm tissues.

Based upon the observed differential preservation of protoplasmic structure, where chloroplasts are the most likely organelle to be found and ER/Golgi bodies are the least likely to be preserved, an index of fidelity (F1-value) to the living condition of each leaf was computed (cf. Table 1 and 2). The maximum possible score for any specimen is 28. The maximum observed score is 15, i.e. *Hydrangea* (Table 1). The lowest scores were tabulated for fossils from the P-40 site. Intrataxonomic comparisons of F1-values for specimens from the different geologic settings of P-33 and P-40 in the same flora indicate that preservational differences are for the most part not due to specific anatomical or chemotaxonomic variations. For example, the F1-values for

TABLE 2. CHI-SQUARE, 2x2 TABLE OF SITE P-33 FOSSIL LEAF CELLS POSSESSING CHLOROPLASTS, MITOCHONDRIA, AND NUCLEI

	Betula nuclei			Hydrangea nuclei			Platanus nuclei			Quercus nuclei		
	+	−		+	−		+	−		+	−	
Chloroplasts												
+	45	45		2	102		30	223		12	144	
−	0	153	$p<.005$	0	30	$p<.70*$	0	24	$p<.25$	2	10	$p<.01$
Mitochondria												
+	45	45		2	102		30	223		12	144	
−	0	810	$p<.001$	0	96	$p<.20*$	0	347	$p<.01$	0	444	$p<.001$
	mitochondria			mitochondria			mitochondria			mitochondria		
	+	−		+	−		+	−		+	−	
Chloroplasts												
+	90	657		104	66		252	324		154	434	
−	0	153	$p<.005$	0	30	$p<.05$	0	24	$p<.01$	2	10	$p<.02$

*Computed by the exact method of variance analysis.

Betula, *Hydrangea*, *Platanus*, and *Quercus* fossils from the Clarkia P-33 locality are 10, 15, 10 and 9, respectively (Table 1)—indicating an intertaxonomic variability of preservation within this locality. Representative fossils of these same taxa from the P-40 locality have virtually identical and low F1-values, i.e. F1=3. (The F1-value of *Betula* is 2 and is due to evidence of excessive biotic degradation—50%.)

Ultrastructural Effects of Dehydration and Compression

To provide a basis for the interpretation of fossil angiosperm leaf ultrastructure, referable modern leaves were examined with the TEM, and subjected to various regimes of dehydration, hydration, and/or compression before their ultrastructure was determined. In addition, the effects of tannic acid fixation were determined, since we had speculated on the ultrastructural fidelity of leaves submerged in an aqueous environment supercharged with this and other naturally occurring phenolics (cf. Niklas et al. 1978; Niklas and Brown 1981). The ultrastructure of unstressed mesophyll cells is well documented in the literature and will not be repeated here; however, in all of the genera we examined, such characteristic protoplasmic features as Golgi apparatus, ER, chloroplasts, mitochondria, and nuclei were evident (Figs. 5-8). As expected, dehydration and compression treatments severely altered the characteristic appearance of the mesophyll tissue and its cellular ultrastructure components.

Dehydration by air-drying results in a random or profuse concertina-like folding of palisade mesophyll cells along their longitudinal axes, while mesophyll cells appear to invaginate and collapse (Figs. 9-12). Attending these deformations, various protoplasmic alterations were observed that are consistent from one genus to another: (1) the loss of observable Golgi apparatus and well-defined ER, (2) the collapse or rupture of the tonoplast membrane(s), (3) vesiculation and/or rupture of the outer chloroplast membrane with a concomitant loss of a well-defined stromagrana orientation parallel to the plastid's major axis, (4) a decrease in the fidelity of mitochondrial cristae, and (5) the loss of the plasmalemma. In addition, demonstrable nuclei were rare or wholly lacking in many of the preparations (e.g. *Betula* and *Quercus*).

In contradistinction to the cell wall deformations induced by dehydration, compression of leaves (buried in a fine-grain sand and mud mixture) resulted in a general vertical collapse of mesophyll cell walls without the formation of concertina-like foldings characteristic of dehydration. Subsequent dehydration of compressed leaves resulted in the aforementioned ultrastructural

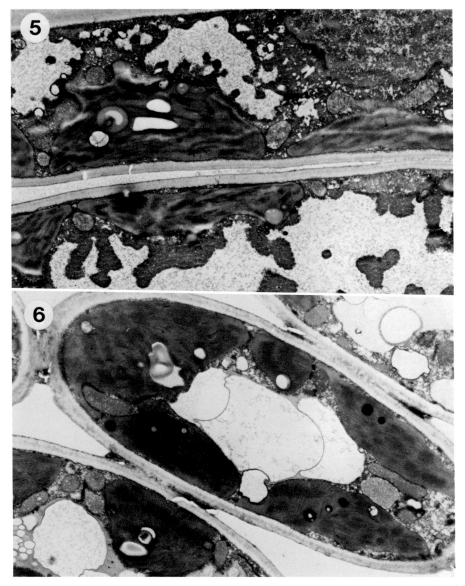

Figures 5 and 6. TEM of extant *Quercus* palisade cells prepared without tannic acid (TA) fixation (Fig. 5) and with TA fixation (Fig. 6). Cell walls are characteristically undistorted in both treatments; however, grana fretwork membrane systems are more pronounced in TA-prepared material. Figure 5 at X 19,000; Fig. 6 at X 11,000.

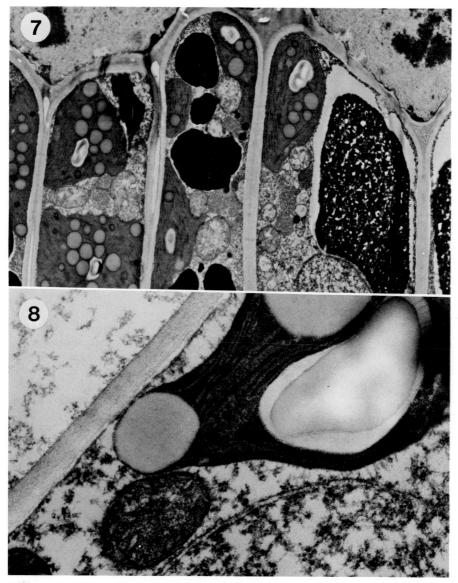

Figures 7 and 8. TEM of extant *Betula* palisade cells, without TA fixation (Fig. 7) and with TA fixation (Fig. 8). Figure 7 at X 8,000; Fig. 8 at X 42,000.

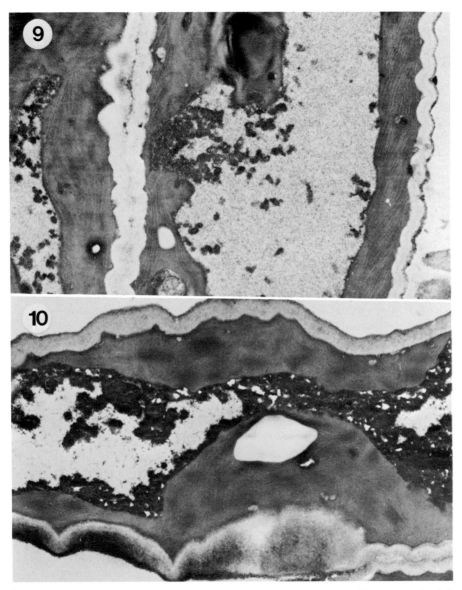

Figures 9 and 10. TEM of air-dried extant *Quercus* palisade cells prepared without TA fixation (Fig. 9) and with TA fixation (Fig. 10). Dehydration results in a concertina-like distortion of the cell walls in both treatments. Tannic acid fixation enhances the contrast seen between protoplasmic features. Figure 9 at X 14,000; Fig. 10 at X 14,000.

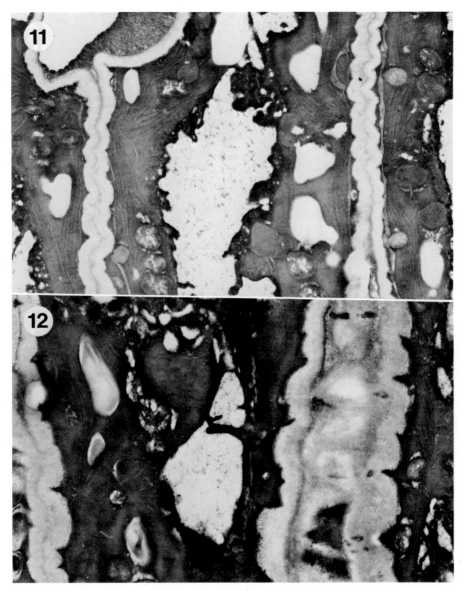

Figures 11 and 12. TEM of air-dried extant *Betula* palisade cells without TA fixation (Fig. 11) and with TA fixation (Fig. 12). Figure 11 at X 14,000; Fig. 12 at X 16,000.

TABLE 3. CHI-SQUARE TABLES OF "OBSERVED" VS. "EXPECTED" CHLOROPLAST/NUCLEUS RATIOS IN FOSSIL LEAF TISSUES

	f^a	fc^b	f-fc	$(f-fc)^2$	$(f-fc)^2/fc$	
Clarkia site P-33	750	250	500	2.5×10^4	1×10^3	$p < .001$
Clarkia site P-40	275	250	25	625	2.5	$p < .15$

[a] f = observed C/N ratio.
[b] fc = expected C/N ratio.

alterations observed in non-compressed, dehydrated mesophyll cells. Subsequent rehydration of air-dried and compressed tissues resulted in a partial re-establishment of the original cell wall configuration.

In at least two genera (*Betula* and *Quercus*), TA fixation of both fresh (Figs. 6, 8) and dried leaves (Figs. 10, 12) resulted in the appearance of "negative contrast" membrane systems, particularly in the infrastructure of chloroplasts. In general, TA fixation resulted in what appears to be the preferential preservation of chloroplast ultrastructure at the expense of the fidelity of other organelles (compare Figs. 9 and 11 with 10 and 12).

DISCUSSION

Ultrastructural examinations of angiosperm leaf fossils, presented here and elsewhere (cf. Niklas et al. 1978; Niklas and Brown 1981), indicate that protoplasmic features may be preserved provided that the physical and biochemical factors attending fossilization are sufficiently mild (Figs. 1-4). Nuclei, nucleoli, chloroplasts, pyrenoids, and even such delicate structures as chromosomes have been described from a wide range of plants and from various geologic periods (Baschnagel 1966; Baxter 1950; Bracken-Hanes and Vaughn 1978; Bradley 1946, 1962; Darrach 1938; Dwivedi 1959; Eisenack 1967; Florin 1936; Gould 1971; Mamay 1957; Millay and Eggert 1974; Schopf 1968, 1974; Stevens 1912; Vishnu-Mittre 1967; see, however, Niklas 1982, 1983). The interpretation of fossil protoplasm is difficult, however, due to: (a) possible pre-depositional alterations of the living state due to necrosis and/or senescence, and (b) post-depositional changes (= diagenesis) attending sediment consolidation and lithification. The present study has attempted to document the intra- and intertaxonomic variation in the ultrastructural preservation of four angiosperm leaf taxa (*Betula*, *Hydrangea*, *Platanus*, and *Quercus*) from two depositional sites of the St. Maries River fossil area of northern Idaho, near the town of Clarkia. The interpretation of these fossils must, however, be placed within the context of the ultrastructure of comparable modern genera, both in their pristine living condition and in states of dehydration and compaction designed to simulate some of the traumatized conditions possibly induced by fossilization (Figs. 5-12). By comparing the ultrastructure of fossil and modern taxa from different depositional environments and under conditions of physiologic stress, respectively, we hope that a more precise quantitative expression of ultrastructural states of preservation may be attained, and inferences concerning the pre- and postdepositional changes in protoplasm may be useful in reconstructing the conditions all ending in fossilization.

Of the 2,300 randomly sampled cells from tissues of *Betula*, *Hydrangea*, *Platanus*, and *Quercus* from site P-33 (cf. Smiley and Rember 1979), 90.1% contain remnants of chloroplasts (C), 26% contain mitochondria (M), and 4.3% have demonstrable nuclei (N) (Table 1). Similarly, a comparable study of tissues from site P-40 indicates that C = 34.6% M < 3%, and N < 1% (Table 1). Statistical analyses of the data from both sites indicate that a well-defined pattern exists in the frequency of preservation of various organelles such that C>M>N regardless of the presumed differences in the environments of deposition and of the taxon selected (cf. Table 2). The significance of the raw data, when tested against the numerical abundance of chloroplasts over nuclei in the cell populations of leaves (C/N was taken as 250), indicates that the statistical probability of these data being the result of chance is less than one in one thousand ($p<.001$; Table 3). The data and the inferred pattern of protoplasm degeneration are consistent with studies based upon other angiosperm leaf tissues of comparable age (e.g. Succor Creek Flora; Niklas 1981, 1982, 1983).

The apparent susceptibility of nuclei to degeneration during fossilization, as opposed to the stability of chloroplasts, is not consistent with some studies of modern plant material subjected to physical or biological stress. From a survey of the literature, the death of a cell appears to follow in most cases a repeatable pattern of protoplasm degeneration, regardless of a variety of causes (Butler and Simon 1971; Israel and Ross 1967; Hoefert 1980; Hurkman 1979; Toyama 1980, and references therein). The fidelity index (F1) provides a method whereby both fossil and modern patterns of degeneration may be placed in a common framework. The first detectable changes in a modern plant cell undergoing necrosis are a decrease in the number of ribosomes and a loss of ER and Golgi body membranes. This is comparable to fossil tissues with an F1-value of 21, and is consistent with the lack of ER and Golgi in all of the fossil cells examined. Chloroplast breakdown is the next most obvious event in cellular necrosis. The stroma disappear, thylakoids swell and disintegrate, and the number and size of osmophilic globuli (= plastoglobuli) increase. While mitochondria are in most cases present until terminal necrosis, they may show a reduction in size, a swelling of their cristae, and a reduction in their number concomitant to chloroplast breakdown. The tonoplast ruptures before most organelles degenerate ($F1 = 18$). The plasmalemma and nucleus are relatively stable during necrosis ($F1 = 11$); death of the cell is marked by the lysis of the plasmalemma and the vesiculation and disappearance of the nucleus ($F1 = 4$). Except for the initial disappearance of ER, Golgi bodies, and the tonoplast, the pattern of modern cell degeneration is basically the reverse of what may be inferred to be the pattern induced by fossilization (cf. Table 2). As will be pointed out later, in some cases chloroplasts have been observed to be the only surviving organelle just before death of the cell (cf. Hurkman 1979).

An appropriate null hypothesis to explain the apparent disparity between these neo- and paleobotanical observations is that the reconstruction of protoplasmic degeneration is biased and ultimately unrepresentative of the phenomenon of fossilization. Given the proposition that all organelles in a cell undergoing fossilization have the same rate of alteration, then inherent differences in the way organelles manifest alteration ultrastructurally would lead to differences in their apparent survival, i.e. organelles with the most diagnostic infrastructure will be diagnosed as present in highly altered cells. The chloroplast, by virtue of its complex grana fretwork system of membranes, may be much more distinctive, even in the altered state, than a nucleus or mitochondrion. This combined with the numerical superiority of chloroplasts over nuclei could account for the low frequency of observed nuclei in our fossil material. A number of observations, however, mitigate this null hypothesis: (a) studies of various modern plant tissues in which nuclei preferentially degenerate (e.g. development and maturation of sieve cells, cf. Evert et al. 1973; sieve elements, cf. Hoefert 1980) provide the ultrastructural context in which to identify even highly altered nuclei; (b) our own studies, based upon dehydrated and/or compacted tissues, indicate that nuclei, while highly altered, are demonstrable even after severe osmotic stress and mechanical trauma; and (c) such techniques as stereological analysis (designed to compute volumetric and frequency distributions) when applied to the fossil material have failed to show a "hidden" population of nuclei in the sample set (data not shown). It is clear, however, from our manipulation of the living material, that chloroplasts are the most distinctive organelle even when the protoplasm is highly altered.

While the inferred degeneration pattern of organelles in the fossils studied cannot be fully explained, information is accumulating as to potential mechanisms. The factors attending fossilization may either be different from those effecting changes in modern cells undergoing necrosis, or the factors may be the same but result in different ultrastructural patterns, due to as yet unknown diagenetic processes. Considering the diversity of exogenous and endogenous factors studied, it is difficult to identify one factor that is unique to the fossilization process. Similarly, the pattern of degeneration observed in the Clarkia fossils is the same for two different environments of deposition, as well as for leaves preserved in pyroclastic (ashfall) deposits of the Succor Creek (Niklas, 1983), indicating that the extent of tissue hydration during fossilization is not as critical as was previously thought. Our preliminary observations do, however, indicate that dehydration and the possibility of "auto-fixation" by naturally occurring plant phenolics before fossilization may play a critical role in both the extent of preservation and in the degenerative pattern seen in organelles. Luzzati and Husson (1962) have shown that at 80% dehydration, the phospholipid bimolecular leaflets of some membranes rearrange into hexagonal arrays that provide few sites for reactions with fixatives such as osmium tetroxide or tannic acid. A "negative staining" results, very similar to the condition seen in many fossil cells (Figs. 10, 12; cf. Niklas et al. 1978), as well as dehydrated modern leaf tissues that have been subsequently fixed with tannic acid. In this regard, we have observed that those taxa characterized by high, naturally occurring concentrations of tannic acids are those often found in the best state of preservation. The negative staining seen in many of these fossils may indicate that a period of dehydration (in which the molecular nature of cellular membranes becomes reorganized) was followed by a period of fixation due to tannic acids or other plant phenolics. The unit membranes of mitochondria and nuclei remain intact even after 95% of their lipids have been removed. In contrast, chloroplast membranes, because of their unusual chemistry ($\cong 80\%$ mono- and digalactosyl diglycerides) are less stable. However, after dehydration and subsequent tannic acid fixation, chloroplast membranes appear to be more stable than those of mitochondria or nuclei. Thus, the differential preservation of the chloroplast over other organelles in some fossils could have resulted from: (a) fully mature leaves undergoing senescence with attending decrease in water content, or dehydration due to premature removal from trees by some mechanical agent; (b) phospholipid-rich bimolecular leaflets of membranes reorganizing into hexagonal arrays that provided few sites for tannic acid fixation, while galactolipid-rich membranes resisted such a rearrangement; (c) subsequent rehydration of leaf tissue in tannic acid rich waters or, in leaves preserved in dry environments, rupture of tonoplasts (containing high concentrations of plant phenolics) resulting in (d) the infiltration of cells by a "natural fixative." Such a simple scenario could explain the diversity of depositional environments in which excellent ultrastructural preservation has been reported, various ultrastructural features suggestive of dehydration (e.g. concertina-like distortion of fossil cell walls, Fig. 2; differential staining of organelles, Fig. 4), and the apparent stability of chloroplasts over nuclei and mitochondria.

Differences in the sequence of chloroplast degeneration due to such factors as senescence, dehydration, and compression are sufficiently pronounced and consistent as to provide some information on the predepositional status of fossil leaves (Table 4). Naturally senescent or detached/aging leaves initially show similarities in the degeneration of chloroplasts. These include the appearance of osmophilic globuli, reorientation of the thylakoidal system, and a distention of the grana-intergrana lamellae (Table 4). However, in detached aging leaves, the chloroplast envelope ruptures before the breakdown of the lamellae, swelling of the grana-intergrana lamellae is not pronounced, and the thylakoidal system of membranes degenerates without vesiculation or the formation of numerous osmophilic globuli. Both types of senescence result in a characteristic negative staining of membranes, much like tissues fixed with tannic acid, perhaps indicative of dehydration-induced reorganization of the unit membranes. Rapid dehydration of leaves results in the formation of irregularly shaped chloroplasts, small numerous osmiophilic globuli, and in advanced plasmolysis, the appearance of myelin-like membrane configurations (Table 4). A nonplastid character of rapid dehydration, previously noted, is a concertina-like failure of cell walls.

TABLE 4. DIAGNOSTIC FEATURES OF ALTERED MESOPHYLL ULTRASTRUCTURE

SENESCENCE

Naturally Senescing

10 days
a. chloroplasts elliptical
b. stroma-grana parallel to major axis
c. stroma have ribosomes and small plastoglobuli

21 days
a. stroma-grana disoriented and swollen
b. contain large plastoglobuli
c. stroma compact to granular

or

a. chloroplasts spherical
b. stroma granular with
c. grana swollen and stacked; appearing as vesicles
d. intergrana lamellae not recognizable

>21 days
a. stroma flocculent
b. no thylakoidal system
c. contains numerous vesicles
d. very large plastoglobuli
e. negative staining of membranes

Detached and Aging

3-4 days
a. chloroplasts elliptical
b. lamellae swollen
c. chloroplast membrane swollen and distended

5 days
a. chloroplast envelope swollen
b. lamellar structure intact
c. stroma flocculent
d. with large plastoglobuli

or

a. chloroplast envelope rupture
b. stroma-grana system still evident
c. very large plastoglobuli
d. parts of the lamellar system and outer membrane appear digested
e. negative staining of membranes

TRAUMA

Dehydration

1-2 days
a. chloroplasts elliptical
b. lamellae swollen
c. small plastoglobuli

3-4 days
a. chloroplasts irregular
b. lamellar structure intact
c. stroma granular
d. small, numerous plastoglobuli

>10 days
a. chloroplast membrane discontinuous or ruptured
b. stroma-grana still evident; or myelin-like stroma
c. negative-staining of membranes

Compression

a. chloroplasts irregular
b. stroma-grana disorganized, undulating
c. small and few plastoglobuli
d. myelin-like stroma

Compression results in irregular chloroplast outlines, an undulated stroma-grana, and myelin-like membranes within the chloroplast (Table 4). Cell walls usually rupture or fail in the axis normal to compression resulting in elliptical wall outlines.

Many of the observed ultrastructural features typically associated with senescence, dehydration, and/or compression have been observed in fossil protoplasm. Negative staining, irregular chloroplast outlines, and myelin-like membrane systems are common in fossils and are indicative of dehydration and/or compression. Similarly, the concertina-like failure of cell walls in some specimens indicates that these ultrastructural features were probably the result of pre-burial dehydration. As noted earlier (cf. Results), rehydration does not totally remove the cell wall failure pattern, perhaps because of a plastic deformation component in failure.

From the foregoing ultrastructural observations, it is possible to identify fossil leaves that had undergone dehydration, senescence (either naturally or due to premature leaf fall), and/or compression. The bulk of the Clarkia leaves examined (63%) appear to have been the result of natural leaf fall due to leaf abscission, and 15% appear to have been detached due to mechanisms other than leaf abscission. Roughly 89% of all the leaves examined showed some evidence of compression cell wall failure, while 17% show concertina failure patterns.

The ultrastructural examination of the Clarkia fossils reveals a wealth of information on vagaries in the preservation of protoplasm, correlative to pre- and postdepositional factors, which collectively define the Clarkia environment of deposition. Perhaps the most surprising conclusion drawn from the statistically well-defined frequency of organelle preservation is the relative stability of chloroplasts over other organelles, which is counterintuitive to neobotanical patterns of protoplasmic degeneration (Butler and Simon 1971). However, in a few cases, the chloroplast has been reported to be the only organelle surviving terminal necrosis (cf. Hurkman 1979). Further research on fossil tissues and on patterns of simulated fossilization must follow if paleo- and neobotanical observations on ultrastructure are to be reconciled.

LITERATURE CITED

Baschnagel, R. A. 1966. New fossil algae from the Middle Devonian of New York. Trans. Amer. Microsc. Soc. 85:297-302.

Baxter, R. W. 1950. *Peltastrobus reedae*: A new sphenopsid cone from the Pennsylvanian of Indiana. Bot. Gaz. 112:174-182.

Brack-Hanes, S. D., and J. C. Vaughn. 1978. Evidence of Paleozoic chromosomes from lycopod microgametophytes. Science 200:1383-1385.

Bradley, W. H. 1946. Coprolites from the Bridger Formation of Wyoming: Their composition and microorganisms. Amer. J. Sci. 244:215-239.

Bradley, W. H. 1962. Chloroplast in *Spirogyra* from the Green River Formation of Wyoming. Amer. J. Sci. 260:455-459.

Butler, R. D., and E. W. Simon. 1971. Ultrastructural aspects of senescence in plants. Pages 73-129 *in* B. L. Strehler, ed. Advances in Gerontological Research, Vol. 3. Academic Press, New York and London.

Croxton, F. E. 1953. Elementary statistics. Dover Publications, Inc., New York, N. Y.

Darrah, W. C. 1938. A remarkable fossil *Selaginella* with preserved female gametophytes. Harvard Bot. Mus. Leaflets 6:113-136.

Dwivedi, J. N. 1959. Fossil thallophytes from Mohgaon-Kalan locality Chhindwara district. M. P. Current Sci. 7:285-286.

Eisenack, A. 1965. Erhaltung von Zellen und Zellkernen aus dem Mesozoikum und Palaozoikum. Natur. und Museum 95:273-477.

Evert, R. F., C. H. Bornman, V. Butler, and M. G. Gilliland. 1973. Structure and development of the sieve-cell protoplast in leaf veins of *Welwitschia*. Protoplasma 76:1-21.

Florin, R. 1936. On the structure of pollen grains in the Cordaitales. Svensk Botanisk Tidskrift 30:624-651.

Gould, R. E. 1971. *Lyssoxylon grigsbyi*, a cycad trunk from the Upper Triassic of Arizona and New Mexico. Amer. J. Bot. 58:239-248.

Hoefert, L. L. 1980. Ultrastructure of developing sieve elements in *Thlaspi arvense* L. II. Maturation. Amer. J. Bot. 67:194-201.

Huggins, L. M. 1985. Cuticular analyses and preliminary comparisons of some Miocene conifers from Clarkia, Idaho. Pages 113-138 *in* C. J. Smiley, ed. Late Cenozoic History of the Pacific Northwest. Pacific Division, Amer. Assoc. Adv. Sci., San Francisco, Calif.

Hurkman, W. J. 1979. Ultrastructural changes in chloroplasts in attached and detached, aging primary wheat leaves. Amer. J. Bot. 66:64-70.

Isreal, H. W., and A. F. Ross. 1967. The fine structure of local lesions induced by tobacco mosaic virus in tobacco. Virology 33:272-386.

Luzzati, V., and F. Husson. 1962. The structure of the liquid-crystal phases of the lipid water system. J. Cell Biol. 12:207-219.

Mamay, S. H. 1957. *Biscalitheca*, a new genus of Pennsylvanian coenopterids, based on its fructification. Amer. J. Bot. 44:229-239.

Millay, M. A., and D. A. Eggert. 1974. Microgametophyte development in the Paleozoic seed fern family Callistophytaceae. Amer. J. Bot. 61:1067-1074.

Niklas, K. J. 1979. Parameters in assessing geochemical studies of fossil angiosperms. IAAP Newsletter 5(2):1-5.

Niklas, K. J. 1981. Ultrastructural-biochemical correlations in fossil angiosperm leaf tissues: Implications on pre- and post-depositional environments. *In*: J. Brooks, ed. Organic Maturation of Sedimentary Organic Matter, Preservation Problems, and Fossil Fuel Exploration. Academic Press, Inc. London and New York.

Niklas, K. J. 1982. Differential preservation of protoplasm in fossil angiosperm leaf tissues. Amer. J. Bot. 69:325-334.

Niklas, K. J. 1983. Organelle preservation and protoplast partitioning in fossil angiosperm leaf tissues. Amer. J. Bot. 70:543-548.

Niklas, K. J., and R. M. Brown, Jr. 1981. Ultrastructural and paleobiochemical correlations among fossil leaf tissues from the St. Maries River (Clarkia) area, Northern Idaho, U.S.A. Amer. J. Bot. 68:332-341.

Niklas, K. J., R. M. Brown, Jr., R. Santos, and V. Bian. 1978. Ultrastructure and cytochemistry of Miocene angiosperm leaf tissues. Proc. Nat'l. Acad. Sci. U.S.A. 75:3263-3267.

Niklas, K. J., and D. E. Giannasi. 1978. Angiosperm paleobiochemistry of the Succor Creek Flora (Miocene), Oregon, U.S.A. Amer. J. Bot. 65:942-952.

Schopf, J. W. 1968. Microflora of the Bitter Springs Formation, Late Precambrian, central Australia. J. Paleont. 42:651-688.

Schopf, J. W. 1974. The development and diversification of Precambrian life. Orig. Life 5:119-135.

Smiley, C. J., and W. C. Rember. 1979. Guidebook and road log to the St. Maries River (Clarkia) fossil area of Northern Idaho. Idaho Bur. Mines Geol. Inf. Circ. 33:1-27.

Smiley, C. J., J. Gray, and L. M. Huggins. 1975. Preservation of Miocene fossils in unoxidized lake deposits, Clarkia, Idaho. J. Paleont. 49:833-844.

Stevens, N. E. 1912. A palm from the Upper Cretaceous of New Jersey. Amer. J. Sci. 34:421-436.

Taylor, T. N., and M. A. Millay. 1977. Structurally preserved fossil cell contents. Trans. Amer. Microsc. Soc. 96:390-393.

Toyama, S. 1980. Electron microscope studies on the morphogenesis of plastids. X. Ultrastructural changes of chloroplasts in morning glory leaves exposed to ethylene. Amer. J. Bot. 67:625-635.

Vishnu-Mittre. 1967. Nuclei and chromosomes in a fossil fern. Pages 250-251 *in* C. D. Darlington and K. R. Lewis, eds. Chromosomes Today, Vol. 2. Oliver and Boyd, Edinburgh.

THE PALEOBIOCHEMISTRY OF FOSSIL ANGIOSPERM FLORAS. PART I. CHEMOSYSTEMATIC ASPECTS

DAVID E. GIANNASI[1] AND KARL J. NIKLAS[2]

[1] Department of Botany, University of Georgia, Athens GA 30602
[2] Division of Biological Sciences, Cornell University, Ithaca NY 14853

"Fossilized" biological compounds, (e.g. steroids, fatty acids, alkanes, cutinic acids, phenolics, etc.) preserved in plant fossils of Paleozoic to Recent ages provide a potential reservoir of chemical markers that can be compared to the same classes of compounds in extant plants. These compounds have undergone predictable patterns of diagenesis (i.e. "fossilization") over geological time and are directly comparable to the same classes of compounds in extracts of extant plants subjected to thermolytic, simulated diagenesis. Chemical congruence between fossil and putative extant relatives is high and often taxonomically confirmatory down to the generic level. The discovery of flavonoid pigments in Miocene angiosperm leaf compression fossils adds a new paleochemical character to studies of angiosperm systematics and evolution, providing a comparison between fossil and extant taxa at the interspecific level and below. Flavonoids do not undergo diagenesis other than total destruction, and thus, in fossils, represent original unaltered constituents directly comparable to those in extant plants. Using both types of paleochemistries, fossil *Zelkova, Celtis, Ulmus* and *Quercus* from the Succor Creek Flora show high chemical congruence with putative extant relatives and a strong affinity with Asian taxa. Fossil Fagaceae, Betulaceae and Juglandaceae from the Clarkia Flora show a closer affinity with extant eastern American species. Paleobiochemistry provides a new data source in studies of fossil/extant plant floras.

The chemistry of fossil plants and fossiliferous sediments has been employed for many years in applied research in locating potential energy sources and in other areas of geological studies of terrestrial and marine environments. The use of fossil chemistry, i.e. paleobiochemistry, for chemosystematic purposes is a more recent phenomenon (Chaloner and Allen 1969). The preservation of biological compounds (albeit "fossilized"), e.g. fatty acids, steroids, alkanes, amino acids, phenolics, etc., in Paleozoic and Mesozoic plant fossils clearly suggests a potential reservoir of chemical markers that can be compared to the same classes of compounds in extant plants (Niklas 1975, 1976a-d, 1980). To be sure, many of these "fossil" compounds have undergone diagenesis ("fossilization") over geological time along with the plant tissue and inorganic matrix in which they occur. However, their modes of diagenesis are predictable and can be duplicated in the laboratory using extant compounds in a suitable matrix that is subjected to controlled conditions of heat and pressure with sequential monitoring (GC-MS, chromatography, etc.) over time (Niklas 1980, Niklas and Gensel 1976, 1977, 1978).

Thus, an extract from an extant plant, suggested to be a putative extant relative of a fossil taxon, is subjected to thermolysis and the reaction mixture sampled for a range of diagenesis patterns and specific altered classes of compounds (e.g. steroids, as *steranes*) that are then compared as an "ensemble" to those found in the fossil (Niklas 1975; Niklas and Gensel 1978). This complex chemical ensemble of data is easily handled by computer analysis. Taxa, both fossil (of known or disputed identity) and extant, can readily be compared using clustering techniques

Copyright ©1985, Pacific Division, AAAS.

(Niklas and Gensel 1976, 1977, 1978). Congruence between chemistry and morphology using such clustering techniques also is quite high, i.e. fossil algae cluster with related extant algae (Niklas 1976a) and putative fossil land plants cluster with related extant land plants (Niklas and Gensel 1977; Niklas 1979). These chemical data also can suggest possible taxonomic relationships for certain problematical fossils as to whether they are vascular or non-vascular, with additional taxonomic specificity under each of these categories. Thus, in Paleozoic plants, for example, *Sciadophyton* and *Eohostimella* cluster close to the rhyniophytes, even though on morphological bases alone the lack of vascular tissue in *Sciadophyton* and the basal rosette habit (rather than bifurcating axial stems) in *Eohostimella* would set them apart from the true rhyniophytes. *Protoaxites* and *Protosalvinia* show definite relationships with brown or other non-green algal groups. However, some of their chemical composition is terrestrial-like, suggesting experimentation with, or exploration of, terrestrial habitats by basically algal groups.

However, the taxonomic level(s) at which these chemical data proves most useful is, in effect, a reflection of the chronological (geological) and/or evolutionary (comparative morphology, anatomy, etc.) "distance" between two taxa as perceived by the taxonomist (systematist, paleobotanist) and reinforced by the relative presence/absence of greater or fewer intermediate forms (Giannasi and Crawford, in press). Thus, in the case of vascular vs. non-vascular (and subdivisions thereof), this perceived distance (lack of intermediates) is greater and hence so too is the taxonomic differentiation between the two (i.e. at the subdivisional to generic levels). Chemical data (with parallel disjunctions) automatically become useful at these higher taxonomic levels. In a more recently evolved extant group, such as the angiosperms, the degree of evolutionary distance between taxa is reduced by the presence of what are perceived as many intermediate and transitional morphological forms both fossil and extant. The ensemble chemistry, with a larger base of extant variability, also mirrors this overlapping series of taxonomic groupings with, in many cases, a concomitant limit to the level of taxonomic utility (discrimination), i.e. the generic level and above. The chemical data become a "captive" of the subjective assignment of taxonomic level to two or more taxa, and thus, additional (chemical) data are required to provide further taxonomic discrimination, e.g. at the species level and below.

The successful use of paleobiochemistry on the chemosystematics of Paleozoic plant fossils foreshadows its potential use in evolutionary studies on the angiosperms as well. The basic criteria for such studies are: (1) the fossil leaf material examined must be preserved such that enough diagnostic chemical constituents remain to be useful for comparative purposes (cf. section on diagenesis)—this is certainly true for the ensemble of compounds used by Niklas (1979, 1980), which show taxonomic specificity down to the generic level; (2) below the generic level, a second set of diagnostic compounds capable of interspecific discrimination is required, which are also preserved in the fossils; (3) these interspecifically discriminatory compounds have, preferably, undergone little or no diagenesis; and most important, (4) are a set of compounds found throughout the angiosperms (and the extant plant kingdom), thereby providing the broadest base for comparative purposes.

Flavonoids provide just such an ideal class of compounds. These low-molecular weight polyphenol aglycones and glycosides have proven systematically useful in extant plant taxa at many taxonomic levels, particularly interspecifically, and are one of the most widely used classes of secondary metabolites used in contemporary chemosystematics (Giannasi 1978; Crawford 1979; Giannasi and Crawford, in press; Harborne et al. 1975; Smith 1976). Uniquely, these flavonoids, as we shall see, do not undergo diagenesis in fossils other than irreversible destruction. Thus, if they are present in fossils, they represent unmodified original constituents and hence are directly comparable to the same compounds found in current extant plants. Using two independent but compatible forms of chemical data, potential chemosystematic comparisons between extant and fossil taxa (depending on conditions of preservation) can be carried out at least to the species level.

Coincidentally, the recent upsurge in interest in angiosperm origin and phylogeny (Beck 1976; Hickey and Doyle 1977; Paeltova 1977; Krassilov 1977) has resulted in a re-examination of many Tertiary floras in response to: (1) controversies in contemporary classifications of the angiosperms (Cronquist 1968; Tahktajan 1969; Thorne 1976); (2) current thoughts on phytogeography based on theories of continental drift (Raven and Axelrod 1974); and (3) re-evaluation of the analytical approaches in identification of leaf remains, the primary fossil source (Dilcher 1974), and thus provides a unique arena for the application of comparative paleobiochemistry to angiosperm systematics and phylogeny.

FLORAS EXAMINED

Two fossil angiosperm floras have been examined thus far, the Succor Creek Flora of Oregon/Idaho and the recently discovered Clarkia Flora in Northern Idaho (near Clarkia), both roughly of the same age, i.e. Miocene, $22\text{-}17 \times 10^6$ years.

Some specimens of the Succor Creek Flora are buried in pyroclastic deposits (tephra). Streams of lava blocked rivers and created numerous lakes. The surrounding vegetation, as well as accumulated aquatic debris (algal blooms in part), was rapidly buried in many places under ash. The rocks containing Succor Creek fossils have the same stratigraphic position as the intermontane beds, which Kirkham (1931) assigns to the Payette formation, while Chaney and Axelrod (1959) confirm their occurrence as interbeds in the Columbia River basalt. Subsequent work suggests that this flora is Oligocene to Miocene in age (Graham 1963). In contradistinction to the coarse sediments common in the Payette Flora (upper Miocene), which are micaceous sands and gravels, the Succor Creek beds are mostly opaline shales with occasional tephra lenses. The Payette strata, weathering to a reddish hue, show alteration due to contact with basaltic lava flows.

The biochemical analyses given in this and earlier reports (Niklas and Giannasi 1977a, 1978) were derived from material preserved in the uppermost pyroclastic ashfall (tephra) lenses, which showed no evidence of basalt contact or igneous intrusion. On the basis of their state of chemical and morphological preservation, as well as available geologic data, the Succor Creek "green leaves" are thought to have been preserved in a unique microgeologic environment that was the result of rapid, cool volcanic ash deposition (Niklas and Giannasi 1978b). The lack of apparent cuticular and chemical decomposition suggests that burial and subsequent fossilization occurred rapidly. Data derived from the thermal decomposition rates and products of selected flavonoids indicate that these fossils did not experience temperatures greater than 80°C or extreme pH shifts (Niklas and Giannasi 1977b).

The Clarkia Flora (Smiley et al. 1975) is also the result of volcanic activity, lava flows cutting off river and stream beds into quiet, relatively undisturbed lakes. The volcanic ash lenses (tephra) typical of Succor Creek, however, are absent, ashfall present normally occurring as continuous layers. Fossil leaf compressions are preserved in the lowermost two silty-clay layers (Zones 1 and 2) representing an "unoxidized" zone where little or no physical or microbial damage has occurred to the leaves (cf. Smiley et al. 1975, p. 834, and this volume). These two layers are covered by an ash zone (3) suggestive of traumatic volcanic activity. Indeed at the interface of Zones 3 and 2 a large number of preserved fish and insects occur. Above the ash layer (3) occur an ash/silt layer Zone 4, and the uppermost silt layer Zone 5, which represent "oxidized" zones. Fossil leaves in these oxidized zones occur only as impressions with replacement of organic material by inorganic substances.

Many (but not all) leaf compression fossils from the unoxidized zones (1-2) are quite well preserved in autumnal colors of red, brown and blackish green and will peel off the matrix upon drying. Immediate field extraction of samples of these compressions in methanol gives a green color typical of chlorophyll phaeophorbides (Niklas and Giannasi 1977a), while HCl/methanol extraction of the same tissues gives red solutions typical of tannins and proanthocyanidins (both confirmed later in the laboratory; cf. Ribereau-Gayon 1969).

Thus, although the Clarkia compressions are preserved in a potentially "extractive" lacustrine environment, deposition must have been rapid with little disturbance or extractive action, since many leaves appear three-dimensionally oriented through several layers of silt with no physical (i.e. shear) damage. The lack of bacterial decomposition in Zones 1 and 2 seems to support a "gentle," biotically sterile, preservational regime for both plant and animal remains, the fish often containing undigested stomach contents (see this volume).

One can only surmise the conditions of preservation. In addition to little upwelling or disturbance of the lake, the water must have been fairly cold to inhibit microbial activity (cf. Smith and Elder, this volume; Lewis, this volume). It is also possible that the fossil leaves themselves, many of which are woody species, may have released considerable quantitites of tannins into the water, which are also toxic to microbial activity. A situation not unlike the toxic, phenolic-laden "black waters" of the tropics may have existed at the lake bottom. Most probably a combination of these, i.e. low temperature, little disturbance, gentle but continuous inflow of sediments from nearby shistose terrain and perhaps chemotoxicity, contributed to the overall preservation of many leaves. Those leaf specimens with poor chemical preservation may have settled near shallower, more disturbed/exposed (shoreline) sedimentation areas more typical of the later oxidized zones (4 and 5), or may represent seasonally senescent leaves washed into the lake (see Part II of this paper).

Thus, fossils of the Succor Creek appear to have been preserved in a mode of desiccation in warm ash, while Clarkia fossils were preserved in a biotically sterile environment capped by an ash layer, much as one would pour a wax layer seal over a jar of strawberry preserves. Regardless of the mode of preservation, these phenomena were most fortunate indeed for the phytochemist.

METHODOLOGY

Since most of the fossil leaf is consumed in analytical procedures, a regular sequence of documentation has been developed to provide as accurate an identification of the fossil taxon as possible. Each fossil leaf is numbered and photographed for a visual record. About a quarter of the leaf is used for GC/MS analysis of the "ensemble" group of compounds. Two smaller portions are cleared for study of fine venation patterns necessary for accurate taxonomic identification (Dilcher 1974) and for ultrastructural analysis (Niklas et al. 1979a; cf. Part II of this study) respectively. The remaining leaf material is degauged from the inorganic matrix for flavonoid analysis. A portion of the matrix is also tested for possible percolative flavonoid contaminants.

Extractive procedures for isolation of the ensemble compounds are as described by Niklas (1979a), Niklas and Gensel (1976, 1977, 1978) and Giannasi and Niklas (1981) and will not be described here. The flavonoids, since they do not undergo diagenesis other than terminal destruction, are extracted from fossil leaf compressions in the same manner as dried leaf material of extant plants (Giannasi 1975, 1978; Giannasi and Mickel 1979; Giannasi and Niklas 1977; Mabry et al. 1970). Chromatographic and spectral analysis of flavonoids is identical in both fossil and extant plant leaf material and follows standard procedures (Giannasi 1975; Mabry et al. 1970).

Since many of the fossil taxa are related to, or placed in, extant genera that often contain many species (e.g. *Quercus*, 300+ species), gross morphology and micromorphology are first used to place the fossil within a smaller subgeneric group of related taxa. This considerably reduces the number of extant taxa that must be examined (Niklas and Giannasi 1978a) for comparison with the fossil. Where possible this is backed up with comparison of putative extant relatives as suggested in previous studies (e.g. Graham 1963). To randomly survey a large extant genus such as *Quercus* without preliminary discriminatory morphology study leads to much waste of time and material. Should putative extant morphotypes lack flavonoid profiles similar to the fossil, the survey can then be expanded to other morphotype groups to make sure nothing has been overlooked. This has not proven necessary in many of our studies, since there has been good congruence between chemistry and morphotype groupings of fossil and extant plants. The presence of rare or unique flavonoid marker compounds in several of the fossils has also helped to place fossil taxa with extant plants quickly.

However, we stress that in any study of this type, a survey of extant taxa is required to a greater or lesser extent and is the most time-consuming portion of the work. The chemical analysis of the fossil is only an "initial" starting point.

DIAGENESIS

As mentioned earlier, the successful use of fossil biochemistry by Niklas and coworkers on studies of Paleozoic plants is based on two basic tenets. First, the compounds used (i.e. steroids, fatty acids, alkanes, phenolics, etc.) all undergo predictable patterns of modification (=diagenesis) during fossilization (simulated or actual) of the plant tissue itself, as documented in thermolytic studies of extant plant tissues (diagenesis of complete chemical ensembles) and individual known standards from each of the classes of compounds employed. In this manner chemical ensembles from fossils may be compared with their putative extant equivalents (cf. Niklas 1980, and Pt. II, this study for a review of diagenesis; also Niklas and Gensel 1976, 1977, 1978) to confirm the identity of a fossil, often down to the generic level with respect to extant taxa. Once the fossil is placed within a specific taxonomic (i.e. generic) frame of reference then the flavonoids, if present, may be used to carry chemotaxonomic comparison to the inter- and intraspecific level.

The use of flavonoids depends on their preservation (at least the aglycone) intact in the plant and deserves further discussion. Since they normally appear as glycosides in plants, thermolytic studies (Niklas and Giannasi 1977b) indicate that the sugar is cleaved first (at least in the case of O-glycosides) during initial diagenesis, at ca. 80°C. Beyond this temperature the triple ring system begins to disintegrate, producing a number of monophenolic acid residues (Niklas and Giannasi 1977b). At the point of ring breakage, the flavonoid data ceases to be an effective, specific set of characters since many of the phenolic fragments are identical to existing simple phenolic residues normally produced from the shikimic acid pathway (as are portions of the flavonoid) or derived from the breakdown of other tissues, i.e. lignin, tannins, etc., during natural diagenesis. Thus, the phenolic fragments of flavonoids may become indistinguishable from the general milieu of ensemble phenolic data, thereby losing their interspecific discriminatory powers.

While this probably restricts the number of fossil floras (i.e. preservational regimes) in which flavonoids may profitably be used, their presence alone does give some indication of the modes of preservation (temperature, etc.). Further, since they represent original unaltered constituents within their limited frame of preservation, comparative diagenetic studies of flavonoids between fossil and extant taxa are not needed.

With the modes of preservation (i.e. diagenesis) established for both flavonoids and ensemble classes of compounds, two different but mutually integrated groups of chemical data are available for accurate and reliable paleochemosystematic studies. Their presence and hence utility is constrained only by the presence/absence of organic material and the degree of trauma involved in the preservation of the fossil itself.

Also within these limitations, this approach probably provides one of the first real, direct approaches to studying a vertical chronology of chemical evolution in plants from Paleozoic to present, as well as the possibility of a horizontal comparative study of taxa between chronologically equivalent fossil floras, much as is carried on in contemporary chemosystematic studies of extant plants.

PREVIOUS WORK

Considering the efficacy of this paleobiochemical approach in Paleozoic and Mesozoic plant studies, it is logical that such techniques be employed to study the paleochemosystematics of the angiosperms, especially considering the present controversy as to their origin, both chronologically and phylogenetically (Takhtajan 1969; Cronquist 1968; Thorne 1976; Kubitsky 1977; Hickey and Doyle 1977; Hickey and Wolf 1975; Beck 1976). The relative growth rate of the angiosperms also provides many leaf megafossils, which are often morphologically attributable to extant genera and in some cases to putative extant species. Further, the megafossils of some floras (e.g. Succor Creek, Clarkia) are well enough preserved as uncarbonized, non-coalified, unsubstituted, organic, cuticular compressions that such paleobiochemical studies are possible.

Thus, Niklas and Giannasi (1977a) examined the chemistry of a fossil leaf of *Zelkova* (Ulmaceae) from the Succor Creek Flora of Oregon. The intact fossil leaves were preserved in a volcanic ash (dated at $22\text{-}16.7 \times 10^6$ K/A years), were a vivid green in color, and could be removed intact from the ash matrix (cf. Niklas and Giannasi 1978b, for documentation). Transmission electron microscope analysis showed preservation of the three-dimensional ultrastructure of the leaf, including photosynthetic and nuclear apparati, cell walls, starch granules and vascular tissue (Niklas et al. 1978; see Part II, this study). As might be expected from such remarkable physical preservation, much of the internal chemistry was retained, including steroids, fatty acids, and alkanes, apparently with little diagenesis.

Most startling was the discovery of flavonoid pigments, in this case anthoxanthins, in the fossil. As related above, these phenolic pigments have been used for many years in biochemical systematic studies of extant plants (Smith 1976) at various taxonomic levels and on plant groups from bryophytes to composites.

The occurrence of flavonoid pigments in the fossil (a flavonol and a dihydroflavonol) not only was a novel discovery, but also provided a new set of chemical markers. Previous work (Niklas and Gensel 1978) had established that a number of compounds, especially steroids, were very effective in taxonomic documentation down to the generic levels. However, once morphological and ensemble chemical data could place the fossil within a restricted taxonomic frame of reference, the flavonoids could be used interspecifically as is done in contemporary studies. Further, analysis of the fossil rock matrix failed to show any flavonoids that would have been expected had simple percolative contamination from other sources occurred. The "fossil" flavonoids then appeared to be original constituents of the fossil *Zelkova* and were directly comparable with the same compounds found in extant species of *Zelkova* (cf. diagenesis section).

To determine if such results were repeatable, two additional fossil leaf taxa from Succor Creek, *Celtis* sp. and *Ulmus* sp., were examined (Giannasi and Niklas 1977). Both fossil taxa possessed mutually exclusive flavonoid profiles. Fossil *Celtis* was characterized by the presence of C-glycosylated flavones, while the fossil *Ulmus* possessed O-glycosylated flavonols. Thus, if the fossil leaves were simply acting as organic "sinks" (Harborne 1977), sequestering phenolics randomly percolating through the matrix, then all of these taxa would have the same general flavonoid profiles (as would the inorganic matrix as well). This was not the case, and in fact, the intergeneric dichotomy was also observed in the steroid, fatty acid and alkane (i.e. ensemble) profiles even with diagenetic parameters considered (Giannasi and Niklas 1977).

The flavonoid dichotomy between the fossil taxa was also found to occur in extant related taxa. e.g. *Celtis occidentalis* and *Ulmus americana*. Indeed, the fossil flavonoid dichotomy prompted a study of flavonoids in extant Ulmaceae to evaluate its potential use in the contemporary systematics of the family (Giannasi 1978). Results showed the flavonoid dichotomy to be absolute among the 19 extant Ulmaceae genera and support the subfamilial alignment of these genera into two groups as proposed by Grudzinskaya (1965), rather than the longstanding arrangements typified by that of Hutchinson (1967; cf. also Sweitzer 1971).

Finally, authentic known flavonoids were subjected to artificial fossilization, i.e. thermolysis, to determine their stability under heat, pressure and pH in a matrix (e.g. bentonite) similar in character to the volcanic tuft in which the fossils were preserved (Niklas and Giannasi 1977b). Results indicated that flavonoid glycosides were stable in neutral or slightly acidic state up to 80°C, after which the O-glycosides began to hydrolyze. As the temperature approached 100°C and above, the aglycone began to disintegrate in predictable modes, depending on the pH. Preservation of the flavonoid glycosides (especially the O-glycosides), therefore, suggested that the volcanic ash matrix probably did not experience a maximum temperature above 80°C. The flavonoids could have easily survived the conditions of preservation peculiar to this matrix and therefore represent original constituents in the fossils. This certainly is supported by the phenomenal preservation of fossil leaf ultrastructure (Niklas et al. 1978) and other chemical constituents (Giannasi and Niklas 1977). Indeed, the recent eruption of Mt. St. Helens and the concomitant ashfall activity provide a graphic, contemporary experience exhibiting actual phenomena like those thought to have occurred in Succor Creek and other pyroclastic-derived fossil floras (Rosenfeld 1980).

With proof of the survivability and retention of flavonoids in fossil leaves, the legitimate use of paleochemotaxonomy in angiosperms was established. Particularly successful was the study of a fossil oak, *Quercus consimilis*, from Succor Creek, which provided the best example of the systematic, evolutionary and phytogeographic potential of such data (Niklas and Giannasi 1978).

Graham (1963) suggested that the fossil, *Q. consimilis*, was related to the extant Asian taxa, *Q. myrsinaefolia* and *Q. stenophylla*. While steroid, fatty acid, and alkane chemistry as well as exomorphology clearly confirmed its identity as an oak, the flavonoid chemistry showed the fossil to be more similar to two other entire-leaved Asian oaks, *Q. acutissima* and *Q. chenii*, rather than the two putative extant relatives suggested by Graham.

Table 1, summarized from Niklas and Giannasi (1978a), shows the distribution and identity of flavonoids in the taxa included in the study. Note that the fossil *Q. consimilis* possesses glycoflavones (Group A), as do *Q. acutissima* and *Q. chenii*. These compounds are absent in samples of *Q. stenophylla* and *Q. myrsinaefolia* examined, which instead possess flavonols (Group C). Also, even though all of the taxa possess flavonols (Group B), even these (with the exception of *Q. variabilis*) fall into two different groups, which further isolate the fossil from the originally proposed extant relatives, *Q. stenophylla* and *Q. myrsinaefolia*. Thus, the flavonoid chemistry proved to be quite discriminatory in placing the fossil systematically within morphologically similar taxa. Indeed, a random survey of 30 other species of oak, lobed and unlobed, Asian and western, failed to turn up a pattern similar to the fossil and its newly proposed relatives (Giannasi, unpubl.).

Perhaps more important, the chemistry suggests a definite taxonomic link between Asian taxa and North American fossil taxa, a relationship long suspected by phytogeographers (cf. Niklas

TABLE 1. COMPARATIVE DISTRIBUTION OF FLAVONOIDS IN FOSSIL AND EXTANT SPECIES OF *QUERCUS* AND RELATED TAXA FROM SUCCOR CREEK FLORA (cf. Niklas and Giannasi 1978a)

Taxon	\multicolumn{5}{c}{Flavonoid distribution[a]}
	A 1 3 5 9 10 11	B 2 4 12 13 16 17	C 14 15	D 7	E 6 8 18 19
1. *Quercus consimilis*	+ +	+ +		+	+ + +
2. *Q. acutissima*	+ + +	+ +		+	+ + +
3. *Q. chenii*	+ + + + +	+ +		+	+ +
4. *Q. myrsinaefolia*[b]		+ + +	+ +	+	
5. *Q. variabilis*		+ + + + +	+ +	tr	
6. *Castanea dentata*		+ +	+ +	tr	+

[a] Flavonoid code: A = glycoflavones; B = flavonols; C = flavanonols; D = ellagic acid; E = unknown. + = presence of compound; tr = trace amounts.

[b] Pattern for *Q. stenophylla* similar to *Q. myrsinaefolia*.

and Giannasi 1978a). Further, these data and those from earlier studies on fossil *Celtis* and *Ulmus* allow some evaluation of relative rates of chemical evolution over actual geological time. In the fossil oak, the biosynthetic capability of glycoflavone production has existed for at least 22-17 million years and apparently is restricted to a small group of Asian species of *Quercus*. The flavonoid dichotomy in the Ulmaceae is broader in distribution within the family, at a higher taxonomic level, and also appears to have been fixed for at least 22 million years. Examination of still older fossil taxa may provide further chronological (=evolutionary) differences.

We have now begun work (Giannasi and Niklas 1981) on a second fossil angiosperm assemblage, the Clarkia Flora, recently exposed and excavated by Dr. C. J. Smiley and coworkers. The cuticular leaf compressions are preserved in lacustrine shales (Smiley and Rember, this volume) and when exposed display autumn leaf colors of brown, red, and in some cases greenish-black. The Clarkia Flora, dated at ca. 22-16.7 KA, is thus of comparable age with the Succor Creek Flora. Of the 12-14 fossil taxa examined so far from Clarkia (site P-33), distinct profiles have been obtained for *Betula, Hydrangea, Acer, Juglans* and a new taxon related to *Fagus* (Smiley and Huggins 1981). The opportunity to examine the Clarkia leaf compressions was both exciting as well as critical, since it was possible: (1) to compare modes of flavonoid preservation in two different fossil angiosperm floras preserved under different conditions, and most important, (2) to examine similar taxa from both floras, i.e. *Quercus consimilis* (Succor Creek) and *Quercus simulata* (Clarkia), to determine if they are distinct taxa or are actually very closely related, or even the same taxon.

Of special interest to us in the Clarkia Flora was a new fossil taxon assigned to the Fagaceae (Smiley and Huggins 1981). While its vegetative characters are clearly similar to extant and fossil *Fagus* sp. (also found at Clarkia), its more robust leaf form coupled with a distinct fruit (found attached to the stem portion) differing from that of known *Fagus* (extant and fossil), supports its assignment to a new genus and species, but still closely related to *Fagus*, hence the name *Pseudofagus idahoensis* (Smiley and Huggins 1981). Comparative chemical studies of *Pseudofagus* and extant related Fagaceae did not provide a quick easy solution to the taxonomic affinities of the fossil as observed in the Succor Creek taxa. Nevertheless, when comprehensive chemical surveys of other extant Fagoid genera besides *Fagus* were finally completed, a quite different pattern of chemical evolution was observed for *Pseudofagus* (as compared to the Succor Creek taxa), which greatly supports the hypothesized origin of *Pseudofagus* as proposed by Smiley and Huggins (1981).

Chemical analyses of representative fossil genera and species of the Fagaceae (Table 2) indicate that the group may be characterized by the possession of oleanane, 3,4-*seco*-oleanane, ursane, glutane (D:B-*friedo*-O), lupane, and friedelane (D:A-*friedo*-O) (Niklas and Giannasi 1981). While these cycloalkanes seem collectively diagnostic for the fossil representatives of the family, and for living taxa studied, oleanane, ursane, and lupane are commonly found in many orders of angiosperms (e.g. Sapindales, Ebenales, Primulales, Caryophyllales, and Geraniales). Similarly, oleanane, glutane, and friedelane are found in the Ericales; oleanane and ursane in the Arales and Rubiales; and oleanane and lupane in the Myrtales, Hammelidales, Cactales and Rosales. Due to the wide phytochemical distribution of sets of these compounds, the steroid chemistry of the Fagaceae must be carefully approached by broad surveys of both living and fossil taxa and qualified within the context of the Fagaceae and accompanying exomorphological studies (Smiley and Huggins 1981).

The total organic chemical profile of *Pseudofagus* is repeatable, however, in several especially well-preserved specimens we collected. The identification of some GC peaks was tentatively accomplished by co-injection of standards (e.g. 5a-cholestane, onocerane, lupane, stigmastane, lanostane, friedelane), while the relative percent abundance of compounds was calculated by the area under a peak. Confirmation of compound identity was by mass spectroscopy (Giannasi and Niklas 1981). In addition to friedelane, lupane, oleanane, ursane, and glutane (mentioned previously), onocerane, 5a-cholestane, stigmastane, and lanostane were also identified. Thus, *Pseudofagus* specimens have high concentrations of: (1) friedelane, a $C_{30}H_{52}$ pentacyclic triterpane, which has

TABLE 2. CYCLOALKANE DISTRIBUTION IN FOSSIL GENERA OF THE FAGACEAE

Genus	Compound[a]							
	O	S	U	L	G	F	C	On
Fagus[c]	+	−	+	+	−	+	−	−
Pseudofagus[b]	+	−	+	+	+	+	+	+
Quercus[c]	+	+	+	+	−	+	−	−
Castanea[c]	+	−	−	+	+	+	−	−
Lithocarpus[c]	−	−	−	+	+	+	−	−
Nothofagus[c]	+	−	−	+	+	+	−	−
Castanopsis[c]	−	−	+	−	+	+	−	−
Trigonobalanus[c]	+	−	−	−	+	+	−	−

[a] O = oleanane; S = 3,4-*seco*-oleanane; U = ursane; L = lupane; G = glutane (D:B-*friedo*-0); F = friedelane; C = 5α-cholestane; On = onocerane.

[b] Fossil taxon.

[c] Living taxon (cycloalkanes inferred from diagenetic steroid composition).

a complex MS fragmentation pattern typical for structures that do not have a high probability for sites of carbon-carbon bond cleavage; and (2) lupane, a pentacylic compound with one 5-membered ring, with a characteristically intense m/e 191 peak (Giannasi and Niklas 1981). Oleanane, ursane and glutane are apparently low in their relative concentrations to other compounds.

Onocerane, 5α-cholestane, stigmastane and lanostane were not isolated from any other taxa known or thought to belong to the Fagaceae. Stigmastane ($C_{29}H_{52}$) and lanostane ($C_{30}H_{54}$) are reported for fossil green algae (Niklas and Giannasi 1977a), while lanostane has been reported from various ascomycetes, in particular the yeasts. Electron microscopy of fossil angiosperm leaves from the Clarkia locality reveals epicuticular fungi and filamentous algae. These taxa may have contaminated the sterane profile of *Pseudofagus* (Giannasi and Niklas 1981). However, we are not aware of any reports of the occurrence in thallophytes of the other "diagnostic" steranes considered here as unique for *Pseudofagus*. On the basis of having various compounds (friedelane, lupane, etc.) shared with the Fagaceae, and having specific steranes (onocerane, 5α-cholestane) thus far unknown for other fagoid taxa, *Pseudofagus* appears to possess a unique and diagnostic profile.

Although in this case the sterane data confirm that the fossil taxon does indeed belong to the Fagaceae, further intergeneric chemical specificity in identification as witnessed in an earlier study on *Quercus* (Niklas and Giannasi 1978) was not obtainable from the *Pseudofagus* material. Therefore, a more general survey of phenolics in extant Fagaceae was carried out to determine if a significant intergeneric chemical marker might be present that might support a closer alliance of *Pseudofagus* with *Fagus* as suggested by morphological studies.

As indicated by Hegnauer (1966), and confirmed by our own extended studies, all genera in the Fagaceae produce the tannin component ellagic acid, except *Fagus* and *Pseudofagus* (Table 3). Specific taxa in this general tannin survey included those used by Smiley and Huggins (1981) in their morphological/anatomical studies and a few extra taxa representative of the genera involved (cf. Giannasi and Niklas 1981). Thus, the absence of ellagic acid in *Pseudofagus* effectively isolates this taxon from other Fagaceae, placing it close to extant *Fagus* ssp., which also lack ellagic acid.

If, as Smiley and Huggins indicate, *Pseudofagus* is most closely related to *Fagus* itself, then a more detailed infrageneric survey of flavonoids in representative *Fagus* species was desirable to determine if further similarity/dissimilarity existed between the fossil and extant *Fagus* species. Identifications of specific flavonoid compounds found in these two genera are given in Giannasi and Niklas (1981), while their comparative distributions are shown in Table 4.

In terms of interspecific flavonoid profiles (Table 4), *Pseudofagus* possesses a much reduced flavonoid profile, more like that of the North American *Fagus grandifolia*, European *F. sylvatica*, or Eurasian *F. orientalis*, rather than the clearly Asian taxa, e.g. *F. engleriana*, *F. sieboldii*, *F.*

TABLE 3. COMPARATIVE CHEMISTRY OF SELECTED FAGACEAE[a]

	Ellagic Acid	Anthochlors	Flavonols	Flavones
Fagus	−	−	+	−
Pseudofagus (fossil)	−	−	+	−
Quercus	+	−	+	+
Castanea	+	−	+	−
Lithocarpus	+	−	+	+
Nothofagus	+	+	+	−
Castanopsis	+	+	+	−
Trigonobalanus	+	−	+	−

[a] Published references consulted include: Harborne 1967; Harborne et al. 1975; Hegnauer 1966; Niklas and Giannasi 1978; as well as original survey work by the authors.

japonica and *F. lucida*. Two additional Asian taxa, *F. crenata* (sometimes included with *F. sieboldii* or vice versa) and *F. hayatae* (from Taiwan) are very similar to *F. engleriana*.

The similarity of *Pseudofagus idahoensis* with North American and/or European *Fagus* species contrasts with an earlier study (Niklas and Giannasi 1978) of the fossil *Quercus consimilis* from the Succor Creek Flora, which showed clear-cut chemical affinities with extant Japanese and Chinese taxa. On the other hand, Bate-Smith and Richens (1973), in their world survey of the genus *Ulmus*, found the Asian species to be rich in flavonoids, while the European and North American taxa showed a more reduced, depauperate flavonoid pool.

Smiley and Huggins (1981; Smiley, Gray and Huggins 1975; Smiley and Rember, this volume) note many similarities between fossil taxa from the Clarkia Flora and extant southeastern U.S. plants, many of which no longer are found in the western U.S. floras. Smiley and Huggins further suggest that *Pseudofagus* occupied a warm-temperate, humid climate period with mild winters and little drought, much like that of extant forests of the Appalachian area and eastern Asia. Based on flavonoid chemistry, *Pseudofagus* clearly has a closer affinity to the Appalachian taxon of *Fagus*; this conclusion agrees with morphological studies. General uplift and drying of the western land areas presumably led to the demise of *Pseudofagus* and its replacement by a cool temperate *Fagus* species, also found in the Clarkia beds. The Appalachian area apparently has retained much of the phytogeographic and ecological characters typical, and possibly supporting the origin, of a taxon like *Pseudofagus*, no matter how short its original "tenure." One can only surmise the possible phytogeographic connection between western and eastern fossil floras that may have existed previously. Indeed, prior to the Eocene-Oligocene climatic deterioration (ca. 34-38 KA; cf. Wolfe 1978 and references therein), North American Tertiary climates were characterized by high equability and, in some cases, high mean annual temperatures (Wolf 1971, 1972). Provided that the ancestral clade to *Pseudofagus* had a fairly wide geographic range prior to the Oligocene (i.e. western to eastern North America), a single climatic vicariant event could have resulted in the production of two residual geographic distributions. The evolution of *Pseudofagus* in the Northwest may have been a unique event in the clade. The post-Miocene extinction of the genus along with the putative *Fagus* or *Fagus*-like ancestor in the Northwest may or may not have had a concomitant extinction event for the ancestral form in the southeast. Thus a relatively conservative phytochemistry for related taxa may have survived in more equitably climatic regions. By analogy, both the Betulaceae and Salicaceae show evidence of considerable geographic expansion and specific diversification during the Neogene. Many similar connections, i.e. east/west correlations, have been observed in contemporary conifer taxa based on terpenoid data (von Rudloff 1975), along with more localized patterns of evolution and phytogeography of western U.S. conifers (Zavarin et al. 1975). Our data may also be interpreted as evidence for mere phytochemical-ecologic convergence between some populations of *Pseudofagus* and contemporary

TABLE 4. DISTRIBUTION OF FLAVONOIDS IN *FAGUS* (extant) AND *PSEUDOFAGUS* (fossil) SPECIES (cf. Giannasi and Niklas 1981)

Taxon	1	2	3	4	5	6	7	8	9	10	11	12	13	14	15
1. *Fagus engleriana*	+	+			+	+	+	+							
2. *F. japonica*		+		+	+	+	+		+						
3. *F. crenata*	+	+	+	+			+	+			+		+	+	+
4. *F. hayatae*	+	+		+		+	+	+		+					
5. *F. sieboldii*	+	+	+	+							+		+		
6. *F. lucida*	+		+	+							+				
7. *F. grandifolia*		+		+								+			
8. *F. grandifolia* var. caroliniana		+		+								+	tr		
9. *F. sylvatica*		+		+											
10. *F. sylvatica* var. asplenifolia		+		+								tr			
11. *F. orientalis*		+		+								tr			
12. *Pseudofagus idahoensis* sp. nov.		+							+				+	+	

Fagus ssp.; however, if such a model is proposed, the selective pressures resulting in this phytochemical convergence remain obscure.

Considering its individual character, *Pseudofagus* produces a species-specific isorhamnetin 3-0-glycoside (a methylated compound), setting it apart from extant *Fagus* species. Such a biosynthetic step is considered an advanced character (Harborne 1977), suggesting a specialized (derived) condition in *Pseudofagus*. This would fit Smiley and Huggins' scenario in which *Pseudofagus* was a localized endemic with robust habit and perhaps a specialized (=derived) fruit differing from those of the more typical fossil *Fagus* species that succeeded it in cooler and drier conditions. The steroid data, while placing the fossil in the Fagaceae, also show a highly distinct (=derived) profile.

We have examined a single specimen of the fossil *Fagus* species contemporary with and succeeding *Pseudofagus*. Unfortunately, we could not obtain a flavonoid pattern. This may not be as important to our discussion as would be assumed, since if the fossil *Fagus* were different from *Pseudofagus* it would simply further strengthen the validity of *Pseudofagus*. Similar or identical profiles between the two taxa might just as easily support Smiley and Huggins' corollary that *Pseudofagus* was simply a derivative of the contemporary but less abundant *Fagus* species. Since we have only begun our paleochemosystematic work on the Clarkia Flora, we look forward to further attempts to re-examine the fossil *Fagus*, as well as a number of other fossil species of *Betula*, *Acer* and *Juglans*, all of which have yielded excellent flavonoid profiles. Indeed, the flavonoid profile of the fossil *Juglans* shows three flavonol glycosides, which in a survey of the Juglandaceae (Manhart, unpubl.), appear only in the North American taxa, *J. nigra* (North American) and *J. major* (Mexico), thus reinforcing the American affinities of the Clarkia Flora.

CONCLUSIONS

Based on the studies of the Succor Creek and Clarkia floras, paleobiochemical systematics of angiosperms offers a new approach for studying the systematics and evolution of angiosperms. Rather than studying only the morphology of fossil and extant plants or the comparative phytochemistry of extant plants, we now can directly compare both forms of data between fossil and extant plants. Inherent in this approach is the potential ability to observe biochemical evolution over geologic time, in addition to documenting taxonomic similarities/dissimilarities between fossil and extant species.

Just as significant as the taxonomic and evolutionary conclusions to be drawn from fossil leaf chemistry are the phytogeographic implications. Fossil taxa from Succor Creek, especially *Quercus consimilis*, show clear affinities with Asian taxa, supporting earlier work suggesting the Pacific Northwest to be a refugium for older Asian taxa (Niklas and Giannasi 1978a). In the Clarkia Flora, in contrast, *Pseudofagus idahoensis*, in addition to its chemical distinctness, does possess a greater affinity with eastern U.S. or European (Eurasian) beech species rather than purely Asian taxa as at Succor Creek. This supports the contention (Smiley and Huggins 1981; Smiley and Rember, this volume) that the Clarkia Flora shares a close affinity with the extant Appalachian flora and its inherent climatology. Recent work on fossil *Juglans* from Clarkia again shows flavonoid affinities to native U.S. and Mexican species of extant *Juglans* (Giannasi et al. in prep.; Manhart unpubl.).

These preliminary chemical data (Niklas and Giannasi 1978; Giannasi and Niklas 1981) suggest the existence of two different, interdigitating, coexistent floral regimes in the western U.S.; one of perhaps an older and Asian origin (Succor Creek) and the other (Clarkia) a younger and perhaps uniquely native American or Eurasian (cf. Raven and Axelrod 1978).

Despite the promising success of these paleobiochemical data, there are limiting caveats as to the application of such data to plant systematics. First, the number of fossil floras preserved as seen in Succor Creek and Clarkia are obviously fewer than those consisting primarily of impression fossils where chemical data cannot be obtained. Nevertheless, our preliminary results are most encouraging and suggest that intensive analysis of the two interfacing floras, Succor Creek and Clarkia, will not only provide taxonomic and evolutionary data, but will also define the parameter of the sensitivity of chemical data in characterizing individual fossil floras.

A second important caveat is that each fossil taxon actually represents a single leaf preserved at a fixed point in time. Does this individual really represent the "norm" for this taxon through vertical time (stratigraphic) as well as horizontal (geographic) displacement? This can be answered to some degree by recent studies of vertical and horizontal samples of a taxon tentatively identified as *Acer bendirei* (Smiley et al. 1975) from Clarkia. Of the four samples examined one lacked flavonoids, while the remaining three showed the same basic flavonoid pattern regardless of vertical or horizontal origin. This tends to support the reliability of our conclusions based on fossil chemistry, although obviously, similar stratigraphic surveys of other individual fossil taxa are desirable. The ease of obtaining such samplings from the soft and easily split layers of the Clarkia Flora makes it an easier sampling regime than the harder rock-like strata of the Succor Creek tephra.

Although fossil taxa from Succor Creek consistently show flavonoid profiles, samples from Clarkia often lack these phenolic pigments (see discussion of *Acer* above). Indeed, some taxa from Clarkia such as *Magnolia, Liriodendron* and *Platanus* have thus far failed to show flavonoids, although phenolic acids and tannins appear to be present. Whether this represents an intrinsic preservational loss in these taxa, or indicates horizontal placement (i.e. disturbed shoreline vs. lake bottom), seasonal senescence, or secondary placement (river flow input), remains to be answered. Despite this, both floras are botanically rich and exciting in their diversity.

These caveats not withstanding, the presence of preserved taxa from two different floras of approximately the same age also allows for the possibility of comparing the same or similar taxa from both locations. Thus, is *Quercus consimilis* really different from *Q. simulans* of the same Succor Creek formation? Is *Q. simulans* of Succor Creek the same taxon as *Q. simulans* of Clarkia, or, is the Clarkia taxon entirely different, or more closely related to another entire margined oak species? Might *Q. simulata* also be another ephemeral, localized endemic as in the case of *Pseudofagus idahoensis*?

Ultimately, it would be highly desirable to examine other well-preserved fossil floras of older (or younger) geographical age to address overall evolutionary questions. Thus, how far back in time does the flavonol/glycoflavone flavonoid dichotomy exist between recognizable Ulmoid taxa? Is there an earlier Ulmaceae taxon containing both flavonoid classes, thereby providing an actual chronological point prior to chemical (and perhaps morphological) divergence leading to Celtoid and Ulmoid lines in the Ulmaceae? How many fossil taxa may be localized "experiments" actually contributing little to the mainstream of evolution of current taxa? The potential ability to answer even a few of these questions provides a forceful impetus for continued pursuit of angiosperm paleobiochemical systematics.

ACKNOWLEDGMENTS

This work was supported by NSF grants DEB-7904551 (DEG), DEB-8120515, and DEB-7822646 (KJN). We are indebted to Dr. P. K. Holmgren and Miss Eileen Schofield, the New York Botanical Garden, and Mr. Thomas Delendick, Brooklyn Botanic Garden, for supplying extant plant materials for these studies.

LITERATURE CITED

Bate-Smith, E. C., and R. H. Richens. 1973. Flavonoid chemistry and taxonomy in *Ulmus*. Bioch. Syst. Ecol. 1:141-146.
Beck, C. G., ed. 1976. Origin and early evolution of angiosperms. Columbia University Press, New York, N. Y.
Chaloner, W. G., and K. Allen. 1969. Paleobotany and phytochemical phylogeny. Pages 21-30 in J. B. Harborne, ed. Phytochemical Phylogeny. Academic Press, London.
Chaney, R. W., and D. I. Axelrod. 1959. Miocene floras of the Columbia Plateau, pt. 2. Systematic considerations. Carnegie Inst. Washington Pub. 617, pt. 1.
Crawford, D. J. 1979. Flavonoid chemistry and angiosperm evolution. Bot. Rev. 44(4):431-456.
Cronquist, A. 1968. The evolution and classification of flowering plants. Houghton Mifflin, Boston, Mass. 396 pp.
Dilcher, D. L. 1974. Approaches to the identification of angiosperm leaf remains. Bot. Rev. 40: 1-157.
Giannasi, D. E. 1975. The flavonoid systematics of the genus *Dahlia* (Compositae). Mem. N. Y. Bot. Gard. 26:1-125.
Giannasi, D. E. 1978. Generic relationships in the Ulmaceae based on flavonoid chemistry. Taxon 27:331-334.
Giannasi, D. E. 1979. Systematic aspects of flavonoid biosynthesis and evolution. Bot. Rev. 44(4):399-429.
Giannasi, D. E., and D. J. Crawford. In press. Biochemical systematics II. A reprise. Evol. Biol.
Giannasi, D. E., and J. T. Mickel. 1979. Systematic implications of flavonoid pigments in the fern genus *Hemionitis* (Adiantaceae). Brittonia 31:405-412.
Giannasi, D. E., and K. J. Niklas. 1977. Flavonoid and other chemical constituents of fossil Miocene *Celtis* and *Ulmus* (Succor Creek Flora). Science 197:765-767.
Giannasi, D. E., and K. J. Niklas. 1981. Comparative paleobiochemistry of some fossil and extant Fagaceae. Amer. J. Bot. 68:762-770.
Graham, A. 1963. Systematic revision of the Succor Creek and Trout Creek Miocene floras of southeastern Oregon. Amer. J. Bot. 50:921-936.
Harborne, J. B. 1977. Flavonoids and the evolution of the angiosperms. Bioch. Syst. Ecol. 5:7-22.
Harborne, J. B., T. J. Mabry and H. Mabry, eds. 1975. The flavonoids. Chapman and Hall, London. 1204 pp.
Hegnauer, R. (1962–present). Chemotaxonomie der Pflanzen. Vol. 1-8+. Birkhauser, Basel (continuing series).
Hickey, L. J., and J. A. Wolfe. 1975. The bases of angiosperm phylogeny: Vegetative morphology. Ann. Missouri Bot. Gard. 62:538-589.
Hickey, J. J., and J. A. Doyle. 1977. Early Cretaceous fossil evidence for angiosperm evolution. Bot. Rev. 43:3-104.
Hughs, N. F. 1977. Palaeo-succession of earliest angiosperm evolution. Bot. Rev. 43:105-127.
Hutchinson, J. 1967. The genera of flowering plants. Vol. 2. Clarendon Press, Oxford. 968 pp.
Kirkham, V. R. D. 1931. Revision of the Payette and Idaho formations. J. Geol. 39:193.
Krassilov, V. A. 1975. The origin of angiosperms. Bot. Rev. 43:143-176.
Kubitski, K, ed. 1977. Flowering plants: Evolution and classification of higher categories. Plant Syst. Evol., Suppl. 1:1-416. Springer-Verlag, Vienna.

Mabry, T. J., K. R. Markham, and M. B. Thomas. 1970. The systematic identification of flavonoids. Springer-Verlag, New York, N. Y. 354 pp.
Niklas, K. J. 1975. The chemotaxonomy of *Parkia decipiens* from the lower Old Red Sandstone, Scotland (K.K.). Rev. Palaeobot. Palynol. 21:205-217.
Niklas, K. J. 1976a. Chemical examinations of some non-vascular Paleozoic plants. Brittonia 28:113-137.
Niklas, K. J. 1976b. Chemotaxonomy of *Prototaxites* and evidence for possible terrestrial adaptation. Rev. Palaeobot. Palynol. 22:1-17.
Niklas, K. J. 1976c. Morphological and chemical examination of *Courvoisiella ctenomorpha* gen. and sp. nov., a siphonous alga from the upper Devonian, West Virginia, U.S.A. Rev. Palaeobot. Palynol. 21:187-203.
Niklas, K. J. 1976d. Organic chemistry of Protosalvinia (=Foerstia) from the Chattanooga and New Albany Shales. Rev. Palaeobot. Palynol. 22:265-279.
Niklas, K. J. 1979. An assessment of chemical features for the classification of plant fossils. Taxon 28:505-516.
Niklas, K. J. 1980. Paleobiochemical techniques and their applications to paleobotany. Pages 143-181 *in* J. Harborne and T. Swain, eds. Progress in Phytochemistry. Vol. 6, Pergamon Press, Oxford.
Niklas, K. J., and P. G. Gensel. 1976. Chemotaxonomy of some Paleozoic vascular plants. Part I. Chemical composition and preliminary cluster analyses. Brittonia 29:353-378.
Niklas, K. J., and P. G. Gensel. 1977. Chemotaxonomy of some Paleozoic vascular plants. Part II. Chemical characterization of major plant groups. Brittonia 29:100-111.
Niklas, K. J., and P. G. Gensel. 1978. Chemotaxonomy of some Paleozoic vascular plants. Part III. Cluster configurations and their bearing on taxonomic relationships. Brittonia 30:216-232.
Niklas, K. J., and D. E. Giannasi. 1977a. Flavonoids and other chemical constituents of fossil Miocene *Zelkova* (Ulmaceae). Science 196:877-878.
Niklas, K. J., and D. E. Giannasi. 1977b. Geochemistry and thermolysis of flavonoids. Science 197:767-769.
Niklas, K. J., and D. E. Giannasi. 1978a. Angiosperm paleobiochemistry of the Succor Creek Flora (Miocene) Oregon, USA. Amer. J. Bot. 65:942-952.
Niklas, K. J., and D. E. Giannasi. 1978b. The green fossils. Garden 1:6-11.
Niklas, K. J., R. M. Brown, Jr., R. Santos, and B. Vian. 1978. Ultrastructure and cytochemistry of Miocene angiosperm leaf tissues. Proc. Nat'l. Acad. Sci. USA 75:3263-3267.
Paeltova, B. 1975. Cretaceous angiosperms of Bohemia-Central Europe. Bot. Rev. 43:128-142.
Raven, P. H., and D. I. Axelrod. 1974. Angiosperm biogeography and past continental movements. Ann. Missouri Bot. Gard. 61:539-673.
Raven, P. H., and D. J. Axelrod. 1978. Origin and relationships of the California Flora. Vol. 72, University of California Press, Berkeley, Calif. 134 pp.
Ribereau-Gayon, P. 1972. Plant phenolics. Hafner, New York, N. Y. 254 pp.
Rosenfeld, C. R. 1980. Observations on the Mount St. Helens eruption. Amer. Sci. 68:494-509.
Smiley, C. J., J. Gray and L. Huggins. 1975. Preservation of Miocene lake fossils in unoxidized lake deposits, Clarkia, Idaho. J. Paleont. 49:833-844.
Smiley, C. J., and L. M. Huggins. 1981. *Pseudofagus idahoensis*, n. gen. et sp. (Fagaceae) from the Miocene Clarkia Flora of Idaho. Amer. J. Bot. 68:741-761.
Smith, P. M. 1976. The chemotaxonomy of plants. Edward Arnold, London. 313 pp.
Sweitzer, E. M. 1971. Comparative anatomy of Ulmaceae. J. Arnold. Arb. 52:523-585.
Takhtajan, A. 1969. Flowering plants: Origin and dispersal. (Trans. C. J. Jeffrey). Oliver and Boyd, London. 310 pp.
Thorne, R.F. 1976. A phylogenetic classification of the angiospermae. Pages 35-106 *in* M. K. Hecht, C. Ch. Steere and J. W. Wallace, eds. Evolutionary Biology, Vol. 9. Plenum Press, New York, N.Y.
von Rudloff, E. 1975. Volatile leaf oil analysis in chemosystematic studies of North American conifers. Bioch. Syst. Ecol. 2:131-167.
Wolf, J. A. 1971. Tertiary climatic fluctuations and methods of analysis of Tertiary Floras. Paleogr. Paleoclimatol. Palaeocol. 9:27-54.
Wolf, J. A. 1972. An interpretation of Alaskan Tertiary flora. Pages 201-233 *in* A. Graham, ed. Floristics and Paleofloristics of Asia and Eastern North America. Elsevier, Amsterdam.
Wolfe, J. A. 1979. A paleobotanical interpretation of Tertiary climates in the Northern Hemisphere. Amer. Sci. 66:694-703.
Zavarin, E., K. Snajberk and J. Fisher. 1975. Geographic variability of monoterpenes from cortex of *Abies concolor*. Bioch. Syst. Ecol. 3:191-203.

THE PALEOBIOCHEMISTRY OF FOSSIL ANGIOSPERM FLORAS PART II: DIAGENESIS OF ORGANIC COMPOUNDS WITH PARTICULAR REFERENCE TO STEROIDS

KARL J. NIKLAS[1] AND DAVID E. GIANNASI[2]

[1] Division of Biological Sciences, Cornell University, Ithaca, NY 14853
[2] Department of Botany, University of Georgia, Athens, GA 30602

Biochemical markers, referable to the diagenetic products of steroids and other organic compounds, have been isolated from fossil angiosperm tissues from the Clarkia Flora. Alterations in steroid chemistry may be caused by thermal, catalytic, and microbial activity. By one or more of these processes, reduction of olefinic double bonds and oxygenated groups may occur. Basically three categories of compounds have been detected in the cyclic alkane fractions that have been extracted from fossil tissues: (a) pentacyclic compounds with one five-membered ring, (b) tri-, tetra-, and pentacyclic compounds with six-membered rings, and (c) tetracyclic compounds with one five-membered ring. By comparing the steroid chemistry of fossil taxa to that of referable modern taxa, a preliminary evaluation of the original steroid composition of a fossil is possible. Combined geochemical and ultrastructural analyses of fossil tissues assist in the interpretation of diagenetic pathways. Leaf tissues isolated from sediments thought to be deposited in a nearshore lake environment characteristically lack a well-defined biochemical profile. In addition these leaves show evidence of ultrastructural modifications due to senescence, caused by leaf abscission and/or detached aging. As a result, the poor chemical profiles of many fossils may have been due to the degradation of organic compounds before burial in sediments rather than to extreme diagenesis.

The chemical composition of plant fossils provides the organic geochemist with data useful in the interpretation of diagenetic phenomena associated with postdepositional factors attending fossilization. Similarly, these data, if sufficiently extensive and placed within the context of living plant biochemical markers, provide the paleobotanist with relevant taxonomic and possibly phylogenetic insights. A biochemical marker is a compound whose structure can be interpreted in terms of previous biological origin. To be useful, the compound must have sufficient stability to survive long periods of time and sufficient complexity of structure (positional, and relative and absolution confirmation) to render it very distinctive. It can be a chemical originally present in the organism, but it is usually a related structure—the latter derived from the original compound by the operation of diagenetic processes in the formation of the sediment, and later by maturation processes in the ancient sediment. We can approach biochemical markers in one of two ways, either by the characterization of the organic chemicals within geologic materials or by the characterization of products of living organisms in terms of what might be expected in a sediment.

This paper deals with some of the biochemical markers found in Miocene fossil angiosperm leaves collected from the Clarkia Flora (cf. Smiley et al. 1975; Smiley and Rember 1979). Preliminary field work indicated that often distinctive chemical profiles from representative genera are preserved despite the hydrated condition of numerous specimens and evidence of various extents of necrosis. The survey of the Clarkia fossil deposits was prompted by earlier studies of material of a similar age from the Succor Creek Flora, Oregon. These studies indicated that chemical profiles of flowering plant fossils may be so distinctive as to permit direct comparisons with referable

Copyright ©1985, Pacific Division, AAAS.

modern taxa (Giannasi and Niklas 1977; Niklas and Giannasi 1977, 1978). The extensive preservation of individual taxa and the taxonomic diversity represented in the Clarkia deposits provide ideal conditions for a broad chemotaxonomic study, and it is the objective of this paper to present preliminary geochemical findings relevant to the interpretation of individual biochemical markers for each fossil species. Emphasis will be placed on diagenetic phenomena attending the fossilization of the Clarkia material as inferred by the biochemical profiles. As an adjunctive technique to biochemical analyses, references will be made to ultrastructural data which form the basis for determining the physical effects of diagenesis and fossilization phenomena (cf. Niklas et al. 1978; Niklas and Brown 1981; Niklas, Brown, and Santos, this volume).

ANALYTICAL TECHNIQUES

The collection and field methodology necessary for an adequate sampling of fossil material and subsequent chemical analyses is reviewed by Niklas (1979) and Niklas and Giannasi (1978). Specimens from freshly exposed rock surfaces are handled as little as possible and wrapped in aluminum foil or are immediately removed from the rock matrix and stored under nitrogen to prevent the oxidation of labile compounds. Three samples from each specimen or suite of specimens are required for proper documentation; these are for (1) morphologic, (2) ultrastructural, and (3) chemical analyses. Since the chemical techniques used are destructive, voucher specimens and/or immediate photography are essential.

The techniques used in the isolation and identification of chemical constitutents are given by Niklas (1976) and Niklas and Giannasi (1977, 1978). Fossil leaf material was removed from associated matrices, pulverized in a disc mill for 5 minutes (min) and extracted ultrasonically (3 times, 30 min each) in benzene/methanol (3:1, v/v). The preparation was then centrifuged (2,500 rpm, 2 min) and the supernatant removed, reduced in volume, and column chromatographed on neutral alumina (100 g) and eluted with n-hexane (75 ml) and benzene (300 ml). The combined eluates were evaporated and monitored by infrared spectroscopy (IR). The alkane fraction was obtained by preparative $AgNO_3/SiO_2$ thin layer chromatography, and was dissolved in iso-octane (20 ml) and heated under reflux with a molecular sieve for 74 hours (hr). The branched and cyclic alkane fractions (85% of the alkane fraction, 0.25% by weight of leaf material) were obtained by washing the sieve in an all-glass Soxhlet. All GLC analyses were performed on a Perkin-Elmer equipped with a flush heater and flame ionization detector; the columns were (a) $200' \times 0.01''$ Apiezon L grease coated and (b) $150' \times 0.01''$ coated with 7-ring polyphenylether. All analyses were isothermal at $255°C$ with a helium flow rate of 2 ml/min. High resolution MS used a modified MS-9. Co-injection of standards, the use of polar and non-polar phases on GC and MS were collectively used to assign tentative structures to various steranes and triterpanes.

CYCLOALKANE AND TETRAPYRROLE CHEMISTRY

Cycloalkanes

The critical stages of biochemical identifications of taxa are based upon steroid chemistry. Most tricyclic, tetracyclic, and pentacyclic terpanes found in our geologic samples are present as saturated hydrocarbons. In a few cases some oxygenated compounds have been found (betulin, allobetulin, oxyallobetulin, friedlin, and ursolic acid). No saturated hydrocarbon steroids have been found in any living organisms. Thus, barring the unlikely possibility that there existed a reductive enzyme system in the plants found as fossils, the steranes and triterpenoids found in our samples must be the result of diagenesis. Various alterations in steroid chemistry may be caused by thermal, catalytic, and microbial activity. By one or more of these processes reduction of olefinic double bonds and oxygenated groups may occur, as well as decarboxylation by bacterial and/or thermal activity. Diagenesis may also cause skeletal rearrangements. Thus, we must not only look for reduced steroids of known structure but also new types of structures.

Table 1 lists some of the major orders of vascular plants, as well as some non-vascular plant groups, encountered in the fossil record. The cycloalkane distribution as inferred from the reduction of known steroids for each taxon is also given. A number of dicotyledon orders show the same cycloalkane distribution (e.g. oleanane and lupane are found in the Myrtales, Hamamelidales, Rosales, Geraniales, and Fagales, while oleanane is found in 10 out of 17 vascular plant groups). Similarly, stigmastane is known from vascular (Leguminales) and non-vascular plants (Chlorophyceae).

The retention characteristics of some known cycloalkane standards (on a $150' \times 0.01''$ coated column with 7-ring polyphenylether) referable to major steroids in plant fossils are shown in Fig. 1. The retention character of any particular compound is based primarily upon molecular structure and is not necessarily reflected by molecular weight (e.g. adeantane, a 29-carbon numbered cycloalkane, has a longer retention time than lanostane and onocerane III, both C_{30} compounds). From these data it appears that for molecules with the same weight the retention time (RT) increases with increasing numbers of fused rings, and that molecules with 6-membered rings have longer RT than those with one 5-membered ring. Based upon these cycloalkane compounds, extracted from the organic material of angiosperm fossils, it is not possible to identify many orders. More detailed analyses of fossil chemical profiles indicate, however, that particular compounds such as triterpenoids are diagnostic of taxa below the order level. For example, of the genera analyzed from the Pinaceae (e.g. *Picea*, *Pinus*), compounds are present having the molecular structure shown in Fig. 1A (e.g. for *Picea* compounds are found with a m.p. = 307, 3β-OCH_3, 21β-OH, Δ^{14}; m.p. = 276, 3α-OCH_3, 21α-OH, Δ^{14}; and m.p. = 194, 3α-OCH_3, 21β-OH, Δ^{13}). Similarly, in *Alnus* spp. a compound with m.p. = 198 (12β, 20(S)-OH, 3-oxo, $\Delta^{25(26)}$) is commonly encountered having a structure shown in Fig. 2B. Some *Betula* spp. also show compounds of the molecular structure shown in Fig. 2B; however, these are identifiable as Folientriol (m.p. = 196, 3α, 12β, 20(S)-OH, Δ^{24}), Folientetrol (m.p. = 168, 3α, 12β, 17α, 20(S)-OH, Δ^{24}), and Triterpene B (m.p. = 197, 12β, 20(S)-OH, 3-oxo, Δ^{24}) diagenetic products. Extant and fossil representatives of some of the legumes reveal molecular structures shown in Fig. 2C, while taxa of the Rosaceae often show Pomolic acid (m.p. = 301, 3β, 19α-OH, Δ^{12}, 28-COOH) (Fig. 2D). In some instances chemical characterization of cycloalkanes can be used to resolve the generic level. For example, distinctive sesquiterpenoid derivatives have been isolated from fossil *Ulmus*. These compounds all have in common a basic hydroxycadalene skeleton. In addition, *Ulmus* spp. release lignans referable to thomasic and thomasidioic acids and lyoniresinol derivatives.

Cycloalkanes play a critical role in the interpretation of diagenetic phenomena in addition to providing phytochemical limits to the generic status of a fossil. Basically three categories of molecular structures are relevant to the definitive analysis of cyclic alkane fractions extracted from geologic samples: (a) pentacyclic compounds with one 5-membered ring (e.g. lupane), (b) tricyclic, tetracyclic, and pentacyclic compounds with only 6-membered rings (e.g. grammacerane), (c) tetracyclic compounds with one 5-membered ring (e.g. lanostane and all the steranes) (see Fig. 1 for molecular structure and GC-retention characteristics). In these cyclic systems, fragmentation must involve the breakage of two or more carbon-to-carbon bonds. Thus, some stability is conferred on these molecular species, and the most abundant molecular form extracted is usually referable to the saturated parent molecule. Other compounds resulting from the fragmentation of the parent molecule have concentration distributions determined by the susceptibility of various C–C bonds to cleavage. Since the cleavage of any particular bond depends upon the activation energy of the bond and the stability of the positive ions and neutral fragments, theoretical predictions of these two factors can provide for an estimate of the fragmentation patterns of the parent molecule. In general, the following order of carbon-carbon bond cleavages is cycloalkanes may be predicted: $-CH_2-CH_2- < -CHR-CH- < -CR_2-CH_2- < -CHR-CHR- < -CR_2-CHR- < -CR_2-CR_2$, where R is some group other than H. Steroids and triterpenoids, by virtue of their functional groups, give rise to fragmentation patterns generally not encountered by their reduced cycloalkane

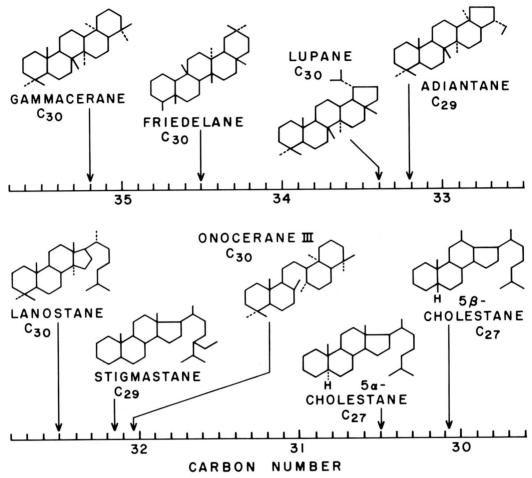

Figure 1. The retention characteristics of authentic steranes and triterpanes on a 7-ring polyphenylether capillary column. For further details, see text.

counterparts. This is particularly important in the interpretation of the former's high intensity ion fragmentation patterns seen in mass spectra, e.g. McLafferty rearrangement in ketotriterpenoids and steroids. In the mass spectra of the cyclalkanes lupane and adiantane, isolated from Clarkia fossil material, the most characteristic (intense) peaks for each is at m/e 191. Lupane shows a P-43 peak that is characteristic of compounds containing an isopropyl group.

The branched and cyclic alkane fraction isolated for each genus recovered at the Clarkia localities was analyzed by means of combined gas chromatography–mass spectroscopy. Compounds such as the isoprenoid alkanes farnesane, pristane, and phytane and a predominance of tricyclic, tetracyclic, and pentacyclic cycloalkanes, and perhydrocarbons were recoverable from most taxa. In general, the cycloclkane distributions for the genera studied were consistent with those given (at the ordinal level) in Table 1. The presence of ergostane and stigmastane in some samples is consistent with an algal source for some organic material in the sediments. Transmission electron microscopy of some angiosperm leaf tissues reveal epiphytic unicellular and filamentous algae.

Tetrapyrrole Pigments

Several homologous series of porphyrins are demonstrable in specimens of *Betula*, *Platanus*, and *Hydrangea*. Deoxophyllerythractioporphyrin (DPEP) and etioporphyrin (EP) were detected in

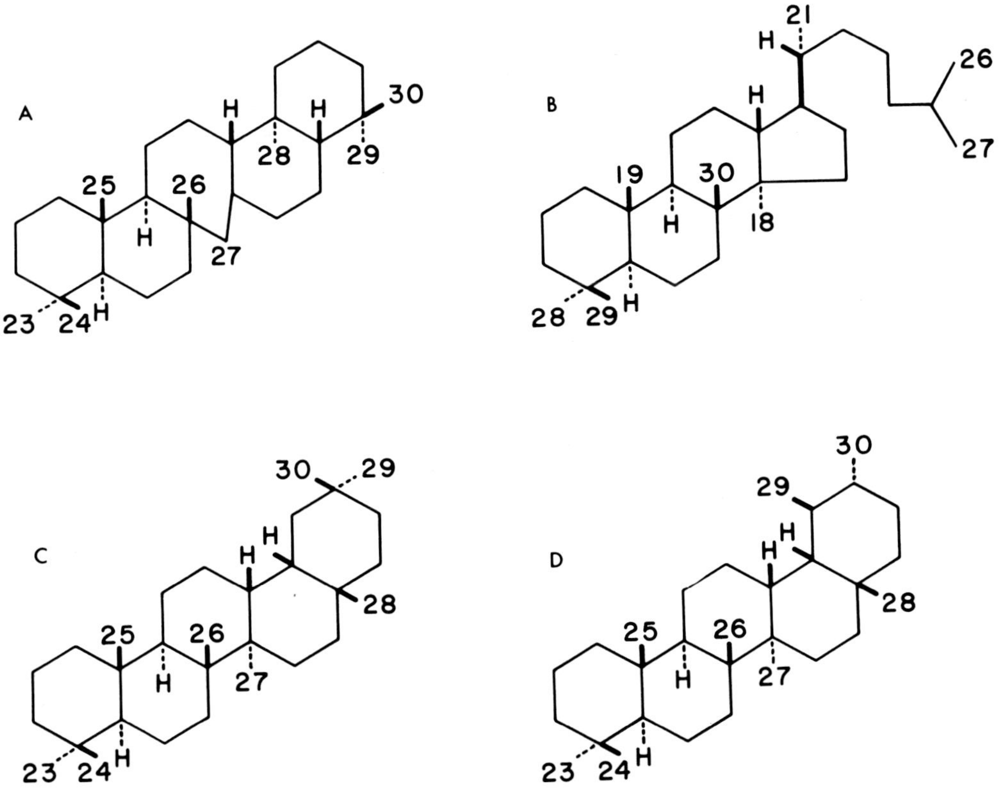

Figure 2. Molecular configurations of various cycloalkanes isolated from fossil leaf tissues of Miocene age.

leaves of *Castanea* and *Persea*. Mass spectra of *Castanea* and *Persea* (acetone-methanol extracts) show m/e peaks at 476 (DPEP) or m/e 478 (EP). Many specimens of leaf tissue have pheophytin *a* (peaks at 278, 459, 461, 516, 592, 870). The m/e 592 peak most probably represents a loss of 278 from pheophytin *a* (mass 870), which is probably phytadiene. The 516 peak is thought to be the result of the loss of the 10-carbomethyoxyl group producing ketene. A major red absorption at 668 nm suggests that metallochlorins are not present as significant components of the extraction.

Other non-tetrapyrrole pigments, such as flavonoids, are discussed elsewhere (Giannasi and Niklas, this volume).

DIAGENESIS AND SEDIMENT MATURATION

Plant material deposited in aqueous environments is subjected to various chemical and physical alterations before burial and post-burial sediment maturation. In some cases microbial decomposition occurs in such significant amounts as to obliterate the chemical profile diagnostic of the plant substrate. Anaerobic microbial activity and sulfate reduction results in extensive hydrolysis, decarboxylation, and deamination of organic material, which eventually results in the formation of n-alkanes, mono-methyl-substituted long chain alkanes, acyclic isoprenoid alkanes, cyclic diterpanes, and cycloalkane derivatives of triterpenoids, steroids, and the degradation of chlorophyll into phytol esters and porphyrins. These compounds are limited in use as biochemical markers at the generic or species level; however, they demonstrate control of positional isomerism in that the carbon atoms are always attached to one another in a precisely determined way. For example, the geologically occurring steranes have their biological counterparts in the steroids and relative stereochemical transformations may be used to reconstruct the precise steroid from which the sterane was derived. If, however, sediment maturation proceeds beyond this point, the cyclic molecular structure of the sterane may be broken, thus making the reconstruction of the parent molecule impossible. Similar obliteration of biochemical markers diagnostic of the generic or even familial ranks may occur even in a relatively dry environment of deposition. Hollerback (1978), in a study of a beech forest leaf litter, has shown that a large variation in the amount of extractable alkanes occurs even in three year old material; e.g. the carbon preference index (CPI), which measures the ratio of the odd- and even-chain lengths of alkanes, varies in green leaves (7.6), recently shed leaves (23), old brown leaves (19.5), and soil samples (5.4). A comparison of these results for different species indicates extreme variability, but in general there exists a rapid mobilization of organic constituents in early diagenetic stages from living matter into the soil. Similarly, phytadienes from green and recent brown leaves are very unstable, disappearing almost entirely in old leaves, and cannot be detected in soil samples.

Assuming that such phenomena as leaf detachment, either by natural senescence or mechanical insult, and microbial degradation have occurred before the leaves found as fossils in the Clarkia Flora were fossilized, then there exists a maximum state of biochemical preservation which does not coincide with an unaltered phytochemical profile for the species level. Such compounds as chlorophylls, steroids, and polyunsaturated acids are unlikely to survive disappearance of α-tocopherol, an antioxidant in plant material (Metzner 1973), from recently shed leaves suggests that saturation due to oxidation of chlorophyll, steroids, and polyunstaurated acids will result in the rapid disappearance of chlorins, steroids, and saturated acids respectively.

Tissues from the Clarkia area contain geochemical constitutents that serve as independent indicators of paleoenvironmental conditions, as well as potential correlates with the potential for the preservation of specific cytologic details. Pristane to phytane ratios (Pr/Ph) and chlorin geochemical profiles are particularly sensitive to redox potentials. Lipid compositions (in particular, unsaturated fatty acids) and cellulose-hemicellulose are good indicators of geothermal and microbial degradation. Pr/Ph ratios determined for the Clarkia P-33 tissue ssamples are equal to or less than unity (Niklas and Brown 1981) and are interpreted as indicative of anoxic paleoenvironmental conditions. The C_{20} isoprenoid skeleton of phytane has a greater potential for preservation under anoxic conditions than does the C_{19} skeleton of pristane. Pheophytin a and dihydropheophytin, both byproducts of chlorophyll degradation, yield dihydrophytol upon hydrolysis. However, oxidation of dihydrophytol produces phytanic acid, which converts to phytane. Similarly, P-33 specimens are relatively rich in biolipids, carotenoids, and chlorins, which is consistent with an anoxic interpretation. In addition to a high organic carbon content (14.2% dry weight), the associated inorganic sediments of the site P-33 have a high percent of HCl-extractable sugars and amino acids. The chlorin content of these tissues and the presence of pheophytin a indicate that chlorophyll degradation is incomplete. These data are interpreted as indicative of mild geophysical gradients during sediment maturation. Tissues from site P-34 (*Castanea* and *Persea*), which are thought to be lake-shore or deltaic in origin, have Pr/Ph ratios of unity and show chlorophyll byproducts reflecting severe oxidative degradation (DPEP and EP). These compounds and the Pr/Ph ratios indicate an oxic or alternating oxic/anoxic environment of deposition. A lake-border or deltaic **environment could have been associated with episodic oxic-anoxic phases, which in turn may have been responsible for the altered chlorophyll profiles in tissues and the low organic carbon** content of the associated inorganic sediments (0.78% dry weight). While thermal diagenesis is also capable of yielding the same geochemical profiles as are reported for site P-34, the presence of native cellulose, albeit altered to some extent, indicates that the thermal gradient was relatively mild, since cellulose is a sensitive indicator of high temperatures.

The presence of tanniferous compounds in most of the P-33 tissues is of particular interest. Some tissues fixed in glutaraldehyde (e.g. *Quercus*) appear to have a "negative staining" in their chloroplasts which is similar to tissues fixed in tannic acid (Niklas and Brown 1981). Tannins

TABLE 1. PALEOCHEMOTAXONOMIC DISTRIBUTIONS OF CYCLOALKANE DISTRIBUTIONS IN REPRESENTATIVE ORDERS OF PLANT FOSSILS

Order	Cycloalkane Distribution	Order	Cycloalkane Distribution
Sapindales	Oleanane	Umbellales	Ursane
Caryophyllales	Oleanane	Rutales	Ursane
Ariales	Oleanane		Arborane
	Ursane	Leguminales	Oleanane
Myrtales	Oleanane		Onocerane
	Lupane		Ergostane
Hamamelidales	Oleanane		Stigmastane
	Lupane	Coniferales	Hopane
Rubiales	Oleanane		Ursane
	Ursane	Filicales	Fernane (E:C-*friedo*-Hopane)
Rosales	Oleanane		Hopane
	Lupane		Adiantane (30-*nor*-Hopane)
Geraniales	Oleanane	Sphagnidae	Ursane
	Ursane		Taraxerane
	Lupane	*Class*	
Fragales	Oleanane		
	3,4-*seco*-Oleanane	Chlorophyceae	Ergostane
	Ursane		Stigmastane
	Lupane		Sitostane
	Glutane (D:B-*friedo*-O)	Basidiomycetes	C_{31} Lanostane
	Friedelane (D:A-*friedo*-O)		Lanostane
Celastrales	Friedelane		Ergostane
Utricales	Lupane		Chlorestane
Rhamnales	Lupane	Ascomycetes	Lanostane
	abeo-Lupane		Chlorestane
Myricales	Taraxerane		Ergostane

occur naturally in many plant tissues and bind strongly to proteins and polysaccharides. Similarly, by complex formation, tannins may inactivate, to varying degrees, enzymatic processes. The ultrastructural appearance of some tissues and the presence of unusually high concentrations of tannins are interpreted as evidence that these tissues may have undergone either autofixation through the necrotic release of cytoplasmically compartmentalized tannins, or were infiltrated gradually by water supercharged with tanniferous products. Peaty or swamp conditions generate high tannic acid concentrations, which inhibit microbial activity. Condensed tannins liberated by the leaching of wood by water, as well as aldehyde-rich lignin degradation byproducts, may have been responsible for the specific states of preservation seen in some tissue samples.

PALEOBIOCHEMICAL AND ULTRASTRUCTURAL CORRELATIONS

The ultrastructural states of preservation seen in the Clarkia material have been presented in detail by Niklas et al. (this volume) and Niklas and Brown (1981). Similarly, the paleobiochemical profiles of referable material used for electron microscopy examination have been reviewed by Giannasi and Niklas (this volume), Niklas and Brown (1981), and the present report. Various correlations among the chemical and structural data provide a basis for explaining some of the apparent inconsistencies observed in our studies. In particular, the variability in the fidelity of biochemical profiles seen in fossil specimens from a single taxon is apparently due to pre-depositional changes in leaf chemistry.

Tissue samples from site P-40 have, in general, poor chemical preservation interpreted to be the result of a relatively oxygenated environment of deposition. In addition, these tissues do not show ultrastructural details as well as their taxonomic counterparts preserved in site P-33, which for biochemical and stratigraphic reasons is thought to represent lake bottom sedimentation and low oxygen concentrations (Niklas and Brown 1981).

Ultrastructural studies of leaf tissues from site P-40 reveal membrane systems that characteristically show a negative staining and chloroplast infrastructure diagnostic of various extents of pre-depositional dehydration, senescence, or aging due to premature removal from the living plant (Niklas et al., this volume). Thus, alterations in the biochemistry of the leaves before burial in the Clarkia Lake sediments may account for the total lack of some chemical components (e.g. steroids, flavonoids, and chlorophyll derivatives) in the P-40 specimens which are characteristically present in the P-33 material. Chloroplast ultrastructure is particularly useful in assessing the predepositional physiologic status of the fossil leaves. Dehydration by air-drying of modern leaf tissues has been shown to result in a profusion concertina-like folding of palisade mesophyll cells along their longitudinal axes (normal to the paradermal plane), while mesophyll cells appear to invaginate and structurally collapse (Niklas et al., this volume). In addition to these cell wall changes, the protoplasm of dehydrated tissues changes: (1) golgi bodies and well defined endoplasmic reticulae disappear, (2) the tonoplast membrane collapses or ruptures, (3) the outer chloroplast membrane vesiculates and/or ruptures and the stroma-grana infrastructure of the chloroplast becomes disorganized, (4) mitochrondria become indistinct, and (5) the plasmalemma is lost. Senescence of leaves brought about by normal aging (e.g. autumal leaf drop), while showing some of these features, is distinctively different in ultrastructural appearance from the effects of dehydration.

The bulk of the leaves preserved at site P-40 show ultrastructural characteristics compatible with those resulting from dehydration and/or senescence. Based upon ultrastructural examinations the leaves preserved at this site most probably had very poor physiologic viability well before their burial in lake shore or shallow water sediments. The apparent lack of such biochemical constituents as flavonoids and chlorophyll in these tissues (Giannasi and Niklas, this volume) is consistent with this observation. The intrataxonomic diversity seen in the chemical profiles of P-40 and P-30 tissue samples may, therefore, be explained by pre-burial vagaries in leaf biochemistry. The excellent ultrastructural and chemical preservation determined for the P-33 material may be due to a differential sedimentation of more viable leaves in these sediments, as well as the apparent anoxic environment of deposition. Collectively these pre- and postburial factors may account for the variability in preservation observed among the Clarkia sites.

LITERATURE CITED

Giannasi, D. E., and K. J. Niklas. 1977. Flavonoid and other chemical constituents of fossil Miocene *Celtis* and *Ulmus* (Succor Creek Flora). Science 197:765-767.

Hollerbach, A. 1978. Early diagenetic alterations of biogenetic organic compounds. Naturwissenschaften 65:202-204.

Metzner, H. 1973. Biochemie der Pflanzen. F. Enke, Stuttgart.

Niklas, K. J. 1976. Chemical examinations of some non-vascular Paleozoic plants. Taxon 28: 505-516.

Niklas, K. J., and R. M. Brown, Jr. 1981. Ultrastructural and paleobiochemical correlations among early fossil leaf tissues from the St. Maries River (Clarkia) Area, northern Idaho, U.S.A. Amer. J. Bot. 68:332-341.

Niklas, K. J., R. M. Brown, Jr., R. Santos, and B. Vian. 1978. Ultrastructure and cytochemistry of Miocene angiosperm leaf tissues. Proc. Nat'l Acad. Sci. 75:3263-3267.

Niklas, K. J., and D. E. Giannasi. 1977. Flavonoids and other chemical constituents of fossil Miocene Zelkova (Ulmaceae). Science 196:877-878.

Niklas, K. J., and D. E. Giannasi. 1978. Angiosperm paleobiochemistry of the Succor Creek Flora (Miocene), Oregon, U.S.A. Amer. J. Bot. 65:943-952.

Smiley, C. J., J. Gray, and L. M. Huggins. 1975. Preservation of Miocene fossils in unoxidized lake deposits, Clarkia, Idaho. J. Paleont. 49:833-844.

Smiley, C. J., and W. C. Rember. 1979. Guidebook and road log to the St. Maries River (Clarkia) fossil area of northern Idaho. Idaho Bur. Mines Geol. Inf. Circ. 33. 27 pp.

INTERPRETATION OF CO-OCCURRING MEGAFOSSILS AND POLLEN: A COMPARATIVE STUDY WITH CLARKIA AS AN EXAMPLE

JANE GRAY
Department of Biology, University of Oregon, Eugene, OR 97403

The Clarkia Lake beds provide an unexcelled opportunity to compare the quantitative and qualitative relationship between pollen and plant megafossils recovered from the same sequence of beds and from the same strata within that sequence. This type of 1:1 comparison, involving close-spaced samples, is particularly valuable in view of the different values claimed for plant megafossils and pollen in paleoenvironmental interpretations.

Taxic inventories indicate that most taxa at Clarkia are represented by co-occurring megafossils and pollen, although the taxonomic resolution is much higher for megafossils than for pollen. In addition to pollen and spores, Clarkia microfossils also include remains of algae and fungi. The fungi are a particularly potent source of paleoenvironmental information that has not been exploited in Tertiary studies.

Quantitative comparisons of pollen and megaremains at Clarkia show both similarities and differences, although the records independently tend to reinforce one another in what appear to be some of the dominant taxa in the catchment area. Quantitatively useful tallies of megaremains are not possible, however, from the close-spaced samples used by pollen analysts; thus information may be lost. Irregularities in the megafossil tallies also suggest the possible influence of sedimentological control on accumulation of leaf remains, which limits the usefulness of quantitative megafossil information in interpreting the source vegetation.

Both methodologies have limitations as well as promise for paleoenvironmental interpretations. Their promise is best exploited when they are used as complementary resources, although pollen data recovered from plant-megafossil beds may be as biased by the immediate shoreline vegetation as the megaremains from the same environment appear to be.

Analysis of samples from the Clarkia leaf locality shows these Tertiary rocks to be extremely rich in the remains of organic microfossils. Samples of a fraction of a gram yield abundant pollen, spores, fungal remains, and microscopic algae, enabling close-spaced sampling from which it is possible to secure details of vegetational history at a scale that would be impossible for most Tertiary localities in the Pacific Northwest. This preliminary account discusses the results of pollen analysis of approximately 360 cm (12 feet) of section (Unit 2) at approximately 30-cm intervals, beginning from the base of Unit 2 to within about 15 cm of the base of Unit 3, in comparison with cooccurring megafossils. Unit 2 is the principal part of the Clarkia sequence at P-33 with both plant megafossils and co-occurring microfossils. Unit 1 is barren of megafossils (Smiley and Rember 1981, Fig. 17-5); I have recovered no pollen from Units 4 and 5.

Although there is increasing interest in using Quaternary pollen and megafossils from the same sediments to provide varied types of complementary information not obtainable from each alone, few such studies have been undertaken with Tertiary materials. Clarkia, however, provides an opportunity to compare the pollen/spores of identified taxa with the co-occurring megafossils recovered from the same unit of sedimentary rock to see how paleoecological reconstructions

Copyright ©1985, Pacific Division, AAAS.

might benefit. Because the mega- and microfossil information was accumulated in close conjunction involving the same or essentially the same strata in Unit 2 at one locality (P-33), it is feasible to look at both the qualitative and quantitative similarities and differences as a possible basis for providing some generalities about the type of "blind spots" that might be expected where only one type of remain is the sole source of information on floristic and climatic history.

In the face of obvious qualitatative and quantitative similarities and differences in the assemblages at P-33, I have used this opportunity to reconsider some theoretical questions related to interpretation of Tertiary megafossil and pollen assemblages that doubtless relate in part to the different circumstances of their accumulation. I have attempted as well to reconcile in a preliminary way the similarities and differences in the two records. I have not attempted a detailed interpretation of the Clarkia pollen flora, nor am I specifically concerned with Clarkia except as a case study of the type of problems arising in environmental interpretation of megafossil and pollen assemblages from the same strata. Although there are a number of leaf-bearing outcrops in the depositional basin of Clarkia Lake, many Tertiary fossil localities consist of isolated outcrops of the type found at P-33, whose contemporaneity with other nearby outcrops is difficult to prove. The megaremains and co-occurring pollen at P-33, in my opinion, thus serve as an acceptable model for addressing problems in interpretation of all Tertiary mega- and microfossil assemblages.

I also see this paper as a vehicle to encourage paleobotanists to ask the types of questions that pollen analysts have asked for many years, and to begin accumulating the hard data needed to put paleobotany on the same type of ecologic basis that pollen analysis enjoys. Only then will it be possible to fully understand the significance of differences and similarities in the records, and to use to their full extent the complementary virtues of each discipline in environmental interpretation.

PLANT MEGAFOSSILS AND FOSSIL POLLEN: THEIR VALUE FOR ENVIRONMENTAL INTERPRETATION

The taphonomic variables that have been suggested to bear on the accumulation of pollen and varied types of megafossils are sufficiently different that they have led to some extreme generalizations about the environmental significance of each type of assemblage. The occurrence of megafossils is said to be "always more or less accidental" with remains mostly from the marginal limnic/fluviatile vegetation "which is frequently climatically indifferent" (Faegri and Iversen 1975:123). Or "land plant megafossils tend to be parochial; . . . certain environments are well represented, most are not represented at all. . . . land plant megafossils are limited in their ability to explicate patterns of community evolution" (Clapham 1970:412-413). The derivation of meaningful quantitative information from megafossils is also questioned. It makes no sense to many paleoecologists to quantify either together or singly structures that are as disparate in size, shape, weight, and weight-area ratio as cone scales, cones, needles, leaves, stamens, fruits, bud scales, bracts, seeds, wood and so on, or that are produced in such different quantities by plants and apt to be deposited largely with regard to their sensitivity to transport and sedimentation, and to treat them as one would a pollen sum (Faegri and Iversen 1975; Watts 1978)—as a basis for providing information about the abundance of their source plants in the vegetation. Many would agree with the following negative assessment of the potential for arriving at a quantitative relationship between plant megafossils and vegetation: ". . . it is not likely that any direct relation will ever be found to link the abundances of the source vegetation component species and their representation in an allochthonous fossil assemblage. The factors affecting litter production alone are sufficiently numerous to introduce significant fluctuations in the source material, which, when mechanisms of transport, deposition, and diagenesis have disproportionately sorted the plant remains, produce such a wide range of possible abundances within the fossil deposit that direct interpretation becomes meaningless" (Spicer 1981:63). One might suspect that such hyperbole was also expressed by many in the early days of pollen analysis with regard to the possible quantification of pollen rains and source vegetation!

Today, after many years of study, pollen analysts know even more about the tremendous number of variables that affect the entrance of pollen into depositional basins and that influence their final sedimentation, many parallel, of course, and of complexity equal to those that affect allochthonous megafossil assemblages. Still the foundations of pollen analysis as established by von Post not only remain intact, but in many respects are strengthened by knowledge of how to deal with many of the variables, even though some of the original assumptions have been challenged. The often regional aspect of pollen rains, representing potentially all environments regionally present, the great abundance and "uniformity" of pollen production, and the "mixing" or "homogenization" in the atmosphere in direct relation to the number of source plants together with the sifting by gravity of pollen over lake and bog surfaces are all assumptions that no longer remain wholly intact, although they form the basis for pollen analysis and are true "in principle." Quantification of pollen, expressed in pollen diagrams to provide an indication of the character of the surrounding vegetation, also makes sense to most paleoecologists because pollen is "comparable" in size, shape, and weight. Thus it follows to many that pollen assemblages are always a far better source of vegetational information and climatic inference than the "climatically indifferent" and "accidentally" accumulated megafossil remains with which they co-occur.

On the basis of such generalities, it might be expected that allochthonously accumulated megafossil and pollen assemblages, as at Clarkia, would scarcely approach one another in taxic diversity or in the diversity of environments represented; that a comparison might point up primarily the varied incongruences between these two sources of information; and that the pollen record would provide an incomparably more reliable guide to the vegetational history and environments of the Pacific Northwest in the middle and late Tertiary.

In recent years, however, there has been renewed and increasing interest in analyzing Quaternary megafossil remains, independently (Dickson 1970; Burrows 1974) or in conjunction with pollen assemblages, because "pollen has limitations which can sometimes be overcome by the combined uses of pollen and macrofossil studies" (Watts 1978:53-54). These limitations have to do principally with the capacity of pollen for transport, often for extended distances from its source vegetation, as well as with the level of taxonomic resolution that is possible for what are pollen-morphologically homogeneous taxa at genus and family level. These two limitations are equally valid whether one is dealing with assemblages from the late Tertiary or the Quaternary. Quaternary fruits and seeds, however, can often be identified to living species; Tertiary leaf fossils are, of course, far less precisely identified, and there are problems of identification within families and genera that are equivalent to those faced by the pollen analyst, though less numerous.

The principal difference in these parts of the geologic column, insofar as the study of co-occurring mega- and microfossils is concerned, is that in the present North Temperate region, Quaternary megafossils are largely the fruits and seeds of obligate aquatics and wetland plants of shoreline or nearshore that have accumulated autochthonously or essentially so in the limnic or bog deposits in which they are found. The occurrence of clearly allochthonous remains from "upland" (i.e. nonlocal) species are always extremely limited in limnic deposits, although needles, seeds, cones and cone scales, and similar fragments of other plants and wood may occur with some frequency. Leaf floras, excluding leaf fragments, tend to be much less common. Exceptional examples are provided by leaves apparently accumulated in proglacial Lake Monongahela (Gillespie and Clendening 1968) and tiny leaves of arctic and alpine species occasionally found abundantly in late-glacial deposits in North America and Europe (Argus and Davis 1962; Baker 1965; Watts 1967, 1979). Fluvial sediments, on the other hand, may have abundant seeds and other remains representing the upland, but still local, forest vegetation (Delcourt and Delcourt 1977). The assemblage described by Delcourt and Delcourt includes heavy seeds of upland plants that appear to have low transport potential, are produced in relatively small numbers, and are seldom found in limnic deposits.

Tertiary megafossil accumulations are overwhelmingly leaf remains that have accumulated through transport just as the pollen with which they co-occur, thus largely complicating the issue of what their presence and abundance means when they are found in limnic deposits. The transport of large plant structures, even those of terrestrial plants obviously not deposited *in situ*, may be minor, however. Thus "allochthonous" as I use it here is a matter of degree. Fine-scale allochthony of the type that operates for both seed/fruit assemblages and leaf accumulations by and large does not bring remains from regions climatically or biogeographically different from those in which the fossils have accumulated. However, the character of the deposit may critically influence the source of the remains. Peat deposits largely accumulate the structures of plants growing on their surface and are obviously more "autochthonous" than leaf-bearing limnic deposits where buoyancy in wind and water plays at least some part in shaping the character of the fossil assemblage. Pollen, on the other hand, *may* have accumulated from hundreds of miles outside of the range of the majority of the fossil assemblage, although the source *may* also be essentially autochthonous.

Examples of incongruences, both qualitative and quantitative, between these two sources of evidence from the fossil and subfossil record, should make one cautious about independent interpretation of either record. Matthews (1975), for example, found autochthonous remains of birch (including samaras, bracts from pistillate "cones" and some leaf fragments) in Quaternary deposits from the Alaskan North Slope, yet pollen of birch, a notoriously large pollen producer in the wind-pollinated category, was scarcely represented in the same deposits. The megafossils appeared to provide direct evidence for the presence of birch; the pollen could be interpreted to indicate that it was windborne into the site of deposition possibly from some distance, although a limited pollen production in an environmentally marginal habitat is an alternative possibility (Matthews 1975). However, in these same deposits, pollen and spores, although less precisely identified than the co-occurring megafossils, exceed them in taxonomic richness, including families and genera not represented by megafossils. Watts and Winter (1966) found that at Kirchner Marsh, the Pleistocene record included 80 identified species of megafossils (largely aquatics), but fewer than 20 of these taxa could be determined to genus or family from their pollen. Grüger (1972), on the other hand, found only a comparatively small proportion of megafossils (predominantly water and swamp plants, together with *Picea* and *Larix* needles, young cones of *Picea* and *Populus* bud scales) relative to the large number of identified pollen taxa (see his Figs. 2, 4, 5) in Wisconsinan samples from Illinois, and change in sediment type from gyttja to clay resulted in the loss of seeds but not pollen. Boulter and Chaloner (1970) from preliminary study of late Tertiary deposits in England, where both mega- and microfossils had been allochthonously accumulated, found that the 6 taxa represented by wood or leaves were all represented by pollen. At least 14 genera and taxa of two additional families were represented only by pollen/spores, together with plant microfossils of unknown affinities. Gray (in Martin and Gray 1962), comparing pollen and megafossils from the Miocene Mascall and associated floras from east central Oregon, found a number of taxonomic discrepancies between them. In the Blue Mountains flora, for example, she found 18 pollen taxa, representing 47% of the total pollen assemblage, that did not occur among the megafossils. Independent tallies of pollen and leaves might also be interpreted to indicate considerable discrepancies in abundance of some taxa in the immediate vegetation. Łancucka-Środoniowa (1979) found even more remarkable disparities in Polish Miocene lacustrine and swamp deposits. Of 68 taxa identified from pollen and 111 identified largely from fructifications, fruits and seeds, only *Selaginella, Glyptostrobus*, Cyperaceae, Graminae, *Acer, Alnus, Betula* and *Carpinus* co-occurred as fossils.

If such examples are typical, it is clear that neither record is without limitations. There are sufficient disparities to suggest that neither should be interpreted within a framework that excludes the other if the objective is to gain a full picture of vegetation and paleoclimate. It is also clear that neither record has a monopoly on greater taxic diversity. Because the potential for transportation, sedimentation, and preservation of varied megaremains and pollen largely determines the abundance and taxic diversity of fossils recovered from sedimentary deposits, the quality of the mega- and microfossil records depends ultimately on such factors as the climatic zone of the depositional site and site size, on water depth, turbulence, chemistry and so on, on the specific structure and composition of the surrounding vegetation, and on the means (aerial, riverine, and so on) by which the remains have accumulated. With regard to preservation potential, Clarkia is a case in point. In oxidized Units 4 and 5 at P-33, leaf remains occur, but I found no pollen. On the other hand, Unit 1 at P-33 contains abundant pollen but no megafossils (Smiley and Rember, 1981, Fig. 17-5, note that Unit 1 is "barren"). At Clarkia, perhaps in part because of the excellent preservation of megafossils and their abundance, the diversity of identified megataxa greatly exceeds that of identified pollen and spores.

PLANT MEGAFOSSILS, FOSSIL POLLEN, AND SOURCE VEGETATION

Attempts to understand the variables and biases that might operate in the accumulation of fossil leaf assemblages are few. Even fewer are attempts to systematically examine the occurrence of leaves, fruits, wood and other megastructures in depositional basins, such as modern counterparts of the Miocene Clarkia Lake, to see how representative accumulating remains might be from other than a very limited set of adjacent riparian or lakeside communities. The question of what affects the opportunity for plant remains to occur in accumulating sediments has seldom been addressed in any type of controlled situation, involving varied vegetation types and varied plant structures. Studies relating to the transportation and preservation of plant remains (potential megafossils) in modern environments are necessary as a model or an "interpretive link" against which to assess the Clarkia fossils. It is necessary to disentangle variables related to: (1) preservation under a variety of circumstances; (2) aerial transport and sorting, which may be limited; and (3) aquatic transport and sorting, which might be more extensive, as in the case of pollen. Few studies attempt to quantify the relation between leaves, fruits, seeds, wood, and other structures accumulating in lakes and source vegetation. Thus we do not know whether it is possible to use megafossils as a basis for determining abundances of source plants. It seems self-evident, however, that the situation will be complex and that sorting prior to final deposition and differential preservation may be critical in shaping the character of a plant megafossil assemblage. Some understanding of these variables is essential in order to reconstruct the fossil vegetation and to understand the vegetational and climatic history of a region.

In contrast, numerous studies over many years have attempted to assess the varied factors that govern the occurrence of pollen in limnic deposits, and to quantify the relations between pollen and its source vegetation (see, for example, Faegri and Iversen 1975; Birks and West 1973; Moore and Webb 1978; Birks and Birks 1980 and references therein). Adding the whole spectrum of questions related to pollen production to those of transport, sedimentation, and preservation, it is clear that many problems relevant to pollen interpretation are still incompletely understood and examined. Nevertheless, the variety of both experimental and observational studies permits, at least in a general way, some feel for the limitations inherent in interpreting pollen assemblages in terms of the surrounding vegetation, and for the sensitivity of pollen in characterizing vegetation. These data accumulated from analyses of surface pollen rains leave little doubt that regional vegetation at biome level can be recognized by characteristic pollen assemblages (for example, Webb 1974; R. B. Davis and Webb 1975; Webb and McAndrews 1976; Delcourt et al. 1982). Pollen analysts are increasingly able to recognize particular plant communities (see, for example, Janssen 1966, 1973; Birks, Webb, Berti 1975; Birks 1976; Caseldine and Gordon 1978).

With pollen deposited in lakes, initial emphasis has been on problems of wind transport. There is increasing evidence, however, that quantities of pollen are delivered to lake deposits aquatically, and this may under some circumstances exceed that derived from an airborne source (McAndrews and Power 1973; Peck 1973; Crowder and Cuddy 1973; Tauber 1977; Bonny 1976, 1978; Crowder and Starling 1980; Starling and Crowder 1981).

One of the few holistic models to challenge the general concept of the airborne "pollen rain" homogenized in the upper atmosphere, and sifted uniformly by gravity over surface waters of lakes, is that proposed by Tauber (1965, 1967a,b, 1977). Tauber's model, in which pollen is trapped in depositional basins in a closed-canopy forest, identifies varied pollen components that may be delivered onto the lake surface from different vegetational sources and from different distances (see also Janssen 1966, 1973). Although the comparison of this model with the situation at Clarkia may be incomplete for varied reasons related to lake size, vegetation type and structure, and the aerodynamics within the Clarkia basin, the model has potentially interesting consequences in the interpretation of pollen samples taken from megafossil-bearing beds with their probability for very localized nearshore accumulation.

Tauber recognized that at least three components comprise the pollen rain delivered to various parts of a lake surface surrounded by heavy forest. The so-called trunk space component, from pollen of ground cover and canopy trees, travels laterally through the trunk space of the forest, is derived from vegetation within a few hundred meters of the depositional basin, and is deposited onto a lake surface within a few hundred meters of the edge of the forest. Depending on the size of the lake, as much as 80% of the pollen may be trunk space component.

That component that travels in low-atmosphere winds just above the canopy and is carried downward to the lake surface largely by eddy diffusion—the so-called canopy component—comes from distances estimated, depending in part on the density of the pollen, the topography, and the lake size, to be within 10 km. Depending on the distance from shoreline (and the forest margin) that the pollen rain is sampled, a sizable proportion of the canopy component may be from distances greater than 10 km and much of it may be more local in derivation (Tauber 1965: Figs. 5, 6). In a small water body, this component may comprise only a minor part or none of the pollen because, according to Tauber, it passes in horizontal wind currents over the lake to be deposited in the vegetation on the opposite side. In a large lake, 70-80% of the pollen rain may be from the canopy component, representing pollen of more regional aspect than trunk space pollen.

The most regional component may come into depositional basins from any distance being washed in from the atmospheric scrubbing by rainfall. This component may be very limited, approximately 5-10% or less of the pollen fallout.

Because each component in this model comes from different source vegetation, each may add a very different pollen assemblage to limnic deposits; the relative amounts of trunk space and canopy components are critical to determining the size of the area from which the pollen has been derived. The variables bearing on the size of each component to a depositional basin will, however, be a function of: (1) the structure and density of the vegetation, both regionally and at lake side, where vegetation margining the lake may be an effective filter to pollen of the trunk space component (Tauber [1965] has suggested, to further complicate the issue, that "area" represented by pollen fallout may change through time from "local" to "regional" by change in density of the forest cover without vegetation change); (2) the rainfall pattern of the region; (3) the size and morphometry of the lake; (4) the regional windspeed; (5) the density of the pollen grains and so on. On the basis of sampling, involving a small lake (200 x 100 m) in Danish closed-canopy temperate deciduous forest, Tauber (1977) has largely confirmed the predicted theoretical model with regard to source of airborne pollen. Thus he found that about 5% of the pollen delivered to the lake surface was rainout component, 35% was canopy component, and about 60% was delivered from within the trunk space. Careful sampling by Currier and Kapp (1974) in and adjacent to a small lake surrounded by temperate hardwood (oak, poplar, maple, plus hickory and linden) and spruce-larch forests confirmed the trunk space and canopy components as contributing to the pollen deposition in the lake. However, possibly related to differences in structure of the local vegetation and in local aerodynamics, they found that the more regional canopy component (their horizontal drift component) delivered more pollen than the trunk space component, thus mitigating to some extent the gross overrepresentation of pollen of extremely local plants found by Tauber.

In individual circumstances involving either a small water body with a limited catchment area and limited or no stream entrance, or a localized embayment of a larger water body (which corresponds to the circumstances for deposition of leaves and pollen at P-33 [Clarkia]), most of the pollen *may* be airborne and much of it as local as the predicted source of megafossils. In an area of some topographic relief (as at Clarkia in the Miocene), surface runoff during times of heavy rain, even in the absence of local inflowing streams, might recruit some previously deposited pollen and spores from catchment-basin soils and forest floor vegetation. In a heavily forested region, however, that pollen is as likely to have had a local source as the airborne pollen.

Although lateral (and vertical) variations in pollen of lake deposits may be reduced by mixing of surface sediments during easonal water circulation (cf. M. B. Davis 1968, 1973), homogenization is incomplete. R. B. Davis et al. (1969) and Sønstegaard and Mangerud (1977) have shown that synchronous pollen assemblages from different parts of the same lake may differ, possibly due to the influence of the vegetation source related to distance from shoreline, among other factors. Bonny (1978:413) found from her studies of Blelham Tarn that surface sediments within an 8-meter water-depth contour had a relatively homogeneous pollen composition, but that pollen assemblages from surface sediments of the shallow-water, near-littoral region varied significantly "from place to place apparently as a result of high but localized pollen input from the nearest stands of emergent or terrestrial vegetation . . ." She noted (1978:413-414) that "resuspension and redistribution of surface sediment is apparently insufficient to obliterate differences in percentage pollen composition . . . despite the facts that (i) the effects of wind-induced turbulence upon the mud surface are likely to be greatest in shallow water, and that (ii) the littoral region down to *c.* 6 m water depth lies above the temperature barrier which the metalimnion presents in summer to the downward transmission of turbulence, so that shallow-water sediment is potentially subject to disturbance all year round." In a permanently or semipermanently stratified lake the limited amounts of resuspension and redistribution of surface sediments in comparison with lakes having seasonal overturn may mean spatial heterogeneity in pollen assemblages comparable to those demonstrated by Turner (1970, 1975) for peats sampled at different distances from the forest border. The bog center received the most regional pollen rain: distance from the source vegetation influenced pollen content in other samples.

In a lake having a large surface area, a large catchment area, and a number of entering streams, a large component of waterborne pollen may mitigate the potentially very local source of airborne pollen, adding complexity to the Tauber model and possibly confusing the issue of local and regional pollen rain. McAndrews and Power (1973) show that while surface pollen spectra from small lakes near Lake Ontario reflect *only* the local forest, spectra from Lake Ontario reflect the varied vegetation of the entire drainage basin related to the main river watersheds, thus implicating rivers as an important source of pollen delivered into the lake. Homogenization was sufficiently limited, however, due to differential vertical and horizontal pollen transport, that the load of specific rivers and the vegetation from which it had come could be identified in surface pollen spectra from offshore sediments (McAndrews and Power 1973). Bonny (1978: 402) felt that as much as 85% of the pollen caught in submerged lake traps in Blelham Tarn (0.67 km length) might have been streamborne, and that that component better reflected the extra-local vegetation than the airborne component alone. Even in a small water body (200 x 100 m) with a minor entering stream, Tauber (1977) found that about 50% of the total pollen *deposited on the lake bottom* had been carried by the incoming stream, thus reducing the significance of pollen input to bottom sediments derived from aerial components, and in this case distorting the bottom spectra with regard to the aerial pollen load and its vegetation source.

In Quaternary studies, where lakes are sampled whose size and catchment area are determinable, both lakes and sampling sites that have the potential for introducing vegetational biases are avoided. This can include very small basins where the pollen of local elements will be overrepresented, as well as marginal deposits of larger lakes, where leaves, fruits, and seeds are apt to be best represented, but where the pollen rain can be heavily influenced by shoreline vegetation as well as depositional basins whose sediments may be locally perturbed by inflowing streams.

In Tertiary studies, one is seldom provided with a choice for which any of the variables that make for an ideal pollen sampling site (see, for example, Faegri and Iversen 1975:86) are determinable. Because of the problems just outlined, however, the use of leaf deposits for pollen samples may build into Tertiary pollen assemblages precisely the kind of vegetational biases which pollen analysis is normally optimally suited to counter, thus abrogating one of its most important aspects in comparison with megafossils—its use to provide information about the regional vegetation and climate.

With regard to megafossils, the same types of questions raised early on by pollen analysts need to be considered, followed by empirical studies to put paleobotany on a footing comparable to pollen analysis. Paleobotanists need to address questions pertaining to: (1) the "localness" or "parochialness" of plant remains found in limnic deposits—the source area from which remains have been derived; and (2) the "accidental" aspect of megafossil accumulations—whether it is possible to quantify the relation between megaremains and source vegetation to give the abundances of remains meaning in terms of vegetational cover.

Studies of fruit/seed accumulations in modern lake deposits (Birks 1973) and in their subfossil Quaternary and Holocene counterparts (for example, Bright 1966; Watts and Winter 1966; Watts and Wright 1966; Rybníček and Rybníčková 1968, 1974; Watts and Bright 1968; Wright and Watts 1969; Watts 1970, 1979; E. Grüger 1972; J. Grüger 1973; Birks 1976; Birks and Mathewes 1978; McMurray et al. 1978, to name a few) have demonstrated the extremely local source of most of the seeds and occasional other remains. Birks (1973) found that few fruits and seeds were moved much beyond the range of their source plants, unless they were adapted for floating or animal dispersal. Wright and Watts (1969:37) have even interpreted some such occasional plant debris in basal limnic deposits as representing forest floor, accumulated *in situ* or only slightly transported. Such assemblages seldom receive much input from "upland" vegetation, although occasionally a small component of the assemblage may reflect the more regional vegetation (see, for example, Wright and Watts 1969; Birks 1976; Watts 1979). In modern Minnesota lake deposits, Birks (1973) finds the remains of "upland" plants or plants from the upland forest, in this context any plant not of aquatic or wetland habitats, as being most common in marginal limnic deposits and as represented primarily by fruits and seeds (such as those of *Betula* and conifers) and conifer needles—both sometimes found in sediments in the centers of deep lakes that are devoid of other plant structures even when the lakes are surrounded by forest. Although Birks found disparities in seed recovery related to seed production, differential potential for transport and so forth, she did find a general trend toward increase in number of fruits/seeds for a taxon as its representation in the ground cover increased. Bright (1966:21) found that where pollen and seeds/fruits of aquatic and semiaquatic plants were apparently tracking changes in lake size and depth, there could be a "remarkable synchronization of the many maxima" in both records, indicating a co-occurrence of varied quantitative changes that suggested a good correlation with vegetational changes. Birks and Mathewes (1978) also noted that pollen and megafossil influx data may occasionally coincide to provide good evidence of local abundance, as in the case of pollen and seeds for the water plant *Nymphaea*.

Because most North Temperate area Quaternary studies deal largely with the fruits and seeds of aquatic plants or those that occur at the interface with the nonaquatic environment, their source would be expected to be "local," and incidence of synchronization in quantitative changes that are of local, *not* regional, significance are more to be expected than in the case of allochthonously accumulated fossils.

However, the limited number of fruit and seed deposits in the Tertiary do not represent plants of marginal aquatic environments. Most Tertiary megafossil assemblages are transport accumulations. Thus, generalizations derived from the study of Quaternary megaremains may not pertain to them. Until studies have been made on the occurrence of allochthonous leaves and seeds derived from diverse vegetation in different topographic and climatic settings, and deposited in lakes of diverse sizes and chemistries, it is too soon to say that localness or parochialism also pertains to them. Limited evidence collected in specialized circumstances suggests, however, that this is the case.

Chaney (1959) discussed from a theoretical point of view the principal "depositional factors" that affect opportunities for the entrance of plant remains into contemporaneous sediments, including:

1. The proximity of the parent plant to the site of deposition.
2. The durability of the organ entering the record. Chaney considered leaf durability empirically only in terms of size and texture. "Texture" is, of course, a composite of varied biological and biochemical factors that influence durability. At one level, it can be demonstrated, for example, that coriaceous leaves, including hard-textured "sun" leaves, are less readily broken down by decomposer organisms than soft-textured leaves (Heath and Arnold 1966; Ewel 1976). Leaf and plant decomposition in general is a complex process (Kaushik and Hynes 1971; Dickinson and Pugh 1974) that needs to be integrated by paleobotanists into hypotheses related to the presence and absence of plant remains.
3. The weight-area ratio of the organ. Those with low weight-to-area ratios will be more easily disseminated than those with high weight-to-area ratios. In addition to inherent genetic variability affecting the weight-area ratios, position of leaves in the crown might also be expected to affect the ratio. The "sun" leaves, which are exposed to greatest solar radiation at the crown periphery, tend to be both smaller and thicker than the "shade" leaves that do not occur in peripheral positions.
4. The height and volume of the parent plant.
5. The deciduous versus evergreen habit of the parent plant. Chaney suggested that seasonality of leaf-drop related to the evergreen-deciduous habit might also be a factor in leaf preservation, depending on whether leaf-drop coincides with times that might be favorable to deposition and burial in any particular environment.

Chaney regarded proximity of the source plant to the depositional basin as the single most important variable in determining the number of leaves deposited as well as the taxa represented. He concluded that fossil leaves and other structures came from source plants that were both common in the vegetation cover and local, and that it was possible to quantify the nature of the vegetation from the relative abundance of leaf types in the deposit, i.e. that a correlation existed between the number of leaves and the numbers of individuals of that taxon in the local vegetation. He saw the relative abundance of fossil leaves as a *direct* clue to the abundance of the source plants.

The first attempt to assess the individual or collective significance of these factors by looking at the accumulation and preservation of modern plant remains in a depositional setting was Chaney's (1924) Muir Woods study in the Coast Ranges of California. This involved an essentially closed-canopy woods in a restricted riverine situation with small, still-water settling pools. In this setting, Redwood Creek is overhung by trees and margined closely by understory vegetation, a situation probably *not* analogous to *most* circumstances, including the situation at Clarkia Lake, in which fossil leaf deposits are formed, although it has provided the basis for quantitative interpretation of most Tertiary leaf floras in Western North America.

Chaney found that none of the plants contributed to his sampling sites from distances of greater than a maximum 152 m (500 feet), and that 91% of the leaves, representing only four species, came from average distances of just over 1.5 m (5 feet) from the depositional sites. The two leaves of oak that had entered Redwood Creek from a minimum distance of 76 m (250 feet) were the only representatives of the bordering and more upland trees. Vegetative organs from plants with a tree habit made up 94% of the individual elements; structures from shrubs and small trees accounted for only 6% of the total; structures that could be attributed to herbs made up only 0.04% of the total.

On the basis of this study, Chaney (1938) concluded that few remains of larger organs of plants are transported *aerially* (he did not address the question of water transport) from more than 30.5 m (100 feet) into basins of deposition. Negligible amounts of coarse airborne particulate matter (mostly leaves) collected from lake traps beyond 10 m of the shoreline and the tendency of fallen leaves to drift shoreward before sinking (Gasith and Hasler 1976 and references therein) reinforce the "localness" of megaremain accumulations. Taxa represented in the fossil record may thus be assumed to be primarily those that lived adjacent to streams or lakes. With regard to upland vegetation at more distant sites, Chaney concluded (1938) that it must be reconstructed by comparison with the vegetation now living on the borders of lowland forests, and from the occasional finds of winged fruits and seeds which, he felt, are more readily transported by wind from upland sites than leaf remains.

Wind, however, is remarkably ineffective in seed transport. Even fruits and seeds that are judged morphologically to be optimal anemochores cannot be depended on to provide information about vegetation growing at greater distances from depositional basins than leaves (Ryvarden 1971; Sheldon and Burrows 1973; and reviews in Levin and Kerster 1974; Harper 1977; Levin 1979). Levin (1979:382), citing relevant literature, noted paradoxically: "The aerodynamic properties of the samaras and single-winged fruits of angiosperm trees will explain why winged fruits do *not* disperse very far beyond the parent trees . . ." Merrill (1928:133) similarly observed the "relative inefficiency of winged fruits" (of the Dipterocarpaceae) for wind dissemination, suggesting that the wings "give the fruit a gyratory motion in falling rather than buoyancy" such that they would probably not be carried "more than a few hundred meters, even in a very heavy wind." Green (1983:357) found that even among single-winged samaras that "spin as they fall and are transported by wind," the morphology may be sufficiently dissimilar that there are distinct differences in travel distances under different circumstances. The more slowly falling maple (*Acer* spp.) samaras "typically disperse farther in the low-speed winds that predominate in forests. . . . ash [*Fraxinus* spp.] and tuliptree [*Liriodendron tulipifera* L.] samaras appear to have greater stability in gusty, high-speed winds . . . and therefore disperse farther under these conditions than maples."

Aerial seed dispersal is strongly leptokurtic, i.e. the frequency distribution falls off rapidly within distances of about 5 to 50 m from the source (Levin and Kerster 1974; Harper 1977) depending on the height of the plant and its areal coverage, seed weight, terminal velocity (settling rate), wind speed, turbulence, and so on. The settling rate is proportional to weight and seems to be the limiting factor in transport (Levin and Kerster 1974). An increase in wind speed results only in a proportional increase in transport distance, and even plumed seeds with low terminal velocities still experience short transport distances (Sheldon and Burrows 1973). Ryvarden's (1971) studies with seed traps suggest that better than 80% of seeds were transported away from source plants less than 5 m; those carried more than 5 m usually had transport distances of 100 m or less. His study dealt largely with herbaceous or shrubby taxa in an exposed area with limited vegetation to filter windborne diaspores. Even in the case of *Salix* (a shrubby species), apparently an optimal anemochore, the seeds are rarely dispersed more than 5 m because they are blown along the ground and readily trapped. Winter-dispersed seeds scraped from snowbeds generally came from distances of up to 2 km, possibly because they were moved over a smooth snow surface (Ryvarden 1975). Seeds of the anemochore *Betula pubescens* had a transport distance of 4 to 6 km in these circumstances, which, are hardly applicable, however, as a basis for establishing transport distances of megaremains in warm-temperate Tertiary forests. Nevertheless, Ryvarden (1971: 225) found that seeds trapped along transects in both aerial and stream traps "quantitatively reflect the composition of the indigenous flora" and that many seeds without apparent morphological adaptation to dispersal were found in both types of traps, as well as in snow beds during the winter.

The relevant literature on forest trees assesses seed transport from isolated trees or from the edge of forest clearings, circumstances not pertinent to evaluating dispersal in closed-canopy forests where there is lower wind speed and where filtering occurs on adjacent plants (see Levin and Kerster 1974; Harper 1977); it is nevertheless very limited even under these circumstances and rarely exceeds 200 m (see, for example, Isaac 1930).

The possibility of generating megafossil deposits by high-energy storm winds is unassessed in terms of type and number of varied plant structures, although some data, largely anecdotal in nature, is available on long-distance storm-wind transport of fruits, seeds, leaves, and other objects (for example, Ridley 1930; van der Pijl 1972, and references therein). The possibility of wind-generated deposits is of interest because high-energy winds may exceed the critical velocity necessary to strip leaves from plants independently of their normal dehiscence habits (Craighead and Gilbert 1962; Scheihing 1980). Thus, one might expect wind-storm deposits to include leaves of evergreen plants whose shedding cycle might not normally predispose them to occur in depositional basins, as well as the leaves of herbaceous taxa that usually dry on the plant. With hurricane-force winds, defoliation may be random, nonrandom or total. Craighead and Gilbert (1962) cite incidences where pine, for example, was wholly stripped of needles and others where little defoliation occurred.

There may be other nonrandom sampling effects. Ridley (1930) and Scheihing (1980) provide examples where high-velocity storm winds were only effective in the canopy of dense forests, leaving the understory vegetation essentially untouched, including shrubs and herbaceous taxa. In such circumstances storm-generated plant deposits would contain leaves, fruits, and seeds heavily biased toward top-canopy members of the plant community. Windborne megastructures are also apparently "sorted out" in falling by weight, and/or aerodynamic properties (Ridley 1930), an additional biasing factor for wind-generated megafossil deposits. Storm-generated deposits could also provide a high potential for scrambling leaves and other plant parts from diverse nonlocal communities under some circumstances. However, it should be noted, as Merrill (1928:134) pointed out, that in circumstances where high winds are accompanied by very heavy rainfall, the elevated humidity effectively "inhibits or prevents the dissemination of most seeds and fruits that have adaptation for distribution through the medium of wind, this being especially true of seeds or fruits that are supplied with hairs . . ."

Despite anecdotal literature on long-distance animal transport of fruits and seeds (van der Pijl 1972), it too is seldom effective beyond 100 m (Howe 1977; Bullock and Primack 1977; Levin 1979), although the capacity of animals for caching seeds that may become fossil or subfossil deposits (Chaney and Mason, 1936, provide such a fossil example) may be enormous.

Dorf (1945, 1951) studied the effects of differential "fossilization" and transportation on plant remains following the eruption of Parícutin in circumstances relevant to the Tertiary of the Pacific Northwest, where many megafossils occur in or associated with ash deposits. Most of the area around the volcano prior to eruption was forested by "an upland, temperate pine-oak association" of approximately 70% pine, 15% oak and minor spruce, madrone, and crabapple (Dorf 1945; Segerstrom 1950) living in a topographic setting not dissimilar to that of Clarkia (Williams 1950). Dorf found incipient fossilization to be highly selective (no herbaceous plants, although herbs were common prior to the eruption; examples of gross overrepresentation and underrepresentation of taxa compared with their original population abundance), and transportation generally deleterious to good preservation. Abundant leaves in an adequate state of preservation were primarily found in subaerial ash falls, where presumably they had been buried immediately upon falling from the trees. In waterlaid deposits, leaves generally were neither well-preserved nor abundant. In limnic beds, Dorf found well-preserved leaves only in marginal sites where they must have fallen directly into the lake and been buried at once. In as little as 30.5 m (100 feet) from the forest edge, only plants with tough, resistant leaves were represented. In general, Dorf (1945:259) found that the "proportions of the species represented by buried plant-remains were in inverse ratio to the distances of the parent trees from the site of burial."

McQueen (1969) sampled leaves and seeds from nine open-water and marginal lake and swamp stations in relation to the surrounding vegetation in the foothills of the Rimutaka Range, New Zealand. Seven of the sampling sites were in Lake Pounui (1.7 x 0.8 km); two were in shallow water of an adjacent swamp. The leaves in both swamp and lake were almost exclusively coriaceous. Potential fossils in the swamp were of plants growing in the swamp or from dry ground nearby. A single leaf from a plant not recorded in the vegetation around the swamp came from at least 2 km distance. Only 9 of the 57 species in communities around and on the swamp were found as "potential fossils." Remains in the lake represented only the dominants of the surrounding vegetation, yielding only 8 of the possible 32 plant species available in the local vegetation. The most inshore sample had the most local contribution of plant remains, although it also contained a few leaves from at least 60 m from the site. The open-water samples had leaves of the more common species in the surrounding hills, i.e. they tended to be less parochial in content although there was no evidence of distant transport. The lake sediments included no remains of woody swamp plants. McQueen hypothesized that swamps on all streams feeding into the lake acted as filters trapping any remains that might have been transported from outside the immediate vicinity.

Roth and Dilcher (1978) examined leaf remains in a small pond (ca. 20 m) surrounded by forest, and found that: (1) leaves from the riparian vegetation were not transported far from shore; (2) fewer than half the woody species in the surrounding forest were represented in the lake bottom sediments; (3) neither of the two woody vines growing at shoreline were represented by leaves; (4) several understory shrubs were not represented, although most of the surrounding trees were; and (5) the foliar physiognomy of leaves obtained from the lake bottom provided a poor match with the physiognomy (size and leaf margin characters) of the leaves of the surrounding vegetation.

Drake and Burrows (1980) provided a semi-quantitative study of plant remains in Lady Lake (1.65 x 0.9 km), New Zealand, in comparison with surrounding predominantly evergreen forest, and scrub, swamp, and wet-meadow vegetation. Lady Lake is 114 m above sea level and 3 km to the west of hills reaching timberline (at a maximum 1156 m), although elevations immediately adjacent to the lake seldom exceed 200 m. Plant litter from three transects into the lake showed surprisingly good qualitative representation of forest taxa and for the most common taxa good quantitative representation as well, although some taxa well represented in the lake litter are poorly represented or are even absent in the vegetation, some taxa prominent in the vegetation are not well represented in the lake, and non-woody ground cover and lake margin swamp plants are both sparse in lake litter. In general, trees and shrubs "are remarkably well represented by the lake litter" (Drake and Burrows 1980:271), with most taxa that are *very important* in the vegetation reflected in the leaf litter. The greatest abundance and diversity of litter together with the "largest and best-preserved leaves, fruit, and seed are found nearest the shore, with a progressive decline in size and preservation . . . towards the centre of the lake" (Drake and Burrows 1980: 272), although entering streams tended to carry large and well-preserved plant remains further into Lady Lake.

From a comparison of megaremains and pollen from the lake sediments (but not the *same* sediments), Drake and Burrows (1980:273) concluded that "Both studies show that most of the taxa which are very important in the vegetation are reflected reasonably faithfully by the pollen and lake litter," while swamp taxa are poorly represented by both types of remains.

In assessing the entry modes of leaf litter into Lady Lake, Drake and Burrows concluded that stream transport, particularly during floods, delivered the richest variety and greatest abundance of megaremains, and that winds, particularly strong winds, are also a very important source of plant debris.

Spicer (1981) has considered some taphonomic variables relating to survival of leaves during fluvial transport and the occurrence of leaf remains in a limited delta front deposit. The type of depositional environment he examined is probably a limited source of Tertiary leaf assemblages and to my knowledge, with the possible exception of the Clarno nut beds (Hansen 1973) which may be fluvial rather than deltaic, has no counterpart in the Tertiary of the Pacific Northwest, where leaf deposits are chiefly limnic. Although it is possible that most leaves delivered to a delta front environment are aquatically transported from some distance, one cannot assume that leaves in lake bottom deposits are derived via fluvial rather than aerial transport from the immediate surrounding vegetation. In the Pacific Northwest, and at Clarkia specifically, most leaf assemblages are comprised largely of what Spicer categorizes as "museum specimens" (with, of course, the usual hash of finely comminuted plant debris) rather than the very fragmental leaf remains that he found in his restricted delta front environment. This fact argues against the application of his model, with its emphasis on turbulent transportation and macerated leaves, to most Tertiary floras. Dorf (1945:259) found at Parícutin that subaerial ash deposits where "it was evident that both ash and plant-materials had been deposited gently, with little or no damaging transportation by strong winds or running water" contained more abundant and better preserved remains than either stream or lake deposits. Nevertheless, the potential for leaves and other remains to float (see Praeger 1913, for example) often for long distances in wholly recognizable condition has been demonstrated, often but *not always* involving more "durable" plant structures such as the coriaceous leaves that McQueen (1969:18-19) found waterborne from at least 30 km. Burrows (1980:321) provides other instances of leaves transported intact by "streams and/or by flotation on lakes and/or by wind" from lake and riverbed sites in New Zealand. Most leaves were isolated specimens moved from distances of less than 10 km. Thus as Burrows suggested, "long-distance transport . . . would not be likely to lead to major contributions to a fossil assemblage as the density of individual items is low and opportunities for preservation are not great." Nevertheless, examples of riverine and flood water transport of isolated plant parts, especially fruits, seeds, and wood, often in tremendous quantities for long distances (see Ridley 1930; Chandler 1964; van der Pijl 1972; Burrows 1980) and of vegetational rafts of the type hypothesized to be responsible for the London Clay flora in the Thames Estuary (Chandler 1964:4-9) attest to the potential for scrambling plant remains from far outside the immediate depositional basin. However, McQueen argues on the basis of his studies in New Zealand that the highly organic fine-grained sediments, such as those in Lake Pounui, apt to have been derived from slow-flowing rivers "at or near base level . . ." and associated with local swamps, which act as filters to floating plant remains, are unlikely to contain exotic (nonlocal) transport elements. Conversely, coarser sediments associated with more rapidly flowing and unimpeded rivers may often contain interbedded leaf deposits having admixtures of ecologically incompatible taxa from higher elevations that are not a part of local plant communities. Chandler (1964:4-9) suggested that in areas of topographic diversity and rapid topographic change, plant beds in lowland sites formed from debris of river transport "must inevitably" include mixtures of remains from outside the immediate basin of deposition. But documented examples in the Pacific Northwest where both sedimentological and floral evidence indicates accumulation of megafossil deposits by large-scale riverine transport are too infrequent to suggest that most lowland leaf deposits were formed in that way. The available information tends to support the generalization, albeit based on studies that are very preliminary and circumstances too specialized for broad speculations, that allochthonous megafossil assemblages are predominantly very "local" in origin, and represent communities and environments that are parochial with regard to the depositional basin in which they are found.

Where one is dealing with the remains of aquatic or wetland plants that are used primarily to provide information about water chemistries, trophic structure of lakes, water fluctuations, hydrosere succession, extinction histories, and the nature of the immediate aquatic vegetation, the "localness" of megaremains does not detract from paleoecological interpretations but can only enhance them (see, for example, Bright 1966; Watts and Winter 1966; Birks and Mathewes 1978; Watts 1978; Birks 1980 and references therein). Such "localness" permits environmental inferences about vegetation seldom obtainable from pollen assemblages (for example, pertaining to small-scale community succession [Rybníček and Rybníčková 1968, 1974]) and documents plant occurrences that cannot be assumed from the presence of pollen even of relatively small pollen producers. Rybníček and Rybníčková (1968:131), for example, found the pollen of *Filipendula, Galium* and *Comarum* to be continuous even where this was not "phytocenotically" possible for the plant communities they were studying.

The presence of some types of plant communities or formations is sometimes ambiguous from pollen data, but readily documented from co-occurring megafossils. For example, tundra in late-glacial pollen spectra is usually denoted by high nonarboreal pollen (NAP) suggesting open, treeless vegetation. This is often associated with tree pollen, however, and sometimes with little or no pollen of arctic herbs and shrubs. The presence of megafossils including leaves and fruits of typical arctic taxa can leave little doubt, however, of a tundra ground cover (see, for example, Argus and Davis 1962; Baker 1965; Watts 1967, 1979; Birks 1976). Megafossils are also a more reliable source of information on plant immigrations into a region than pollen, even data derived from pollen influx, since pollen must inevitably include some transported component. Matthews (1975) for example, found remarkable discordance between the presence of megafossils (indicating presence of the taxa) and their pollen frequencies in considering dispersal histories of birch and alder on the North Slope.

To extrapolate some sense of the regional vegetation and catchment environments from allochthonously accumulated leaf, fruit, and seed deposits, it is essential to be aware of biases that could be ecologically misleading. Certain environments may be represented to the exclusion of others. McQueen (1969) and Drake and Burrows (1980) found no or limited evidence of swamps from adjacent lake deposits. Megaremains may or may not represent the dominants even of the local vegetation (Dorf 1945, 1951; McQueen 1969; Roth and Dilcher 1978). Remains may be very disproportionate relative to plant abundance. Dorf (1945) found that the co-dominants pine and oak occurred at Parícutin in the ratio of about 4:1. Pine in comparison to all dicot leaves was recovered from incipient fossil deposits in the ratio of 1000:1. Leaves of oak were most numerous of the dicots; they were grossly underrepresented compared to pine, but overrepresented compared to the local abundance of *Quercus*. All other taxa were greatly underrepresented relative to their occurrence in the vegetation. Biasing toward the remains of trees is evidenced in all studies (Chaney 1924, 1938; Dorf 1945; Roth and Dilcher 1978; Drake and Burrows 1980); the absence or limited representation of herbs and the leaves of small trees and woody shrubs is notable.

Such examples should make one cautious about megafossil counts as a basis for arriving at conclusions about the significance of taxa in plant assemblages. The use of small numbers of remains and certain types of remains for predicting the composition of the regional vegetation also should be approached with caution for allochthonous assemblages. Their use requires that several questions be addressed through empirical studies when possible: (1) Are taxa that are represented by small numbers of structures, or by certain types of structures, to be construed automatically as those plants having limited opportunities to become fossil because of distant growth sites from basins of accumulation? Their occurrence may have less to do with their growth sites than with their capacity for transportation and durability. (2) Are perceptions for dispersal of structures based solely on their morphology correct? Ryvarden (1971, 1975) found that seeds *without* obvious adaptation for dispersal were transported greater distances in higher proportion than the obvious anemochores. (3) Have the tolerances of specific taxa remained the same so that they have been similarly segregated in their growth sites through time, some automatically having more limited opportunities to become a part of the depositing sediments than others?

QUANTITATIVE COUNTS IN PALEOBOTANY

Leaf counts have been a feature of floristic studies of western American Tertiary leaf assemblages for many years (see Chaney 1924, 1938, 1959, etc.) where they have been used to suggest which taxa were abundant, common, uncommon, and rare in the lowlands and which were restricted to more distant sites ("slopes and ridges") from depositional basins. Chaney (1959:46), for example, used the *recorded* abundance of megafossils as "suggestive of their *actual* abundance" and in conjunction with the type of structure or organ represented (leaf, winged fruits, branchlet, etc.) and the habitat occupied by the "living equivalent taxa," segregated the Mascall flora into broad "associations" or "communities." He did not address the appropriateness of quantifying as unit items (e.g. Faegri and Iversen 1975:123) a variety of disparate megafossil structures and organs (although he dealt mainly with leaf remains) to arrive at his results. If, however, fruits and seeds can be considered sufficiently "comparable" (e.g. Watts and Winter 1966) to give the interpretation of such counts validity in the Quaternary, one might suspect that it makes as much sense *theoretically* to quantify leaves that are perhaps no more variable among themselves than these other structures.

Chaney and Sanborn (1933) suggested some arbitrary "correction factors" to adjust for inequities in leaf production and delivery. It doubtless will be appropriate *in conjunction with vegetational studies* to consider this question in detail, as pollen analysts have for pollen, in order to better understand the relation between leaves in depositing sediments and the character of the source vegetation. Critical to the use of leaf abundances for environmental interpretation is some understanding of whether the fossil frequencies reflect on differences in production, transport potential, durability, and so on, or commonness of the source plant. A single pollen grain of *Ilex* may reflect more on the local occurrence and commonness of holly than 100 grains of *Pinus* on its local commonness; a single leaf of one species may have the significance of 100 leaves of another in reconstructing the vegetation.

Of far greater significance to the use of relative frequencies of megafossils in vegetational analysis is the potential for "homogenization" or random mixing of organs and structures in an allochthonously derived deposit. The analogous problems in dispersal and mixing of pollen cannot be minimized. But the potential for aerial mixing of pollen in proportions related in a complex way to the numbers of source plants in the vegetation cover forms in large part the theoretical basis for pollen analysis, even though the original model of the "pollen rain" as a random mixture of the pollen of the regional vegetation sifted uniformly over the landscape is too simplistic to have credibility in modern pollen analysis. This has not deterred the use of pollen analysis in vegetational reconstruction. It *has* meant the application of more sophisticated, often numerical, techniques to help understand the significance of pollen assemblages, and detailed study of modern pollen rains and their vegetation source as a basis for extrapolating to fossil information. Sedimentary facies control has not been investigated by pollen analysts to any extent, in part because of the emphasis on the airborne pollen rain. For Cretaceous rocks, however, Batten (1969, 1972, 1973, 1975) demonstrated differences in pollen and spore (micro- and megaspore) assemblages with regard to spore diversity, preservation, absolute numbers, and such morphological attributes as size and exine sculpture and thickness in relation to variations in lithology (grain size, sorting, and so on). He suggested that changes in assemblages reflect both depositional and source environments, although in his examples (Batten 1975:452) the distribution of fossils is "probably more closely related to sedimentary facies than to the distribution of the parent plants."

The possibility for mixing of leaves by atmospheric turbulence has not been examined, but the evidence pertaining to the aerodynamics of fruit and seed dispersal does not suggest confidence in the potential *for any type* of "natural megafossil rain" in which megastructures are mixed in proportions related to the population abundance of the source plants. On the contrary—all evidence tends to suggest the absence of such atmospheric mixing. Thus as Levin and Kerster (1974:164-165) note: "The settling rate seems to be the limiting factor in dispersal. Most seeds either are not airborne long enough to experience the turbulence that will carry them to high altitudes or are simply too heavy to respond dramatically to the turbulence they experience." Examination of seeds accumulated in soils attests to the veracity of this observation. Sanford (pers. commun. 1982) found no evidence for mixing of seeds in soil samples from a wind-swept plateau in eastern Oregon. Seeds of a single plant or a single group of plants (e.g. grasses) dominated each sample. These seeds, including even those of anemochores, obviously never became a part of the atmospheric particulate matter, but accumulated so locally that they provided no information about the regional vegetation or even the immediate vegetation where they were collected. Pollen analyzed from the same seed-bearing soil samples, on the other hand, provided a reasonable depiction of the immediate vegetation. The extremely localized nature of seed assemblages in topsoil samples demonstrated by Keepax (1977) and Minnis (1981) similarly supports the idea of limited mixing of seeds in "natural seed rains." There is no reason to suppose that the terminal velocities of leaves and other structures will not similarly restrict both their transport and mixing in the atmosphere, particularly in a closed-canopy forest.

At the same time, practical experience with fossil deposits and varied observations reinforces the idea that the depositional environment exercises considerable selectivity on megastructures with widely different physical characteristics in relation to transportation, sorting, and sedimentation that will result in nonrandom accumulations of varied types of structures. Using a small aquarium, I found that all leaves continued to float long after all seeds, except maple samaras, had sunk. In circumstances where our perceptions of the presence or absence of a plant in the reconstructed vegetation might depend on finding only its seeds in a leaf deposit, the co-occurrence of these structures in the same strata cannot be assured. The flat, thin, maple samara, the similarly constructed conifer seeds, and seeds of *Cedrela*, for example, are among the more consistently occurring seeds in Tertiary leaf assemblages. Hydrodynamically, and perhaps aerodynamically, they must behave as leaves. Birks (1973) regularly notes only the needles of pine and winged seeds in the centers of lakes to the exclusion of other organs and structures of either indigenous water plants or those from "upland" habitats. McQueen (1969) found only the coriaceous leaves of *Nothofagus menziesii* transported in free-flowing streams below stands of this species in New Zealand. Roth and Dilcher (1978) found small, coriaceous sun leaves better represented in lake center (where they may have been either wind- or waterborne) than the shade leaves from the same plants. Plants with only small, coriaceous leaves will be disproportionately represented in some depositional environments compared with those producing only large thin leaves.

The often lenticular and localized occurrence of megafossil deposits, with remains largely accumulated in certain types of rocks, indicates the sensitivity of megaremains to sedimentological factors such as change in sediment lithology. Sedimentological control will be most obvious in continuous stratigraphic sequences where changes in abundances of ecologically compatible taxa appear to relate to lithologic variations rather than environmental causes. The consistent occurrence of remains of only some taxa excluding other ecologically expectable species must suggest ". . . a preservational or sedimentation mechanism that favours these species and rejects others"

Watts 1978:60). The "patchy" occurrence and differential preservation of plant remains in alluvial plain and valley-fill deposits at Parícutin appeared to correlate at least partly with sediment variations. Dorf noted (1945:259), for example, that while most pockets of coarse ash were barren of plant remains, "occasionally they were filled with numerous highly abraded and broken leaf fragments, twigs, and acorns." That fossil wood, seeds, and leaves occur in largely discrete deposits is well known. Even where disparate types of remains occur together in the same rock unit, they are seldom intermixed. In the Miocene Mascall Formation of eastern Oregon, for example, hackberry (*Celtis*) fruits/seeds are locally common in the riverine tuffs containing mammal remains, but hackberry is represented by only two leaves in leaf-bearing lacustrine horizons together with hackberry pollen. Barbour's (1925) Hackberry Conglomerate from western Nebraska—typical of the incidence of these seed accumulations, often in the millions over a wide area of the northern High Plains (see Chaney 1925) without any evidence of co-occurring plant remains, including hackberry leaves—indicates the potential for extreme localization of some types of plant remains in obvious relation to selective sedimentological factors. *Celtis*, incidentally, occurs in few *fossil assemblages*, and is never abundant. The London Clay flora and others in that area of early Tertiary age are largely segregates of seeds. The same is true of the Eocene Clarno flora of eastern Oregon where the "nut-bed" deposit is segregated from the leaf deposits in the same formation.

These obvious examples of nonrandom megafossil accumulations involving fluvial transport suggest that it would be unwise to overlook the more subtle possibilities for sedimentological control in lake basins such as might be occasioned by: (1) differential flotation of some types of structures involving different taxa; (2) differential preservation of taxa and structures in some sediment types and under some conditions of sedimentation; (3) changes in the source of organic-laden sediments entering lakes and shifts in sites of deposition; and (4) nonrandom changes in rates of sedimentation that could result in considerable variability in the number of structures being deposited from different taxa. The latter two factors might be especially critical where it is necessary to tally megaremains from large increments of deposits as at Clarkia, in order to obtain statistically valid numbers of fossils.

If the aerodynamic and hydrodynamic properties of single types of structures, such as seeds, leaves, or wood, are sufficiently similar, they will lead to mixed allochthonous accumulations; if they are different, individually, they will lead to allochthonous accumulations of single types of structures. In either case, the occurrence of plant remains in allochthonous deposits principally because of aerodynamic and hydrodynamic properties by which they are either mixed or segregated effectively negates tallies of their relative frequencies for providing any direct measure of the number of their source plants. Changes in relative abundances of structures through time related to vagaries of sedimentation including chance physical factors, such as changes in sediment types and rates of sedimentation, will give changes in abundance of plant parts in deposits that are independent of changes in taxic abundance in the vegetation cover. With regard to sedimentological control, only if the accumulation of fossiliferous sediments has occurred randomly at a reasonably constant rate is there any possibility for suggesting that frequency changes of fossils reflect changes in the vegetation, apart from all the other variables that affect entry of megastructures into depositional basins.

PLANT MEGAFOSSILS AND POLLEN AT CLARKIA, P-33: A COMPARISON

Qualitative Results

Taxa represented both as megafossils and pollen/spores. A large number of plant taxa are represented in Unit 2 by both megafossils and pollen/spores. This list includes: (1) *genera* that have been identified both from pollen and megafossils; (2) *families* for which genera have been identified from megafossils, but for which the pollen record is now indeterminate below the family level, or where there are several genera whose pollen is morphologically sufficiently close to preclude determination with a light microscope to a single genus; and (3) *family complexes* whose pollen is not determinable at the level of a single family but for which morphologies are represented that could belong to one or more families also represented by megafossils. This category includes: *Osmunda* (Osmundaceae); *Polypodium* and others (Polypodiaceae) (Plates 1, 2, 3, 4, figures 1-8; 1,2; 1-9; 4-6); Taxodiaceae (megafossils include *Metasequoia, Taxodium, Sequoia, Glyptostrobus, Cunninghamia*) (Plate 5, figures 1-3); Cupressaceae and/or Taxaceae (megafossils include *Chamaecyparis, Thuja, Amentotaxus*); *Abies* (Plate 6, figure 6), *Pinus* (Plates 5,7, figures 4-6; 1,2) (Pinaceae); Gramineae; *Pterocarya* (Plate 8, figures 7,8), *Carya* (Plate 9, figures 10,11) (Juglandaceae); *Alnus, Betula* (Plate 4, figures 1,2), *Ostrya* (or *Carpinus* from pollen), *Corylus* (Betulaceae); *Fagus* (Plates 8,10, figures 1-6; 4,5), *Quercus* (Plate 11, figures 1-4), *Castanea* (and/or *Castanopsis* and *Lithocarpus*, from pollen) (Fagaceae); *Celtis* (Plate 4, figure 3), *Ulmus-Zelkova* (Ulmaceae, all 3 genera represented as megafossils); *Liriodendron, Magnolia* (Plate 10, figure 6) (Magnoliaceae); *Liquidambar* (Plate 12, figures 9,10) (Hamamelidaceae); *Platanus* (Platanaceae); Rosaceae (megafossils include *Amelanchier, Crataegus, Prunus, Rosa*), *Rhus* (Plate 12, figures 1-3) (Anacardiaceae); *Ilex* (Plate 9, figures 1-3) (Aquifoliaceae); *Acer* (Figure 1) (Aceraceae), *Aesculus* (Plate 7, figures 4-7) (Hippocastanaceae); Rhamnaceae (megafossils include ?*Rhamnus, Berchemia, Paliurus*); *Parthenocissus, Vitis* (Plate 6, figures 7,8) (Vitaceae, megafossils include also cf. *Amelopsis*); *Tilia* (Plate 10, figures 1-3) (Tiliaceae); *Nyssa* (Plate 6, figures 3-5) (Nyssaceae); Ericaceae (Plate 6, figures 1,2; megafossils include *Vaccinium*, cf. *Rhododendron*, cf. *Pieris*); *Symplocos* (megafossils attributed to Symplocaceae or the closely allied Styracaceae); *Fraxinus* (Oleaceae).

Although a large number of genera and families can be independently identified from co-occurring megafossils and pollen and spores, the taxonomic resolution is much higher with megafossils than pollen. Leaf taxa in the late Tertiary generally can be compared favorably with one or

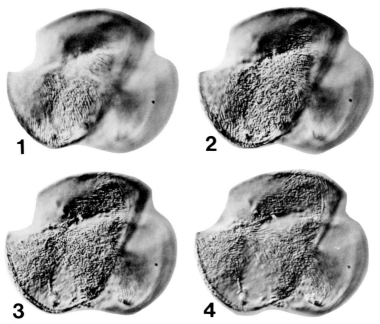

Figure 1. *Acer* sp. (x 1000) Nomarski interference photography. Oblique polar view of tricolpate grain showing details of exine sculpture including the fine striae in high focus characteristic of maple pollen. All photomicrographs taken by J. Gray.

two modern species. The impreciseness with which pollen taxa can be identified, together with their scarcity in some cases (Table 1), has greatly complicated the issue of their presence or absence in comparison with megafossils in Unit 2. For example, the families Taxodiaceae, Rosaceae, Rhamnaceae, Vitaceae, and Ericaceae together represent from megaremains possibly as many as 18 genera. This spectrum of taxa cannot be distinguished in the pollen record. Some other examples exist, e.g. *Ulmus* and *Zelkova*, *Ostrya* and *Carpinus*, *Castanea*, *Castanopsis*, and *Lithocarpus*, where pollen cannot be discriminated at the generic level. Thus one pollen taxon may equal several megafossil taxa (see Faegri and Iversen 1975:46). The megafossil assemblage unquestionably provides a fuller picture of the local vegetation from presently identified taxa than the pollen analysis. The variety of unidentified pollen types, if eventually placed into families and genera, may alter this taxonomic imbalance.

Megafossil taxa without pollen/spore representation. A number of genera are represented by megafossils in Unit 2 that are not represented by pollen. Taxa whose families are represented by pollen/spores but where the generic identification is now indeterminate are included in category 1. Category 2 includes: *Equisetum* (Equisetaceae); ?*Keteleeria* (Pinaceae); *Smilax*, cf. *Heterosmilax* (Liliaceae); *Caldesia* (Alismataceae); *Salix*, *Populus* (Salicaceae); *Cercidiphyllum* (Cercidiphyllaceae); *Cocculus* (Menispermaceae); cf. *Asimina* (Annonaceae); *Lindera*, *Persea*, *Sassafras* (Lauraceae); *Hydrangea* (Saxifragaceae); *Exbucklandia*, cf. *Parrotia* (Hamamelidaceae); *Robinia*, *Amorpha*, cf. *Derris*, cf. *Zenia*, *Gleditsia*, *Gymnocladus* (Leguminosae); cf. *Gordonia* (Theaceae); *Diospyros* (Ebenaceae); cf. *Halesia* (Styracaceae); *Paulownia* (Scrophulariaceae); *Viburnum* (Caprifoliaceae); cf. *Malva* (Malvaceae); *Cedrela* (Meliaceae).

The taxa on this list include a number that have *never* been recorded as pre-Quaternary fossil pollen at the generic level, or in some cases generic *or* familial levels (see Muller's [1981] compilation based on a worldwide survey of angiosperm fossil pollen records). These include: (1) *Smilax*, *Heterosmilax* (other genera in the family Liliaceae have a few pollen records beginning in the upper Eocene); (2) *Caldesia* (the family Alismataceae has acceptable Pleistocene records, cf. Godwin, 1975, not recorded by Muller); (3) *Cocculus or* Menispermaceae; (4) *Asimina* (three other genera have scarce records beginning in the Upper Cretaceous); (5) *Lindera*, *Persea*, *Sassafras* (there are apparently two records from the Upper Paleocene that *may* be acceptable for the family, although as Muller notes, the low pollen production and the thin, poorly resistant exine make for poor fossilization—I personally doubt that they are authentic); (6) *Hydrangea or* Saxifragaceae (there are, however, Pleistocene pollen records for the Saxifragaceae, cf. Godwin, 1975, not recorded by Muller); (7) *Exbucklandia*; (8) *Robinia*, *Amorpha*, *Derris*, *Zenia*, *Gleditsia*, *Gymnocladus* (the Leguminosae have a number of widely scattered pollen records, confined almost exclusively to the Mimosoideae and the Caesalpinioideae, which Muller treats as families; the Papilionoidae—the Fabaceae of Muller—have been identified reliably only in the Pliocene of North Africa); (9) *Halesia or* Styracaceae; (10) *Paulownia* (the family Scrophulariaceae has no acceptable records in the Tertiary, although there are a number of Pleistocene pollen records assigned to various genera, cf. Godwin, 1975, not recorded by Muller); (11) *Malva* (there are about a half-dozen records of *Hibiscus*-type pollen, basically undifferentiated to genus, beginning in the Upper Eocene); and *Cedrela*.

The following have Tertiary fossil pollen/spore records: *Equisetum*, *Keteleeria*, *Salix*, *Populus*, *Cercidiphyllum*, *Parrotia*, *Gordonia*, *Diospyros* and *Viburnum*. All are based on scattered, sometimes single, records of varied degrees of reliability. *Populus*, for example, is notorious for its low preservation potential (see Sangster and Dale 1961). The frequency of its pollen in surface sediments is always at variance with its atmospheric frequency and with the local population abundance of poplar (for example Davis and Goodlett 1960; Ritchie and Lichti-Federovich 1963). It seems doubtful that poplar pollen would survive in recognizable form into the Tertiary. The only one of these genera that I have previously found in pollen assemblages of the Pacific Northwest is *Cercidiphyllum*

TABLE 1. IDENTIFIED TAXA AND RELATIVE FREQUENCY COUNTS OF POLLEN
AND SPORES IN SAMPLES C-1 THROUGH C-11, UNIT 2, CLARKIA, P-33[1]

Taxa	C-1	C-2	C-3	C-4	C-5	C-6	C-7	C-8	C-9	C-10	C-11
Fungi (hyphae, spores, conidia)	p	p	p	p	p	p	p	p	p	p	p
Dinoflagellates?	3.8	21.2	30.4	47.6	22.2	9.2	18.1	33.8	8.6	6.0	7.2
Anthoceros?											p
Lycopodium selago-type				0.2							
Lycopodium sp.		p									
Isoetes	p	p	p	p	p	p	p	p	p	p	p
Osmunda	p	p									
Polypodiaceae	0.9	1.1	0.5	0.5	0.8	0.6	0.5	0.6	0.8	0.6	0.7
Polypodium vulgare-type		p	0.2	p				p			0.2
Pinus	4.2	3.4	2.1	6.2	7.9	4.0	4.5	4.6	3.3	6.7	11.3
Abies	p	0.1		0.2	0.2	0.1					
Cedrus	1.2	1.1	1.1	2.7	3.1	4.4	0.6	1.3	1.6	1.1	3.1
Picea	0.3	0.4	0.3	0.2				0.1		p	0.2
Pseudotsuga-type		p	0.2	p				p			
Tsuga heterophylla-type	p	0.1	p	0.2							
Taxodiaceae-Cupressaceae-Taxaceae	39.8	35.9	28.2	27.7	23.7	28.6	25.1	31.2	31.6	28.6	28.0
Ephedra	0.2	p									
Typha-type			0.3	p	p	p		p	0.4		p
Gramineae		0.1	p								
Cyperaceae	0.3			0.3							
Myrica	p	0.1	p								
Juglans	0.1	0.1		p							
Carya	p	0.1	p	0.5	1.4	0.7	0.9	0.9	0.2	0.4	0.5
Pterocarya	6.1	2.2	1.8	1.6	0.6	1.0	0.4	1.2	p		0.2
Engelhardia-type	p										
Alnus	4.5	0.9	0.6	1.8	0.8	1.7	0.8	0.7	1.8	0.8	1.9
Betula	3.8	2.3	2.6	2.7	2.2	3.0	2.1	3.0	3.5	1.4	1.2
Ostrya-type	0.1	8.8	17.3	7.6	7.7	4.0	5.6	3.1	0.6	1.0	0.5

TABLE 1. CONTINUED

Sample Numbers

Taxa	C-1	C-2	C-3	C-4	C-5	C-6	C-7	C-8	C-9	C-10	C-11
Corylus	0.1	0.4	0.8	0.2	p	0.1	0.4	0.3		0.1	0.2
Castanea-type	5.1	1.8	2.4	2.2	2.8	2.3	4.1	4.3	3.7	6.3	7.4
Fagus	0.9	2.4	6.8	9.2	9.2	9.5	13.2	9.5	11.1	8.6	7.4
Quercus	0.6	2.4	4.5	6.2	6.8	6.8	5.2	6.8	7.4	8.1	6.0
Celtis	0.1	0.4	0.9	1.3	2.0	0.4	1.0	0.4	p	1.0	0.2
Ulmus-Zelkova	0.7	0.8	2.6	2.2	3.1	3.4	3.6	1.9	3.9	1.3	1.7
Chenopodiaceae	p	0.3	0.2	p	0.5	0.1	0.1	p		p	p
Liriodendron	0.1	0.4	p								
Magnolia	0.1	0.8	2.0	p	p	p	0.1	0.1	p	p	p
Liquidambar	1.4	3.5	2.1	1.9	2.2	1.7	1.7	0.9	0.8	1.1	1.5
Platanus	0.3	p	0.3	1.1	1.2	2.6	1.7	1.5	1.8	1.1	0.9
Rosaceae		p	0.2	0.2	0.2	p	p			p	0.2
Rhus		p	0.2	p		p		p	p	p	
Ilex	0.1	p			p		p			p	
Acer	p	0.4	0.2	0.6	0.6	0.4	0.1	0.4	0.2	0.2	
Aesculus	0.6	0.6	0.2	0.2	0.5	p	p	0.3	p	0.6	0.3
Rhamnaceae				p			p				0.2
Vitis-type	p	p	p	p			p	p	p	p	p
Parthenocissus	0.1						0.1	p			
Tilia	p	p	p	0.2	p				p	p	
Shepherdia	p	p		p						0.1	
Nyssa	p	0.1	0.2	p	0.2	0.1	0.1	p		0.1	p
Ericaceae	p					p	p				
Symplocos?	p										
Fraxinus	3.0	3.0	0.9	1.0		0.1	0.1		0.2	0.6	0.2
Unknown/Unidentifiable	24.9	26.5	20.9	21.7	22.7	24.8	27.8	26.7	27.3	30.3	26.2
N	691	800	664	628	649	703	774	674	513	717	583

1. The pollen sum on which the frequencies are based includes all pollen and spores (N) for each spectrum, but excludes the "dinoflagellates." "Dinoflagellate" frequencies are given in terms of total pollen and spores. P indicates pollen or spores found on scanned slides not included in the counts of that sample. The Unknown/Unidentifiable category includes pollen and spores of unknown affinities as well as poorly preserved pollen and spores.

(Gray *in* Martin and Gray 1962, Figure 4a) based on a few grains from the Mascall Formation.

Many of the taxa on this list are zoophilous with a limited pollen production by comparison with many wind-pollinated taxa. Zoophilous species are always underrepresented or absent in pollen diagrams of both Quaternary and Tertiary ages. That they were locally present in Tertiary floras of the Pacific Northwest, where all but *Heterosmilax, Caldesia, Derris* and *Zenia* have previous records, is clearly demonstrated by megafossils from Clarkia and elsewhere. Understanding of their vegetational history in this region necessarily depends on their megafossil remains.

Pollen/spore taxa without megafossil representation. There are a number of genera as well as families whose records in Unit 2 at Clarkia depend on pollen/spores only: *Isoetes* (Plate 2, figures 7,8) (Isoetaceae); *Lycopodium selago*-type and other *Lycopodium* spp. (Lycopodiaceae); *Cedrus* (Plate 2, figures 5,6), *Picea* (Plate 3, figure 10), *Pseudotsuga*-type (Plate 4, figure 7), *Tsuga heterophylla*-type (Plate 11, figures 5,6) (Pinaceae); *Ephedra* (Plate 2, figures 3,4) (Ephedraceae); Cyperaceae; *Juglans, Engelhardtia*-type (Juglandaceae); *Myrica* (Myricaceae); *Sarcobatus*-type (Plate 9, figures 8,9) (Chenopodiaceae); *Shepherdia* (Elaeagnaceae).

This list is diminutive compared with the megafossils that have no corresponding pollen/spore records. Most taxa listed have extremely limited occurrence, including most of the wind-pollinated conifers (Table 1). Pollen of ecologically expectable taxa such as *Myrica, Juglans* and *Engelhardtia* are as rare as pollen probably added to the assemblage by regional or long-distance transport. The list does not, of course, include unidentified taxa, of which there is some abundance (Table 1).

Two taxa, *Isoetes* and *Engelhardtia*, are represented by megafossils at Clarkia, but not in Unit 2 at P-33; *Juglans may* be represented in the Clarkia megaflora. *Isoetes* microspores are most common in the lower 4 samples at Unit 2 and become rare in the upper part of the sequence. This trend may relate to a change in trophic structure of the lake during deposition of Unit 2, since *Isoetes* is often common in oligotrophic lakes in circumstances of limited competition with other plants. It is also an early colonizer of nutrient-poor substrates in some types of lakes (Sculthorpe 1967; Hutchinson 1975). Because *Isoetes* is often but not exclusively a water plant or one of marginal wet habitats (Pfeifer 1922), it is unlikely that its spores were airborne into Unit 2 (although they might have been waterborne). Rather it seems probable that *Isoetes* was a common plant at this site despite the absence of co-occurring megafossils.

Engelhardtia has seldom been recorded as a megafossil in the Pacific Northwest, although a single distinctive fruit has been found at Clarkia P-38 (Smiley, pers. commun., 1982). Its pollen (including any of the complex of genera with similar pollen) occurs sporadically in Tertiary rocks of Washington and Oregon (Gray, unpublished data). Leopold and McGinitie (1972) record pollen of *Alfaroa-Engelhardtia* in the Rocky Mountain region from the Upper Cretaceous to the Oligocene, in Alaska in the late Oligocene–earliest Miocene (cf. Wahrhaftig et al. 1969) and throughout the Miocene of the Pacific Northwest and Idaho. They provide no documentation, however. The presence of *Juglans* megafossils at Clarkia is based on a few leaves that are of "questionable identification" without co-occurring fruits for verification (Smiley, pers. commun., 1982). Juglandaceous pollen is rare (Table 1). *Myrica* now grows in coastal bogs and thickets in the Pacific Northwest. Megafossil records of *Myrica* are infrequent, and it has not yet been found at Clarkia; its pollen is scarce at P-33 (Table 1), and it is equally scarce in other Tertiary pollen assemblages of this region (Gray, unpublished data).

The other taxa represented are a disparate group including pollen of: (1) a few herbaceous and shrubby elements that may be better represented in pollen than in megafossil assemblages; (2) genera potentially important in the regional environment, mainly gymnosperms; and (3) taxa of drier, more open environments than are otherwise represented by megafossils or most of the pollen taxa.

Pollen of herbaceous and shrubby taxa (the NAP of pollen analysts) may represent local, wet-ground plants and regional or even extra-regional taxa, e.g. Chenopodiaceae, *Shepherdia* and *Ephedra*. Sporadic grass and sedge pollen (Table 1) could have had a dry-ground, regional source, though it is also potentially local.

Herbaceous taxa are often poorly documented in Tertiary floras compared with woody plants, consistent with the possibility that herbs were not yet common in the vegetation. However, leaves and pollen bias the record toward woody taxa in a closed forest. Fruits and seeds of herbs, on the other hand, may be common and represent taxa otherwise unknown in the Tertiary (Łan-cucka-Środoniowa 1977, 1979). Low NAP at Clarkia, by analogy with modern circumstances, suggests a closed canopy forest where shrub and ground-cover pollen was suppressed. Hansen (1949) found pollen from the lowest arborescent and shrubby layers to be largely absent in moss polsters from evergreen forests in western Oregon. Faegri and Iversen (1975:135) cite a figure of 10% or less of the arboreal pollen for the NAP in limnic deposits in a heavily forested region, related in part to the low wind speeds in dense vegetation. Pollen released low to the ground, where it has essentially no opportunity through updraft to become a part of the canopy or general atmospheric component, may fall vertically to the ground (Handel [1976] noted that the dispersal distance of pollen varies inversely with the height of the pollen source) or if it is moved horizontally, it may be filtered out on the local vegetation (decreased movement of pollen in tree stands being well documented). Handel (1976) demonstrated empirically the limited movement of NAP in a forest situation. He found that pollen of two wind-pollinated species of *Carex* moved only 1 or 2 meters from the plants; pollen of one of the species was largely confined within 1 meter. Some NAP may enter limnic environments from ground surface runoff, although the amount will be tempered by the structure of the vegetation and the character of the local rains.

It is difficult to say what part wind or water transport might have played in the pollen occurrence of taxa not represented by megafossils. Big, heavy grains like those of *Abies, Picea, Pseudotsuga* (or *Larix*) and *Tsuga* have the reputation for settling rapidly in the air, and thus comparatively limited transport potential (Potter and Rowley, 1960, and Janssen, 1966, review much of the literature related to pollen production, settling velocity and dispersal potential of these and other taxa; see also Gregory, 1973). However, Hansen (1949) found hemlock and fir pollen in limited numbers in surface samples in western Oregon, where they must have been carried from at least several kilometers, and Kapp (1965) found low quantities of spruce and fir carried across the High Plains from the Front Range of the Rockies to the west. McAndrews and Wright (1969) in a similar transect from the Rocky Mountains to the North Dakota Plains showed that spruce, fir, and Douglas fir pollen were carried at least a "few tens of kilometers" beyond their source areas. Hansen (pers. commun. 1982) nevertheless confirms my impression from study of surface samples in the Pacific Northwest that many of the big-grained conifers have comparatively limited dispersal potential, especially in forest situations. This, together with the relatively low pollen production of some taxa (King and Kapp, 1963, for example, found spruce pollen frequency well below pine in surface spectra from Ontario, even though the trees were equally abundant in the forest) doubtless accounts for their infrequency at Clarkia.

Megafossil records indicate that *Picea, Tsuga*, and *Pseudotsuga* were established regionally in the Pacific Northwest in the late Tertiary. Megafossils of *Tsuga* and *Pseudotsuga*, however, have been found only in "floras of distinctly upland aspect" including the Blue Mountain and Trout Creek floras of Oregon and the Thorn Creek flora of Idaho (Chaney and Axelrod 1959:144). Whatever transport has occurred has not been extensive.

Ephedra pollen on the other hand (as well as pollen of the Chenopodiaceae) is widely disseminated away from its source areas (see, for example, King and Kapp 1963; Ritchie and Lichti-Federovich 1963; Maher 1964; and Janssen 1966, who provide examples of extreme long-range drift for *Ephedra* and Chenopodiaceae, including *Sarcobatus* pollen). However, *Ephedra* is also widely recorded among pollen assemblages in the Pacific Northwest during the Tertiary (Gray, unpublished data). Although *Ephedra* grows in extreme southeastern Oregon, I have never found it in modern pollen rains either to the east or west of the Cascades. Without confirmation of radical changes in wind patterns for this area, it seems reasonable to assume that *Ephedra* may have grown more regionally in the Pacific Northwest in the Tertiary.

The microfossil assemblage includes a variety of other types of fossils. There are, for example, abundant fungal remains, kindly examined by Dr. Martha Sherwood, who finds a diversity of spores (Plate 7, figure 3), hyphae, fruit-bodies, and so on (see Appendix). Many of these can be identified to familial and generic levels, and most appear to have environmental indicator value that provides an important supplement to interpretations derived from other remains.

Fungi have the potential to add to paleoenvironmental interpretations in other contexts. Although they possess a variety of adaptations to facilitate aerial takeoff of their spores, and once subjected to atmospheric turbulence they can be carried unlimited distances (Gregory 1973), many fungal remains found in the decaying plant litter near, at, or beneath the soil surface will endure only limited aerial or aquatic transport beyond the immediate environment of their occurrence (Muller 1959). The apparent correlation between high influx values of fungal remains in limnic sediments and increased levels of erosion and surface runoff demonstrated by Bradbury and Waddington (1973) shows that abundant fungal remains may also provide a useful index for tracking disturbances in the vegetation cover of the catchment area. Under those circumstances, fungal remains may represent largely transport assemblages of variable areal extent. In undisturbed forest, however, limited transport will be the rule for hyphae produced within a thick mat of decaying plant material as well as for large spores released into the air below a dense vegetation; their distribution will be as circumscribed as the pollen of herbs and shrubs that fail to escape from the low wind environment within the canopy space. Fungal remains either deposited with plant litter on which they were growing or airborne short distances into depositional basins can thus provide an effective index to autochthonous or essentially autochthonous deposits (for example, Muller 1959; Cohen 1970). Depending on the type of remains, fungi will tend to have a more local source than the pollen rain, which invariably includes a transported component. At Clarkia, the abundant, relatively large spores of the so-called impaction type (those that would be readily filtered out on the leaves, twigs and trunks of immediate vegetation), the high incidence of fungi characteristic of humid environments (M. A. Sherwood, pers. commun., 1982) which would tend to be damped out rather than transported aerially, together with the incidence of fungal hyphae and fruitbodies, evokes a limited transport assemblage.

The Clarkia fungal remains are of interest in still another frame of reference. Cohen (1973, 1974) found fungi to be consistently more common in the Okefenokee Swamp in microenvironments associated with forest swamp (*Taxodium* swamps with *Persea, Magnolia, Gordonia, Nyssa, Ilex*, all occurring at Clarkia) with their abundant surface litter of leaves and wood and their drier environments, than in open-water, unforested marshes dominated by floating and emergent, herbaceous aquatics. At Clarkia, the high-diversity fungal assemblage which is more characteristic of woody than herbaceous swamps (Sherwood, pers. commun., 1982), and the prominence of taxa indigenous to woody plants and evergreen leaves (Appendix) in conjunction with abundant Taxodiaceae and scarce remains of herbs reinforce the prominent part of tree swamps or perhaps of very humid lake-margin taxodiaceous woodlands adjacent to P-33 rather than open-water swamps.

Some "algae" (Figure 2) are common enough to provide relative counts (e.g. dinoflagellates?). Although it is not yet possible to interpret their significance, their potential for adding to the limnological history at Clarkia is high. *Ovoidites* is found preferentially, where its modern habitat has been examined, in freshwater environments (Cohen and Spackman 1972, Figure 11; Rich et al. 1982) and within the freshwater habitat in open-water marshes with emergent and floating aquatic herbs of the type termed "prairies" in the Okefenokee Swamp. Its occurrence is either diminished or it disappears entirely in sediments dominated by the pollen of shrubs and

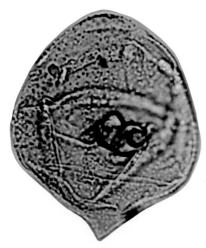

Figure 2. Dinoflagellate? (x 1000). This taxon, of unknown affinities (Rex Harland [pers. commun. 1982] suggests that these specimens may be dinoflagellates; R. Castenholz indicates [pers. commun. 1980] that they could be the encysting stage of a number of organisms, including dinoflagellates) is common throughout Unit 2, where it shows several major fluctuations in abundance that may relate to the chemistry of the lake and could be seasonal. Many dinoflagellate blooms occur in the early spring and summer.

trees (*Taxodium, Ilex, Cephalanthus*) (Rich et al. 1982). The scarcity of *Ovoidites* (Plate 13, figures 1-4) at Clarkia, where its record is confined to a single stratum (C-2), concurs with other evidence from the local vegetation that suggests nonoptimal circumstances for the occurrence of the *Ovoidites*-producing organism reported by Rich et al. (1982) to be *Spirogyra* or some closely related zygnemataceous algae.

A few "leaves" of mosses have also been recovered in the microfossil assemblages from sediments at P-33. No doubt bulk macerations of samples would add to our knowledge of this component at this locality.

Quantitative Results

There is much current enthusiasm among Pleistocene workers for using quantitative seed data, based on relative frequency counts or particularly seed influx, in conjunction with pollen counts from the same core to derive environmental information (see references, p. 192, herein; Watts 1978; Birks 1980 and references therein). In the Pacific Northwest, few Tertiary localities have proved to be rich enough in megafossils to compare megafossil counts from closely spaced samples with corresponding pollen counts from the same beds. Gray (*in* Martin and Gray 1962) compared pollen and leaf tallies from the Mascall and related deposits, but the information was not directly comparable in that the spot samples used for pollen were not closely correlated with leaf counts from any narrowly defined part of the Mascall Formation.

At Clarkia, because of abundant megafossils—predominantly leaves, but including a variety of other structures and organs—these can be counted in stratigraphic sequence, as can Quaternary fruit and seed remains, and used to make a "leaf diagram" of the pollen-diagram type. This diagram shows variations in the relative abundance of taxa through time that can be compared more or less directly with counts obtained from pollen (Figure 3). There are, however, some differences in the way the counts were obtained and the information that can be derived from them. Because of scarcity of megafossils at Clarkia, *in comparison* with pollen, leaf counts were secured from remains in contiguous 30-cm intervals of the sedimentary column (Figure 3). The pollen counts come from discrete samples representing not more than a 1-cm sample thickness

(Figure 3). This means that a direct comparison at a 1:1 level is not possible for the leaf and pollen counts. The possibility of obtaining "continuous" statistically meaningful counts from cm collecting intervals is one of the major advantages of pollen analysis in tracing vegetational history in comparison with megafossils, which usually are only locally abundant. This means that pollen can provide information on a scale for the Clarkia locality that is not possible with megafossils. The counts of leaves in cm intervals, based on from 13 to 56 specimens (Smiley and Rember 1981, Figure 17-6) provide interesting but not statistically meaningful information in comparison with pollen counts from similar units, in view of the size of the flora. Minor fluctuations in the vegetation that may have occurred during deposition of the 30 cm of sediments are lost in the leaf counts. With pollen it might be possible to trace vegetational changes related to the numerous ash falls that occur throughout Unit 2, because close-spaced samples can be taken both above and below the ash layers.

It is possible to examine the pollen content of the thin layers or laminations that occur in certain parts of the sequence to determine whether they have been deposited on a seasonal basis and represent nonglacial varves. Some information collected by Wu Zuo-ji and myself from a limited number of laminae suggests that this may be the case because there are taxic differences, and differences in absolute abundance of pollen in the light and dark laminae. Thus the light layers contain a much higher proportion of pollen of taxa that flower in the early spring than the dark laminae where the ratio of pollen of summer/autumn-flowering taxa to spring-flowering taxa is higher. In my opinion, laminae at P-33 are more easily accounted for as seasonal deposits than by the storm hypothesis advanced by Smiley and Rember (1981:556-557). I know of no similarly laminated deposits where storms are invoked in their explanation either in the marine or the nonmarine environment. The example cited by Smiley and Rember to support their hypothesis of rapid storm burial involves vertically and horizontally oriented fern fronds whose burial Brown and Gow (1976:279) likened to "bar deposition in rivers today," circumstances hardly analogous to the laminae at Clarkia Lake. Additionally, I do not regard the occurrence of single leaves crossing laminae to be preserved on two bedding planes as necessarily indicating more rapid deposition than implied by seasonal laminae. In an anaerobic environment where there is no evidence for bioturbation, it seems probable that a partially buried leaf could easily survive intact to have been wholly buried at a later time, although obviously we need to know from modern circumstances how long unburied organic material will survive in such an environment. Modern seeds and fruits in almost perfect condition are retrieved in shallow dredge sample from the Puerto Rico Trench, attesting to the potential for preservation of organic material in an anaerobic environment in the absence of both bioturbating and biodegrading organisms (Gunn and Dennis 1976).

It is relevant to note that the low total leaf counts from the 30-cm intervals (Figure 3, left column) leave room for statistical error in dealing with taxa with low-frequency counts. This is also a factor for pollen, but because of its abundance, it is possible to tally enough grains (literally thousands) to ensure statistical reliability for all the taxa that are present. Influx (grains/cm^2/yr) values can also be more readily secured with pollen than with leaves, although at Clarkia because of abundant megafossils and some knowledge of the sedimentation rate, it would be possible to look at leaf influx as Quaternary workers have secured such information for fruit and seed assemblages.

Finally, because of the greater abundance of pollen/spores produced by plants and their greater abundance in sediments at Clarkia, as compared with leaves and other structures, the opportunity for recovery of pollen of any nearby wind-pollinated or zoophilous taxon with relatively large pollen production is much greater than for the recovery of co-occurring megafossils of that taxon.

Comparative quantitative counts at Clarkia. Figure 3 is a diagram that combines relative frequency data for both megafossils and microfossils, predominantly pollen and spores, from Unit

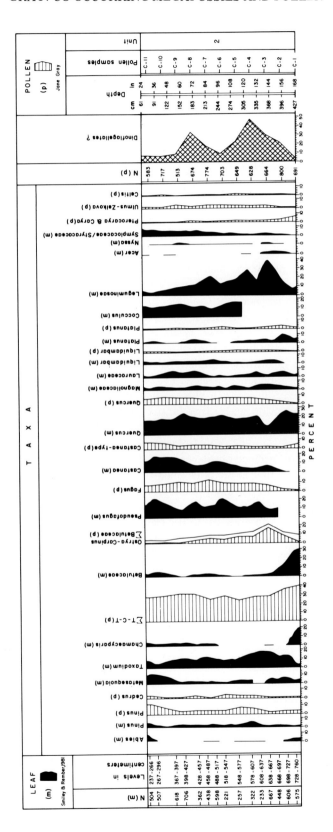

Figure 3. Comparative percentage diagram for megafossils (m) and pollen (p), for Unit 2, P-33, Clarkia Lake, Idaho. Percentages plotted against time. The basis of calculation for the pollen curves is the total pollen and spores [N(p)] in 1-cm increments of sediment at C-1 through C-11. The dinoflagellate? curve is calculated in terms of the pollen sum. The basis of calculation for the leaf curves is the total megaremains [N(m)] in 30-cm increments of sediment indicated at "levels in centimeters." For further explanation see text.

2 at Clarkia. Leaves are entered onto this diagram if they are represented by greater than 5% of the "leaf sum," which includes all leaves counted from each 30-cm unit of rock. Pollen is entered if it is represented by greater than 2% of the "pollen sum," which includes all pollen and spores from each 1-cm unit of rock. The pollen sum excludes the "dinoflagellates" total, which has been calculated as a percentage of the pollen sum.

Several genera and families are represented by sufficient leaf and pollen frequency to co-occur on the diagram.

Pinaceae. Two genera, *Pinus* and *Abies*, are represented by megafossils on the diagram. *Pinus* and *Cedrus* are the members of this family represented by pollen. The minor rise in pine pollen indicated at the top of Unit 2 may correspond to the increased abundance of pine megafossils found in overlying Unit 5 (from which pollen has not been recovered). The megafossil evidence, based on needles, cone scales, and seeds, indicates local occurrence of pine. Pine pollen representing more than one species is sufficiently uncommon to suggest that *Pinus* may not have been growing in the immediate vicinity of the depositional site. Ten percent pine pollen is frequently cited as the cutoff point in suggesting local presence or absence of pine (see Faegri and Iversen 1975:153). In low pollen-producing plant communities, which Clarkia was not, surface samples may contain as much as 15-25% pine pollen even in the absence of pine, and many kilometers distant from its nearest occurrence (for example, Bright 1966; McAndrews 1966; McAndrews and Wright 1969). Even though the seeds and needles may have been wind- or waterborne into Clarkia Lake, it is unlikely that they would have come from greater distances than could readily have accommodated aerial or aquatic transport of pine pollen.

Pine produces great quantities of pollen; its pollen is usually believed to be overrepresented relative to its population abundance. Janssen (1966) suggests, however, that most examples of pine pollen overrepresentation occur in vegetational assemblages where pine is *absent* and not where it is found in association with other large or relatively large pollen producers. In other words, its pollen records must be considered within the context of the particular source vegetation in which it occurs. Swamping out of pine pollen by the T-C-T count (p. 213 herein) is an alternative, because local occurrence of pine cannot be discounted in view of its megafossils. Potter and Rowley (1960) showed that in the pollen rain of a southwestern pine-juniper (a typical T-C-T pollen type) woodland, pine was strongly underrepresented and juniper pollen overrepresented in relation to all phytosociological measures of the vegetation. Whitehead and Tan (1969) found that in the southeastern Evergreen Forest of North Carolina, a vegetation type more comparable to Clarkia in taxic diversity and composition, pine pollen was more or less proportionately represented to pine abundance in the forest, but *Taxodium* plus Cupressaceae pollen was overrepresented. Thus, examples of overrepresentation of pine pollen in what are obvious transport circumstances may be due, as Janssen suggests, as much to its good dispersal potential as its excessive pollen production. At Clarkia, the combination of only a limited number of local pine trees and the abundance of T-C-T may account for the low pollen record.

The presence of *Abies* needles and cone scales, without the co-occurrence of large numbers of fir pollen, is less of a puzzle, because the large, heavy grains (*Abies* appears to have a terminal velocity several times that of other pollen and spores, according to information given in Gregory [1973]) seldom seem to be distributed any distance from the parent plant; this would especially be the case in dense vegetation.

Cedrus has never been identified from megafossils in Tertiary floras of the Pacific Northwest, although its pollen occurs regularly (Gray, unpublished data). Whether this means its megafossils have been consistently misidentified over the years, or that the trees were growing in sites which would have made the possibility of transport into depositional basins prohibitive cannot be determined. *Cedrus* pollen is more common than that of the other Pinaceae (*Tsuga, Pseudotsuga, Picea*) which have no co-occurring megafossils. The relative frequency difference is minor, but significant: cedar, with one exception, always exceeds 1% of the pollen sum and reaches a maximum of over 4%. None of the other three conifers exceeds a maximum 0.4% of the tally. Cedar pollen occurs in all samples; pollen of the others has sporadic occurrence (Table 1). This may mean that cedar pollen is more readily transported, aerially and aquatically, than the others, that it is produced in greater abundance, or that *Cedrus* was more common or more locally distributed than the other taxa. Some information relevant to the pollination ecology of cedar may help to resolve the question. Unlike other north temperate trees, cedar sheds its pollen in the fall over a 10-14 day period (Wright 1952). Wright (1952:28) has noted, however: "The male strobili are erect and so stiff that little pollen is shed unless the branches are shaken by strong winds." Most pollen is thus dispersed on cold, fall days by winds of greater-than-average velocity when cedar's deciduous associates have lost their leaves or are in the process of leaf-loss. Dispersal distances for cedar pollen are more or less comparable to those of *Pseudotsuga* and *Picea* pollen (Wright 1952), which is shed over a few days' period before or simultaneously with leafing of deciduous associates in the spring, and probably on windy days as well.

The sum of this information, including the difficulty of getting cedar pollen airborne, suggests that the more common occurrence of *Cedrus* pollen may relate to the presence of more trees and/or their more local occurrence in comparison with *Tsuga, Pseudotsuga* and *Picea*. Nevertheless, the minor pollen frequency of all these taxa without co-occurring megafossils is expectable for plants apparently not growing in the immediate vicinity of the depositonal basin. *Pinus* and *Abies* were clearly local plants whose low pollen frequency is possibly anomalous.

Taxodiaceae-Cupressaceae-Taxaceae (T-C-T). In Unit 2, the common megafossils that may correspond to the pollen count are *Metasequoia, Taxodium*, and *Chamaecyparis* (Cupressaceae). Pollen in Unit 2 potentially represents the Cupressaceae, Taxaceae, and Taxodiaceae, all of which are found as megafossils in the same unit. T-C-T pollen dominates the pollen tally with relative frequencies between 24 and 40%. That much of this pollen is taxodiaceous is indicated by the common incidence of papillate grains. The presence or absence of a papilla and variation in pollen size and surface pattern suggests, however, a number of genera and families. Of the megafossils present in sufficient abundance to be represented in the tallies (Figure 3), only *Taxodium* occurs in frequencies approximating the T-C-T count. It is tempting, because of this and its continuous occurrence in Unit 2, to assume that *Taxodium* has also contributed most of the pollen. However, the buoyancy and easy dispersal of pollen in all three families should urge caution in making this assumption.

T-C-T pollen dominates the pollen counts. In the megafossil diagram, co-dominance is shared with various members of the Fagaceae, and locally with Leguminosae and *Cocculus*, neither of which appear in the pollen diagram.

Betulaceae. The frequency curves based on megafossils and pollen show some significant differences. The Betulaceae curve for the megafossils is based mainly on *Betula*, the most common leaf taxon of this family at P-33 (Smiley, pers. commun., 1982). The pollen curve, however, is mainly *Ostrya-Carpinus* (Figure 3). Megafossil frequency increases precipitously in the lower 60 cm of Unit 2. The betulaceous pollen curve shows a corresponding abrupt decline. It also shows a gradual upward increase reaching a maximum at C-3. Thus the interval of maximum pollen frequency does not correspond to the interval of maximum megafossil frequency.

Some betulaceous taxa, e.g. *Betula, Alnus* and *Corylus*, are among the most prolific pollen producers, whose pollen is readily transported (Faegri and Iversen 1975). The limited representation of *Betula* and *Alnus* pollen might be interpreted as due to aerial or aquatic transport, if both taxa were not locally represented by megafossils. Like pine, the pollen of both taxa are probably responding to the large representation of T-C-T pollen. *Corylus* is scarce as leaves at P-33, but considered "common" elsewhere at Clarkia (Smiley, pers. commun., 1982). Its low pollen occurrence (Table 1) may have very little to do with the abundance of hazel in the immediate vegetation. *Corylus* only flowers abundantly in open, disturbed situations (Ritchie and Lichti-Federovich 1963). As a common understory shrub in closed-canopy deciduous forests, *Corylus* flowers infrequently or not at all (Ritchie and Lichti-Federovich 1963; Faegri and Iversen 1975).

Fagaceae. Fagaceous pollen (Figure 3) includes *Fagus, Castanea*-type, and *Quercus*. Corresponding megafossils are *Pseudofagus, Castanea*, and *Quercus*. If the pollen of the extinct *Pseudofagus* is morphologically close to that of *Fagus*, it may be included in that pollen curve. Fagaceous leaves are a much more significant proportion of the total leaf remains than the co-occurring pollen of the pollen sum. There are additional differences: leaves of *Pseudofagus* and *Castanea* are missing from the base of Unit 2—the corresponding pollen curves go to the base of the Unit, although the pollen of *Fagus* and *Quercus* decrease and pollen of *Castanea* increases slightly in abundance; the diminution of *Quercus* pollen is not matched by change in frequency of its leaf remains.

***Liquidambar* and *Platanus*.** The two remaining taxa with sufficient frequency to co-occur on the diagram are *Liquidambar* and *Platanus*, both represented by a small number of mega- and microfossils. Their megafossils are discontinuous compared with their pollen. *Liquidambar* megafossils disappear toward the base of Unit 2; *Platanus* has a disrupted record near the middle of Unit 2 and a significant decrease toward the top of the Unit. *Platanus* megafossils show an increase in frequency toward the base of the section that is not reflected in the pollen counts.

Hereafter, taxa abundant enough to be represented on the diagram by either leaves or pollen diverge. The following taxa with at least 5% of the leaf count in some 30-cm intervals of Unit 2 have either limited pollen records or none (see Table 1): *Liriodendron, Magnolia*, Lauraceae, *Cocculus*, Leguminosae, *Acer, Nyssa*, Symplocaceae/Styracaceae. The pollen diagram has minor frequencies without corresponding leaf counts of *Pterocarya* plus *Carya* (both are represented by pollen, but most of it in the early part of the sequence is *Pterocarya*), an undifferentiated *Ulmus/Zelkova* curve (both taxa are represented as megafossils and both may be involved in this curve), and a small but consistent frequency of *Celtis*, which has only sparse megafossil representation.

The pollen counts confirm the megafossil predominance of taxodiaceous and fagaceous taxa, although they do not duplicate in detail information implied from the leaf frequency counts. The pollination ecology of the taxodiaceous genera has not been examined in detail compared with other taxa. The size of the trees, the small, buoyant pollen grains, seemingly optimally adapted for wind-pollination, and the predominantly lowland, often wetland, habitat of many of the genera, would suggest that they might be among the group of taxa whose pollen is "overrepresented" in pollen diagrams from limnic sediments, i.e. plants whose pollen tends, in part because of the distribution of the trees, to swamp out many of the others when growing in association with them. Potter and Rowley (1960) in fact found Cupressaceae pollen, and Whitehead and Tan (1969) found *Taxodium* plus Cupressaceae pollen to be overrepresented in two different vegetation types, both of which included other large pollen producers such as pine. In addition, assuming no change in phenology over time, the early spring pollination of many of these taxa, including *Taxodium*, before leafing out of their deciduous associates, would assure unobstructed aerial transport of their pollen into depositional basins.

Modern species of oak and beech tend to be among the small-to-moderate pollen producers. Oak pollen especially may be "overrepresented" or "underrepresented" depending on its immediate associates. The abundant pollen of this group, particularly in association with T-C-T, would tend to confirm that fagaceous taxa, and members of the T-C-T complex (most of which may have been taxodiaceous) were among the dominants in the catchment vegetation around Clarkia Lake, even though the mega- and microfossil records do not conform in detail.

The implications of the other frequencies are difficult to assess. The pollen counts depart most significantly from the megafossil tallies where zoophilous taxa are concerned, many of which have provided abundant leaves but low pollen counts.

While there are large disparities in the independent leaf and pollen tallies, sufficient similarities exist in abundances to suggest that the megafossil record is not *merely* "happenstance." It reflects on the abundance of taxa in the vegetation cover just as pollen does. There are features related to the leaf tallies, however, that tend to support aspects of the "accidental" argument with regard to megafossil accumulations. Conspicuous features of the megafossil counts (Figure 3) are abrupt discontinuities (e.g. *Metasequoia, Chamaecyparis, Castanaea, Pseudofagus, Cocculus*), large-scale fluctuations within continuous records (e.g. *Quercus*), and major changes in relative abundances of taxa suggesting change in their frequency through time (e.g. *Chamaecyparis*, Betulaceae, Leguminosae). *Cocculus* ends abruptly at an ashfall layer (not indicated on the diagram). The presence of abundant leaves in laterally equivalent beds about 100 m away (Smiley, pers. commun., 1982) only reinforces the accidental aspect of their accumulation. In the absence of this information, it might appear that *Cocculus* was eradicated by the ash fall independently of the other genera, which show no disruption at this horizon. The record of *Pseudofagus* also ends abruptly, and independently of other taxa or of ash fall layers. Betulaceous leaves are diminished abruptly while the pollen record continues largely unabated. Did a sudden flush of betulaceous leaves unmixed with remains of other taxa swell their presence at the base of Unit 2 out of proportion to their local occurrence? Or is there a "real" change in the abundance of betulaceous taxa at Clarkia for which the pollen record appears discordant? The significance of these megafossil discontinuities is a major concern for interpreting leaf counts, either in stratigraphic section, as at Clarkia, or where comparing local florules that appear to represent the contemporaneous, regional vegetation such as the Miocene floras of the Columbia Plateau discussed by Chaney and Axelrod (1959). The concern at Clarkia pertains particularly to taxa whose megaremains end abruptly with an abundance that suggests their continued presence and whose pollen also continues with a frequency consistent with previous co-occurrence.

The pollen curves clearly are more continuous than those of the megafossils and tend to show relatively minor fluctuations. This type of difference has been noted by those comparing seed/fruit and pollen data in the Quaternary. Birks and Mathewes (1978:480), for example, attribute it to the greater opportunities to find pollen because of the abundance in which it is produced, compared with megastructures. Some of the abrupt discontinuities and fluctuations at Clarkia may be the result of the necessity to lump leaves and other structures from 30-cm intervals, thereby masking changes in leaf abundance through time. Additionally, or alternatively, sudden losses and extreme fluctuations in remains of plants that are ecologically compatible may reflect vagaries in the potential for mixing or "homogenization" of different plant organs and structures of different weight, size, and shape dispersed into a water body under different circumstances of leaf fall, wind velocity and/or changes in sediment type or other sedimentological factors. The "accidental" aspect imparted to the leaf counts independently and in part because of the co-occurring pollen records may well reflect leaf deposition in response to physical parameters.

Discontinuities of the type illustrated at Clarkia effectively mean that the absence of leaves in parts of a stratigraphic section or in a local florule cannot necessarily be attributed to the absence of the source plant in the vicinity of the depositional basin, even though the presence of megaremains generally provides incontrovertible evidence of plant occurrence near depositional sites (except in the case of fluvial deposits!). The anomaly of *Cocculus*, and others at Clarkia, points up the critical sampling problem for those who wish to use megafossils in making community interpretations. *Unless* it is possible at fossil sites to acquire data from a number of closely spaced sections where megafossils can be carefully sampled, there is always the possibility of being misled into believing that changes in abundance demonstrated for any single section may be ecologically related. A study by Christophel (1976) provides an early Tertiary example of small-scale horizontal discontinuities in plant megaremains that illustrates several points raised with regard to the Clarkia megafossils, including how the extent of sampling can influence the numbers and types of taxa seen to make up a plant community. Christophel's map (1976: Text-Fig. 6) of the lateral distribution of megafossils within a tuff horizon demonstrates that remains contributed by different genera to the depositional site were not randomly mixed prior to settling. Rather, with the exception of the uniformly distributed leaf-bearing branchlets of *Glyptostrobus*, remains of other taxa occur in restricted non-overlapping patches where apparently they were sorted out selectively prior to final deposition.

Pollen presence, unlike megaremains, may or may not carry the significance of local occurrence. Pollen absence, however, may well mean the local and regional absence of that taxon, particularly, but not exclusively, in the case of wind-pollinated species.

SUMMARY AND CONCLUSIONS

Because of the abundant megaremains in comparatively small units of the beds, and the excellent preservation of the fossils, the Clarkia Lake beds provide an opportunity, not previously available for Tertiary floras in the Pacific Northwest, to look at the qualitative and quantitative records of pollen and megaremains, principally leaves, and to compare them within the framework of the theoretical expectations said to affect accumulation of plant mega- and microfossils. Some expectations of what the differences would be are borne out by this case study, but by no means all of them. The possibility of assessing the records for closely comparable units of rock strengthens the significance of both the differences and the similarities. Information derived within the context of the Clarkia flora may provide for some generalizations applicable within the larger sphere of Tertiary floras of the Pacific Northwest.

Unlike comparative studies of the Quaternary that involve autochthonously accumulated fruits and seeds and allochthonous pollen, the mega- and microfossils at Clarkia have both accumulated allochthonously. Although transport may be limited, we can assume that the occurrence of plant remains is the result of at least some aerial and aquatic transport and physical sorting prior to sedimentation independently affecting each set of remains.

Taxic inventories make it clear that the leaf record better represents zoophilous plants with small pollen production and taxa whose pollen is fragile and thus poorly preserved. These taxa will always be "blind spots" in pollen analyses. The leaf record affords to Tertiary assemblages, just as fruit/seed studies afford to North Temperate Late Quaternary assemblages, a quality of taxonomic resolution that is impossible with pollen. Yet it may be possible, as in the Quaternary, to use the megafossil taxonomy where fossil taxa are closely related to living species and closely co-occur with pollen, to better elucidate the specific relations of some pollen taxa. For now, the quality of the identifications means that Tertiary megafossils can provide a great deal more specific environmental information than pollen. This also means that Tertiary megafossils are a much better source of information concerning *some* details of vegetational history than pollen whose taxonomic resolution is often at the generic and familial levels, including migrational history of taxa and information on migration rates, biogeographic information, and aspects of speciation and extinction. Axelrod (1983), for example, is able from fossil cones to trace the distributional history and evolution of coastal conifer forests in California, providing information that is impossible to obtain from pollen analysis.

However, megafossils are always far more localized in occurrence and less abundant than pollen. This will continue to be their principal limitation in the elucidation of vegetational history, except in small regions with abundant fossil localities, such as the Pacific Northwest. Also, it is impossible to derive quantitatively significant information from megafossils in small units of rock of the type available from pollen, which means the potential loss of small-scale vegetational information.

At Clarkia, as elsewhere, the pollen and spore record favors the wind-pollinated taxa including those whose growth sites may be prohibitive to the entry of large plant structures into depositional basins except by chance. The occurrence of some pollen types in the Clarkia assemblage not represented by megafossils suggests that the size of the area contributing to the pollen rain is less circumscribed, i.e. more regional, than that contributed to by megafossils. This judgment is provisional, however, because it is difficult to determine the meaning of megafossil absence. At Clarkia the "regional" or potentially long distance component is very small, qualitatively and quantitatively, and may be related to the large proportion of pollen delivered via the trunk space to nearshore deposits where leaf remains were also accumulating. Were these deposits farther from shore, one might expect that the rain-out and canopy components would have represented a larger share of the pollen rain relative to the pollen derived from local parts of the forest.

Microfossils also include the remains of fungi and algae and spores of bryophytes and lower vascular plants that may be poorly or not at all represented by fossil leaves or other structures. At Clarkia because of excellent megafossil preservation, taxa such as *Isoetes*, bryophytes, and ferns are better represented as megafossils than is normally the case. Thus disparities normal to the occurrence of certain types of spores relative to their megafossils are less than usual.

The quantitative aspects are difficult to assess because so little is known of the controls on the occurrence of leaves and other structures in fossil deposits. There are nevertheless a number of similarities as well as dissimilarities between the pollen and leaf counts in Unit 2 at P-33. Those similarities that tend to reinforce the dominance of some key taxa make it reasonable to assume that their leaf number is in part a function of their commonness in the vegetation cover of the catchment area, *even though* there are a variety of other types of controls operating on leaf occurrence (cf. Chaney 1959). This suggests that leaf abundance may indicate the dominants of communities and that leaf counts have the *potential* to provide effective quantitative information. Paleobotanists need to take the extra steps taken by pollen analysts and sample occurrences of leaves and other megastructures allochthonously accumulated in varied types of lakes in a variety of topographies where the vegetation is analogous to fossil assemblages. This information will provide a frame of reference against which to assess presence and absence and abundance data and a far better guide for reconstructing former vegetation than a *direct* comparison between fossil leaf remains and modern vegetation of the type now used by paleobotanists. From such studies, which involve comparison between modern leaf accumulations and modern vegetation on the one hand, and fossil leaf accumulations on the other, paleobotanists will be provided with "interpretive links" between megafossils and analogous modern vegetation of the type provided by surface pollen rains. Chaney (1924) took the first preliminary steps toward interpreting megafloras by comparing fallen leaves and vegetation. If the raw data of such comparative studies are subject to refinements of the types made by pollen analysts with pollen, and if the data are used conservatively, paleobotanists should be able to impart more specific interpretations to fossil leaf counts and to link them, albeit in a complex way, to numbers of source plants. This approach will be more fruitful in providing useful information than the experimental approach of Spicer (1981) or the preliminary experiment performed on leaves by Ferguson (1971). These experiments involve highly specialized circumstances which may only poorly duplicate nature and that probably have little to do with conditions that lead to most fossil leaf accumulations.

Because mixing of megastructures may be dependent on their physical characters in conjunction with sedimentological factors, and because leaves, fruits and seeds and so on are produced in comparatively low frequency, the presence or absence of taxa and changes in their relative abundances from florule to florule, or within a single stratigraphic sequence, may be misleading as a basis for suggesting floral change. Absence of megafossils carries no greater significance than occurrences of some pollen types. The nonrandom mixing of megaremains is probably a major factor influencing the character of the Clarkia megafossil counts at P-33 and doubtless accounts for some of the discontinuities, disruptions, and fluctuations that are not documented by pollen counts.

If the megafossil accumulation at Clarkia is vegetationally biased in terms of local lakeside plants, whatever biases exist cannot be demonstrated to be potentially more significant than have affected the pollen accumulation where a large number of megafossil taxa are not represented. The advantage of pollen—the addition of a large number of regional elements that would "correct for" the supposedly parochial nature of the megafossils and add evidences of communities not represented by them—is not a major feature at Clarkia in Unit 2, P-33. The regional component is very limited. Because megaremains appear to accumulate in nearshore situations where the immediate vegetation may be overrepresented, pollen accumulated in these same deposits may be equally poorly mixed and represent predominantly vegetation within a few tens of meters of the shoreline. In other words, serious sampling errors in terms of regional vegetation may attend the collections of pollen and spores from megafossil-bearing deposits because the distribution of pollen is not uniform over synchronous surfaces, especially where these surfaces impinge on the shore. In such circumstances, overrepresentation of local plant remains may be a problem that is not overcome by the pollen rain, and similarities in megafossil and pollen tallies that tend to reinforce one another may only indicate biases in the same direction.

Considering the taxonomic limitations, the lower taxic diversity, and the variety of unidentified taxa, pollen analysis at Clarkia gives a less complete picture of the immediate vegetation of the catchment area around Clarkia Lake than the megafossils and thus, potentially, a less complete picture of the paleoecology.

ACKNOWLEDGMENTS

Several years ago, C. J. Smiley asked me to examine pollen from the Clarkia locality. I would especially like to thank him for the opportunity to take part in this study, for his patience with regard to my long delay in birthing this paper, for answering my numerous questions with regard to unpublished information on the Clarkia plant megafossils, and for reading a first draft of this ms. and offering many editorial suggestions. I would also like to thank Martha A. Sherwood for her identification of the fungal material attributed to *Helicoön* (*Helicodendron*) and A. J. Boucot for reading several versions of the ms. and asking a number of annoying questions, which, together with Smiley's comments, forced me to become even more obdurate about many of the ideas expressed in this paper. Frances Duryee, my faithful laboratory technician, can be thanked for her preparation of the sediment samples.

LITERATURE CITED

Argus, G. W., and M. B. Davis. 1962. Macrofossils from a late-glacial deposit at Cambridge, Massachusetts. Amer. Midl. Nat. 67(1):106-117.

Axelrod, D. I. 1983. New Pleistocene conifer records, coastal California. Univ. Calif. Pub. Geol. Sci. 127:1-108.

Baker, R. G. 1965. Late-glacial pollen and plant macrofossils from Spider Creek, southern St. Louis County, Minnesota. Geol. Soc. Amer. Bull. 76:601-610.

Barbour, E. H. 1925. Hackberry conglomerate. Nebraska State Museum Bull. 1(8):87-90.

Batten, D. J. 1969. Some British Wealden megaspores and their facies distribution. Palaeontology 12(2):333-350.

Batten, D. J. 1972. Recognition of the facies of palynologic assemblages as a basis for improved stratigraphic correlation. Proc. 24th Int'l. Geol. Congr., Montreal, Canada, Sect. 7:367-374.

Batten, D. J. 1973. Use of palynologic assemblage-types in Wealden correlation. Palaeontology 16(1):1-40.

Batten, D. J. 1975. Wealden palaeoecology from the distribution of plant fossils. Proc. Geol. Assoc. 85(4):433-458.

Birks, H. H. 1973. Modern macrofossil assemblages in lake sediments in Minnesota. Pages 173-189 in H. J. B. Birks and R. G. West, eds. Quaternary Plant Ecology. John Wiley & Sons, New York, N. Y.

Birks, H. H. 1980. Plant macrofossils in Quaternary lake sediments. Archiv. für Hydrobiologie Beiheft: Ergebnisse der Limnologie 15:1-60.

Birks, H. H., and R. W. Mathewes. 1978. Studies in the vegetational history of Scotland. V. Late Devensian and Early Flandrian pollen and macrofossil stratigraphy at Abernethy Forest, Inverness-Shire. New Phytol. 80:455-484.

Birks, H. J. B. 1976. Late-Wisconsinan vegetational history at Wolf Creek, central Minnesota. Ecol. Monogr. 46(4):394-429.

Birks, H. J. B., and H. H. Birks. 1980. Quaternary palaeoecology. University Park Press, Baltimore, Md. 289 pp.

Birks, H. J. B., T. Webb III, and A. A. Berti. 1975. Numerical analysis of pollen samples from central Canada: A comparison of methods. Rev. Palaeobot. Palynol. 20:133-169.

Birks, H. J. B., and R. G. West, eds. 1973. Quaternary plant ecology. John Wiley & Sons, New York, N. Y. 326 pp.

Bonny, A. P. 1976. Recruitment of pollen to the seston and sediment of some Lake District Lakes. J. Ecol. 64(3):859-887.

Bonny, A. P. 1978. The effect of pollen recruitment processes on pollen distribution over the sediment surface of a small lake in Cumbria. J. Ecol. 66(2):385-416.

Boulter, M. C., and W. G. Chaloner. 1970. Neogene fossil plants from Derbyshire (England). Rev. Palaeobot. Palynol. 10:61-78.

Bradbury, J. P., and J. C. B. Waddington. 1973. The impact of European settlement on Shagawa Lake, northeastern Minnesota, U. S. A. Pages 289-307 in H. J. B. Birks and R. G. West, eds., Quaternary Plant Ecology. John Wiley & Sons, New York, N. Y.

Bright, R. C. 1966. Pollen and seed stratigraphy of Swan Lake, southeastern Idaho: Its relation to regional vegetational history and to Lake Bonneville history. Tebiwa Misc. Pap. Idaho State Univ. Mus. Nat. Hist. 19(2):1-47.

Brown, J. T., and C. E. Gow. 1976. Plant fossils as indicators of the rate of deposition of the Kirkwood Formation in the Algoa Basin. So. Afr. J. Sci. 72:278-279.

Bullock, S. H., and R. B. Primack. 1977. Comparative experimental study of seed dispersal on animals. Ecology 58(3):681-686.

Burrows, C. J. 1974. Plant macrofossils from Late-Devensian deposits at Nant Ffrancon, Caernarvonshire. New Phytol. 73:1003-1033.

Burrows, C. J. 1980. Long-distance dispersal of plant macrofossils. New Zealand J. Bot. 18:321-322.

Caseldine, C. J., and A. D. Gordon. 1978. Numerical analysis of surface pollen spectra from Bankhead Moss, Fife. New Phytol. 80:435-453.

Chandler, M. E. J. 1964. The Lower Tertiary floras of Southern England. IV. A summary and survey of findings in the light of recent botanical observations. Brit. Mus. Nat. Hist., London. 151 pp.

Chaney, R. W. 1924. Quantitative studies of the Bridge Creek flora. Amer. J. Sci., 5th Ser., 8(44):127-144.

Chaney, R. W. 1925. Studies on the fossil flora and fauna of the western United States. III. Notes on two fossil hackberries from the Tertiary of the western United States. Carnegie Inst. Wash. Pub. 349:49-56.

Chaney, R. W. 1938. Paleoecological interpretations of Cenozoic plants in western North America. Bot. Rev. 9(7):371-396.

Chaney, R. W. 1959. Miocene floras of the Columbia Plateau, Part I. Composition and interpretation. Carnegie Inst. Wash. Pub. 617:1-134.

Chaney, R. W., and D. I. Axelrod. 1959. Miocene floras of the Columbia Plateau, Part II. Systematic considerations. Carnegie Inst. Wash. Pub. 617:135-229.

Chaney, R. W., and H. L. Mason. 1936. A Pleistocene flora from Fairbanks, Alaska. Amer. Mus. Novit. 887:1-17.

Chaney, R. W., and E. I. Sanborn. 1933. The Goshen flora of west central Oregon. Carnegie Inst. Wash. Pub. 439:1-103.

Christophel, D. C. 1976. Fossil floras of the Smoky Tower locality, Alberta, Canada. Paleont. Abt. B, Bd. 157(1-4):1-43.

Clapham, W. B., Jr. 1970. Evolution of Upper Permian terrestrial floras in Oklahoma as determined from pollen and spores. Proc. No. Amer. Paleontol. Convention, Chicago, 1969, E:411-427.

Cohen, A. D. 1970. An allochthonous peat deposit from southern Florida. Geol. Soc. Amer. Bull. 81:2477-2482.

Cohen, A. D. 1973. Petrology of some Holocene peat sediments from the Okefenokee swamp-marsh complex of southern Georgia. Geol. Soc. Amer. Bull. 84:3867-3878.

Cohen, A. D. 1974. Petrology and paleoecology of Holocene peats from the Okefenokee swamp-marsh complex of Georgia. J. Sediment. Petrol. 44(3):716-726.

Cohen, A. D., and W. Spackman. 1972. Methods in peat petrology and their application to reconstruction of paleoenvironments. Geol. Soc. Amer. Bull. 83:129-142.

Craighead, F. C., and V. C. Gilbert. 1962. The effects of Hurricane Donna on the vegetation of southern Florida. Quart. J. Florida Acad. Sci. 25(1):1-28.

Crowder, A. A., and D. G. Cuddy. 1973. Pollen in a small river basin: Wilton Creek, Ontario. Pages 61-77 in H. J. B. Birks and R. G. West, eds. Quaternary Plant Ecology. John Wiley & Sons, New York, N. Y.

Crowder, A., and R. N. Starling. 1980. Contemporary pollen in the Salmon River Basin, Ontario. Rev. Palaeobot. Palynol. 30:11-26.

Currier, P. J., and R. O. Kapp. 1974. Local and regional pollen rain components at Davis Lake, Montcalm County, Michigan. Mich. Acad. 7(2):211-225.

Davis, M. B. 1968. Pollen grains in lake sediments: Redeposition caused by seasonal water circulation. Science 162(3855):796-799.

Davis, M. B. 1973. Redeposition of pollen grains in lake sediment. Limnol. Oceanog. 18(1):44-52.

Davis, M. B., and J. C. Goodlett. 1960. Comparison of the present vegetation with pollen-spectra in surface samples from Brownington Pond, Vermont. Ecology 41(2):346-357.

Davis, R. B., L. A. Brewster, and J. Sutherland. 1969. Variation in pollen spectra within lakes. Pollen et Spores 11(3):557-571.

Davis, R. B., and T. Webb III. 1975. The contemporary distribution of pollen in eastern North America: A comparison with the vegetation. Quatern. Res. 5:395-434.

Delcourt, P. A., and H. R. Delcourt. 1977. The Tunica Hills, Louisiana-Mississippi: Late glacial locality for spruce and deciduous forest species. Quatern. Res. 7:218-237.

Delcourt, P. A., H. R. Delcourt, and T. Webb III. 1982. Mapping and calibration of modern pollen and vegetation in eastern North America as a quantitative basis for Quaternary paleoecological reconstructions. Geol. Soc. Amer. Abstr. with Prog. 14(7):474-475.

Dickinson, C. H., and G. J. F. Pugh, eds. 1974. Biology of plant litter decomposition. Vols. I and II. Academic Press, London. 146 pp., 175 pp.

Dickson, C. A. 1970. The study of plant macrofossils in British Quaternary deposits. Pages 233-254 in D. Walker and R. G. West, eds. Studies in the Vegetational History of the British Isles. Cambridge University Press, Cambridge.

Dorf, E. 1945. Observations on the preservation of plants in the Parícutin area. Trans. Amer. Geophys. Union 26(2):257-260.

Dorf, E. 1951. Lithologic and floral facies in the Parícutin ash deposits, Mexico. Trans. N. Y. Acad. Sci. 13:317-320.

Drake, H., and C. J. Burrows. 1980. The influx of potential macrofossils into Lady Lake, north Westland, New Zealand. New Zealand J. Bot. 18:257-274.

Ellis, M. B. 1976. More dematiaceous Hyphomycetes. Commonwealth Mycological Institute, Kew. 507 pp.

Ewel, J. J. 1976. Litter fall and leaf decomposition in a tropical forest succession in eastern Guatemala. J. Ecol. 64(1):293-308.

Faegri, K., and J. Iversen. 1975. Textbook of pollen analysis. 3rd ed. Hafner Press, New York, N.Y. 295 pp.

Ferguson, D. K. 1971. The Miocene flora of Kreuzau, western Germany, 1. The leaf-remains. (Verh. Koninkl. Nederlandse Akad. Wetensch. afd. Natuurkunde, Tweede Reets, 60:1). North-Holland Publishing Co., Amsterdam, London. 297 pp.

Gasith, A., and A. D. Hasler. 1976. Airborne litterfall as a source of organic matter in lakes. Limnol. Oceanog. 21(2):253-258.

Gillespie, W. H., and J. A. Clendening. 1968. A flora from proglacial lake Monongahela. Castanea 33:267-300.

Glen-Bott, J. I. 1951. *Helicodendron giganteum* n. sp. and other aerial-sporing Hyphomycetes of submerged dead leaves. Trans. Brit. Mycol. Soc. 34(3):275-279.

Godwin, H. 1975. The history of the British flora: A factual basis for phytogeography. 2nd ed. Cambridge University Press, Cambridge. 541 pp.

Green, D. S. 1983. The efficacy of dispersal in relation to safe site density. Oecologia 56:356-358.

Gregory, P. H. 1973. The microbiology of the atmosphere. 2nd ed. John Wiley & Sons, New York, N. Y. 377 pp.

Grüger, E. 1972. Pollen and seed studies of Wisconsinan vegetation in Illinois, U. S. A. Geol. Soc. Amer. Bull. 83:2715-2734.

Grüger, J. 1973. Studies on the late Quaternary vegetation history of northeastern Kansas. Geol. Soc. Amer. Bull. 84:239-250.

Gunn, C. R., and J. V. Dennis. 1976. World guide to tropical drift seeds and fruits. Times Books, New York, N. Y. 233 pp.

Heath, G. W., and M. K. Arnold. 1966. Studies in leaf-litter breakdown. II. Breakdown rate of 'sun' and 'shade' leaves. Pedobiologia 6:238-243.

Handel, S. N. 1976. Restricted pollen flow of two woodland herbs determined by neutron-activation analysis. Nature 260:422-423.

Hansen, C. B. 1973. Geology and vertebrate faunas in the type area of the Clarno Formation, Oregon. Geol. Soc. Amer. Abstr. with Progr., Cordilleran Section, 5(1):50.

Hansen, H. P. 1949. Pollen content of moss polsters in relation to forest composition. Amer. Midl. Nat. 42(2):473-479.

Harper, J. L. 1977. Population biology of plants. Academic Press, New York, N. Y. 892 pp.

Howe, H. F. 1977. Bird activity and seed dispersal of a tropical wet forest tree. Ecology 58(3): 539-550.

Hutchinson, G. E. 1975. A treatise on limnology. Vol. III. Limnological botany. John Wiley & Sons, New York, N. Y. 660 pp.

Isaac, L. A. 1930. Seed flight in the Douglas fir region. J. Forest. 28:492-499.

Janssen, C. R. 1966. Recent pollen spectra from the deciduous and coniferous-deciduous forests of northeastern Minnesota: A study in pollen dispersal. Ecology 47(5):804-825.

Janssen, C. R. 1973. Local and regional pollen deposition. Pages 31-42 *in* H. J. B. Birks and R. G. West, eds. Quaternary Plant Ecology. John Wiley & Sons, New York, N. Y.

Kapp, R. O. 1965. Illinoian and Sangamon vegetation in southwestern Kansas and adjacent Oklahoma. Contr. Mus. Paleont., Univ. Michigan XIX (14):167-255.

Kaushik, N. K., and H. B. N. Hynes. 1971. The fate of the dead leaves that fall into streams. Archiv für Hydrobiologie 68(4):465-515.

Keepax, C. 1977. Contamination of archaeological deposits by seeds of modern origin with particular reference to the use of flotation machines. J. Archaeol. Sci. 4:221-229.

King, J. E., and R. O. Kapp. 1963. Modern pollen rain studies in eastern Ontario. Can. J. Bot. 41:243-252.

Łańcucka-Środoniowa, M. 1977. New herbs described from the Tertiary of Poland. Acta Palaeobotanica 18(1):37-44.

Łańcucka-Środoniowa, M. 1979. Macroscopic plant remains from the freshwater Miocene of the Nowy Sacz Basin (West Carpathians, Poland). Acta Palaeobotanica 20(1):3-117.

Leopold, E. B., and H. D. MacGinitie. 1972. Development and affinities of Tertiary floras in the Rocky Mountains. Pages 147-200 *in* A. Graham, ed. Floristics and Paleofloristics of Asia and Eastern North America. Proc. Symposia, Systematics Section, XI Int'l. Bot. Congr., Seattle, Wash., and Japan-U.S. Coop. Sci. Program, Corvallis, Ore., 1969. Elsevier Pub. Co., New York.

Levin, D. A. 1979. The nature of plant species. Science 204:381-384.
Levin, D. A., and H. W. Kerster. 1974. Gene flow in seed plants. Evol. Biol. 7:139-220.
Maher, Jr., L. J. 1964. *Ephedra* pollen in sediments of the Great Lakes region. Ecology 45(2): 391-395.
Martin, P. S., and J. Gray. 1962. Pollen analysis and the Cenozoic. Science 137:103-111.
Matthews, J. V., Jr. 1975. Incongruence of macrofossil and pollen evidence: A case from the late Pleistocene of the northern Yukon Coast. Geol. Surv. Can., Pap. 75-1 (Pt. B):139-146.
McAndrews, J. H. 1966. Postglacial history of prairie, savanna, and forest in northwestern Minnesota. Mem. Torrey Bot. Club 22(2):1-72.
McAndrews, J. H., and D. M. Power. 1973. Palynology of the Great Lakes: The surface sediments of Lake Ontario. Can. J. Earth Sci. 10(5):777-792.
McAndrews, J. H., and H. E. Wright, Jr. 1969. Modern pollen rain across the Wyoming basins and the Northern Great Plains (U.S.A.). Rev. Palaeobot. Palynol. 9:17-43.
McMurray, M., G. Kloos, R. O. Kapp, and K. Sullivan. 1978. Paleoecology of Crystal Marsh, Montcalm County, based on macrofossil and pollen analysis. Mich. Acad. 10(4):403-417.
McQueen, D. R. 1969. Macroscopic plant remains in Recent lake sediments. Tuatara 17(1):13-19.
Merrill, E. D. 1928. Flora of the Philippines. Pages 130-167 *in* R. E. Dickerson, ed. Distribution of Life in the Philippines. Monogr. 21, Bureau of Science, Bureau of Printing, Manila, Philippines.
Minnis, P. E. 1981. Seeds in archaeological sites: Sources and some interpretive problems. Amer. Antiquity 46(1):143-152.
Moore, P. D., and J. A. Webb. 1978. An illustrated guide to pollen analysis. John Wiley & Sons, New York, N. Y. 133 pp.
Muller, J. 1959. Palynology of Recent Orinoco delta and shelf sediments: Report of the Orinoco Shelf Expedition: Vol. 5. Micropaleontology 5(1):1-32.
Muller, J. 1981. Fossil pollen records of extant angiosperms. Bot. Rev. 47(1):1-142.
Peck, R. M. 1973. Pollen budget studies in a small Yorkshire catchment. Pages 43-60 *in* H. J. B. Birks and R. G. West, eds. Quaternary Plant Ecology. John Wiley & Sons, New York, N. Y.
Pfeiffer, N. E. 1922. Monograph of the Isoetaceae. Ann. Mo. Bot. Garden 9:79-232.
Potter, L. D., and J. Rowley. 1960. Pollen rain and vegetation, San Augustin Plains, New Mexico. Bot. Gaz. 122(1):1-25.
Praeger, R. L. 1913. On the buoyancy of the seeds of some Britannic plants. Sci. Proc. Royal Dublin Soc. 14(n.s.):13-62.
Rich, F. J., D. Kuehn, and T. D. Davies. 1981. *Ovoidites ligneolus* R. Pot. as an indicator of ancient freshwater marsh habitats. Palynology 5:242.
Rich, F. J., D. Kuehn, and T. D. Davies. 1982. The paleoecological significance of *Ovoidites*. Palynology 6:19-28.
Ridley, H. N. 1930. The dispersal of plants throughout the world. L. Reeve & Co. Ltd., London. 744 pp.
Ritchie, J. C., and S. Lichti-Federovich. 1963. Contemporary pollen spectra in central Canada. I. Atmospheric samples at Winnipeg, Manitoba. Pollen et Spores 5(1):95-114.
Roth, J. L., and D. L. Dilcher. 1978. Some considerations in leaf size and leaf margin analysis of fossil leaves. Courier Forschungsinstitut Senckenberg 30:165-171.
Rybníček, K., and E. Rybníčková. 1968. The history of flora and vegetation on the Bláto mire in southeastern Bohemia, Czechoslovakia (palaeoecological study). Folia Geobot. Phytotaxonom. 3(2):117-142.
Rybníček, K., and E. Rybníčková. 1974. The origin and development of waterlogged meadows in the central part of the Sumava foothills. Folia Geobot. Phytotaxonom. 9(1):45-70.
Ryvarden, L. 1971. Studies in seed dispersal I. Trapping of diaspores in the Alpine Zone at Finse, Norway. Nor. J. Bot. 18:215-226.
Ryvarden, L. 1975. Studies in seed dispersal II. Winter-dispersed species at Finse, Norway. Nor. J. Bot. 22:21-24.
Sangster, A. G., and H. M. Dale. 1961. A preliminary study of differential pollen grain preservation. Can. J. Bot. 39:35-43.
Scheihing, M. H. 1980. Reduction of wind velocity by the forest canopy and the rarity of non-

arborescent plants in the Upper Carboniferous fossil record. Arg. Palaeobot. 6:133-138.
Sculthorpe, C. D. 1967. The biology of aquatic vascular plants. St. Martin's Press, New York, N. Y. 610 pp.
Segerstrom, K. 1950. Erosion studies at Parícutin, State of Michoacán, Mexico. U. S. Geol. Surv. Bull. 965-A:1-164.
Sheldon, J. C., and F. M. Burrows. 1973. The dispersal effectiveness of the achene-pappus units of selected Compositae in steady winds with convection. New Phytol. 72:665-675.
Smiley, C. J., and W. C. Rember. 1981. Paleoecology of the Miocene Clarkia Lake (Northern Idaho) and its environs. Pages 551-590 (Chapter 17) in J. Gray, A. J. Boucot, and W. B. N. Berry, eds. Communities of the Past. Hutchinson and Ross Publishing Co., Stroudsburg, Pa.
Sønstegaard, E., and J. Mangerud. 1977. Stratigraphy and dating of Holocene gully sediments in Os, western Norway. Norsk Geol. Tid. 57(4):313-346.
Spicer, R. A. 1981. The sorting and deposition of allochthonous plant material in a modern environment at Silwood Lake, Silwood Park, Berkshire, England. U. S. Geol. Surv. Prof. Pap. 1143: 1-77.
Starling, R. N., and A. Crowder. 1981. Pollen in the Salmon River system, Ontario, Canada. Rev. Palaeobot. Palynol. 31:311-334.
Tauber, H. 1965. Differential pollen dispersion and the interpretation of pollen grains. Danmarks Geol. Under. II Raekke, Nr. 89:7-69.
Tauber, H. 1967a. Investigations of the mode of pollen transfer in forested areas. Rev. Palaeobot. Palynol. 3:277-286.
Tauber, H. 1967b. Differential pollen dispersal and filtration. Pages 131-141 in E. J. Cushing and H. E. Wright, Jr., eds. Quaternary Paleoecology. Yale University Press, New Haven, N. J.
Tauber, H. 1977. Investigations of aerial pollen transport in a forested area. Dansk. Bot. Ark. 32: 1-121.
Turner, J. 1970. Post-Neolithic disturbance of British vegetation. Pages 97-116 in D. Walker and R. G. West, eds. Studies in the Vegetational History of the British Isles. Cambridge University Press, Cambridge.
Turner, J. 1975. The evidence for land use by prehistoric farming communities: The use of three-dimensional pollen diagrams. Pages 86-95 in J. G. Evans, S. Limbrey and H. Cleere, eds. The Effect of Man on the Landscape: The Highland Zone. Res. Rep. 11, Council for British Archaeology.
van der Pijl, L. 1972. Principles of dispersal in higher plants. 2nd ed. Springer-Verlag, Berlin, Heidelberg, New York. 162 pp.
Wahrhaftig, C., J. A. Wolfe, E. B. Leopold, and M. A. Lanphere. 1969. The coal-bearing group in the Nenana coal field, Alaska. U. S. Geol. Surv. Bull. 1274-D:D1-D30.
Watts, W. A. 1967. Late-glacial plant macrofossils from Minnesota. Pages 89-97 in E. J. Cushing and H. E. Wright, Jr., eds. Quaternary Paleoecology. Yale University Press, New Haven, N. J.
Watts, W. A. 1970. The full-glacial vegetation of northwestern Georgia. Ecology 51(1):17-33.
Watts, W. A. 1978. Plant macrofossils and Quaternary paleoecology. Pages 53-67 in D. Walker and J. C. Guppy, eds. Biology and Quaternary Environments. Australian Academy of Science, Canberra.
Watts, W. A. 1979. Late Quaternary vegetation of central Appalachia and the New Jersey Coastal Plain. Ecol. Monogr. 49(4):427-469.
Watts, W. A., and R. C. Bright. 1968. Pollen, seed, and mollusk analysis of a sediment core from Pickerel Lake, northeastern South Dakota. Geol. Soc. Amer. Bull. 79:855-876.
Watts, W. A., and T. C. Winter. 1966. Plant macrofossils from Kirchner Marsh, Minnesota: A paleoecological study. Geol. Soc. Amer. Bull. 77:1339-1360.
Watts, W. A., and H. E. Wright, Jr. 1966. Late Wisconsin pollen and seed analysis from the Nebraska Sandhills. Ecology 47(2):202-210.
Webb, T., III. 1974. Corresponding patterns of pollen and vegetation in lower Michigan: A comparison of quantitative data. Ecology 55(1):17-28.
Webb, T., III and J. H. McAndrews. 1976. Corresponding patterns of contemporary pollen and vegetation in central North America. Geol. Soc. Amer. Mem. 145:268-299.

Whitehead, D. R., and K. W. Tan. 1969. Modern vegetation and pollen rain in Bladen County, North Carolina. Ecology 50(2):235-248.

Williams, H. 1950. Volcanoes of the Parícutin region, Mexico. U. S. Geol. Surv. Bull. 965-B: 165-279.

Wright, H. E., Jr., and W. A. Watts. 1969. Glacial and vegetational history of northeastern Minnesota. Minnesota Geol. Surv. Spec. Pub. SP-11:1-59.

Wright, J. W. 1952. Pollen dispersion of some forest trees. Northeastern Forest Experiment Station Paper 46, U. S. Department of Agriculture, Forest Service. 42 pp.

PLATES

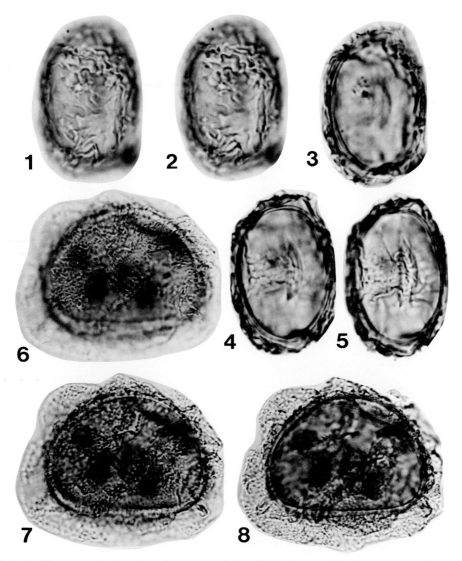

Plate 1. Figures 1-5. Monolete fern spore 1 (x 1000). Unidentified fern spore with perispore showing details of surface pattern and character of the scar. Figures 6-8. Monolete fern spore 2 (x 1000). Unidentified fern spore with perispore showing details of surface pattern. The number of distinctive fern spore types, though not here defined in terms of genera, exceeds distinctive fern megafossils at Clarkia.

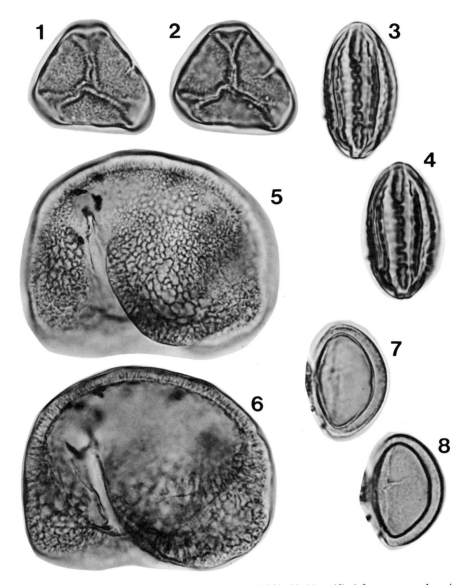

Plate 2. Figures 1, 2. Trilete fern spore 1 (x 1000). Unidentified fern spore, showing details of surface pattern and scar. Figures 3, 4. *Ephedra* sp. (x 1000). Figures 5, 6. *Cedrus* sp. (x 1000). In lateral view showing body and bladder pattern, bladder attachment and cap thickness. Figures 7, 8. *Isoetes* sp. (x 1000). Microspore in oblique polar view.

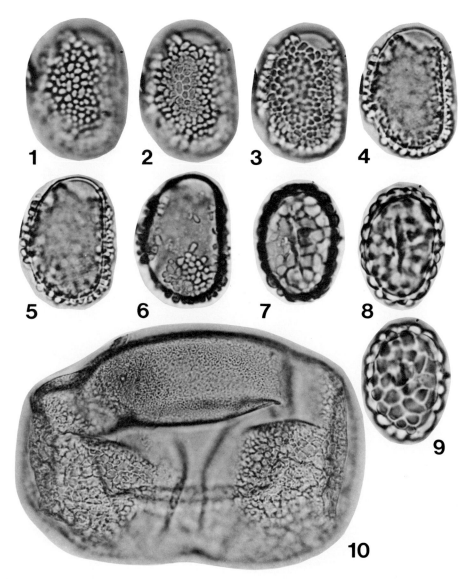

Plate 3. Figures 1-6. Monolete fern spore 3. (x 1000). Unidentified fern spore. In lateral view showing the patchy loss of surface elements and the possibility for obtaining a smooth-walled spore. Many such "naked" spores are found in the Clarkia samples; some may result from the type of degradation shown here. Figures 7-9. Monolete fern spore 4. (x 1000). Unidentified fern spore. In polar view showing details of scar, and surface pattern. Figure 10. *Picea* sp. (x 750).

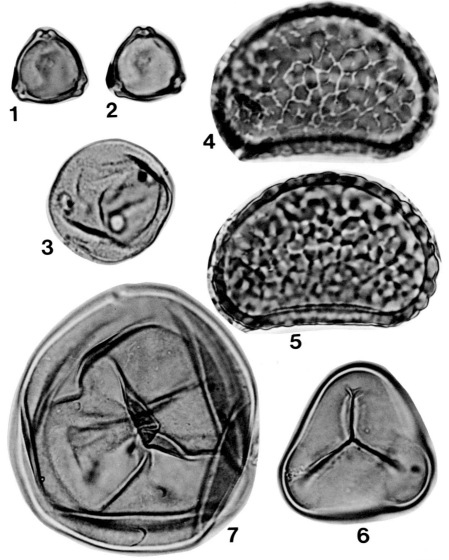

Plate 4. Figures 1, 2. *Betula* sp. (x 1000). Polar view. Figure 3. *Celtis* sp. (x 1000). Figures 4, 5. *Polypodium vulgare*-type (x 1000). Figure 6. Trilete fern spore 2 (x 1000). Unidentified, smooth-walled fern spore. Figure 7. *Pseudotsuga*-type. (x 750). A morphologically similar pollen grain is produced by *Larix*.

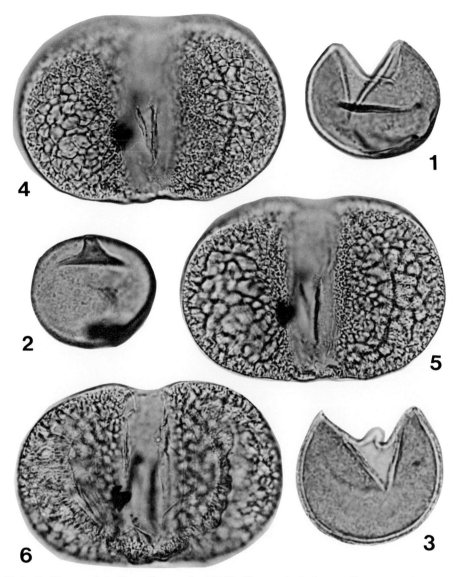

Plate 5. Figures 1-3. Taxodiaceae (x 1000). Figures 1, 3 show split open grains, the common configuration, with intact papilla characteristic of the Taxodiaceae, in Figure 3. Figure 2 is a whole (unsplit) grain, with large papilla of the *Sequoia* type. Figures 4-6. *Pinus* sp. 1 (x 1000). In polar view, showing broad, unconstricted attachment of bladders.

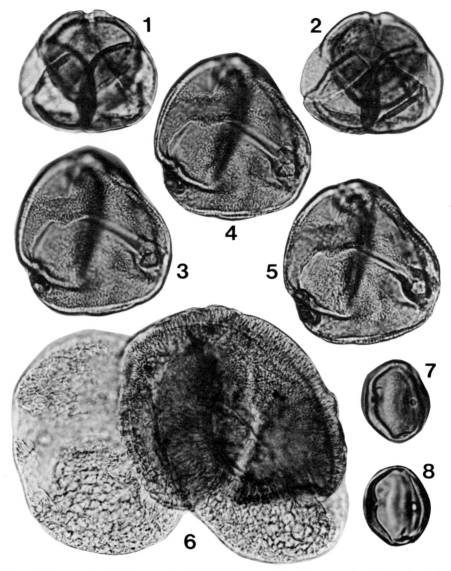

Plate 6. Figures 1, 2. Ericaceae (x 1000). Figures 3-5. *Nyssa* sp. (x 1000). In oblique polar view showing details of surface pattern and apertures. Figure 6. *Abies* sp. (x 750). Figures 7, 8. *Vitis* sp. (x 1000). Equatorial view.

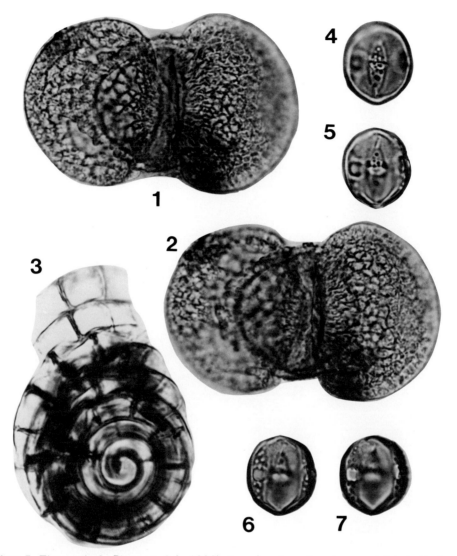

Plate 7. Figures 1, 2. *Pinus* sp. 2 (x 1000). In polar view, showing constricted attachment of bladders. This morphologic type of grain is sometimes attributed to the Podocarpaceae (cf. Boulton and Chaloner 1970). Figure 3. *Helicoön-Helicodendron*-type conidium (x 1000: identification by Martha Sherwood). Helicosporous Hyphomycetes such as *Helicoön* and *Helicodendron* are characteristic of decaying wood from very damp places and of submerged wood and leaves "decaying under relatively anaerobic conditions" (Glen-Bott 1951:276; Ellis 1976). Although mycelia are present on submerged plant material, the helicoid conidia, which temporarily trap air in the coils to float on the water surface, form only when the fungal-bearing material is exposed to air (Glen-Bott 1951). Figures 4-7. *Aesculus* sp. (x 1000). Equatorial view of tricolporate grain, showing heavily textured furrows and large pores.

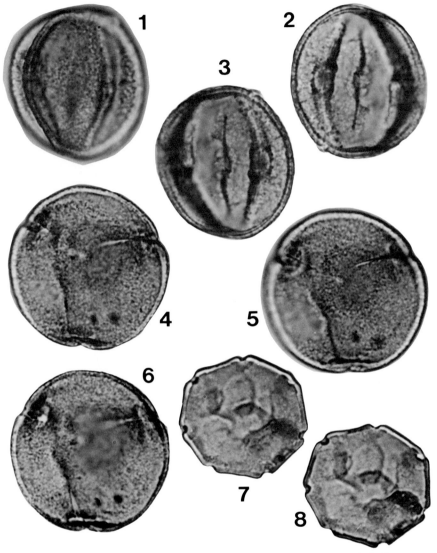

Plate 8. Figures 1-3. *Fagus* sp. (x 1000). Equatorial view. Figures 4-6. *Fagus* sp. (x 1000). Oblique polar view. Figures 7, 8. *Pterocarya* sp. (x 1000). Polar view of 7-pored specimen.

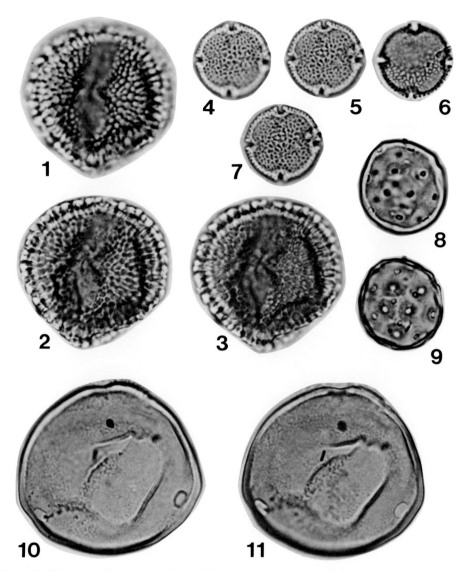

Plate 9. Figures 1-3. *Ilex* sp. (x 1000). Equatorial view showing the characteristic clavate-baculate elements. Figures 4-7. Unidentified pollen type (x 1000). Polar view. Pollen of this unknown taxon occurs in a number of the late Tertiary pollen assemblages in the Pacific Northwest, but is generally uncommon. The four apertures and uneven reticulum are characteristic. This fossil resembles pollen of *Reevesia*, an Asiatic member of the Sterculiaceae comprised of shrubs and generally small trees, found most commonly in the Evergreen Sclerophyllous Broad-leaved and Rain Forests of the Insular, Maritime and Southwestern Provinces of China. The distinctly aspidate apertures, with protruding thickened endexine, of pollen of *R. pubescens* Mast. and *R. thyrsoidea* Lindl. are not, however, sufficiently clearly defined in the fossil material to warrant the generic identification. Figures 8, 9. Chenopodiaceae cf. *Sarcobatus*-type. (x 1000). This relatively few-pored pollen type occurs regularly in late Tertiary pollen assemblages of the Pacific Northwest. Figures 10, 11. *Carya* sp. (x 1000).

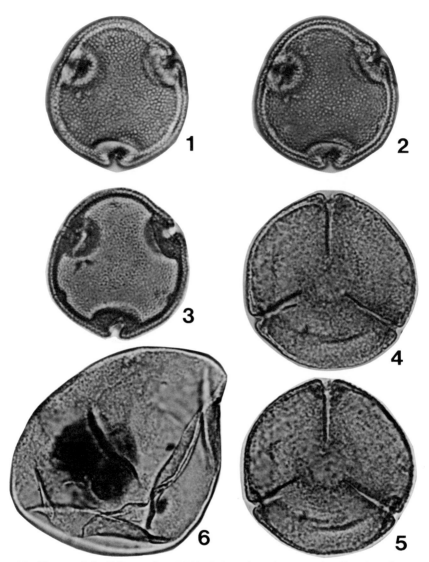

Plate 10. Figures 1-3. *Tilia* sp. (x 1000). Polar view showing details of surface pattern and apertures. Figures 4, 5. *Fagus* sp. (x 1000). Polar view. Figure 6. *Magnolia* sp. (x 1000). A monocolpate grain in oblique polar view.

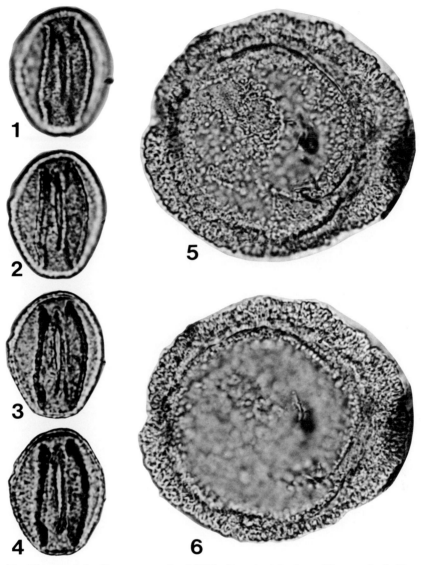

Plate 11. Figures 1-4. *Quercus* sp. (x 1000). Equatorial view. Figures 5, 6. *Tsuga heterophylla*-type (x 1000). Polar view. Note the small spines that distinguish pollen of the western American *T. heterophylla* and the Appalachian *T. caroliniana*, as well as that of most Asiatic hemlocks from the common North American *Tsuga canadensis*.

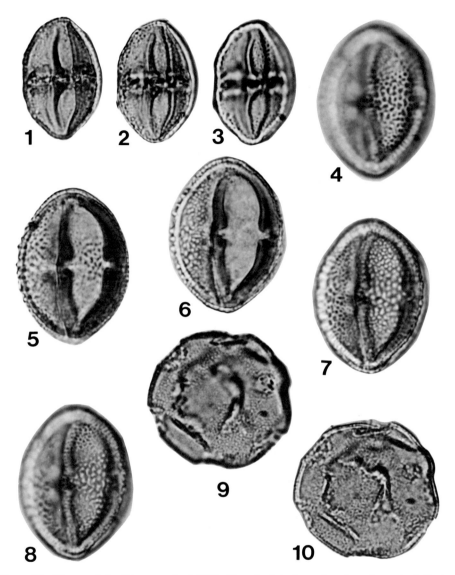

Plate 12. Figures 1-3. *Rhus* sp. (x 1000). Equatorial view showing characteristic transverse furrows and thickenings at the margin of the furrow. Figures 4-8. Unidentified tricolporate pollen type (x 1000). Figures 9, 10. *Liquidambar* sp. (x 1000).

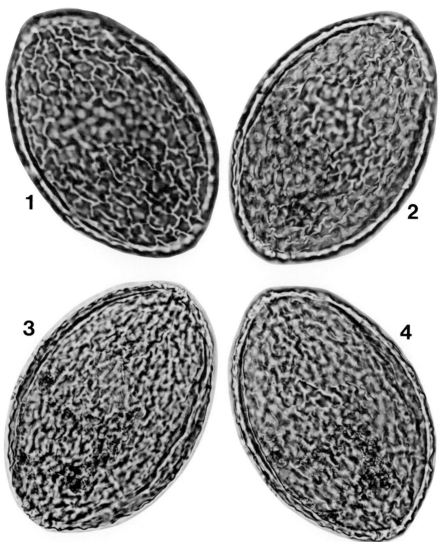

Plate 13. Figures 1-4. *Ovoidites* cf. *ligneolus* (× 600). Microfossil of unknown affinity, possibly an algal cyst. A common taxon in some Tertiary assemblages (at Clarkia recovered only in C-2); found also in freshwater marsh peat from Georgia and Florida (Rich et al. 1981). Characteristic features are the coarse reticulum in upper focus and the "dehiscence" line which extends around the long dimension (see Figures 2-4 near specimen margin).

APPENDIX:
PRELIMINARY CHECKLIST OF THE FUNGI OF THE CLARKIA FLORA, UNIT 2, P-33

MARTHA A. SHERWOOD
Department of Geology, University of Oregon, Eugene, OR 97403

The following checklist is based on isolated spores and fragments of fruitbodies found in palynological preparations from Unit 2, P-33. It is by no means an exhaustive list, since it includes only those forms which could be readily identified by reference to modern forms. No formal taxonomy has been attempted. We hope that in the future a more detailed and diagnostic paper, with illustrations and formal descriptions, will be prepared.

The fungi were identified using Ellis's (1971) keys and descriptions of dematiaceous hyphomycetes, and Sutton's (1980) monumental account of coelomycetes. Reference is made to fossil fungi figured by Sheffy and Dilcher (1971) and Pirozynski and Weresub (1979) for comparative purposes. No attempt has been made at this time to compare the fungi in the Clarkia flora to all of the relevant described fossil fungi. Our flora has a number of the same spore types as the Eocene assemblage described by Sheffy and Dilcher. It does appear to have a larger number of distinctive spore types.

The heading "Saccardo spore group" refers to P. A. Saccardo's 19th-century artificial classification of fungi according to the shape, color, and septation of the spore. As has been pointed out (Pirozynski and Weresub 1979), this system is particularly useful for classifying fossil fungi.

The majority of the spores observed belonged to the Fungi Imperfecti, and are presumed to be the asexual stages of Ascomycetes. Many conidia (asexual spores) can be distinguished from ascospores by the presence of a scar at one or both ends of the spore. Both the sexual and the asexual (e.g. uredospores of rusts) spores of Basidiomycetes also bear attachment scars. Isolated basidiospores can generally be distinguished from conidia by their shape and ornamentation. During this study, one possible rust teliospore and two possible basidiospores were observed. Judging from the spora, the fungal flora of this community was overwhelmingly Ascomycetous.

Although the presence of an attachment scar denotes an asexual spore among the Ascomycetes, its absence is not necessarily diagnostic of an ascospore. I have listed as Ascomycetous only those spores which are unambiguous. Xylariaceous ascospores, dark unicellular spores with a lateral germ slit, are abundant in the material. Spores of *Melannoma, Leptosphaeria*, and an unidentified, possibly microthyraceous, fungus were rare.

Spores in the sample were all pigmented to some degree. Judging from their morphology most of the spores belong to genera which are dark-spored today. In some cases the pigmentation may be a secondary result of the preservation environment. It is also likely that spores with thick dark walls were selectively preserved.

Among the genera of Hyphomycetes represented, there is a preponderance of forms which occur on dead wood, and a lack of characteristic grassland fungi. This suggests that the fungal flora came from a wooded region, probably one which was humid and temperate or tropical. The presence of Microthyriales and abundance of Xylariaceae would be unlikely in a frigid climate. Beyond this it is difficult to make ecological statements, since comparatively little is known of the biogeography of fungi.

Suprageneric Taxon	Genus	Saccardo Spore Group	Fossil Name	Typical Substrate
ASCOMYCOTINA				
Euascomycetadae				
Xylariaceae	?	Phaeosporae	*Inapertisporites* Hammen cfr. *I. disciformis* Sheffy and Dilcher	Usually on wood, esp. warm-temperate and tropical
Sordariaceae	*Podospora* Ces.	Phaeosporae	?	Dung
	? (ornamented)	Phaeosporae	?	Dung, burnt debris
Sphaeriales		Phaeodidymae	?	?
Loculoascomycetadae				
Pleosporaceae	cfr. *Melanomma* Nits. ex. Fuckel	Phaeophragmae	*Chaetosphaerites* Felix	Wood
	Leptosphaeria Ces. and de Not.	Phaeophragmae	cfr. *Multicellaesporites fusiformis* Sheffy and Dilcher	Herbaceous debris
Microthyriales	?	?	cfr. *Callimothallus* Dilcher	Evergreen leaves
Microthyriales?	?	Hyladodidymae		Evergreen leaves
FUNGI IMPERFECTI				
Hyphomycetes				
Dematiaceae	*Acrogenospora* M. B. Ellis (*Domingoella* Cifferi)	Phaeosporae	*Lacrimasporonites* (Clarke 1965) Elsik 1968	Wood
	Bactrodesmium Cooke	Phaeophragmae		Bark
	Berkleasmium Zobel	Phaeodictyae		Wood
	Bispora Corda	Phaeodidymae	cfr. *Multicellaesporites bigeminatus* Sheffy and Dilcher	Wood
	Brachysporiella Bat.	Phaeodictyae	?	Wood, tropics
	Brachysporium Sacc.	Phaeophragmae	?	Rotten wood, cosmop.

Cordana Preuss	Phaeodidymae		Wood and bark
Curvularia Boedijn	Phaeophragmae or Helicosporae	cfr. Pirozynski (1979, fig. 26.11)	Pluriverous, esp. on grasses
Dictyosporium Corda	Staurosporae	?	Rotten wood and herbaceous debris
Exosporium Link ex. Schlecht. (*Corynesora* Güssow)	Scolecosporae	cfr. Pirozynski (1979, fig. 26.14)	?
Helicöon Morgan (*Helicodendron* Peyron.)	Helicosporae	?	Wet wood (*Helicöon*) or debris in stagnant water (*Helicodendron*)
Rhinocladium Sacc. and Marsh(?)	Phaeosporae	?	Wood
Septonema Corda	Phaeophragmae	*Multicellaesporites* Elsik cfr. Sheffy and Dilcher (1971, fig. 16.57)	Wood and bark
Spiropes Cifferi	Phaeophragmae		On *Meliola* (fungi)
Sporidesmium Link ex Fr.	Phaeophragmae		
Spondycladiella Linder	Phaeophragmae	?	On *Corticium* (fungi)
Taeniolella Hughes	Phaeophragmae		
Xylohypha (Fr.) Mason	Phaeosporae	cfr. *Dyadosporonites didymus* Sheffy and Dilcher	On wood
Coelomycetes			
Cheirospora (Moug.) Fr.	± Staurosporae	?	Twigs, bark
Diplodia Fr. apud Moug. s. l.	Phaeodidymae	cfr. *Didymoporisporonites normalis* Sheffy and Dilcher	?
cfr. *Staganospora* (Sacc.) Sacc.	Hyalophragmae	*Multicellaesporites* Elsik	Grass leaves

LITERATURE CITED

Ellis, M. B. 1971. Dematiaceous Hyphomycetes. Commonwealth Mycological Institute, Kew. 608 pp.

Pirozynski, K. A., and L. K. Weresub. 1979. The classification and nomenclature of fossil fungi. Pages 653-688 *in* B. Kendrick, ed. The Whole Fungus. Volume 2. National Museum of Natural Sciences, National Museums of Canada for the Kannanaskis Foundation, Ottawa.

Sheffy, M. V., and D. L. Dilcher. 1971. Morphology and taxonomy of fungal spores. Paleontographica Abt. B. 133(1-3):34-51.

Sutton, B. C. 1980. The coelomycetes. Commonwealth Mycological Institute, Kew. 696 pp.

MIOCENE INSECTS FROM THE CLARKIA DEPOSITS OF NORTHERN IDAHO

STANDLEY E. LEWIS
Dept. of Biological Sciences, St. Cloud State University, St. Cloud, MN 56301

A taxonomically diverse collection of insects has been examined and identified from a Miocene locality near Clarkia, Idaho. The collection includes one new order and eight new families of insects previously not recorded from Miocene deposits of the region. Several trace fossils, representing damage done to plant material by Hymenoptera and Lepidoptera, have also been examined and identified.

Insects have been collected and identified from the freshwater sediments of "Latah"[1] Formation since the work of Cockerell (1924). Additional studies of collections and descriptions were published by Berry (1928), Carpenter et al. (1931), Lewis (1969, 1970, 1973) and most recently by Barr and Gillespie (*in* Smiley et al. 1975). Prior to 1975 specimens were obtained from sites near Spokane, Washington, and Juliaetta, Idaho. Recently a new Miocene site (Clarkia locality P-33) in northern Idaho has yielded 135 additional specimens (Smiley et al. 1975).

This preliminary study is based upon the fossil insect material collected by Smiley and co-workers during the 1970s. The 135 Clarkia specimens include 8 orders and 24 families; several genera have been identified, but further research will be required for more precise taxonomic treatment. From this study one specimen representing the Ephemeroptera adds this order to the list of Miocene insects from this region. The cumulative total of insects recovered from the Miocene sediments over the past 56 years now includes 14 orders and 34 families.

Table 1 summarizes, by investigator, year, and area, our current knowledge of the fossil insects from the Miocene deposits of Washington and Idaho. Information used in the preparation of this table is from Cockerell (1924), Berry (1938), Carpenter et al. (1931), Lewis (1969, 1970, 1973), Barr and Gillespie (*in* Smiley et al. 1975), and the present investigation.

METHODS

All specimens were studied with an A/O Spencer and/or Wild M-5 stereoscopic microscopes. Measurements were made using an ocular micrometer in the A/O Spencer microscope. All photographs were taken with a 35 mm Nikon camera, with extension bellows for smaller specimens. Because these insect specimens are so similar to modern forms, it was possible to use existing references, keys, etc. for their identifications.

All specimens discussed in this study will be deposited in the paleontological collections at the University of Idaho College of Mines Museum (UIMM). All specimens examined are from locality UIMM P-33. Terminology used for anatomical parts such as wing veins and cells follows the convention format set forth in Borror, Delong and Triplehorn (1976).

SYSTEMATICS

A preliminary, annotated list of specimens found at Clarkia site P-33 represents determinations to the level of order or family, with certain fossils given tentative generic status. Such a diverse insect fauna as exists in the Clarkia deposits makes it difficult for one entomologist to

---

[1] "Latah"—name applied to the sediments in which these fossil insects have been collected.

Copyright ©1985, Pacific Division, AAAS.

identify all taxa to the level of species. It is hoped that this list will spur other specialists in entomology to study these specimens further and eventually place some of them into more specific categories.

Class INSECTA
Order EPHEMEROPTERA
Family Siphlonuridae
Genus? *Ameletus* Eaton or *Parameletus* Bengtsson
(Figure 1)

Diagnosis: A single specimen represented by an incomplete nymph with portions of the head, thorax, legs and abdomen missing. Body length is 13.7 mm; lateral cercus is 5.4 mm in length; wing pads are present, but not distinct, abdomen is long, slender, flattened and somewhat teleoscopic; posterolateral projections of terga are present, especially prominent on the apical segments; gills are visible on abdominal segments 1-7, appearing single, ovoid and small; terminal filament is incomplete and fringed on both sides; cerci are fringed on the medial side only with a short darkened area at midpoint.

Discussion: Although this specimen lacks critical morphologic features for identification, existing body form and morphological characters indicate a tentative placement in the family Siphlonuridae, probably close to the genus *Ameletus* Eaton or *Parameletus* Bengtsson of the subfamily Siphlonurinae (Edmunds et al. 1976:131-132, 134-136).

Nymphs of the genus *Ameletus* are usually found in small, rapidly flowing streams among pebbles, near banks or among vegetation and debris; certain species can be found in large rivers, lakes, and ponds. Distribution of this genus is holarctic, with nearctic species most abundant and diverse in the north, extending along the mountains south to such states as Georgia, Illinois and California.

Nymphs of the genus *Parameletus* are found primarily in quiet swamps and forest pools with emergent vegetation. This genus is holarctic and generally confined to northern regions with high elevations.

Specimen No.: UIMM T-0052
Locality: UIMM P-33

Order ORTHOPTERA
Suborder ENSIFERA
Family Tettigoniidae
Barr and Gillespie, 1975
(*in* Smiley et al. 840, Pl. 1, Fig. 5)
(Figure 2)

Diagnosis: Two specimens represent metathoracic legs (part of the femora, tibia and tarsi) and a faint impression of the basal half of the forewing. Only the Sc, R, and MA veins are discernible from the forewing of specimen T-0019 (Fig. 2). The forewing preservation measures 29.3 mm. The overall length of the leg segments measures 43.9 mm to 45.8 mm. Metafemora are slender and incomplete. The metatibia, which are uniformly slender and elongate, measure 24.4 mm in length. The metatarsi measure approximately 4.9 mm in length, with segments 1-3 short and globular; the fourth (4th) metatarsal segment is elongate and extended. Measurements are based upon the incomplete impressions of the body parts.

Specimen No.: UIMM T-0019
Locality: UIMM P-33.

TABLE 1. MIOCENE INSECT FAUNA FROM SPECIFIC LOCATIONS IN EASTERN WASHINGTON AND NORTHERN IDAHO

Taxonomic Categories	Cockerell 1924	Berry 1928	Carpenter et al. 1931	Lewis 1969	Lewis 1970	Lewis 1973	Barr & Gillespie 1975	Lewis 1980
Ephemeroptera								
Siphlonuridae								C
Odonata								
Libellulidae			BY	MC				
Orthoptera								
Suborder Ensifera								
Tettigoniidae							C	
Suborder Blattaria								
Blattellidae								C
Suborder Caelifera								
Family (?)								C
Plecoptera								
Nemouridae				J				
Hemiptera								
Pentatomidae			BY	BY				C
Saldidae				J				
Homoptera (?)								
Psocoptera				J				
Coleoptera								
Carabidae	DC		BY	J,BY				C
Dytiscidae			BY					
Elateridae							C	C
Meloidae								C
Dascillidae						BY		
Lucanidae								C
Scarabaeidae							C	C
Chrysomelidae (?)								C
Anthribidae							C	
Isoptera								
Rhinotermitidae			BY	J				
Mecoptera (?)				J				
Trichoptera								
Limnephilidae		BY	BY	BY,J,DC	BY		C	C
Phryganeidae			BY	J				C
Leptoceridae (?)							C	
Lepidoptera								
Nepticulidae				BY				
Gracilariidae								C
Diptera								
Tipulidae				BY				
Sciaridae				J				
Mycetophilidae				J,BY				
Bibionidae				J,BY				C
Anisopidae				J				
Bombyliidae				DC				
Cecidomyiidae								C
Hymenoptera								
Tenthredinidae				J				
Ichneumonidae				J,BY				C
Formicidae			BY	BY,J,DC				C
Cynipidae								C
Apidae			BY					C

*Sites: BY = Brickyard near Spokane, WA; DC = Deep Creek near Spokane, WA; C = Clarkia, ID; J = Juliaetta, ID; HC = Hangman's Creek near Spokane, WA; MC = Marshall Creek near Spokane, WA.

Suborder BLATTARIA
Family (?) Blattellidae
Genus (?) *Parcoblatta* Hebard
(Figure 3)

Diagnosis: This specimen represents a dorsal view of portions of the thorax and a complete abdomen of a cockroach, and it is the first cockroach recorded from Miocene deposits of the region. The specimen measures 29.3 mm in length, from the prothoracic segment ot the tip of the abdomen. The abdomen measures 11.7 mm at the widest part. No portions of the appendages or wings are preserved.

Discussion: This determination is based upon the estimated overall size and basic configuration of the thoracic and abdominal segments. The anatomy of this specimen most closely resembles that of the wood cockroach (*Parcoblatta* Hebard). Several species in this group are presently found in the southern United States near fallen trees and in ground litter.
Specimen No.: UIMM T-0053.
Locality: UIMM P-33.

Suborder CAELIFERA
Superfamily (?) Locustopseidea
Family unknown
(Figure 4)

Diagnosis: One orthopteran specimen appears to be one or two wings overlaid one upon another. The wing length is 39.0 mm.

Discussion: The anterior wing margin of one portion of this fossil resembles wing venations of members of the superfamily Locustopseidea (Sharov 1968:88-99). It is too incomplete for reference to a family at this time.
Specimen No.: UIMM T-0054.
Locality: UIMM P-33.

Order HEMIPTERA
Suborder GEOCORIZAE
Family Pentatomidae
Genus (?) *Chlorochroa* Stal. 1872
(Figure 5)

Diagnosis: A dorsal view of a complete insect, except for wings and portions of the legs, is represented. The body, which measures 16.6 mm in total length, is broadly ovoid in shape.

Head: Oval in shape, occupying a recessed position in the pronotum; width of head from one outereye margin to another, 3.4 mm; length of head, 2.9 mm; juga approximately the same length as tylus; anterolateral margins gently rounded; dorsal surface punctate; eyes prominent and extended laterally; ocelli present; eye facets not visible; antennae slender, with individual segments not readily discernible.

Thorax: Width of thorax at widest part, 8.3 mm; pronotum and scutellar region coarsely and unevenly punctate; width and length of pronotum, 7.8 mm and 3.9 mm respectively; scutellum basically triangular in shape, with sides slightly bowed outward; moderately pronounced apical lobe rounded posteriorly, length of scutellar base and sides, 4.9 mm and 3.9 mm, respectively; appendages incomplete; wings absent.

Abdomen: Six abdominal segments visible; length and width of abdomen 8.8 mm and 9.3 mm respectively; distinct alternating light and dark patterning visible along side of abdominal segments; genitalia present.

Discussion: According to available references on fossil hemipterans of the family Pentatomidae and comparison with current museum specimens, this specimen is tentatively placed in the genus *Chlorochroa*. It closely resembles *Chlorochroa uhleri* Stal., described from the Heteroptera of Eastern North America (Blatchely 1926:111-112).

The genus *Chlorochroa* is holarctic, with nearctic species ranging from Quebec and New England across the northern United States and as far south as Southern California. Members of this genus are phytophagous and have been found in sandy-dry places and occasionally on various trees and bushes, especially willows and junipers.

Specimen No.: UIMM T-0055.
Locality: UIMM P-33.

Order COLEOPTERA

Approximately 35% of the fossils found at the Clarkia site are members of the order Coleoptera. Of this percentage, approximately 23% belong to the family Elateridae. The other families identified are Carabidae, Meloidae, Lucanidae, Scarabaeidae, Chrysomelidae, and Anthribidae. Some specimens are identified only to the ordinal level.

Family Carabidae
(Figure 6)

Diagnosis: Two specimens are represented by elytra only. One specimen (Fig. 6) is a complete view of the right elytron, 14.1 mm in length, 5.9 mm and 4.9 mm in width at the anterior and posterior ends, respectively. The elytron has nine long rows of punctations, with the five punctate striae on the sutural side less crowded than those next to the outer margin. The wing punctations are much coarser on the outer margin than those near the sutural side.

Discussion. This fossil resembles the genus *Carabus* Linne as described in Tertiary Coleoptera of North America (Scudder 1900:14-16). Study of the gross morphology of the wing indicates that the specimen could also be associated with species of the genus *Pterostichus* Bonelli (Dillon and Dillon 1961:85). Because of the lack of definitive morphologic characters, this specimen is not identified beyond the family level at this time.

Both genera (*Carabus* and *Pterstichus*) are holarctic, with species found throughout nearctic regions. Species of both genera occur beneath logs and stones of woodland areas.

Specimen No.: UIMM T-0056.
Locality: UIMM P-33.

Family Elateridae
Genus *Ctenicera* Latreille
(Figure 7)

Diagnosis: Eleven specimens represent the family Elateridae. Seven of these are partially or nearly complete dorsal or ventral views of the adult, with elytra sometimes visible. Two specimens represent only the prothoracic tergite. The overall body lengths range from 10.7 to 31.0 mm.

Discussion: Four of the more complete specimens resemble members of the genus *Ctenicera* Latreille in size range, and in general body, prothoracic, and elytral form. One specimen appears to be quite similar in the above-mentioned characters to *C. aethiops* (Herbst) (Dillon and Dillon 1961:321). No further taxonomic placement for this specimen will be made at this time.

Members of the genus *Ctenicera* are commonly found in the eastern and southeastern United States. Species can be found on trees and shrubs such as walnut, hickory, mustard and rhubarb, and close to the ground under stones and rubbish.

Specimen No.: UIMM T-0057.
Locality: UIMM P-33.

Figures 1-6. Fig. 1. Genus? *Ameletus* Eaton or *Parameletus* Bengtsson. Fig. 2. Undetermined (family Tettigoniidae). Fig. 3. Genus (?) *Parcoblatta* Hebard. Fig. 4. Undetermined (superfamily? Locustopseidae). Fig. 5. Genus (?) *Chlorochroa* Stal. 1872. Fig. 6. Undetermined (family Carabidae).

Genus *Alaus* Esch.
(Figure 8)

Diagnosis: This specimen is represented by a portion of the prothorax and a major portion of one elytron. The prothoracic segment measures 7.3 mm in length, while the elytron measures 18.5 mm in length and 4.9 mm in width. The estimated length of the insect is 27.3 mm.

Discussion: This specimen is placed in the genus *Alaus* Eschscholtz with no species designation given at this time. It definitely represents a member of this genus, which is so readily distinguished from other elaterids by the eyelike maculae on the pronotum. The adults of this genus are commonly found beneath the bark of dead pines and are numerous in the southern United States.

Specimen No.: UIMM T-0058.
Locality: UIMM P-33.

Family Meloidae
(Figure 9)

Diagnosis: One specimen of the family Meloidae represents a dorsal view of the adult form, measuring 34.1 mm in total body length. The head is somewhat circular in shape, with eyes placed forward and lateral. Several short antennal segments are also preserved. The elytra measures 18.0 mm in length and covers approximately one-half of the abdomen.

Discussion: This fossil resembles members of the genus *Meloe* Linne as described by Dillon and Dillon (1961:299) and Jacques (1951:148), but no taxonomic placement below the family level will be attempted at this time.

The meloids are widespread throughout the United States and are commonly found associated with flowers, especially the composites. The adults are phytophagous, while the larval stage may be carnivorous, feeding on eggs of grasshoppers.

Specimen No.: UIMM T-0059.
Locality: UIMM P-33.

Family Lucanidae
Genus *Lucanus* Scopoli
(Figure 10)

Diagnosis: A single impression of a lucanid beetle shows only the fore and hind wings plus a small portion of the thorax and legs. The forewing, which measures 11.7 mm in length and 5.9 mm in width at the widest part, appears black, non-punctate and non-striated. The hindwing measures 16.6 mm in length.

Discussion: The hind wings are sufficiently well preserved to aid in its family and tentative generic placement, as the hind wings of beetles are often useful for identification. The dark strong venation is most reminiscent of members of the genus *Lucanus* Scopoli (Dillon and Dillon 1961: 568-569). No taxonomic placement below the generic level will be made at this time.

The genus *Lucanus*, which commonly occurs in the southern United States, can be found in wooded areas around oak stumps. The larvae occur in decaying wood and look very much like white grubs that can be found in a garden.

Specimen No.: UIMM T-0069.
Locality: UIMM P-33.

Family Scarabeidae
Genus *Osmoderma* Le Pel.
(Figure 11)
Barr and Gillespie, 1975 (*in* Smiley et al., 840, pl. 1, Fig. 4)

Diagnosis: This specimen is represented by a dorsal-ventral preservation of an adult, which measures 21.5 mm in total body length. Head length is 2.9 mm and width is 3.4 mm. Eyes occupy a latero-posterior position. Antennae are lamellate in form. The prothorax is globular in shape, with a width of 4.9 mm and a length of 6.3 mm. Forewings measure 9.8 mm and 4.4 mm, respectively. The elytra are not punctate nor striated. The hind wing is 11.7 mm in length, with venation visible. A distinct triangular scutellar area is visible. Hind legs are present.

Discussion: Because of the antennal shape, wing venation and overall body form, this specimen is placed in the family Scarabeidae as identified previously by Barr and Gillespie (1975). Upon comparative studies (Dillon 1961:555-557), using criteria such as thoracic, elytra and general body form, this specimen closely resembles members of the genus *Osmoderma* LePeletier Serville. Further taxonomic placement will be conducted at a later time.

The genus *Osmoderma* is commonly found in Eastern North America. The adults are nocturnal and are commonly found near the edge or in open woods.
Specimen No.: UIMM T-0018.
Locality: UIMM P-33.

Family Scarabeidae
Genus *Geotrupes* Lat.
(Figure 12)

Diagnosis: This specimen is represented by a ventral view of an incomplete adult form, which measures 17.6 mm in body length. Head length is 5.9 mm and width is 4.4 mm. The antennae are visible and show a lamellate form. Forewing length is 14.1 mm and its width is 6.8 mm. Wings are strongly striated and punctate.

Discussion: Because of antennal shape, elytral characters and general body form, this specimen is placed in the family Scarabeidae. Upon comparative studies (Dillon 1961:523, 527), using such criteria as head and thoracic shape and elytral characters, this specimen closely resembles members of the genus *Geotrupes* Latreille. Further taxonomic placement will be conducted at a later time.

Members of the genus *Geotrupes*, referred to as earth-boring dung beetles, are commonly found throughout eastern North America. These beetles are found beneath dung or carrion and often occur in open, somewhat moist habitats.
Specimen No.: UIMM T-0061.
Locality: UIMM P-33.

Family(?) Chrysomelidae
(Figure 13)

Diagnosis: Seven specimens resemble members of the family Chrysomelidae. Five are single or paired elytra, and two are represented by elytra and portions of the thorax. Most specimens show shiny black punctate elytra that measure 4.4 to 9.8 mm in length.

Discussion: With minimal morphologic characters available to use in the identifications, these specimens are tentatively placed in this family. Further taxonomic placement should be conducted by a specialist currently working on this group of coleopterans.
Specimen No.: UIMM T-0062.
Locality: UIMM P-33.

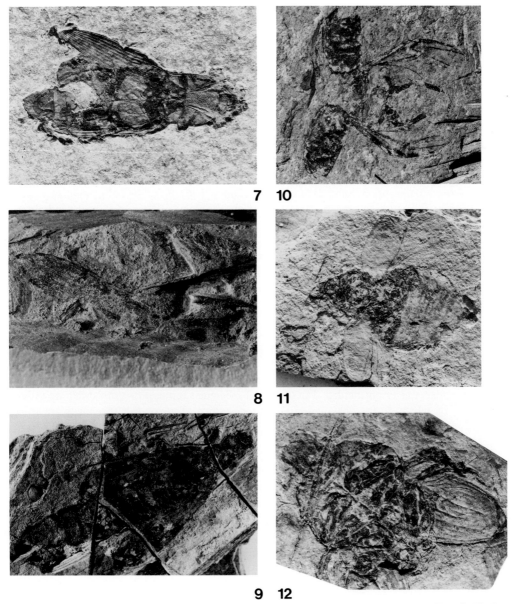

Figures 7-12. Fig. 7. Genus *Ctenicera* Latreille. Fig. 8. Genus *Alaus* Esch. Fig. 9. Undetermined (family Meloidae). Fig. 10. Genus *Lucanus* Scopoli. Fig. 11. Genus *Osmoderma* Le Pel. Fig. 12. Genus *Geotrupes* Lat.

Family Anthribidae
Genus *Eurymyeter* Le Conte
(Figure 14)
Barr and Gillespie, 1975 (*in* Smiley et al., 843, Pl. 1, Fig. 2)

Diagnosis: This specimen is a dorso-lateral view of a complete adult form, which measures 29.0 mm in total body length. The head, which tapers anteriorly, measures 4.9 mm in both length and width. The mouth parts are extended and are included in the measurement of the head length. The antennae are many-segmented and filiform. The prothorax appears shield-like with small protuberances. The prothorax length is 3.9 mm and the width is 8.3 mm. The elytron, which does not cover the last abdominal segments, measures 17.6 mm in length. The legs are visible with only portions preserved. Five abdominal segments are visible.

Discussion: Based upon comparative studies (Dillon 1961:737,740-742), using such criteria as general body form, tarsal segment arrangement, morphology of broad-flat beak, and antennal and elytral length, this specimen closely resembles members of the genus *Eurymyeter* Le Conte. It was previously identified to the family level by Barr and Gillespie (1975). Further taxonomic placement will be conducted at a later time.

Anthribids, or fungus weevils, are found throughout North America. In eastern North America, adult members of the genus *Eurymyeter* are found on dead twigs and on fungus growing on dead trees, such as beech.

Specimen No.: IUMM T-0016.
Locality: UIMM P-33.

Order TRICHOPTERA

Three families of caddisflies have been identified from Clarkia locality P-33: Limnephilidae, Lepidostomatidae and Phryganeidae. Ten of the 12 specimens representing this order are caddisfly larval cases, of which 6 are here identified and described. Two of the specimens are isolated forewings whose characters match species previously described by Carpenter et al. (1931). Four of the larval cases have not yet been identified.

Family Limnephilidae
(Figure 15)
Miopsyche martynovi Carpenter Ann. Ent. Soc. Amer.
24(2):307-322, Fig. 5, 1931.

Diagnosis: A single incomplete forewing measures 12.0 mm. Anterior margin of wing is convex, with apical border rounded. Subcoastal space is narrower than in *M. alexanderi*, also from Latah Formation (Carpenter 1931).

Discussion. The size, shape and venation refer this fossil to the species *M. martynova*, which was described from other Miocene deposits of the region. This is an extinct genus and species.

Specimen No.: UIMM T-0063.
Locality: UIMM P-33.

Platycentropus Ulmer
(Figure 16)

Diagnosis: Two specimens of caddisfly larval cases measure 16.6 mm in overall length and 3.9 mm in width at the widest part. The cases are constructed from what appear to be segments of grasses and sedges arranged in no regular pattern.

Discussion: These specimens closely resemble cases constructed by living members of the genus *Platycentropus* (Wiggins 1977:276-277) in case construction, size and general form. Placement below the generic level will not be attempted at this time.

Platycentropus is a nearctic genus confined to the eastern half of North America (Manitoba to Newfoundland, south to Louisiana). Habitats for this genus range from cool streams to warm ponds. Species are very tolerant to warm quiet waters near the margin of marshes and lakes.
Specimen No.: UIMM T-0064.
Locality: UIMM P-33.

Genus (?) *Limnephilus* Leach
(Figure 17)

Diagnosis: A single specimen of a larval case measures 25.5 mm in overall length. This case is constructed in a random fashion from bits of leaves, pieces of wood and other materials.

Discussion: The specimen closely resembles cases constructed by living members of the genus *Limnephilus* (Wiggins 1977:254-255). This specimen is tentatively placed in this genus based on its size, case construction, and general form. Further taxonomic placement will be done at a future time.

Limnephilus is a holarctic genus found over a large area of North America. Members of this genus inhabit ponds, lake margins and marshes. Certain species occupy temporary pools and streams.
Specimen No.: UIMM T-0065
Locality: UIMM P-33.

Family (?) Lepidostomatidae
(Figure 18)

Diagnosis: A single larval case measures 12.7 mm in overall length. The general form shows a slight taper from the anterior opening (2.9 mm in width) to the posterior attachment (2.0 mm in width). The case is constructed from small sand pebbles and some unidentifiable organic debris.

Discussion: This larval closely resembles cases made by modern species of the genus *Lepidostoma* Rambur (Wiggins 1977:156-157, Fig. F). This fossil specimen resembles the larval case construction that is so characteristic of the early instar cases of this genus. Members of this genus are widespread over North America, especially in the western United States.

Lepidostoma is a holarctic genus that is widespread over much of North America. Members of this genus inhabit lakes and streams.
Specimen No.: UIMM T-0066.
Locality: UIMM P-33.

Family Phryganeidae
Phryganea spokanensis Carpenter
(Figure 19)
Phryganea spokanensis Carpenter. Ann. Ent. Soc. Amer.
24(2):307-322, Fig. 3, 1931

Diagnosis: This specimen represents a well-preserved forewing. The wing measures 21.5 mm in length and 8.0 mm in width. The anterior margin is gently curved. The wing venation is very similar to *P. spokanensis*. as described by Carpenter (1931).

Discussion: The genus *Phryganea* is widely distributed in Europe, Asia, North America and parts of South America. There are two species of this genus that are presently found in North America. Larval forms are often found on the edges of marshes and lakes.

Genus (?) *Phryganea* Linnaeus
(Figure 20)

Diagnosis: Two larval cases measure 29.3 mm in overall length and 6.8 mm in width at the

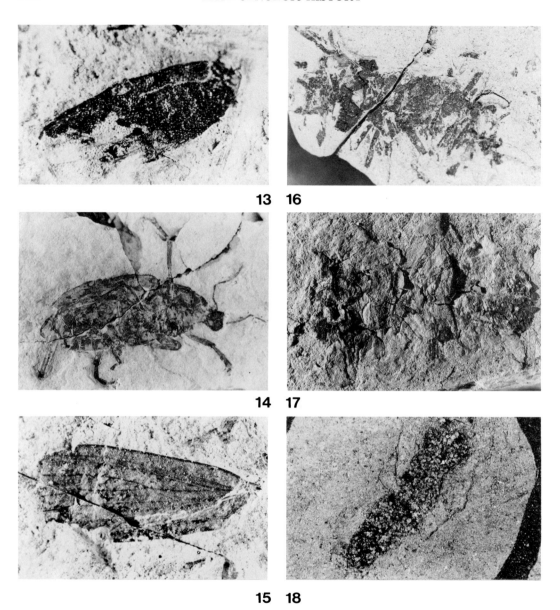

Figures 13-18. Fig. 13. Undetermined (family [?] Chrysomelidae). Fig. 14. Genus *Eurymyeter* Le Conte. Fig. 15. *Miopsyche martynovi*. Fig. 16. Genus *Platycentropus* Ulmer. Fig. 17. Genus (?) *Limnephilus* Leach. Fig. 18. Undetermined (family [?] Lepidostomatidae).

widest part. Conifer needles and miscellaneous plant parts were used in the construction of this case. The conifer needles were placed in a parallel position to one another and oriented in spiralling bands around the case.

Discussion: This particular orientation pattern of plant parts used in the larval case construction is found in several genera of the family Phryganeidae: *Agrypnia, Banksiola, Oligotrichia*, and *Phryganea* (Wiggins 1977:322-323, 324-325, 332-333, and 334-335, respectively). The fossil specimens more closely resemble members of living species of *Phryganea*. This association also supports the previous findings of forewings of this genus in these Miocene deposits (Carpenter 1931:320-322; Lewis 1969:99).

Specimen No.: UIMM T-0068.
Locality: UIMM P-33.

Order LEPIDOPTERA

Two specimens showing lepidopterous damage to plant leaves are recorded from Clarkia site P-33. One specimen definitely represents insect damage (blotch mining) by a member of the family Gracilariidae. Another, not discussed at this time, represents blotch mining by a member of either Nepticulidae, Gracilariidae, Colephoridae, Heliozelidae or Lavernidae (Needham, Frost and Tothill 1928:279-294).

Family Gracilariidae
Genus (?) *Lithocolletis* Clemens
(Figure 21)

Diagnosis: A single specimen shows blotch mines on the leaf of a live oak (*Quercus*).

Discussion: The larval damage resembles that of living members of the genus *Lithocolletis*. The blotch mine looks similar to damage by the solitary oak leaf miner, *Lithocolletis hamadryadella* (Clem.), which occurs on oaks throughout a large area of the eastern United States (Baker 1972:406-407). The amount of damage to this leaf has caused extreme distortion, giving it an unusual shape.

Specimen No.: UIMM T-0069.
Locality: UIMM P-33.

Order DIPTERA

Two families of the order Diptera are identified in this study: Bibionidae and Cecidomyiidae. The bibionids are represented by either complete body and wing impressions or just wing impressions. The cecidomyiids are represented by galls formed on the stems of bald cypress (*Taxodium*). This is the first time any gall-forming trace fossils have been described from Miocene deposits of the region.

Family Bibionidae
Genus *Bibio* Latreille
(Figures 22, 23)

Diagnosis: Seven specimens are represented by either body and wing or isolated wing impressions. Total body length measures 9.8 mm. The body is slender and somewhat fusiform. Other specimens show more robust bodies, which may be indicative of differences between species. The wing is long and narrow with an overall length of 6.3 mm.

Discussion: All specimens are placed in the genus *Bibio* because of venational characters (Curran 1934:128-129). The specimens will be taxonomically placed below the generic level at a future time.

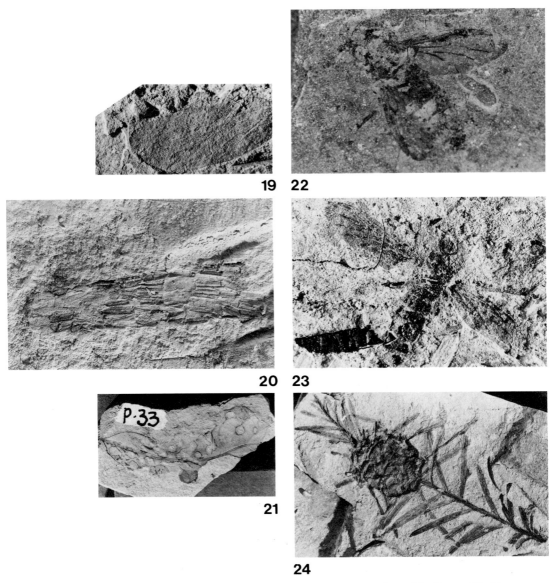

Figures 19-24. Fig. 19. *Phryganea spokanensis* Carpenter. Fig. 20. Genus (?) *Phryganea* Linneaus. Fig. 21. Genus (?) *Lithocolletis* Clemens. Figs. 22, 23. Genus *Bibio* Latreille. Fig. 24. Genus (?) *Thecodiplosis* Kieffer.

The genus *Bibio* is holarctic in its distribution. The adults are commonly found on flowers and are extremely abundant during the spring and early summers.
Specimen No.: UIMM T-0070, T-0071.
Locality: UIMM P-33.

Family Cecidomyiidae
Genus (?) *Thecodiplosis* Kieffer
(Figure 24)

Diagnosis: The shape of the galls ranges from small, smooth, spherical structures to more elongate robust structures with scale-like surfaces that may superficially resemble in prolific cones. They range in length from 3 to 17 mm and in width from 3 to 10 mm. Conifer needles commonly protrude beyond the gall's surfaces.

Discussion: Thirteen stem galls closely resemble stem galls formed on the living bald cypress *Taxodium* by members of the genus *Thecodiplosis*, family Cecidomyiidae (Riley 1870:244). The galls occur at various locations on the stems of modern trees of this conifer, which is common in the swampy areas of the southeastern United States. The galls are not to be confused with the cypress flower gall described by Felt (1940:46).
Specimen No.: UIMM T-0072.
Locality: UIMM P-33.

Order HYMENOPTERA

Four families of the order Hymenoptera, representing 17 specimens, are presently identified from the Clarkia deposits. They include the families Ichneumonidae (ichneumons), Cynipidae (gall wasps), Formicidae (ants), and Apidae (bees). The formicids are represented by 14 of the 17 specimens, and a single specimen is representative of each of the other families. The cynipid is identified from galls on oak leaf.

Family Ichneumonidae
(Figure 25)

Diagnosis: A single specimen has an overall body length of 15.9 mm. The head is circular in shape, with a diameter of .73 mm. Antenna is incomplete with only one or two segments present. The thorax is 2.77 mm in length. Forewings, which measure 6.8 mm in length, are overlapping and disarticulated. The legs are missing.

Discussion: The general body shape and wing venation are used to place this specimen at the family level. More specific taxonomic placement will be conducted at a later time.

Ichneumonids are holarctic in distribution. Many ichneumons are internal parasites of immature stages of their host (i.e. other insects).
Specimen No.: UIMM T-0073.
Locality: UIMM P-33.

Family Cynipidae
(Figure 26)

Diagnosis: Several galls are present on what appears to be a red oak leaf. The galls are circular in vertical view and discoid in lateral view. They vary slightly in size but are typically 2.9 mm in diameter. Several galls have a slightly convex upper surface, and a zone of lighter color around the margin. The gall color is reddish brown on a black leaf compression. Some of the galls have irregular or sinuate margins.

Discussion: These galls closely resemble present-day members of either the genus *Andricus* or *Neuroterus* (Darlington 1968:198-203, Fig. 188; Johnson and Lyon 1976:386-387). Many modern gall wasps form galls on oak leaves, and commonly are restricted to particular host species.
Specimen No.: UIMM T-0074.
Locality: UIMM P-33.

Family Formicidae
(Figures 27, 28)

Diagnosis: Fourteen specimens are mostly impressions of forewings, with only a few body parts preserved. The wing length varies from 7.8 to 10.7 mm, with typical formicid venation.

Discussion: No generic determination can be made from available fossil material, although there is some resemblance to the genus *Messor* Forcel (Carpenter 1930: Plate 11; Fig. 5). Additional specimens are required to show body forms and venational characters of the Clarkia formicids.
Specimen No.: UIMM T-0075, T-0076.
Locality: UIMM P-33.

Family Apidae
(Figure 29)

Diagnosis: A single specimen is represented by a dorsal view of a poorly preserved body and wings. The overall length is 14.6 mm. The head is flattened from the anterior to the posterior margin, having a length of 1.7 mm and a width of 3.7 mm. The wing length is 8.9 mm, and three submarginal cells are present. Five abdominal segments are discernible.

Discussion: This specimen closely resembles living members of the family Apidae (Borror and DeLong 1976:551, Fig. D), and the wing venation compares closely with members of the genus *Apis*. If this generic reference proves to be correct, it will be significant in tracing the ancestry of the apids in North America.
Specimen No.: UIMM T-0077.
Locality: UIMM P-33.

DISCUSSION

The importance of the Clarkia insect assemblage cannot be overemphasized from the standpoint of the number of specimens discovered and the diversity of the species. The present investigation supports the findings of previous studies on Miocene insects of this region. This Clarkia study has added one new order and eight new families of insects to the existing list (Table 1). Fourteen orders and 34 families of insects have now been identified from the Miocene deposits of eastern Washington and northern Idaho.

Upon reviewing this insect assemblage, one finds that the terrestrial forms outnumber the aquatic forms by a ratio of approximately 9:1. This is to be expected when one considers the fossilization process and its effect on soft-bodied forms, such as the aquatic larval stages of insects. The orders represented by aquatic insects in Miocene deposits of this region include the Ephemeroptera, Odonata, Plecoptera, Trichoptera and Coleoptera. The types of preservation include adult body parts (such as wings), immature body forms, and indirect fossil evidence (such as caddisfly cases). The Trichopterans (caddisflies) represent approximately 80% of the aquatic fossils discovered to date, and most of these are larval cases.

The total number of specimen terrestrial insects is dominated by the orders Coleoptera (35%), Diptera (17%), and Hymenoptera (14%). The fossil coleopterans (beetles) are represented largely by adult body parts such as wings. The most common family of coleopterans includes the

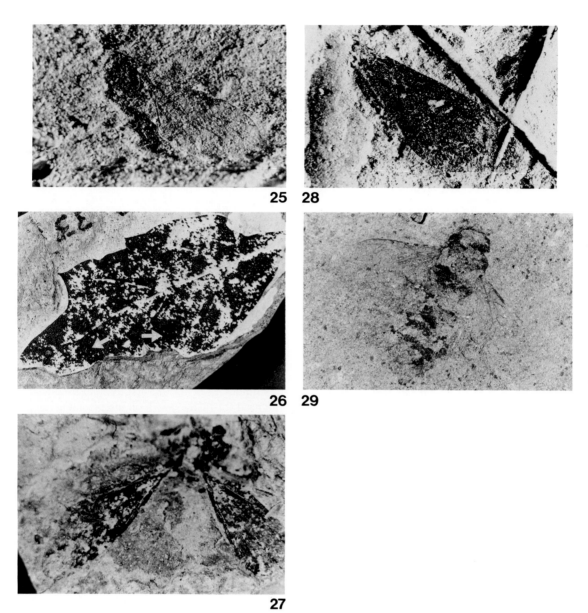

Figures 25-29. Fig. 25. Undetermined (family Ichneumonidae). Fig. 26. Undetermined (family Cynipidae). Figs. 27, 28. Undetermined (family Formicidae). Fig. 29. Undetermined (family Apidae).

click beetles (Elateridae), representing 23% of the total beetle population. This family of beetles is presently widespread, occupying numerous niches on flowers, under bark or on vegetation. It appears that a suitable number of similar niches for such a family of beetle existed during the Miocene in this region. One of the most outstanding specimens of beetles belongs to the family Anthribidae, represented by a complete dorso-lateral view of the entire adult form (Fig. 14). Not only are the morphological features well preserved, but a clear image of what the insects must have been like 15-20 million years ago is portrayed.

The fossil dipterans are represented mainly by members of the families Bibionidae and Cecidomyiidae. The bibionids (March flies) have been recorded in previous studies of the region, and at this time do not add significantly to our understanding of the insect fauna. With more discoveries of different generic representatives, it may be possible to piece together their role in this paleoenvironment.

The cecidomyiids comprise approximately 56% of the dipterans found. All specimens represent galls preserved on stems of bald cypress (*Taxodium*). In southeastern North America the habitat of *Taxodium* is a mesic bottomland region with a swampy component. As indicated by Smiley et al. (1975:8), a marginal swamp developed during the infilling of the Miocene Clarkia Lake.

The fossil hymenopterans are dominated mainly by wing impressions of ants. Although numerous specimens have been discovered in both this and prior studies, the fossil material is inadequate for identification below the family level.

Significant discoveries representing the order Hymenoptera include single fossils of the families Cynipidae, Ichneumonidae, and Apidae. This represents the first cynipid and the second apid recorded from Miocene deposits of the region. The cynipid is represented as a gall formation on what seems to be red oak leaf. This trace fossil is unique in being the first cynipid-produced gall found in the region. This discovery points out the need for further work on insect galling (and mining) of leaves in Miocene floras of the region.

The other hymenopterans recovered from the Clarkia deposits are represented by an impression of an ichneumonid wasp (family Ichneumonidae) and an impression of an adult apid (bee). The apid is possibly a significant discovery in that few bees have been found in these sediments. Members of the genus *Bombus* were recorded by Cockerell (1931:309-311). The Clarkia specimen closely resembles members of the genus *Apis* in the venation pattern of the wings, but this relationship must remain tentative for the present.

Lepidopterans normally are not found in large numbers in any fossil deposit. If fossils are present, they usually are represented by indirect evidence such as leaf damage. One interesting impression of an oak leaf with insect mining closely resembles leaf-mining damage by members of the family Gracilariidae. This specimen represents the first fossil evidence of such damage by this family in Miocene deposits of the region. Again, this fossil points out the need to closely observe fossil plant material for insect damage.

CONCLUSIONS

In this preliminary report, the Clarkia insect fauna has been reviewed and given taxonomic status at the familial or generic level. Preliminary investigation on modern affinities of the Clarkia insects indicates that many of the identified fossils (such as representative of the orders Orthoptera, Hemiptera, Coleoptera, and Diptera) resemble living forms presently found in the southeastern United States, around the southern Appalachian Mountain area. This similarity in fossil and modern insect groups correlates well with floral comparisons. Smiley (1979:8-10) has noted that the Clarkia floral assemblage closely resembles modern forest associations of the southeastern United States. Further information on the degree of similarity between the Clarkia insect fauna and living taxa of the southeastern U. S. will require more specific determination of the fossils.

Future studies will be conducted on the Clarkia insects, including a re-evaluation of past investigations (Lewis 1969, 1970, 1973). New fossil collections and additional studies will ultimately clarify the Miocene-Recent relationships in this complex yet exciting field.

ACKNOWLEDGMENTS

The author wishes to express his deepest appreciation to his wife, Carol, for her help and encouragement during the preparation of this paper. Thanks are also extended to the following people who helped type the manuscript: Ann Miller, Barb Raymond, and Elaine Thrune. Special thanks go to Dan Marek and his staff at the Production Services Unit, St. Cloud State University, for the excellent photographs.

LITERATURE CITED

Baker, W. L. 1972. Eastern forest insects. U.S. Dep. Agric. Misc. Pub. No. 1175. 642 pp.

Barr, W. F., and J. M. Gillespie. 1975. Systematic Paleontology Insecta. *In* Smiley, C. J., et al., Preservation of Miocene Fossils in an Oxidized Lake Deposit. J. Paleont. 49(5):833-844.

Berry, E. W. 1928. A caddis case of leaf pieces from the Miocene of Washington. J. Wash. Acad. Sci. 18(3):60-61.

Blatchley, W. S. 1926. Heteroptera or true bugs of eastern North America. Nature Publishing Co., Indianapolis, Ind. 1116 pp.

Borror, D. J., Delong, D. M., and Triplehorn, C. A. 1976. An introduction to the study of insects. Holt, Rinehart and Winston. 852 pp.

Carpenter, F. M. 1930. The fossil ants of North America. Bull. Mus. Comp. Zool. Harv. Univ. 70(1):3-66.

Carpenter, F. M., Cockerell, T. D. A., Kennedy, C. H., Snyder, T., and Wickham, H. F. 1931. Insects from the Miocene (Latah) of Washington. Ann. Ent. Soc. Amer. 24(2):307-322.

Cockerell, T. D. A. 1924. Fossil insects in the United States National Museum. Section 4. A fossil beetle from Washington State. Proc. U. S. Nat'l. Mus. 64:14-15.

Curran, C. H. 1934. The families and genera of North American Diptera. Ballon Press, New York, N. Y. 512 pp.

Darlington, A. 1968. The pocket encyclopedia of plant galls. Blandford Press, London. 191 pp.

Dillon, E. S., and Dillon, L. S. 1961. A manual of common beetles of eastern North America. Row, Peterson, and Co. Pub. 884 pp.

Edmunds, G. F., Jensen, S. L., and Berner, L. 1976. The mayflies of north and central America. University of Minnesota Press, Minneapolis, Minn. 322 pp.

Felt, E. P. 1940. Plant galls and gall makers. Comstock Pub. Co., Inc. 364 pp.

Jacques, H. E. 1951. How to know the beetles. Wm. C. Brown Co. Pub. 372 pp.

Johnson, W. T., and Lyon, H. H. 1976. Insects that feed on trees and shrubs. An illustrated practical guide. Cornell University Press, Ithaca, N.Y. 464 pp.

Lewis, S. E. 1969. Fossil insects of the Latah Formation (Miocene) of eastern Washington and northern Idaho. Northwest Sci. 43(3):99-115.

Lewis, S. E. 1969. Lepidopterous larval-mining of an oak(?) leaf from the Latah Formation (Miocene) of eastern Washington. Ann. Ent. Soc. Amer. 62(5):1210-1211.

Lewis, S. E. 1970. Fossil caddisfly (Trichoptera) cases from the Latah Formation (Miocene) of eastern Washington and northern Idaho. Ann. Ent. Soc. Amer. 63(2):621-622.

Lewis, S. E. 1973. A new species of fossil beetles (Coleoptera: Dascillidae) from the Latah Formation (Miocene) of eastern Washington. Ann. Ent. Soc. Amer. 66(3):697.

Needham, J. G., Frost, S. W., and Tothill, B. H. 1928. Leaf-mining insects. Wms. & Wilkins Co., Baltimore, Md. 351 pp.

Riley, C. 1870. Cypress-gall. Amer. Ent. 2:244.

Scudder, S. H. 1900. Adephagous and clavicorn Coleoptera from the Tertiary deposits at Florissant, Colorado. Monogr. U. S. Geol. Surv. 60:148 pp.

Sharov, A. G. 1968. Phylogeny of the Orthopteroidea. Acad. Sci. USSR. Trans. Instit. Paleont. 118:88-99.

Smiley, C. J., Gray, J., and Huggins, L. M. 1975. Preservation of Miocene fossils in unoxidized lake deposits, Clarkia, Idaho. J. Paleont. 49(5):833-844.

Smiley, C. J. 1979. Guidebook and road log to the St. Maries River (Clarkia) fossil area of northern Idaho. Idaho Bur. Mines Geol. Inf. Circ. 33:1-19.

Wiggins, G. 1977. Larvae of the North American caddisfly genera (Trichoptera). University of Toronto Press. 401 pp.

EVOLUTION OF FRESHWATER DRAINAGES AND MOLLUSCS IN WESTERN NORTH AMERICA

DWIGHT W. TAYLOR
Department of Geology, Oregon State University, Corvallis, OR 97331

Mesozoic and Cenozoic freshwater molluscs in the western United States include 54 family and subfamily groups. The total number has remained fairly constant through the Cenozoic, progressively changing in composition. Early Tertiary faunas have a clearly tropical character and lived in relatively few large basins of integrated drainage.

Later Tertiary faunas are more diverse because of the preservation of small shells and a wider range of habitats than known earlier. Progressive subdivision of drainage promoted vicariant speciation, and lakes including a variety of habitats supported faunas locally much richer than those of today. In Miocene and Pliocene faunas a climatic gradient like that of the present is evident, with sparser faunas to the south; Pliocene climate was generally less humid than in the Miocene. Most if not all living species differentiated in the Tertiary: there is no evidence of Pleistocene evolution, and modern distribution is correlated with Tertiary tectonic features.

Numerous Pleistocene lakes in the Great Basin had faunas richer than at present, but few had localized species. Quaternary extinct species are most numerous in the late Pleistocene—early Holocene. Regional climatic change brought arid to semi-arid climate in the latest Quaternary, causing both regional changes in range and absolute extinctions.

The modern fauna is distinctive from its localized endemic forms, both lacustrine and nonlacustrine, but is numerically unimpressive by contrast with that of the more humid east. Present composition of the fauna and distribution of species are a product of Tertiary evolution of species, Tertiary and Quaternary extinctions, some Quaternary range extensions, and latest Quaternary development of arid climate. Exotic species due to human importation have had little effect on natives, as they mostly fill niches left vacant by Tertiary extinctions.

Drainage history of some local areas is discussed, with speculative interpretations to account for both past and modern species distributions, under the following headings: Northern Continental Divide; Snake River, Idaho; Green River, Wyoming; Former Snake River, California-Oregon; Lahontan Basin; Bonneville Basin; Central California; Southern Continental Divide; Former Gulf of California; Lower Colorado River; and Rio Grande.

The Cenozoic history of western North America has included tectonic and climatic changes that rearranged the surface drainage repeatedly. Elements of freshwater fauna that are closely tied to their habitat have been affected in various ways—by enlarging geographic range, evolving into new species, reducing geographic range, or becoming extinct. Those freshwater molluscs that are limited to perennial waters at all stages of their life cycle are the principal source of interpretations herein. Emphasis is on later Cenozoic times.

Molluscs of perennial waters differ from fishes in several ways that affect zoogeographic interpretations. They can move only slowly and are dependent on local substratum, current, oxygen content, and temperature of the water, as well as protection from flood scour. The smaller species can persist in individual springs or drainages too small to support fishes. These attributes mean that more local and precise inferences of drainage may be made from molluscs, and in some cases no other source of such evidence is available.

An exception to the foregoing generalizations is the freshwater mussels with their fish-borne parasitic larva. As adults the mussels are nearly sessile, but in the larval state are as mobile as their individual fish hosts. Consequences of this special life history are evident in distributions of the species: they occupy characteristically more individual drainages than prosobranch snails. From Alaska to the southern end of the Plateau of Mexico, the *Anodonta* species are most closely related to those of Eurasia, not of eastern North America. No gastropod group has such a distribution. Most likely this is an effect of the mid-Tertiary spread of modern minnows (Cyprinidae) in times when drainage divides were less pronounced and perhaps individual basins fewer than at present.

The interpretations of past drainages are the simplest accounting for distribution and inferred relationships of the species. A fundamental principle is that spread of the individual animals is entirely within their habitat as part of a biota. The state of taxonomic knowledge varies from group to group, hence future refinements are certain in understanding both relationships and histories.

Within the time span, region, and groups considered, there is no evidence of detailed phylogeny. Subdivision of an inferred ancestor into daughter-species, extinctions, fluctuations of range, and occasional bursts of expansion are the broad features of the record. Most of these were determined by changes in climate and landscape that affected quality and extent of habitat, with no known biological interactions. Less often, especially in small habitats such as springs and minor drainages, competitive exclusion seems a characteristic determinant of the snails present.

Phylogenetic interpretations thus play no part in these inferred drainage histories. There is no element of cladistic analysis, and none whatever of that abhorrence, "vicariance biogeography." I have previously (Taylor 1960) acknowledged, and do so once more herein, my great debt to the ideas of Léon Croizat. The interested reader can compare my understanding of his ideas as assessed by Croizat (1962:vii) with his appraisal of "vicariance biogeography" (Croizat 1982), ostensibly derived in part from his works.

Throughout the entire Mesozoic and Cenozoic record, all of the faunal changes can be accounted for by slow climatic changes (either global or regional), by rearrangement of drainages due to tectonic activity, and by differentiation of genera and species within families already established. Extinctions can be documented, but the evidence for organic evolution or faunal immigration is indirect. There is no evidence of the catastrophes that have been invoked to explain extinctions or range changes in other organisms (Hsü et al. 1982).

Neither is there evidence of great lateral fault movement, not even those documented for the San Andreas Rift in California. Perhaps this is due partly to the arid climate of much of the coastal ranges in California, precluding nearly all molluscs of permanent fresh water. Late Tertiary assemblages from west of the San Andreas are few (Figs. 4,5), as well as sparse, and provide no evidence of significant movement.

GENERAL FEATURES:
THE FAMILY/SUBFAMILY LEVEL

The living families and subfamilies of freshwater molluscs make a fully modern first appearance in the fossil record, whether Mesozoic or Cenozoic. Some extinct groups show character combinations not known in the modern fauna, but in no case is an ancestral-descendant relationship plausible. Figure 1 lists the 54 families and subfamilies known, fossil or living, in the western United States, with their stratigraphic ranges. The total number of groups has been fairly constant throughout the Cenozoic, but with progressive change in composition through absolute and regional extinction (Fig. 2).

Distinctive features of the fauna are a combination of generally North American elements, those shared with Eurasia, with the Old World or New World tropics, and some attributable to ancient (presumably Paleozoic) wider distribution around an ancestral Pacific Ocean. Five groups of snails (*Reesidella*-group, Zaptychiinae, Lancidae, Payettiidae, Pliopholyginae) are restricted to western North America, of which only the Lancidae survive to the present.

The Mesozoic and Tertiary record gives practically no hint of ancient stocks that appear in astonishing diversity in the rich Pliocene lake faunas of Oregon, Idaho, and Utah. These include such varied lineages as the following: Among the Diotocardian or Archaeogastropod snails, an effusion of freshwater Neritacea (Payettiidae), filling the spectrum from *Crepidula*-like shells, through *Septaria*-like forms to even fully *Patella*-form shells, a wider range of shell form than in any other family of Neritacea. Among the lower Monotocardian or Megastropod snails, a flowering of the filter-feeding Pliopholygidae, apparently with both nearshore and deep-water types, and ecologically perhaps equivalent to both Viviparidae and Bithyniidae. In the Cerithiacea a local family possibly related to the Potamididae; and Pleuroceridae attaining enormous size. In the Rissoacea not only the Hydrobiidae one would expect, but seemingly the Old World Pyrgulidae as well. Among the lower pulmonate gastropods, surprises are even greater: a local family related to the Chilinidae of southern South America; and Amphibolidae (as well as Salinatoridae?—I still remain undecided), otherwise known only from New Zealand.

Seemingly there was a faunal reservoir ready to help fill the large lakes of Oregon and Idaho when they came into existence. None of the groups mentioned has been found in regions east of the continental divide, nor in the Miocene or Pliocene lakes of western Nevada or California. Hence more northern sources, consistently west of the continental divide, are indicated. Likely possibilities are the mountain streams of central Idaho and eastern Oregon. Field and laboratory studies in progress should cast more light on these interpretations. At present, the almost unheralded appearance of such faunal diversity serves as further testimony to the sketchiness of the fossil record.

THE EARLY TERTIARY

Intermontane basins from northern Utah through Colorado and Wyoming to Montana contain a rich record in deposits of ancient lakes, alluvial plains, and marshes (see map of lake deposits, Feth 1963). To the west and south the record is spotty, consisting of fossils from inland freshwater deposits, and, along the Pacific Coast, freshwater fossils sometimes as nonmarine assemblages and sometimes mixed with marine faunas. Minute and small shells are notably less well represented than larger specimens, through accidents of preservation and collectors' interests. Thus two common, widespread groups in the late Tertiary and modern faunas are poorly known: small clams of the family Sphaeriidae, and the presently abundant and diverse Rissoacean snails. In the Rocky Mountains larger forms—freshwater mussels (Unionidae) and two groups of perennial water snails (Pleuroceridae, Viviparidae)—are common as fossils. They contrast sharply with both the late Tertiary and living faunas of the region, where these families are unknown or poorly represented.

Six families known from Cretaceous rocks in the western United States are as yet unrecorded from early Tertiary faunas (Fig. 1). These include clams of the extinct family Neomiodontidae, and five extant families of snails. Freshwater Neritidae, the Melanopsidae and Pyrguliferidae may have permanently disappeared in the region by the Tertiary, but the Pomatiopsidae and Lancidae reappear later, so their absence probably reflects the faulty record. Three groups widespread in Cretaceous rocks of the Rocky Mountains persist in the early Tertiary only along the Pacific Coast: both subfamilies of Thiarid snails, and clams of the family Corbiculidae. All three are characteristically tropical groups, so that a change towards more continental interior climates with cooler winters may have been responsible. There is no evidence of a catastrophic change in fauna at the close of the Cretaceous.

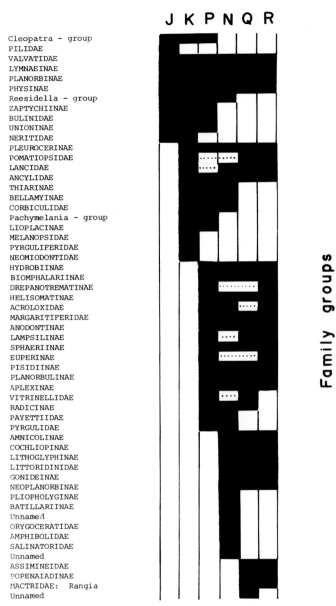

Figure 1. Stratigraphic range of families and subfamilies of freshwater molluscs in western United States and Canada. A few groups that might deserve subfamily status are listed informally. J, Jurassic; K, Cretaceous; P, Paleogene; N, Neogene; Q, Quaternary; R, Recent.

Living genera recognized in early Tertiary faunas amount to 28, of which only 13 survive in the western United States (Table 1). Many genera are now found only in tropical or warm-temperate areas of the Old World; only *Campeloma* and *Elliptio* are now restricted to eastern North America, only *Hemisinus* to tropical America. A continuity of favorable habitats around the northern Pacific Ocean is indicated, with barriers of climate, drainage divides, and other environmental features to the south and east. Early Tertiary deposits of the northern Rockies and plains have generally similar faunas spread across the present continental divide, but contain no trace of

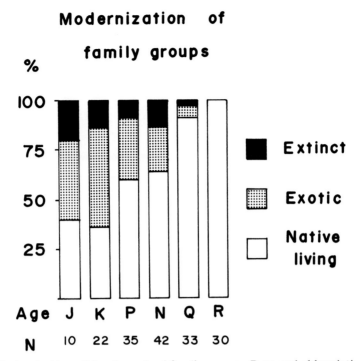

Figure 2. Modernization of family and subfamily groups. Data and abbreviations as in Fig. 1.

the characteristically east American Viviparidae–Viviparinae, and nothing like the diversity of living Unionidae and Pleuroceridae found in the southeastern United States.

Another noteworthy aspect of early Tertiary extant genera is the small number that is endemic (*Lanx*, and perhaps *Juga*). To some extent this can be explained by the general lack of small, well-preserved shells, and by the limited diversity of habitats represented. Large areas of integrated drainage also precluded isolating mechanisms that operated in the later Cenozoic.

The tropical character of Eocene climates in the northern Rocky Mountains has been documented by paleobotanical studies and is likewise shown by fossil molluscs. The clearest evidence comes from the planorbid *Biomphalaria pseudoammonius*, widespread in Eocene rocks of the Rocky Mountains, western Europe and Asia (Fig. 3). These fossils may even represent the living tropical American *B. glabrata*, a species that receives much study because it is a vector of human schistosomiasis. The refined taxonomy of recent years has combined a number of nominal living species and increased the known range of variation in shell features. A consequence is that distinctiveness of the Eocene species is doubtful.

Field and laboratory studies of *Biomphalaria glabrata* indicate optimum temperature for oviposition is 25° C., diminishing to zero at extremes of 20 and 30° C. Laboratory maintenance at 5°C. results in death of the snails within a week. Survival and slow growth is possible at 15°C. though reproduction is suppressed. Seasonal desiccation may prove advantageous to a population, enabling it to postpone reproduction to a time when temperatures are more favorable.

Application of these data concerning the living *Biomphalaria glabrata* to the Eocene *B. pseudoammonius* provides interpretations fully consistent with those of paleobotany. Tropical climate during the middle Eocene in Wyoming has been inferred by MacGinitie (1969, 1974). Such climate appears to have extended much farther north, into Montana and southernmost British Columbia, than previous data for the Rocky Mountains have shown. On the Pacific Coast, tropical and paratropical forests are known as far north as Alaska (Wolfe 1978, and references therein).

TABLE 1. EXTANT GENERA OF FRESHWATER MOLLUSCS
IN EARLY TERTIARY OF WESTERN UNITED STATES

	Geographic regions					
	WNAm	ENAm	TropAm	EAs	Afr	Eur
Valvata	X	X		X	X	X
Juga	X			?		
Hemisinus			X			
Pachymelania				F	X	
Bellamya				X	X	
Campeloma		X				
Clenchiella				X		
Lanx	X					
Lymnaea	X	X	X	X	X	X
Radix				X	X	X
Planorbis					X	X
Gyraulus	X	X	X	X	X	X
Anisus					X	X
Biomphalaria	X	X	X		X	
Drepanotrema			X			
Helisoma	X	X				
Bulinus					X	X
Physa	X	X	X	X	X	X
Acroloxus	X	X				X
Margaritifera	X	X		X	X	X
Elliptio		X				
Anodonta	X	X	X	X	X	X
Lampsilis		X				
Corbicula				X	X	X
Batissa				X		
Sphaerium	X	X	X	X	X	X
Eupera		X	X	F	X	F
Pisidium	X	X	X	X	X	X

X, living occurrence; F, early Tertiary fossil. Geographic regions: WNAm, North America west of continental divide; ENAm, eastern North America; TropAm, tropical America; EAs, eastern Asia; Afr, Africa; Eur, Europe.

In the Rocky Mountains at least, the northward extension of tropical climate was greater in the Eocene than at any other interval in the Tertiary. *Biomphalaria pseudoammonius* appears abruptly in the late early Eocene (post-Graybull), with no evident precursor in older rocks. After a period of abundance in middle and late Eocene times, it vanished. By analogy with modern *B. glabrata*, the spread of *B. pseudoammonius* into suitable habitats could have been rapid. The snails can self-fertilize in isolation to found a colony; they survive desiccation in seasonal water bodies, live in a variety of habitats, and can be transported passively into new situations. A northward expansion of tropical climates in the late early Eocene, with recession of winter temperature lows, seems a well-founded interpretation.

Diverse evidence for global cooling (Lillegraven 1979, Wolfe 1978) is consistent with the disappearance of *Biomphalaria*. Tropical climate in the continental interior never again reached so far north as in the Eocene. Hence this climatic change cannot be ascribed solely to the temporary formation of a planetary ring (O'Keefe 1980), nor to meteoritic impact (Alvarez et al. 1982; Ganapathy 1982).

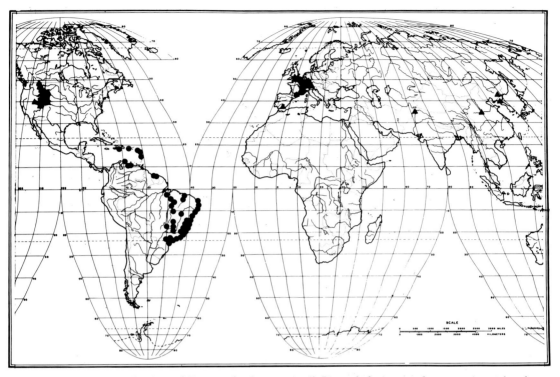

Figure 3. Distribution of Eocene freshwater snail *Biomphalaria pseudoammonius*, triangles, and of living *B. glabrata*, solid dots (family Planorbidae).

Although there are few early Tertiary freshwater molluscs known from the southwestern United States, and none from northwestern Mexico, available information suggests a climatic gradient with more arid areas to the south. Freshwater genera such as *Bellamya*, and some tolerant of slightly brackish water such as *Batissa, Corbicula*, and *Pachymelania*, would be expected to survive in the middle American tropics had they reached so far south. Apparently rainfall was inadequate to support coastal marshes and mangrove swamps along much of the coast of northwestern Mexico, just as today. Similar, but more firmly based, conclusions were drawn by MacGinitie (1969:55) from Eocene floras: "Evidence on the location of the tropical high-pressure system of the Tertiary is given by a group of genera common in western Tertiary floras and in the living flora of central and southern China, but now extinct in America . . . . The list strongly suggests that some climatic barrier prevented the dispersal of such genera from the western Tertiary south into Mexico and the southern Appalachians. The barrier was probably a dry area or belt along the southwestern border of the United States under a tropical high-pressure area."

THE LATE TERTIARY

Late Tertiary freshwater molluscs west of the Rocky Mountains are known principally from lacustrine deposits in the Great Basin, Snake River drainage, and adjacent areas to the west (Figs. 4,5). Contrasts with the early Tertiary record are: (1) westward displacement of the fossil localities, (2) smaller size of the basins, (3) greater number of basins represented, (4) more common occurrence of small shells, and (5) relatively smaller number of mixed freshwater and marine assemblages.

Early and late Tertiary faunas contain practically the same number of family and subfamily groups, but with changes in composition (Figs. 1, 2). Survivorship of extant genera in the western

United States reveals the faunal modernization more clearly (Tables 1, 2). Of 28 extant early Tertiary genera, only 13 persist in the region; of 35 extant late Tertiary genera, 26 survived. The nine exotic genera survive in the Old World (5), eastern North America (3), and in tropical America (1). The contrast between early and late Tertiary faunas is heightened by a gap in the fossil record. There are practically no Oligocene or early Miocene freshwater molluscs known in the region.

The greater diversity of late Tertiary faunas is influenced substantially by the better record of small shells, and by a wider range of habitats than known in the early Tertiary. Both immigration and evolution are also plausible factors, but permitted by ignorance rather than substantiated by knowledge.

Extant species, not recognized with certainty in the early Tertiary, appear first in the Miocene. Table 3 lists 41 known in the late Tertiary, with the stratigraphic units in which they have been found. The data in that list are plotted (Fig. 6) as taxonomic-ecological groups of species according to the number of stratigraphic units in which they have been found. The most widely

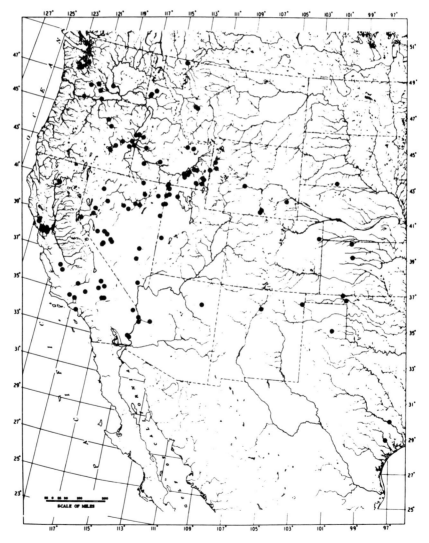

Figure 4. Localities of Miocene freshwater molluscs in western United States.

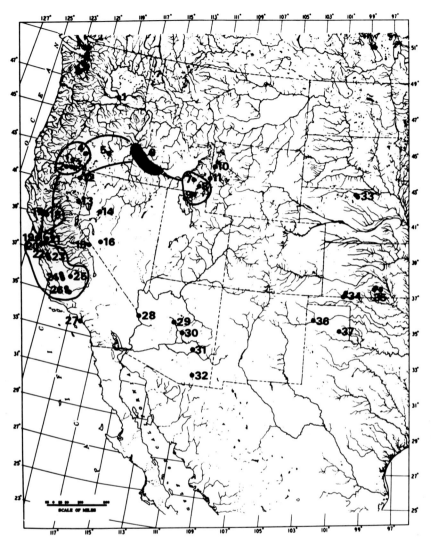

Figure 5. Localities of Pliocene freshwater molluscs in western United States. Solid lines enclose localities linked by common occurrences of at least one species of snail of larger perennial waters. Dotted lines connect groups of localities that share clams, but no snails, of larger perennial waters.

1, Ringold formation, Wash. 2, Butte Valley, Calif. 3, Yonna formation, Ore. 4, Summer Lake, Ore. 5, "Danforth formation," Ore. 6, Glenns Ferry formation, Oregon-Idaho, the younger deposits of Lake Idaho. 7, Salt Lake group, Marsh Creek valley, Ida. 8, Cache Valley formation, Gentile Valley, Ida. 9, Cache Valley formation, Cache Valley, Utah. 10, Jackson Hole, Wyo. 11, Star Valley, Wyo. 12, Alturas formation, Calif. 13, Honey Lake, Calif. 14, Mopung Hills, Nev. 15, Mono Lake basin, Calif. 16, Sodaville, Nev. 17, Cache formation, Calif. 18, Tehama formation, Calif. 19, Santa Clara formation of San Andreas rift, Calif. 20, lower Santa Clara formation, Calif. 21, Livermore gravels, Calif. 22, Purisima group, Calif. 23, San Benito gravels, Calif. 24, Tulare and San Joaquin formations, Kreyenhagen and Kettleman Hills, Calif. 25, Tulare formation, Lambertson's Well, Calif. 26, Tulare and San Joaquin formations, Lost Hills, Buttonwillow, McKittrick and Elk Hills areas, Calif. 27, San Pedro sand, Calif. 28, older Colorado River alluvium, Ariz. 29, Verde formation, Ariz. 30, Payson basin, Ariz. 31, San Carlos basin, Ariz. 32, St. David formation, Ariz. 33, Sand Draw local fauna, Keim formation, Nebr. 34, Saw Rock Canyon, Fox Canyon, Rexroad and Bender local faunas, Rexroad formation, Kans., and Spring Creek and Sanders local faunas, Ballard formation, Kans. 35, Dixon local fauna and Swingle locality, Kans. 36, Red Corral local fauna, Tex. 37, Whitefish Creek, Tex. For more details see Taylor (1966), Taylor and Smith (1981).

TABLE 2. EXTANT GENERA OF FRESHWATER MOLLUSCS IN LATE TERTIARY OF WESTERN UNITED STATES

	Geographic regions					
	WNAm	ENAm	TropAm	EAs	Afr	Eur
Valvata	X	X		X	X	X
Juga	X			?		
Batillaria		X	X	X		
Thiara				X	X	
Bellamya				X	X	
Fontelicella	X					
Savaginius	X					
Amnicola	X	X				
Lithoglyphus	X					X
Tryonia	X	?	X			
Zetekina			X			
Lanx	X					
Lymnaea	X	X	X	X	X	X
Bulimnea		X				
Bakerilymnaea	X	X	X	X		
Fossaria	X	X	X	X	X	X
Pseudosuccinea		X	X			
Radix				X	X	X
Ferrissia	X	X	X	X	X	X
Gyraulus	X	X	X	X	X	X
Anisus					X	X
Helisoma	X	X				
Planorbella	X	X	X			
Vorticifex	X					
Planorbula	X	X				
Menetus	X					
Promenetus	X	X				
Physa	X	X	X		X	X
Sibirenauta	X	X		X		
Margaritifera	X	X	X	X	X	X
Anodonta	X	X	X	X	X	X
Gonidea	X			X		
Corbicula				X	X	X
Sphaerium	X	X	X	X	X	X
Pisidium	X	X	X	X	X	X

X, living occurrence. Geographic regions: WNAm, North America west of continental divide, ENAm; eastern North America; Trop Am, tropical America; EAs, eastern Asia; Afr, Africa; Eur, Europe.

distributed species are snails of seasonal or small perennial waters. Of those living in larger perennial waters, the Sphaeriidae are clearly more widespread than snails. These conclusions from the Tertiary record are in accord with the modern distributions of the species.

Late Cenozoic tectonic activity has brought rearrangement of drainage patterns, and climatic changes have altered habitats drastically in large areas. Nonetheless, the Tertiary records of extant species of larger perennial waters are practically all within, or adjacent to, the local drainages in which they now live. This observation emphasizes the conservative spread, or lack of spread, of the species and their zoogeographic value. Hence, present-day local endemic forms are thought to be

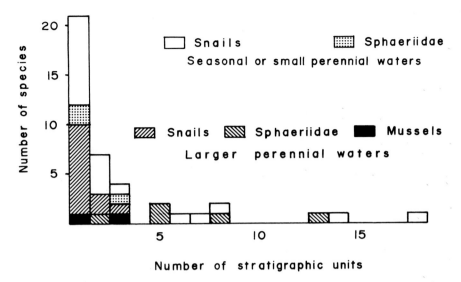

Figure 6. Distribution of extant species in late Tertiary stratigraphic units. Data from Table 3.

Tertiary origin in most if not all cases. The geographic distribution of living local species (Fig. 33) is conspicuously limited to the west of the present Wasatch Front, correlated with the late Tertiary development of basin-range structure. This correlation provides further suggestive evidence that the modern fauna had already differentiated in the Tertiary.

Latitudinal changes in diversity of the late Tertiary faunas are impefectly recorded. The most diverse faunas are not the most northern, but those of the Snake River drainage and northern Great Basin. Three groups of snails have been mapped (Figs. 7-9) to provide more detail. Their record is significant because their shells are relatively large and persistent as fossils, and though all three are restricted to perennial water bodies, they have different habitats.

The pleurocerids *Juga* (subgenus *Calibasis*) (Fig. 7) live in rivers and large springs where they browse on the organic film (aufwuchs) on stones or vegetation. They are characteristic of pure, well-oxygenated waters. Tertiary occurrences in west-central Nevada and the San Joaquin Valley, California, indicate a former lacustrine habitat not known for the group today.

Viviparidae of the genus *Bellamya* (Fig. 8) now live in the Old World in lakes, marshes, and slow streams. They may feed by the selection of fine particles, or by muco-ciliary trapping of suspended material. Their extinction in the western United States (the youngest range in the New World) by the end of the Miocene testifies to the regional disappearance of large areas of suitable habitat. Like *Juga* (*Calibasis*), they formerly occurred from Washington to western Nevada, but ranged slightly further south and substantially further east.

Bulimnea (Fig. 9) is a giant of the pulmonate family Lymnaeidae. The single living species is found in southeastern Canada and the adjacent United States, where it lives in lakes, creeks, rivers, and marshes. It may browse on vegetation or even the flesh of dead animals. Diet, distribution, and habitats are wider than in the previous two genera. The distribution of *Bulimnea* shows the effects of progressive aridity and also the development of local areas of relict habitat. The genus was widespread in the west during the Miocene, but only two Pliocene occurrences are known (Fig. 9). These are both adjacent to, or within, areas of present or Holocene local endemism in northeastern California and southeastern Idaho. Again, a late Tertiary origin of local modern endemism is suggested.

Miocene drainages cutting across present major divides are indicated by the distributions of some species. The evidence is especially strong that the Snake River of southern Idaho formerly

TABLE 3. LATE TERTIARY RECORD OF EXTANT SPECIES OF FRESHWATER
MOLLUSCS IN UNITED STATES WEST OF CONTINENTAL DIVIDE[1]

Group A. Mussels
    Anodonta californiensis Lea. Mio., Muddy Creek Fm., Calif.
    A. wahlamatensis Lea. Plio., Cache Fm. (17), Tehama (18) and lower Santa Clara Fm. (20), all Calif.

Group B. Sphaeriidae of larger perennial waters
    Sphaerium simile (Say). Plio., Rexroad Fm. (34), Kans., but not known fossil west of continental divide.
    S. striatinum (Lamarck). Mio., Bidahochi Fm., Ariz. Plio., Glenns Ferry Fm. (6), Ida.; Salt Lake Group, Marsh Creek Valley (7), Ida.; Cache Valley Fm., Gentile Valley (8), Ida., and Cache Valley (9), Ut. Purisima Group (22), San Joaquin Fm., Reef Ridge (24), and Tulare Fm., Kettleman Hills (24), Calif.
    Pisidium compressum Prime. Mio., Bidahochi Fm., Ariz. Plio., Summer Lake basin (4), Ore.; Glenns Ferry Fm. (6), Ida.; Salt Lake Group, Marsh Creek Valley (7), Ida.; Cache Valley Fm., Gentile Valley (8), Ida., and Cache Valley (9), Ut.; Jackson Hole (10), Wyo.; Alturas Fm. (12), Calif.; Honey Lake (13), Calif.; Cache Fm. (17), Calif.; Tehama Fm. (18), Calif.; Santa Clara Fm. of San Andreas rift (19), Calif.; Mopung Hills (14), Nev.
    P. punctatum Sterki. Plio., Glenns Ferry Fm. (6), Ida.; Salt Lake Group, Marsh Creek Valley (7), Ida.; Cache Valley Fm., Gentile Valley (8), Ida., and Cache Valley (9), Ut.; Tulare Fm., Kettleman Hills (24), Calif.
    P. supinum Schmidt. Plio., Glenns Ferry Fm. (6), Ida.; San Joaquin Fm., Reef Ridge (24), and Tulare Fm., Kettleman Hills (24), Calif.
    P. ultramontanum Prime. Mio., Deer Butte Fm., Ore. Plio., Butte Valley (2), Calif.; Yonna Fm. (3) and Summer Lake basin (4), Ore.; Honey Lake (13), Calif.

Group C. Snails of larger perennial waters
    Valvata tricarinata (Say). Mio., Laverne Fm., Okla., but west of continental divide only Pleistocene and living.
    V. utahensis Call. Plio., Honey Lake (13), Calif.
    "*Fluminicola*" *avernalis* (Pilsbry). Mio., Muddy Creek Fm., Nev.
    Fontelicella idahoensis (Pilsbry). Plio., Glenns Ferry Fm. (6), Ida.
    Lithoglyphus columbianus (Pilsbry). Mio., Deer Butte Fm., Ore. and Chalk Hills Fm., Ida.
    L. hindsi (Baird). Plio., "Danford Fm." (5), Ore.
    L. virens (Lea). Mio., Yakima Basalt, Wash.
    Savaginius yatesianus (J. G. Cooper). Plio., Tehama Fm. (18), Santa Clara Fm. (20), and Tulare Fm., in subsurface northeast of Kettleman Hills (25), Calif.
    Tryonia imitator (Pilsbry). Plio., San Pedro sand (27), Calif.
    Lymnaea hinkleyi F. C. Baker. Plio., "Danford Fm." (5), Ore.
    L. kingi Meek. Plio., Cache Valley Fm., Cache Valley (9), Ut.
    Helisoma anceps (Menke). Plio., "Danford Fm." (5), Ore.
    H. newberryi (Lea). Plio., Honey Lake (13), and Mono Lake basin (15), Calif.

Group D. Sphaeriidae of seasonal or small perennial waters
    Pisidium casertanum (Plio). Mio., Laverne Fm., Okla.; ?Barstow Fm., Calif. Plio., Butte Valley (2), Calif. Jackson Hole (10), Wyo.; St. David Fm. (32), Ariz.
    P. insigne Gabb. Plio., Tulare Fm., Kettleman Hills (24), Calif.
    P. ventricosum Prime. Plio., Jackson Hole (10), Wyo.

Group E. Snails of seasonal or small perennial waters
    Valvata humeralis Say. Mio., Yakima Basalt, Wash.; Neeley Fm., Ida.; Starlight Fm., Ida.; Teewinot Fm., Grand Valley, Ida.-Wyo. and Jackson Hole, Wyo.; Humboldt Gr., Nev.; Salt Lake Gr., Ut.; Bidahochi Fm., Ariz. Plio., Glenns Ferry Fm. (6), Ida.; Salt Lake Gr., Marsh Creek Valley (7), Ida.; Cache Valley Fm., Gentile Valley (8), Ida.; Jackson Hole (10) and Star Valley (11), Wyo.; Alturas Fm. (12), Calif.; Honey Lake (13), Calif.; Cache Fm. (17), Calif.; Purisima Gr. (22), San Joaquin Fm., Reef Ridge (24), and Tulare Fm., Kettleman Hills (24), Calif.
    Lymnaea caperata Say. Mio., Ogallala Fm., Okla. Plio., St. David Fm. (32), Ariz.
    L. montanensis (F. C. Baker). Plio., Jackson Hole (10), Wyo.
    L. palustris-group. Mio., Teewinot Fm., Grand Valley, Ida.-Wyo. Plio., Butte Valley (2), Calif.; Jackson Hole (10) and Star Valley (11), Wyo.; Alturas Fm. (12), Calif.; Payson basin (30) and St. David Fm. (32), Ariz.
    Bakerilymnaea cubensis (K. G. L. Pfeiffer). Plio., St. David Fm. (32), Ariz.
    B. techella (Haldeman). Plio., Glenns Ferry Fm. (6), Ida.
    Fossaria modicella (Say). Plio., Jackson Hole (10), Wyo.; Verde Fm. (29), Ariz.
    F. parva (Lea). Mio., Starlight Fm., Ida. Plio., Verde Fm. (29) and St. David Fm. (32), Ariz.
    Ferrissia rivularis (Say). Plio., Glenns Ferry Fm. (6), Ida.
    Gyraulus parvus (Say). Mio., Green Valley Fm., Calif. Plio., Glenns Ferry Fm. (6), Ida.; Cache Valley Fm., Gentile Valley (8), Ida.; Alturas Fm. (12), Calif.; Honey Lake (13), Calif.; San Pedro sand (27), Calif.; Verde Fm. (29) and St. David Fm. (32), Ariz.
    Planorbella subcrenata (Carpenter). Mio., Bidahochi Fm., Ariz. Plio., Alturas Fm. (12), Calif.
    P. tenuis (Dunker). Plio., Tulare Fm. of McKittrick area (26), Calif.; San Pedro sand (27), Calif.
    Planorbula campestris (Dawson). Mio., Teewinot Fm., Grand Valley, Ida.-Wyo.
    Menetus centervillensis (Tryon). Plio., Tulare Fm., Kettleman Hills (24), Calif.
    Promenetus exacuous (Say). Plio., Alturas Fm. (12), Calif.

[1] Numbers of Pliocene localities correspond to those in Fig. 5.

TABLE 3. Continued.

Group E. Continued.
  P. umbilicatellus (Cockerell). Mio., "Danforth Fm.", Ore.; Neeley Fm., Ida; Starlight Fm., Ida.; Glacier Park, Mont.; Teewinot Fm., Grand Valley, Ida.-Wyo. and Jackson Hole, Wyo.; Bidahochi Fm., Ariz. Plio., Glenns Ferry Fm. (6) Ida.; Cache Valley Fm., Gentile Valley (8), Ida. and Cache Valley (9), Ut.; Jackson Hole (10), Wyo.; Alturas Fm. (12), Calif.; Payson basin (30) and St. David Fm. (32), Ariz.
  Physa gyrina Say. Plio., Glenns Ferry Fm. (6), Ida.; Alturas Fm. (12), Calif.
  P. virgata Gould. Plio., Tulare Fm., Kettleman Hills (24), Calif.; San Pedro sand (27), Calif.; Verde Fm. (29), Payson basin (30), San Carlos basin (31), and St. David Fm. (32), Ariz.
  Sibirenauta elongatus (Say). Plio., Jackson Hole (10), Wyo.

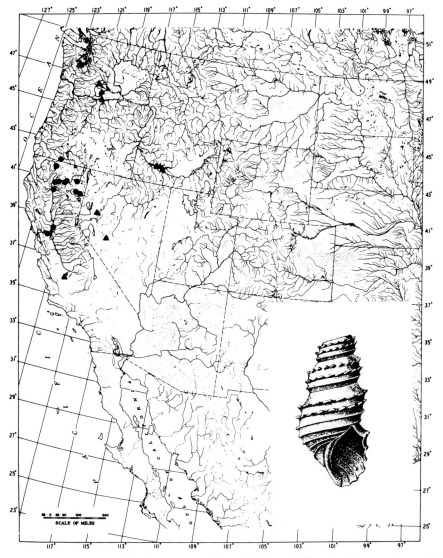

Figure 7. Distribution of freshwater snail *Juga* (*Calibasis*) (family Pleuroceridae). Solid dots, modern; triangles, Miocene and Pliocene. Inset: type of *J. acutifilosa* (Stearns 1890) from original illustration.

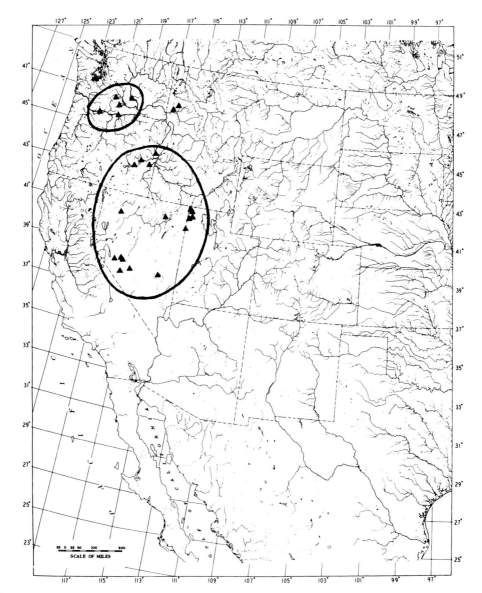

Figure 8. Distribution of Miocene species of freshwater snail *Bellamya* (family Viviparidae). Northern enclosed area, undescribed species; southern enclosed area, *B. turneri* (Hannibal). Poorly preserved specimens from northern Idaho lack the distinctive spiral sculpture of the undescribed species, but appear shorter and more globose than *B. turneri*.

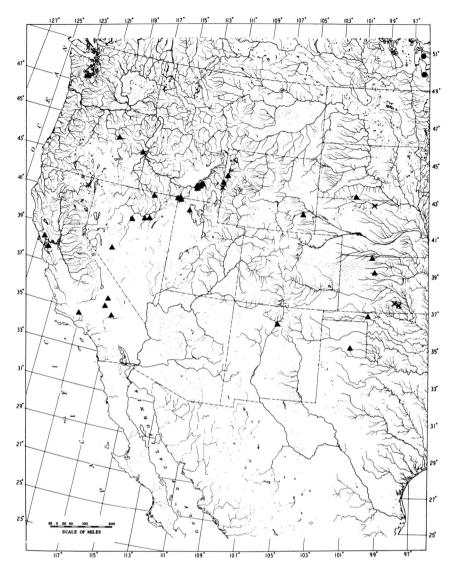

Figure 9. Distribution of freshwater snail *Bulimnea* (family Lymnaeidae). Triangle, Miocene; X, Pliocene; solid dots, southwesternmost occurrences of *B. megasoma* (Say), the sole living species of the genus.

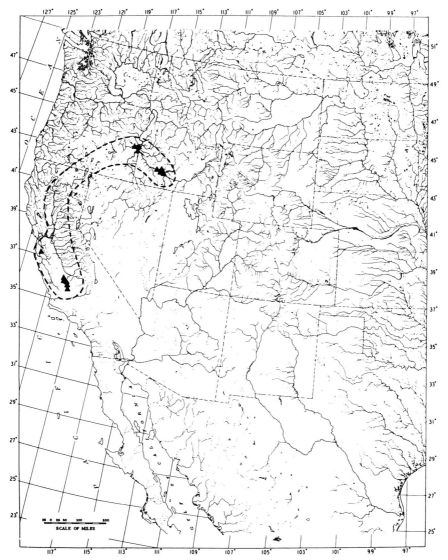

Figure 10. Distribution of Miocene-Pliocene freshwater mussel *Gonidea coalingensis* Arnold (family Unionidae).

flowed westward, independent of the Columbia, to which it is now tributary. Separate species-pairs of the stream and lake snails *Juga* (subgenus *Calibasis*) and *Bellamya* (Figs. 7, 8) are known in the middle Columbia basin and the northwestern Great Basin. The distribution of *Bellamya turneri* is further instructive in cross-cutting the modern divides of the Snake River, Oregon basins of internal drainage, Lahontan basin, Bonneville basin, and internal basins south of the Lahontan basin. It evidently records a period of integrated drainage, prior to the development of modern isolated basins, but separate from the middle Columbia River. Miocene interconnections between the present Snake River drainage and central California are shown by the distributions of the extinct mussel *Gonidea coalingensis* (Fig. 10) and the extinct snail genus *Scalez* (Fig. 11), as well as fossil fishes (Smith 1975). Precise interpretations of locations and time of the inferred connection are not yet possible.

In other parts of the west the fossil molluscs hint at quite different drainages, but provide no precise data of timing or connection. Figures 12-14 show the distributions of three groups of snails that seemingly antedate the development of the present Colorado River and Great Basin drainages. These and the preceding data indicate major reaarrangement of drainage patterns during or after the Miocene, when many extant species were already in the region.

A diagram of middle to late Miocene drainages (Fig. 15) shows some generalized interpretations based on freshwater molluscs. The separate drainages of the middle Columbia and regions to the south (dashed lines) are based largely on the groups previously discussed (Figs. 7, 8). Miocene faunas of southeastern Idaho are sharply distinct from those to the west, but distribution of

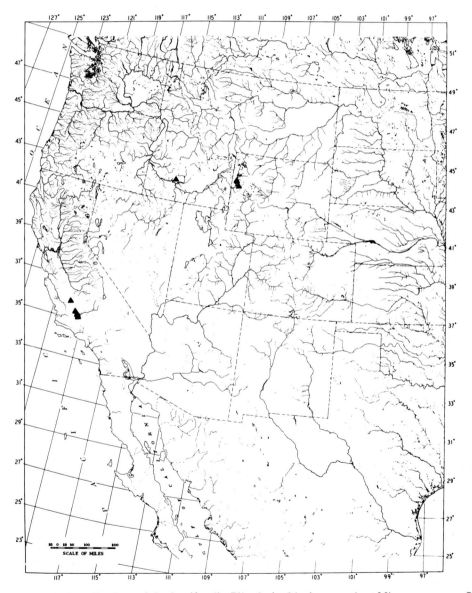

Figure 11. Distribution of *Scalez* (family Pliopholygidae), an extinct Miocene genus. Occurrences in California are *S. petrolia* Hanna and Gaylord; those in Idaho and Wyoming possibly but not surely conspecific.

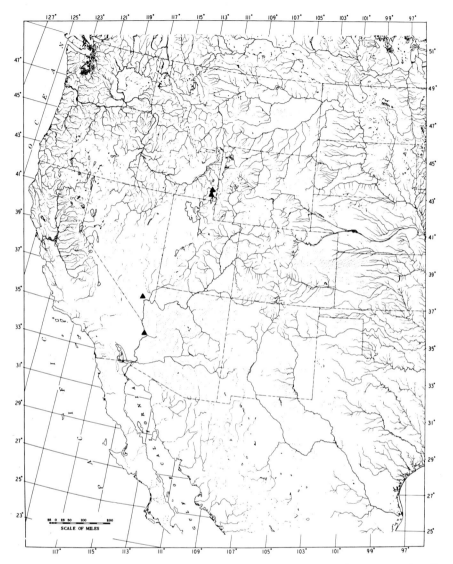

Figure 12. Distribution of Miocene freshwater snail *Valvata idahoensis* Taylor (family Valvatidae).

Valvata idahoensis (Fig. 12) shows relationship to the south. No Miocene fossils clearly indicate drainage from western Montana southward into Idaho, but the later record includes species in southeast Idaho thought to be so derived.

The inferred drainage shown cutting across bold mountain ranges from Montana southward in the Miocene and Pliocene (Figs. 15, 25) represents a choice among implausibilities. No geologic evidence supports such an interpretation (Robinson 1963 and references therein). A late elevation of the Centennial Mountains (separating the present streams of southwestern Montana and southeastern Idaho) is implicit in the diagrammatic drainages. For this there is no direct evidence, and one must appeal for a parallel to the documented post-Miocene uplift of the Teton Range on the eastern Idaho border.

Biogeographic evidence from the molluscs is Miocene and younger, and particularly from southeastern Idaho. Regardless of drainage interpretation, the evidence is that molluscan history in southeastern Idaho is distinctive. The Miocene assemblage described by Yen (1946) from the Salt Lake group is unlike any other of that age. The remarkable genus *Papyrotheca* is otherwise unknown in North America, found elsewhere only in southeastern Europe. Characteristic and conspicuous genera of Miocene assemblages to the west and southwest are lacking—*Bellamya turneri*, all Pleuroceridae, larger Hydrobiidae, all bivalves. Southward drainage through the Rocky Mountain trench, and Miocene interconnections of streams with the region of southeastern Idaho, are inferred to have introduced *Papyrotheca* and other groups of the modern fauna with Old World elements derived from a former trans-Siberian fauna.

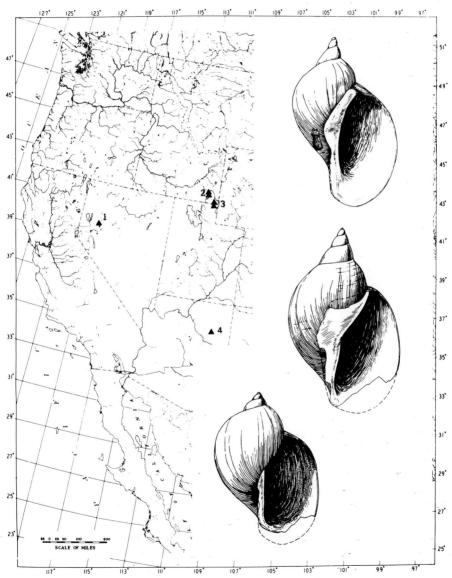

Figure 13. Distribution of Miocene-Holocene freshwater snail *Lutrilimnea* (family Lymaeidae). 1, *L. polyskelidis* Taylor, Pliocene; 2, *L. gentilis* Taylor and *L. ursina* Taylor, Pleistocene; 3, *L. ursina* Taylor, Pleistocene-Holocene; 4, *L. dineana* (Taylor), Miocene. Insets: original illustrations of *L. dineana*, from Taylor (1957).

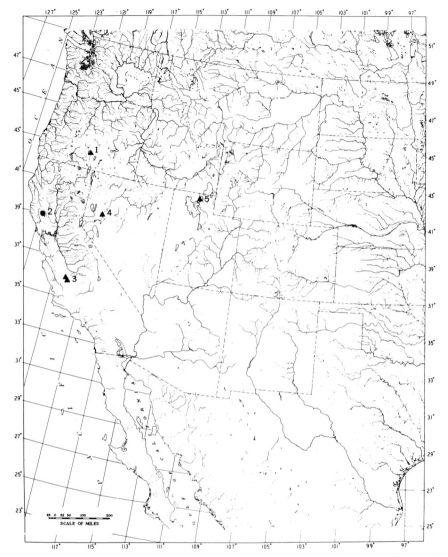

Figure 14. Distribution of freshwater snail *Physa* (subgenus *Costatella*) (family Physidae). 1, *P.* sp., Pliocene; 2, *P. costata* Newcomb, living; 3, *P. wattsi* Arnold, Pliocene; 4, *P. humboldtiana* Taylor, Pliocene; 5, *P. harpa* Taylor, Pliocene.

Post-Miocene evidence of eastern American species in southeastern Idaho, and others occurring more widely, is discussed under "Northern Continental Divide."

Miocene faunas were richer than now not only in fresh waters but in brackish, marginally marine habitats. Brackish-to-fresh-water clams of the genus *Corbicula* are native to the Old World. They were widespread in the Rocky Mountains during the Cretaceous, but the youngest New World occurrence is a few Miocene species in central California. *Batillaria*, a brackish-water snail (family Potamididae), is represented by two Miocene species in California that belong to widely different geographic groups. The species of the southern San Joaquin Valley (then the site of a Pacific Coast embayment) is related to the living East Asian forms. The species of the former embayment in the lower Colorado River Valley is most similar to one living in the Gulf of Mexico and Caribbean (Taylor 1983).

Pliocene freshwater molluscs are known from fewer localities than those of the Miocene (compare Figs. 4,5). The contrast is partly due to the shorter duration of the Pliocene, but also to environmental change. Extensive areas of Miocene lake deposits in the northern Great Basin (Esmeralda and Humboldt Groups, Truckee Formation) and Colorado Plateau (Bidahochi Formation) have sparse Pliocene successors. A further contrast is the generally better preservation of Pliocene fossils, permitting more secure species identifications. More detail in stratigraphic correlations and in interpretations of distribution is thus possible.

The more southern Pliocene localities have no species of larger perennial fresh waters. The interpretation that Pliocene climate to the south was more arid, and faunas sparser, is in accord with earlier and modern faunas (Fig. 34).

Localities of richer fauna have been grouped (Fig. 5) according to shared occurrences of snails of larger perennial waters. These groups are thought to represent separate Pliocene drainages

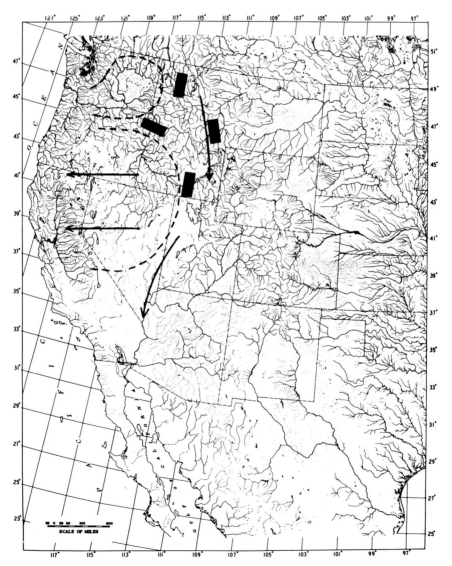

Figure 15. Middle-late Miocene drainages. Black bar: inferred divide crossing present drainage.

in most cases. Shared occurrences of clams are more widespread, interpreted as an inheritance from earlier times, due to their slower differentiation. These areas of localized fauna imply the subdivision of the Miocene river systems into smaller drainages. Clearly, the area of widespread drainage, or interconnected drainages, shown by the Miocene distribution of *Bellamya turneri* (Fig. 8) had been broken up. Local differentiation of many modern species may date from this period of more local drainages, though as yet there are no detailed sequences of fossils to document time of origin.

During latest Pliocene or earliest Pleistocene time, the areas of Pliocene endemic faunas (Fig. 5) were largely eradicated. The rich faunas in the San Francisco Bay area and southern San Joaquin Valley, California; Klamath Lake area, California-Oregon; southwestern Idaho; southeastern Idaho-northern Utah; and western Nevada were substantially eliminated. A new pattern of molluscan distribution came into being, with fewer species, all extant, spread in and around the northern and western Great Basin (Fig. 33). This pattern is shown by the overlapping, disjunct ranges of lake and river species (e.g. *Valvata utahensis*, Fig. 16; *Pyrgulopsis*, Fig. 17; *Lymnaea kingi*, Fig. 18; *Helisoma newberryi*, Fig. 19; *Pisidium punctatum*, Fig. 20; *P. ultramontanum*, Fig. 21); by marginal ranges of others (e.g. *Vorticifex effusus*, Fig. 22), and includes numerous localized forms. This pattern is presumably due to interconnections of drainage at different times, none precisely dated. It is older than known late Pleistocene drainage, older than some early Pleistocene drainage of some areas, and younger than some Pliocene faunas. As it cuts across all of the principal areas of internal drainage, these basins in their modern detail are younger than late Pliocene or early Pleistocene.

In now-arid southeastern California, Pleistocene faunas were notably enriched by the occurrence of species now restricted far to the north and far to the southeast. The association of such species as *Valvata utahensis* (Fig. 16), *Vorticifex effusus* (Fig. 22), *Sphaerium triangulare* (Fig. 23) and *Musculium transversum* (Fig. 24) is unknown elsewhere. External drainage to the lower Colorado River of several now-internal basins has been postulated previously, and would account for the distributional data. No precise date is possible, but as the molluscs represent freshwater drainages, the events are evidently younger than the latest Miocene brackish-water to marine Bouse embayment formerly in the lower Colorado River Valley.

A diagram of Pliocene drainage (Fig. 25), derived principally from interpretation of mollusc distribution, shows the great extent to which internal drainage and rearrangement of external drainages are thought to be of Pliocene and post-Pliocene age.

QUATERNARY

Pleistocene and Holocene shells are widespread through the western United States and represent practically all modern habitats. Most of the record is either Holocene or very late Pleistocene, with fewer early Pleistocene localities (Fig. 26). None of the extinct families of Neogene faunas persists into the Quaternary (Fig. 1), and only three extinct genera are known.

The geographic distribution of extinct species (Fig. 27) is generally correlated with latitude. Drainages in southern California and Texas have the majority of extinct forms, though the highest total for a single basin is that of Lake Thatcher, Idaho. The regional development of arid climate was probably a major cause of the southern extinctions, and habitat changes were directly and indirectly caused by local volcanism in the case of Lake Thatcher.

In addition to these absolutely extinct species, three others are regionally extinct: *Probythinella lacustris* (Fig. 28), *Pisidium cruciatum* (Fig. 29), and *Rangia cuneata* (Fig. 30). Their fossil occurrences in the western United States are hundreds of kilometers from the nearest modern localities. *Probythinella* in the Thatcher basin, Idaho, and *Rangia* in the Pecos River valley, New Mexico, add to the already notable extinct element in the faunas of the two areas.

The Quaternary was predominantly a time of changes in distribution and of extinctions. The

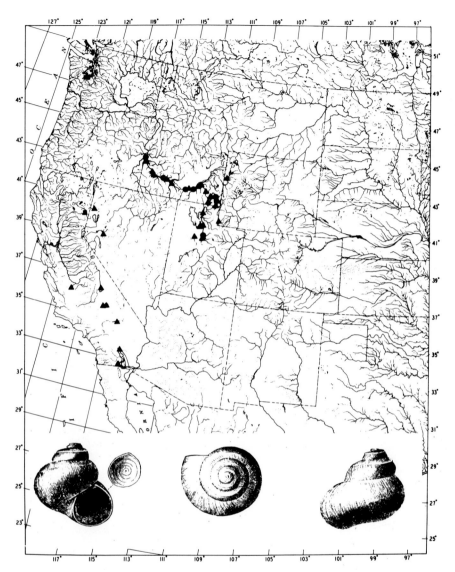

Figure 16. Distribution of freshwater snail *Valvata utahensis* Call (family Valvatidae). Solid dots, modern; triangles, fossil. Inset: original illustrations of the species, from Call (1884).

fossil record provides no clear evidence that any species evolved within this brief interval of about 1.5 million years. Furthermore, extinctions are heavily concentrated in the last tenth of this time span—latest Pleistocene and early Holocene. The number of early Pleistocene extinct species is far less than that in the latest Pleistocene (Fig. 27).

The latest Pleistocene—early Holocene extinctions are only one aspect of regional change in faunas. Throughout the western United States and northern Mexico, in areas of now arid or semi-arid climate, youngest Quaternary faunas are characteristically richer than those living locally (Fig. 31). Radiocarbon dating has gradually revealed that the modernization of faunas was approximately contemporaneous from the Pacific Northwest and northern Great Plains to the southern Plateau of Mexico, in late Wisconsin and early Holocene times. Such a widespread change is most probably correlated with changes in planetary circulation associated with deglaciation.

Climatic change toward more arid climate was evidently more severe in the latest Pleistocene-early Holocene than at any time earlier in the Quaternary. This conclusion follows from the greater number of extinct species in young faunas (Fig. 27), together with the interpretation that extinction was due to climatic change. It follows that present climates are atypical of the Pleistocene, and that the present is not typical of past interglacial times. A similar conclusion was drawn previously from faunas in the High Plains, even though there are scarcely any latest Quaternary extinct species of molluscs there (Taylor 1965).

These data thus indicate that aquatic habitats were not subject to cyclical fluctuations of

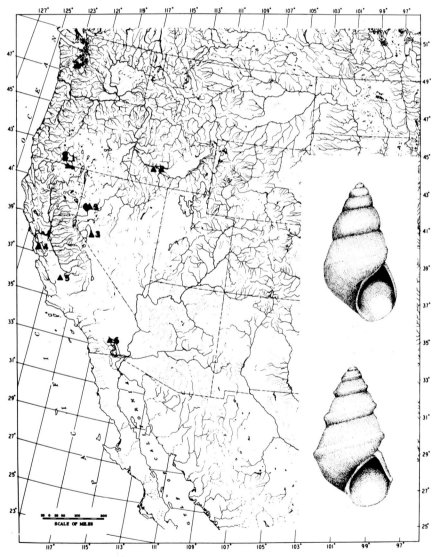

Figure 17. Distribution of freshwater snail *Pyrgulopsis* (family Hydrobiidae) in western United States. 1, *P. archimedis* Berry; living, Klamath Lake; Holocene, Tule Lake. 2, *P.* sp., Pliocene, Glenns Ferry formation. 3, *P. nevadensis* (Stearns), living, Pyramid Lake; Pleistocene-Holocene, Winnemucca Lake and Walker Lake. 4, *P. tropidogyra* Pilsbry, Pliocene, Santa Clara formation. 5, *P. vincta* Pilsbry, Pliocene, Tulare formation. 6, *P. imminens* Taylor, Pleistocene, Lake Cahuilla. Insets: smooth and carinate forms of *P. nevadensis*; drawn by A. D'Attilio.

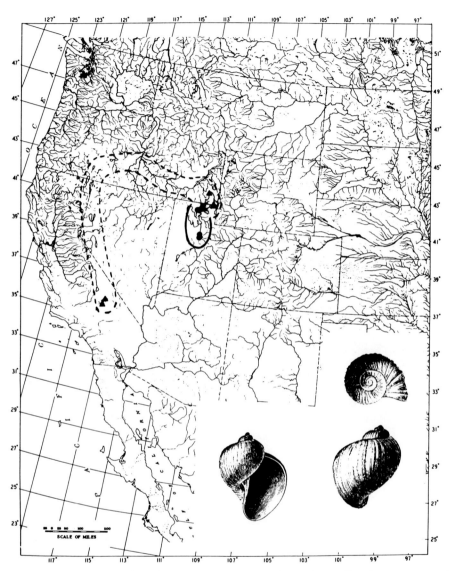

Figure 18. Distribution of freshwater snail *Lymnaea kingi* Meek (family Lymnaeidae). Solid dots, modern; triangles, fossil. Insets: original illustrations of the synonymous form *Radix utahensis*, from Call (1884).

the same intensity throughout the Pleistocene. Evolutionary conclusions drawn from an opposite assumption (Smith 1981b) are, I believe, fundamentally unsound. This is not to deny the abundant evidence of Pleistocene climatic changes, at least some of which were cyclical.

A prominent feature of the Great Basin in late Pleistocene times was the development of many lakes in basins now dry or nearly dry (Snyder et al. 1964). Scarcely any of these lakes contained localized species (Fig. 32), another indication that evolution of species requires more time than only part of the Quaternary. Faunal diversity was not correlated with lake size. Vast Lake Bonneville had only one species restricted to it, and a total fauna far smaller than that of nearby Lake Thatcher.

290 LATE CENOZOIC HISTORY

MODERN FAUNA

The modern fauna in the western United States (estimated at about 300 species in 65 genera and 24 families) is numerically unimpressive by contrast with that in the more humid east. Similar contrasts apply to freshwater fishes (Miller 1958:191). The distinctiveness of the fauna is due to endemic genera and species, partly in lakes (Fig. 32). These local forms are separated by a broad area of mountains and plains lacking such endemics (Fig. 33) from the rich localized faunas of the east.

Species and even genera characteristic of large streams are a conspicuous element in the southeastern United States, but in the west are rare. Only about 20 species and 2 genera are so restricted, found in the following major drainages:

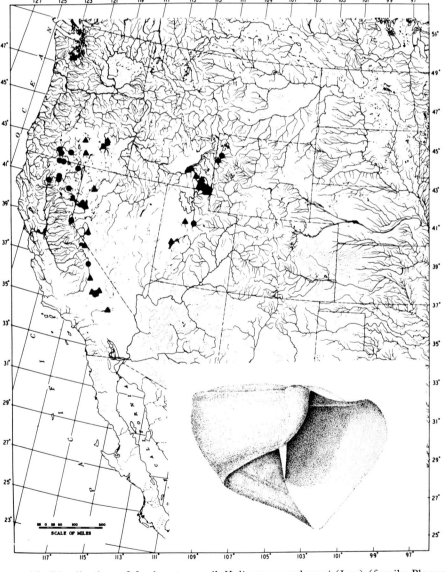

Figure 19. Distribution of freshwater snail *Helisoma newberryi* (Lea) (family Planorbidae). Solid dots, modern; triangles, fossil. Inset: drawn by A. D'Attilio.

	Columbia-Salmon	Upper Snake	Sacramento	Other
Species	9	6	3	3
Endemic genera	0	1	1	0

The fossil record of large-stream environments is scanty, but all of the species are probably of Miocene or Pliocene origin. Those with a known Tertiary record, listed in Table 3, are *Fontelicella idahoensis, Lithoglyphus columbianus, Savaginius yatesianus,* and *Lymnaea hinkleyi*. The species cited as occurring in the "Upper Snake" (the Snake River in southern Idaho, above Hells Canyon) include some that are known from the deposits of Pliocene Lake Idaho: *Fontelicella idahoensis* and an undescribed genus (Taylor, in prep.). These may have evolved originally in a lacustrine environment, although they are known only from Snake River in modern times.

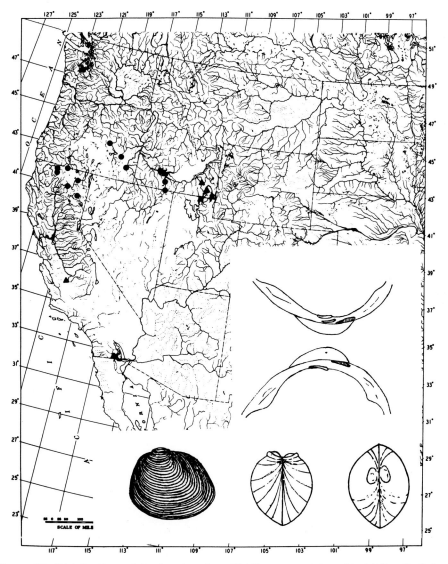

Figure 20. Distribution of freshwater clam *Pisidium punctatum* Sterki (family Sphaeriidae). Solid dots, modern; triangles, fossil. The species is found living also in eastern North America outside the map area. Insets: original illustrations of the species, from Sterki (1895).

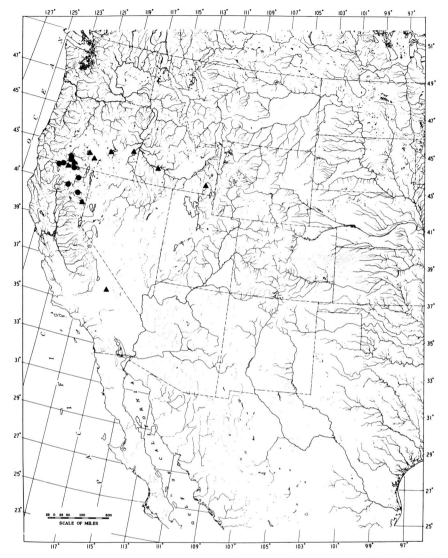

Figure 21. Distribution of freshwater clam *Pisidium ultramontanum* Prime (family Sphaeriidae). Solid dots, modern; triangles, fossil.

In the Sacramento River drainage, California, one of the three river species may also be of lacustrine origin. *Savaginius yatesianus* is known widely in central California in Pliocene deposits of lakes or sluggish lowland streams. It was apparently restricted to the Sacramento River delta before being exterminated (for further data on distribution of the three species see Taylor, 1981).

Bar graphs of native species diversity (Fig. 34) show the combined effects of present-day climate (sparse fauna in more arid areas) together with inheritance from the Tertiary (local evolution of species).

The influence of man has increased species diversity by accidental or intentional importations. In contrast to some other parts of the world, ill effects have been minimal. A probable cause is that the introduced forms are merely filling niches left vacant by late Tertiary extinctions. Practically all of the genera formerly lived in the western United States (Table 4).

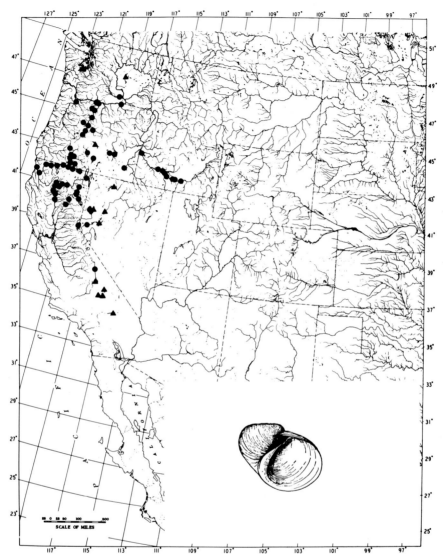

Figure 22. Distribution of freshwater snail *Vorticifex effusus* (Lea) (family Bulinidae). Solid dots, modern; triangles, Pleistocene. Inset from Keep and Baily (1935), ca. 3x, by permission of Stanford University Press.

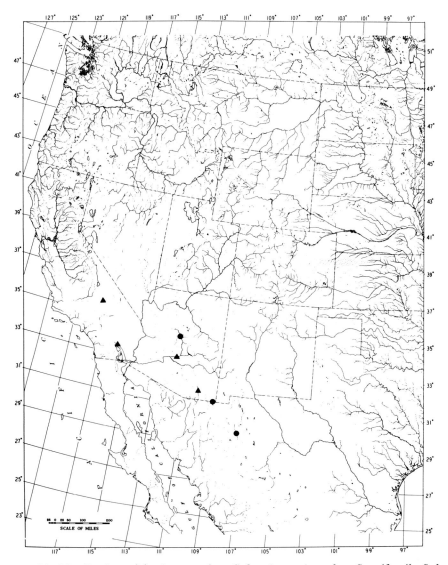

Figure 23. Distribution of freshwater clam *Sphaerium triangulare* Say (family Sphaeriidae). Solid dots, modern; triangles, Pleistocene-Holocene. The species occurs widely on the Plateau of Mexico south of the map area.

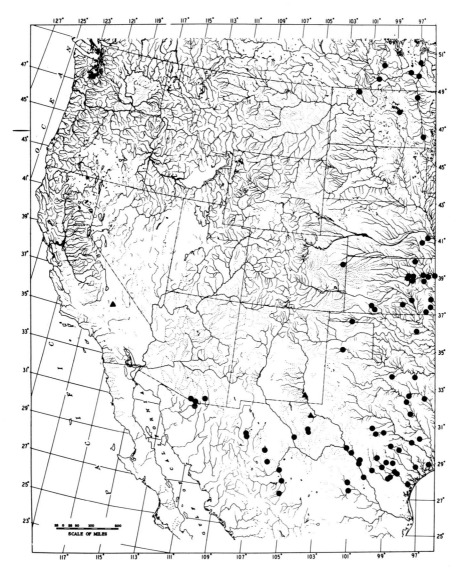

Figure 24. Distribution of freshwater clam *Musculium transversum* (Say) (family Sphaeriidae). Solid dots, modern; triangles, Pleistocene-Holocene. The species occurs widely to the east and south of the map area.

296　　　　　　　　　　　LATE CENOZOIC HISTORY

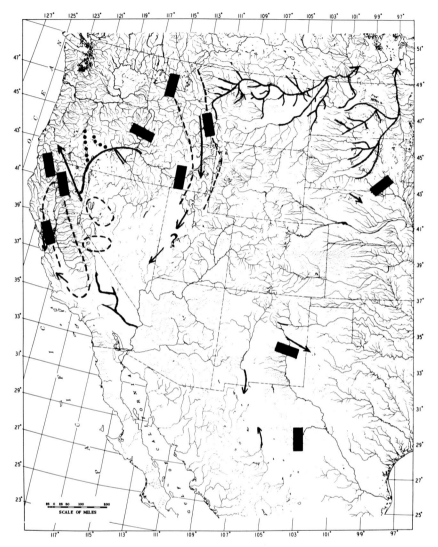

Figure 25. Pliocene drainages. Black bar: inferred divide crossing present drainage. Dotted lines: drainage connections younger than the Pliocene drainage (solid line). Drainage in northeastern part of map area from geological evidence in Lemke et al. (1965), Skinner and Hibbard (1972).

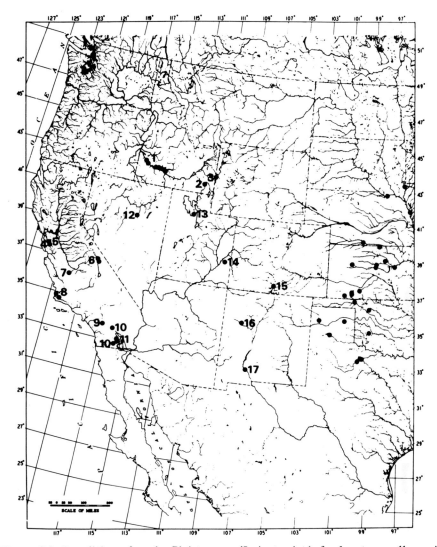

Figure 26. Localities of early Pleistocene (Irvingtonian) freshwater molluscs in western United States.

1, Bruneau formation, Ida. 2, Thatcher basin, Ida. 3, Star Valley, Wyo. 4, upper Santa Clara formation, Calif. 5, Irvington gravels, Calif. 6, Owens Valley, Calif. 7, upper Tulare formation, Calif. 8, Paso Robles formation and Careaga sandstone, Calif. 9, Bautista formation, Calif. 10, Palm Spring formation, Calif. 11, Brawley formation, Calif. 12, Hay Ranch formation, Nev. 13, Saltair core, Utah. 14, San Juan Co., Utah. 15, Alamosa formation, Colo. 16, Valencia Co., N. Mex. 17, Camp Rice formation, N. Mex.

Unnumbered dots to the east are localities known or inferred to be associated with the Pearlette "O" ash, the Cudahy fauna in a broad sense.

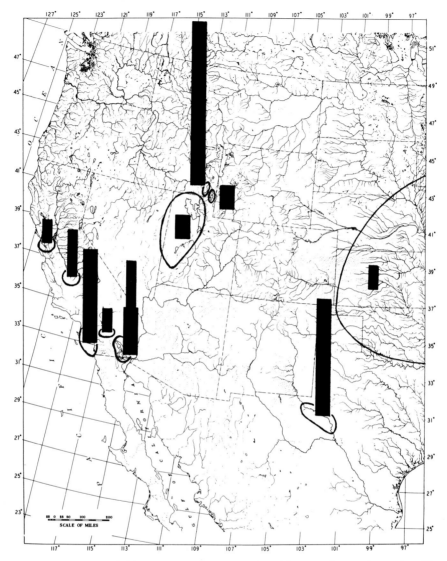

Figure 27. Numbers of Pleistocene extinct species. Shortest bar = 1 species. Narrow bar: early and middle Pleistocene. Wide bar: late Pleistocene-Holocene.

LOCAL SUMMARIES

The sequence of areas discussed is generally north-south and west-east.

Northern Continental Divide

Freshwater molluscs characteristic of larger perennial waters are practically all different species on either side of the continental divide in the northern United States. The few species found on both sides of the divide are in all cases far more widespread on one side or another, and are interpreted as indicating derivation through stream capture. A minimum of two drainage transfers is indicated, each one-way.

1. West to east. Two principally western species occur in the tributaries of the upper

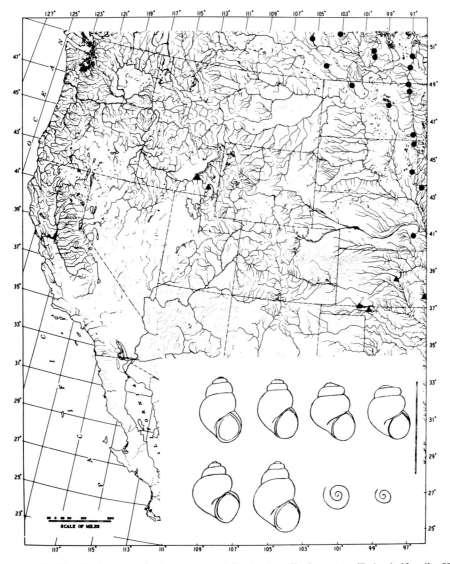

Figure 28. Distribution of freshwater snail *Probythinella lacustris* (Baker) (family Hydrobiidae). Solid dots, modern; triangles, Pleistocene. Insets from Baker (1928); scale line 1 mm.

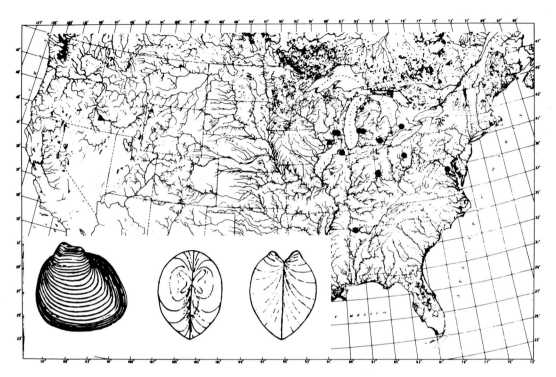

Figure 29. Distribution of freshwater clam *Pisidium cruciatum* Sterki (family Sphaeriidae). Solid dots, modern; triangles, fossil. Insets: original illustrations of the species, from Sterki (1895).

Missouri: the gastropod *Lymnaea hinkleyi*, and the pearl-mussel *Margaritifera falcata*. Both were probably in glacial ice-dammed waters of the Yellowstone Plateau that were diverted across the present continental divide during the last glaciation. To the north, in Canadian drainages tributary to Hudson Bay, Clarke (1973) identified the otherwise Pacific drainage species *Anodonta kennerlyi* (= *A. oregonensis*), *Lymnaea bulimoides*, and *Menetus cooperi* (= *M. callioglyptus*). All are misidentified, in my opinion.

 2. *East to west.* Far more species have most of their range east of the continental divide, with only a small part to the west. Some are known from only one or two western populations, whereas others are common in several states.

 The extreme of the disjunct western occurrences is shown by species such as *Probythinella lacustris* (Fig. 28) and *Pisidium cruciatum* (Fig. 29). These never became widespread in the west, and are known only as Pleistocene fossils remote from the living eastern range.

 The gap between eastern and western range is less in the case of *Amnicola limosa* (Fig. 37). It is widespread in eastern North America, but known living west of the continental divide only in Flathead Lake, Montana, and Utah Lake. Other species widespread in the east, but in Pacific drainage only in western Montana, are *Valvata tricarinata*, *Planorbella campanulata*, and *Pisidium adamsi*. All are known from only a few western localities, and lack a fossil record in the northwestern United States.

 Sphaerium simile has a broadly similar distribution. It is widespread east of the Rocky Mountains, but with a limited range in the west: western Montana and northeastern Washington, northwards into central and northern British Columbia. An isolated occurrence in Bear Lake valley, southeastern Idaho, is reminiscent of the nearby fossil occurrences of *Probythinella lacustris* and *Amnicola limosa*. No fossil occurrences of *Sphaerium simile* are known in the west.

The marsh and pond snail *Planorbula campestris* is found principally in the plains of southern Canada and the northernmost U.S., east of the Rocky Mountains. To the west it is known in the Flathead Lake area of western Montana, and Jackson Hole, Wyoming; like *Sphaerium simile*, it ranges northward into British Columbia. Late Miocene occurrences in the Teewinot formation of Grand Valley and Star Valley, Idaho-Wyoming, are the oldest fossil record of the species. East of the continental divide it is known first in early Pleistocene faunas: the Cudahy fauna of southwestern Kansas, and in the Alamosa Formation of the Rio Grande valley of Colorado.

The small clam *Pisidium punctatum* (Fig. 20) has an extensive east American distribution, but in the west is known living only in California, Idaho, and Oregon. It has an extensive Pliocene

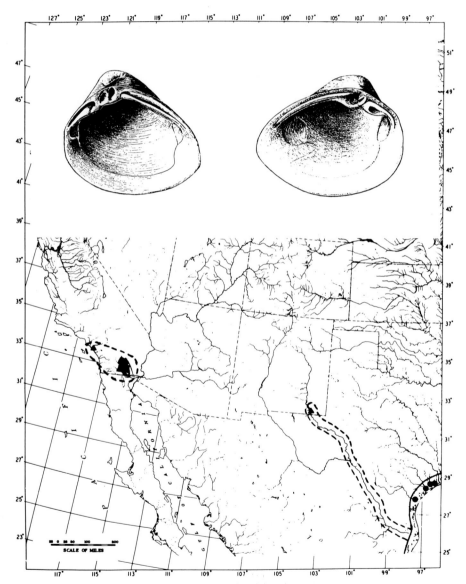

Figure 30. Distribution of brackish-water clams *Rangia cuneata* (Gray) and *R. lecontei* (Conrad) (family Mactridae). Solid dots, modern; triangles, Pleistocene. Insets from Dall (1894); left, *R. lecontei*; right, *R. cuneata*.

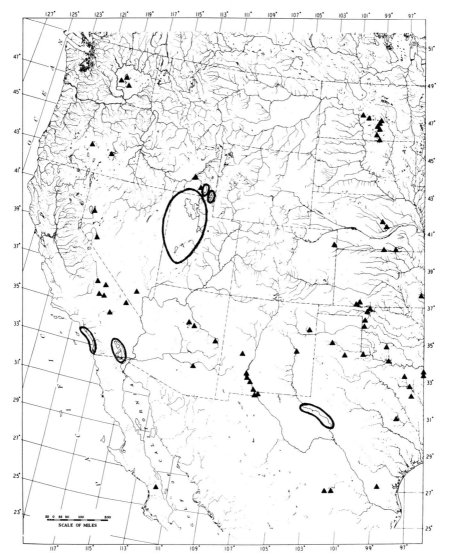

Figure 31. Latest Pleistocene-Holocene occurrences of locally extinct (triangles) and absolutely extinct (closed lines) species of freshwater molluscs.

record. Two species still more widespread in the west are *Helisoma anceps* and *Sphaerium striatinum*; they too are known from the Pliocene.

The geographic and stratigraphic ranges of all these predominantly eastern species can be accounted for most readily by a derivation from western Montana. That is shown diagrammatically as both Miocene (Fig. 15) and Pliocene (Fig. 25). Molluscan evidence is entirely inconsistent with the bizarre interpretation that the Bonneville basin was ever tributary to Hudson Bay (Smith 1981a).

Snake River, Idaho

That the Snake River formerly flowed westward from southern Idaho and was not part of the Columbia drainage is inferred from such evidence as distinct species of Miocene *Bellamya* in the region (Fig. 8), distribution of the Miocene-Pliocene mussel *Gonidea coalingensis* (Fig. 10), and of fishes (Smith 1975). Traces in the modern fauna remain in such snails as the large-river *Lymnaea emarginata* group. *L. apicina* is limited to the Columbia and its major tributary Salmon River. The related species *L. hinkleyi* of the Snake River and adjacent drainages on the east is known also from the Pliocene of eastern Oregon. Most of the evidence of the former course of the Snake River comes from Pliocene and Pleistocene fossils of such species as *Valvata utahensis* (Fig. 16), *Pyrgulopsis* (Fig. 17), *Lymnaea kingi* (Fig. 18), *Helisoma newberryi* (Fig. 19), *Pisidium punctatum* (Fig. 20), and *P. ultramontanum* (Fig. 21). Details of the western course remain uncertain, but both molluscs and fishes show evidence of the Snake River in northeastern California (Taylor and Smith 1981).

The present course of the Snake from southern Idaho is through the deep gorge of Hells

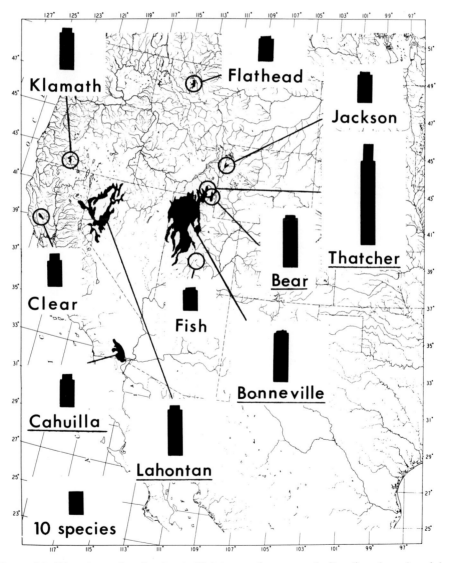

Figure 32. Diversity and endemism in Pleistocene (names underlined) and modern lakes. Wide bar, non-endemic; narrower bar, species endemic to lake. All lakes with localized species are shown.

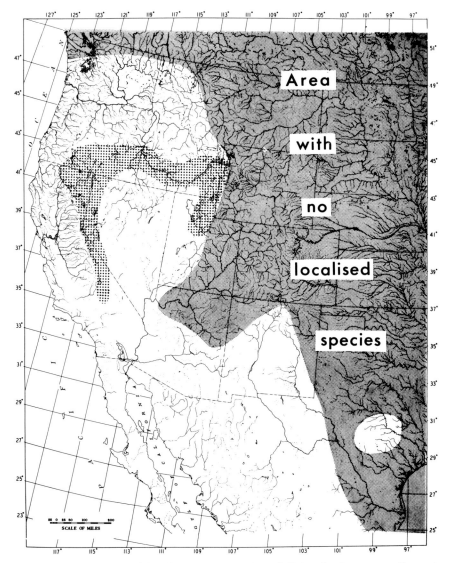

Figure 33. Distribution of local endemic species of living freshwater molluscs (unstippled area). Coarse stipple: area with local endemic species as well as numerous disjunct populations of others.

Canyon northward to the Columbia River in Washington. The canyon is thought to have been cut before deposition of the early Pleistocene Bruneau Formation from interpretation of the history of *Fisherola nuttalli* (Fig. 36).

The sister-group of *Fisherola* is *Lanx*, the only other genus of the Lancidae, a family restricted to the Pacific Northwest. *Lanx*, but not *Fisherola*, is found in the Pliocene Glenns Ferry formation of southwestern Idaho, and the two genera are thought to have been analogous in distribution to the two species of *Bellamya* (Fig. 8). Both are restricted to large streams and lakes and provide some of the firmest molluscan evidence for former drainage connections. *Fisherola* is first recorded as a fossil in the early Pleistocene Bruneau Formation. It is believed to have spread upstream through Hells Canyon (Fig. 36, large arrow) into southern Idaho. Its limpet-like shell is streamlined, and the snails can be found on rock in swift current where no other molluscs survive. Thus an up-current migration seems plausible even though no other molluscs are thought to have spread similarly.

The eastern headwaters of the Snake River in southeastern Idaho and western Wyoming seem also to be a late addition to the main river. Miocene faunas in southeastern Idaho are sharply different from those to the west, thus hinting at different drainages (Fig. 15). The strongest evidence comes from the Pliocene faunas of the Glenns Ferry and Cache Valley formations (Nos. 6-9, Fig. 5). Numerous related but distinct species in a number of genera indicate vicariant speciation, attributed to a drainage divide cutting across the present Snake River Plain. The northern locality of Cache Valley fauna (No. 7) is in Marsh Creek valley, now tributary to the Snake River;

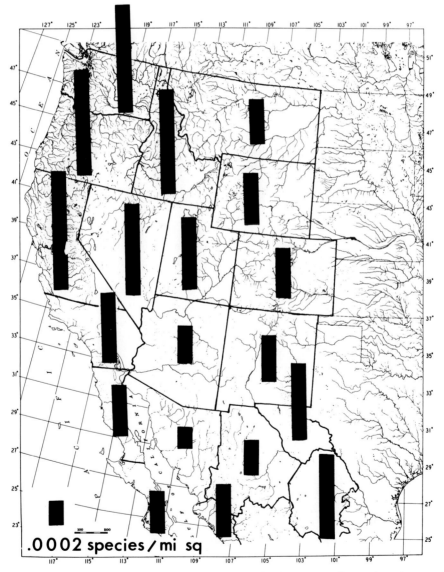

Figure 34. Normalized species diversity of living freshwater molluscs, based on preliminary taxonomic studies. 1 mi^2 = 2.59 km^2.

TABLE 4. INTRODUCED SPECIES

Introduced species	Source	Youngest fossil record of genus in western North America
Bellamya japonica	Japan	Miocene
Cipangopaludina chinensis malleata	Japan	None
Bithynia tentaculata	Europe	None
Batillaria attramentaria	E. Asia	Miocene
Thiara granifera mauiensis	Hawaii	Pliocene
T. tuberculata	E. Asia	Pliocene
Lymnaea columella	E. U. S.	Miocene
Radix auricularia	Europe	Miocene
Planorbella duryi	Florida	(Recent)
Anodonta corpulenta	E. U. S.	(Recent)
Corbicula fluminea	E. Asia	Miocene

nevertheless the fauna is like that of the Cache Valley lake in the Bonneville basin (No. 9; Taylor 1966). Fishes of the Cache Valley formation likewise differ from those of Lake Idaho and support inference of a drainage divide between the two areas (Smith 1981a).

Southeastern Idaho is also a region where several predominantly eastern American species were found, such as *Probythinella lacustris* (Fig. 28) and *Amnicola limosa* (Fig. 37). These have already been mentioned as evidence for spread across the continental divide (p. 302). Their occurrences in Idaho are Pleistocene, but they are attributed to a Tertiary drainage from what is now eastern Montana southward, before development of the Snake River Plain, into the Bonneville basin.

Eruption of vast lava sheets over the eastern Snake River Plain created a surface-water desert with no runoff from large areas. Presumably at one time all the streams draining southward from the Rocky Mountains of central Idaho flowed into the Snake. The names of some—Lost River, Little Lost River—indicate their present fate, to sink into the lava plain. Only in Birch Creek is there molluscan evidence of former connection with the Snake River. Here are a species of *Fontelicella*, of southern affinity; *Lymnaea hinkleyi*, elsewhere in Idaho only in the main-stem Snake River and in Henry's Fork; and the regional river and creek snail *Lithoglyphus hindsi*.

Green River, Wyoming

The Green River in western Wyoming, the northernmost part of the Colorado drainage, is adjacent to both the Snake River (of the Columbia system) and to the Bear River (of the Bonneville basin). The living mollusc fauna of the Green River shares a number of species with its neighbors, but has some anomalous differences. It lacks freshwater mussels and small Hydrobiidae, but nevertheless does have *Lithoglyphus hindsi* and *Lymnaea hinkleyi*. A former drainage connection was evidently selective in the transfer of species across the present drainage divide.

A plausible site for glacial diversion of waters from the Snake River drainage into the Green River was shown to me by J. D. Love in 1959. It is in sec. 36, T. 40 N., R. 111 W., and sec. 1, T. 39 N., R. 111 W., Uinta County, Wyoming. Here the divide between Kinky Creek, running into the Gros Ventre River of the Snake drainage, and Tepee Creek of the Green River drainage, is a morainal valley with scattered ponds. An ice-dammed lake in the Gros Ventre River valley could easily have been diverted into the Green River. Habitats of the species in the Jackson Hole region, Wyoming, are such as to make likely the selection of the *Lithoglyphus* and *Lymnaea*, but not *Margaritifera*, by such a mechanism.

The Green River, and even more so the Bear River, has an entirely western molluscan fauna, with no trace of any Mississippi Valley elements from the east. The drainage divide between Bear River and Green River, though locally low, has seemingly been stable since mid-Tertiary times. The only two species hinting at derivation from a neighboring drainage can be accounted for by the simple glacial diversion mentioned above. This is in stark contrast to the faunas of adjacent southeastern Idaho, with elements derived from disparate and remote sources.

Former Snake River, California-Oregon

The westward course of Snake River from southern Idaho through the Harney Lake Basin is believed to be indicated by fossils. A critical occurrence is Pliocene *Lymnaea hinkleyi* in what

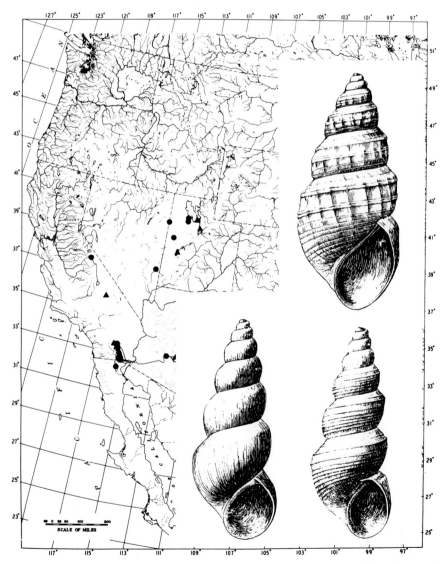

Figure 35. Distribution of freshwater snail *Tryonia protea* (Gould) (family Littoridinidae). Solid dots, modern; triangles, fossil. Insets from Stearns (1901).

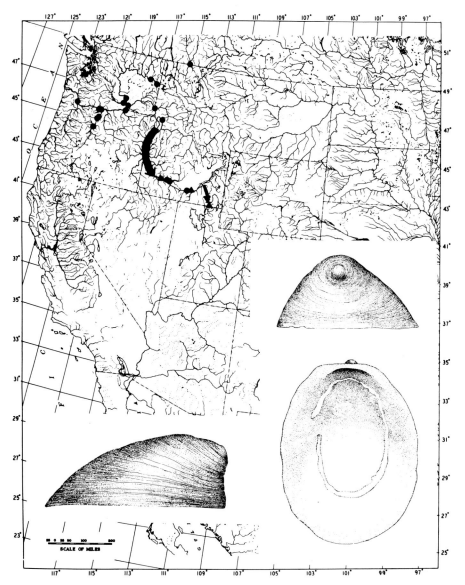

Figure 36. Distribution of freshwater snail *Fisherola nuttalli* (Haldeman) (family Lancidae). Solid, dots, modern; triangles, fossil. Arrows show inferred upstream spread after cutting of Hells Canyon (larger arrow) by early Pleistocene; and spillover of Lake Bonneville (smaller arrow) in latest Pleistocene. Insets drawn by A. D'Attilio.

was formerly Danforth Formation (5, Fig. 5). This is a large-stream snail, in southern Idaho found only in the Snake River and Henry's Fork, but not in any of their large tributaries.

The general westward and southwestward course of the Snake into the Pit River region of northeastern California is shown by the disjunct occurrences of such lake and stream species as *Helisoma newberryi* (Fig. 19) and *Pisidium ultramontanum* (Fig. 21). Other evidence from both molluscs and fishes has been summarized by Taylor and Smith (1981).

Only one living species is now restricted to both Snake River and Modoc Plateau drainages of Oregon and California. This is the spring and stream snail *Fontelicella intermedia* (Tryon), found in the Pit River below Fall River; in springs tributary to Lower Klamath Lake, California; in Crooked Creek, a tributary of the Owyhee River of the Snake drainage in southeastern Oregon; and in Barren Valley, a tiny basin of internal drainage southeast of the Harney basin, Oregon. The last two occurrences are referred to further in this section. *F. intermedia* is one of the species that gives a clue to the type of stream that may have flowed last between southeastern Oregon and northeastern California. It can live in springs or rivers, but is absent from the slow-moving upper Pit River and all of its tributaries. At some localities it lives with *Helisoma newberryi*. Although *Pisidium ultramontanum* is partly similar in distribution, this small clam lives in slower streams or in lakes and has not been found in association with *Fontelicella intermedia*.

With uplift of the Columbia Plateau, the westward flow of the Snake River was broken, and block faulting created the various independent Oregon lake basins. During disruption of through-flowing drainage, there seems to have been short-lived interconnection with the Deschutes River of the Columbia drainage. This interpretation is based principally on distribution of *Anodonta wahlamatensis* (Fig. 38), a mussel of lakes and larger rivers. Its range is strikingly disjunct but as yet without fossil records bridging the gap. The same interconnection is a likely explanation for the creek, river, and lake snail *Vorticifex effusus* (Fig. 22). Its patchy distribution in the Columbia River drainage would seem to indicate it did not spread downstream through the Snake River, that it spreads only slowly upstream, and is stopped by falls. *Vorticifex effusus*, but not *Anodonta wahlamatensis*, is known from the Fossil Lake basin of southeastern Oregon. Associated fossils include salmon, interpreted by Allison (1966) as indicating an overflow stage connected at one time with Pacific drainage.

An implication of this interpretation is that *Vorticifex effusus* is a relatively recent immigrant into the Columbia drainage. The only other living species of the genus, *V. neritoides* of the lower Columbia, would then form with *V. effusus* a pair similar in distribution to the Miocene *Bellamya* (Fig. 8).

Other areas in southeastern Oregon where now-internal basins had former external drainage include the Harney Lake basin, Catlow Valley (on the western side of the Steens Mountains), and Barren Valley (a tiny basin southeast of Harney Lake).

Harney Lake and its larger neighbor Malheur Lake were on the course of the former Snake River as it flowed westward from Lake Idaho. Presumably, it was in latest Pliocene times (before diversion of the Snake River through Hells Canyon to the Columbia) that such large-stream and lake species as *Helisoma newberryi* (Fig. 19) and *Pisidum ultramontanum* (Fig. 21) became distributed through the various now-internal drainages of southeastern Oregon. With regional uplift Harney and Malheur Lakes drained via Malheur River into the Snake, the former westward drainage being reversed. Their outlet was later blocked by a Pleistocene flow, the Voltage lava flow (Gehr 1980, Gehr and Newman 1978) over which the lake spilled at times in the late Pleistocene. Holocene lowering of the lake has eliminated the lake and large-stream elements of the fauna.

Catlow Valley is a small internal basin on the west side of the Steens Mountains, south of the Harney Lake basin. The only permanent-water mollusc known here is *Lithoglyphus turbiniformis*, indeed this is its type locality. This small snail is characteristic of springs and spring-brooks, and entirely unknown from lakes or slow streams. The nearest localities to Catlow Valley are to the southwest, in the Warner Lakes basin of south-central Oregon; immediately north in the headwaters of the Donner und Blitzen River (tributary to Malheur Lake); and far to the northwest on the eastern slope of the Cascade Range in the Deschutes River drainage (of the Columbia system). Here again there are indications of former stream connections across the desert of south-central Oregon, but with little indication of precise routes or timing.

The high desert of southeastern Oregon is an arid region of sagebrush and basalt, with few habitats for aquatic molluscs and only rare localized endemic species. Even so there are hints of former different drainage. *Fontelicella intermedia* (Tryon) is found in Crooked Creek (a tributary

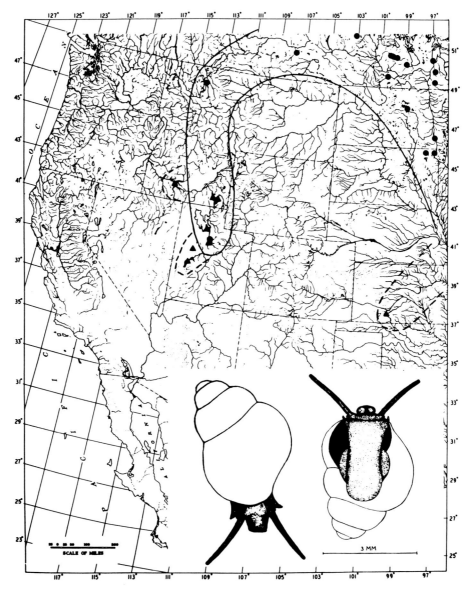

Figure 37. Distribution of freshwater snail *Amnicola limosa* (Say) (family Hydrobiidae). Solid dots, modern; triangles, Pleistocene. Insets from Baker (1928).

of the Owyhee River) and also in the small internal basin of Barren Valley, immediately northwest. This occurrence is accounted for as follows: Regional uplift, Snake River diversion, and cutting of Hells Canyon controlled the formation of the lower Owyhee River canyon. When the Pliocene Snake River flowed westward from Lake Idaho, the Owyhee River continued northwestward on the trend of its upper course, to the vicinity of the Malheur basin (Fig. 25). Crooked Creek likewise flowed northwestward through the present separate basin of Barren Valley.

Building of the High Cascades has obliterated fossil and living molluscan evidence of the ancient course of the Snake River to the sea. The course shown diagrammatically (Fig. 25) is inferred from sketchy fossil evidence and from distribution of a few modern species. It is also worth noting that the Rogue and Umpqua Rivers of Oregon have sparse faunas considering their size, and scarcely any endemic forms. Possibly this reflects widespread molluscan extinction due to volcanic activity.

A close similarity of the upper Klamath and upper Pit River faunas is evident from species in common. The richer Klamath fauna has a number of lacustrine species not in the Pit; virtually all of the faunal differences can be accounted for on ecological grounds. There are no species pairs suggesting a common ancestor gave rise to representative forms in the two drainages. Stream connections of the now-independent basins were therefore relatively recent as well as large. Pliocene faunas of the Glenns Ferry formation (6, Fig. 5) and Cache Valley formation (7-9, Fig. 5) by contrast have species-pairs in various families.

External relations of the lacustrine species in the Klamath fauna that do not occur also in the Pit River are to both east and west. *Pyrgulopsis archimedis* in Klamath Lake is most like *P. nevadensis* of Pyramid Lake-Walker Lake, less so to a Pliocene Lake Idaho species (Fig. 17). Another living species, *Amnicola*, is close to a Pliocene form from Butte Valley (2, Fig. 5), *A. margaretana*. *Lanx klamathensis* of the Klamath drainage is most like a Pliocene form of the Glenns Ferry formation, Idaho (6, Fig. 5).

One species found in lakes and slow rivers that might be expected in Klamath Lake is the freshwater mussel *Anodonta wahlamatensis* (Fig. 38). It is found in the Sacramento drainage upper Pit, and lower Columbia Rivers. Perhaps the critical factor is distribution of fish hosts, not modern habitat.

Faunal history in the Klamath region can be improved with further investigation of fossils in the area. At present the fauna seems most likely a composite of older lacustrine and large-stream endemic elements of Miocene age, and younger stream and spring elements shared with the Pit River in late Pliocene time.

The southernmost trace of the ancient Snake River in California is the Pliocene fauna of Honey Lake (13, Fig. 5; Taylor and Smith 1981). Modern remnants on the Modoc Plateau of northeastern California include the snail *Juga acutifilosa* and the small clam *Pisidum ultramontanum* (Fig. 21). Still further south there is one hint of possible drainage connection with the Honey Lake basin. The freshwater mussel *Anodonta oregonensis* is found in Sierra Valley, now drained by the Feather River (of the Sacramento system). This is its southernmost occurrence anywhere, and the only one south of the Upper Pit and upper Klamath drainages. If this one occurrence is to be accounted for by stream connections, the implication is that Sierra Valley was once tributary to an area with Modoc Plateau fauna—and Honey Lake is the next basin to the north. Ecological factors are an adequate explanation for the gap in range between the Pit River and Sierra Valley.

Lahontan Basin

This major internal basin of western Nevada (with small areas in adjacent California and Oregon) has several similarities to the Bonneville basin. Both are thought to be post-Miocene in age, and to be composites of several formerly separate drainages; both had their latest external drainage to the south; both had a Tertiary lake with a distinctive molluscan fauna; and both had late Pleistocene lakes with a non-distinctive molluscan fauna. In a geological sense both basins can be considered as still widening and deepening, acquiring larger drainage basins and more diverse fauna.

The distribution of the viviparid snail *Bellamya turneri* (Fig. 8) indicates a Miocene (Barstovian-Clarendonian) interval of integrated external drainage over a large area including the present Lahontan basin. Pliocene faunas of the basin include several discrete assemblages. The fauna of the Honey Lake basin, northeastern California (13, Fig. 5) shows marked influence of the former Snake River. Although now in the Lahontan basin, it was evidently part of the former Snake drainage (Taylor and Smith 1981). The Pliocene fauna of the Mopung Hills, Nevada (14,

312 LATE CENOZOIC HISTORY

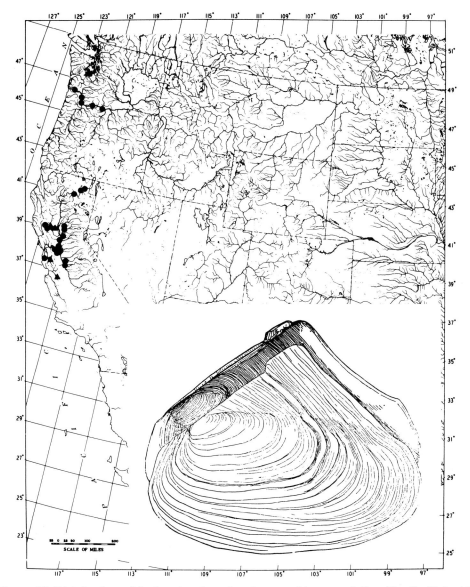

Figure 38. Distribution of freshwater clam *Anodonta wahlamatensis* Lea (family Unionidae). Solid dots, modern; triangles, Pliocene. Inset from Keep and Baily (1935), approximately natural size, by permission of Stanford University Press.

Fig. 5) records a lake with a substantial independent history; the fauna is not large but distinctive, with most species endemic (Taylor and Smith 1981). The one Pliocene species known from farther south in the Lahontan basin (16, Fig. 5) is otherwise found not at more northern localities, but outside the modern Lahontan basin in the Mono Lake basin (15, Fig. 5).

Bonneville Basin

This largest interior basin of the west is thought to be post-Miocene, and a combination of originally separate drainages. The northwestern part of the basin includes localities of *Bellamya turneri* (Fig. 8), suggesting westward drainage. This large-shelled species has never been found in the Salt Lake group of southeastern Idaho, where the faunas are sharply distinct from those of Nevada and Oregon. *Valvata idahoensis* of the Salt Lake Group, or a closely similar species, is found also in the Muddy Creek Formation of the lower Colorado River area (Fig. 12), hinting at southward drainage in the late Miocene. The distinctive Pliocene fauna of the Cache Valley Formation (9, Fig. 5) implies a long-enduring freshwater lake habitat, hence one with exterior flow. Distribution of the living *Tryonia protea* (Fig. 35), a snail tolerant of somewhat saline springs, provides a further hint of exterior drainage. The species occurs south of the Bonneville basin in the lower Colorado River drainage, and in Meadow Valley Wash, Nevada, perhaps connected with a late southern outflow, or at least a stream interconnection. Pleistocene Lake Bonneville was by far the largest lake in the west (Fig. 32), but had no such impressive fauna. Only one species is restricted to that lake, the extinct *Lymnaea bonnevillensis*.

Lymnaea bonnevillensis is of both biogeographic and ecological interest. It belongs to the *Lymnaea emarginata* group, a subgroup of *Lymnaea* (subgenus *Stagnicola*) restricted to larger streams and lakes. Its relatives are all more northern and eastern. *L. bonnevillensis* is known only from late Pleistocene Lake Bonneville, but its close relative *L. hinkleyi* is known from Pliocene rocks. An age for both of at least Pliocene, probably Miocene, is thus plausible. This is the upper limit for derivation of the ancestor of *L. bonnevillensis* from some northern source. Throughout the latest Tertiary and nearly all of the Pleistocene, there was enough water flowing into the Bonneville basin to provide a suitable habitat. The species died with Lake Bonneville: the last episode of aridity was evidently more severe than any previous.

Fossil *Fisherola nuttalli* in Cache Valley are interpreted as indicating that the snails spread upstream in the Lake Bonneville overflow (Fig. 36, small arrow), just as they had spread earlier up Hells Canyon (Fig. 36, large arrow). This is the only addition to the Bonneville fauna that can be attributed to the spillover of the lake; there are no plausible additions to the Snake River fauna due to that spillover.

The close faunal similarities between the upper Snake River and the Bonneville basin are due to repeated drainage changes prior to the spillover of Lake Bonneville. In Pliocene time there was a drainage divide on what is now the Snake River Plain, with drainage southward into the Bonneville basin from the upper Snake River. In early Pleistocene times the upper Snake River had been added to the lower Snake, and the Bear River of the present Bonneville basin was one of the main tributaries of that river. The rich fauna of Pleistocene Lake Thatcher in southeastern Idaho was part of this Bear River drainage.

The marked diversity of the Lake Thatcher fauna (Fig. 32) is due to several factors. One was a local diversity of habitats that persists to some extent today; despite massive extinctions the region is one of the richer areas in the western United States. Second was the persistence of lake and larger stream habitats continuously since the Tertiary. Finally, the complex drainage history of the area brought the evolution of local species as well as the local persistence of forms from diverse drainages, and the immigration of species from the Snake River. Thus the fauna included such Great Basin genera as *Lutrilimnea* (Fig. 13) and *Tryonia*, with endemic species of both. The almost exclusively east American snails *Probythinella lacustris* (Fig. 28) and *Amnicola limosa* (Fig. 37) are thought to have been inherited from former drainage from what is now western Montana. Species derived from the Snake River included the small clam *Pisidium ultramontanum* (Fig. 21), the mussel *Gonidea angulata*, and most remarkably a pleurocerid snail *Juga*, now regionally extinct. Other species of the Lake Thatcher fauna included a series of forms distributed in the Snake River and around the margin of the Great Basin, such as *Valvata utahensis* (Fig. 16), *Lymnaea kingi* (Fig. 18), and *Helisoma newberryi* (Fig. 19). Most species now living in the area lived also in Lake Thatcher.

Volcanism built lava dams that finally raised the level of Lake Thatcher so high that it began to spill into the Bonneville basin, thereby draining the lake and diverting the Bear River from

Snake River drainage to its present-day course into the Bonneville basin (Bright 1967). Yet there is no firm evidence that any species of Lake Thatcher survived for a time in Lake Bonneville. After Lake Bonneville rose to its highest level, with an arm filling part of the Thatcher basin, the local fauna was entirely of the relatively poor Bonneville type.

Bear Lake, in the Bear River drainage and in the valley next east from the Thatcher basin (Fig. 32), is unique among the larger freshwater lakes of the western United States in that it has no living molluscs whatever. Deficiency in some elements, rarity of aquatic vegetation, and perhaps food deficiency are some of the probable reasons. In latest Pleistocene and early Holocene times the lake basin supported a rich fauna, though not so diverse as that of Lake Thatcher. Two species were endemic: *Lutrilimnea ursina* (Fig. 13) and *Fontelicella pilsbryana*. Earlier, before the draining of Lake Thatcher had extinguished its fauna, *Lutrilimnea ursina* had lived also in Lake Thatcher as well as Bear Lake. No species are restricted both to the Bonneville basin and the two Pleistocene lakes of the Bear River drainage. This observation emphasizes that stream connections do not necessarily lead to spread of the aquatic fauna: continuity of habitat is also necessary. Diversion of Bear River from the Snake River into the Bonneville basin brought only one faunal change: extinction of Lake Thatcher and its fauna.

Central California

Miocene drainage connections between the Snake River of southwestern Idaho and the Sacramento-San Joaquin River of California are indicated by such species as the freshwater mussel *Gonidea coalingensis* (Fig. 10), related species of *Juga* (subgenus *Calibasis*) (Fig. 7) and freshwater fishes (Smith 1975). Already by Pliocene times the species of gastropods of the two regions were distinct, those in common being bivalves (Fig. 5). At present the fauna of the northeastern part of the Sacramento drainage (the Pit River above its falls) is distinct from that below. It is to this Pit River fauna that the modern and fossil faunas of the Snake River and Great Basin show special relation (Fig. 33, coarse stipple). Thus it is probable that the Pit River has been added to the Sacramento drainage in post-Pliocene times, from the former Snake River (Fig. 25).

Earlier Pliocene faunas in central California (those of the upper San Joaquin Formation in the San Joaquin Valley, and of the lower Santa Clara Formation in the San Francisco Bay trough) are notably distinct from one another. Yet in later Pliocene times the distinctions were much reduced. A sparse and nearly uniform fauna characterized by the still-living *Savaginius yatesianus* was found from the Sacramento Valley (Tehama Formation; 18, Fig. 5) through the San Francisco Bay trough (Santa Clara Formation; 20, Fig. 5) into the drainage of Monterey Bay (San Benito Gravels; 23, Fig. 5). Remarkably, the same *Savaginius* is recognized also in the southern San Joaquin Valley (lower, but not basal, Tulare Formation; 25, Fig. 5).

The early Pleistocene upper Tulare Formation of the San Joaquin Valley has a still different fauna, most notably including *Valvata utahensis* (Fig. 16) and *Lithoglyphus seminalis*. These are thought to have been species of the Pit River on the Modoc Plateau, indicating that the Pit River had been added to the Sacramento drainage, and that the Sacramento-San Joaquin system had been unified, since Pliocene times.

The Pliocene-Pleistocene faunal changes are combined with radiometric dates and geological interpretation (Sarna-Wojcicki 1976, and references therein; Taylor 1966) in the following scenario:

1. In early Pliocene times (about 3.5-4 million years [my]) was the maximum number of separate drainages. The Pit River was tributary to a Snake River that flowed northwestward to the Pacific. The Sacramento River (without the Pit) flowed south to a marine to freshwater embayment in the southern San Joaquin Valley that connected with the sea by a strait in the southern Coast Ranges. The San Francisco Bay trough drained southward to Monterey Bay; the Golden Gate and Carquinez straits were not yet in existence.

While the southern San Joaquin Valley had been filled by a marine embayment during the

Miocene, separate freshwater tributaries developed faunas with distinctive species in many cases. As the embayment filled and finally became fresh, the barriers between separate streams disappeared and the result was a locally rich lacustrine fauna, especially in Hydrobiidae. This fauna became extinct by early Pleistocene time with few exceptions, as changes in rainfall and the rise of the Coast Ranges eliminated the stable lacustrine and river habitats.

2. In approximately mid-Pliocene times the distinctive fauna of the basal Tulare Formation had become extinct. The strait through the southern Coast Ranges permitted a last brief incursion of seawater, with a marginal marine and brackish-water fauna preserved in the lower but not basal Tulare Formation.

3. In later Pliocene time (prior to about 1.5 my), in the San Francisco Bay trough the localized fauna of the Santa Clara Formation was extinguished by continued tectonic changes and by the cutting of the Carquinez Straits. The Sacramento River now flowed through the San Francisco Bay trough into Monterey Bay, with a sparse but widespread fauna. The common species *Savaginius yatesianus* reached the mouth of the river, spread into the lowland confluence of other streams including the ancestral Salinas River, and ultimately spread upstream in that river to what is now part of the southwestern San Joaquin Valley. The southern San Joaquin Valley drained to the Pacific Ocean through the remains of the former marine strait in the Coast Ranges.

4. In early Pleistocene times there is evidence of ponding of waters in the San Joaquin Valley. An extensive lake is indicated by the Corcoran Clay, about 0.7-0.75 my old (Davis et al. 1977) with the implication that westward drainage through the southern Coast Ranges had been sealed off. The San Joaquin River now joined the Sacramento River in flowing through Carquinez Straits into San Francisco Bay. At or before this time the bay had come into existence as an estuary open to the sea. The trough no longer drained south, but received drainage of the San Benito River from the south.

With the unification of the San Joaquin and Sacramento rivers, northern species reached the San Joaquin Valley for the first time. *Valvata utahensis* and *Lithoglyphus seminalis* in the Tulare Formation above the Corcoran Clay are interpreted as indicating the Pit River had been added to the Sacramento drainage by this time.

A sparse fauna in the Santa Clara Formation of the San Andreas Rift (19, Fig. 5) is entirely distinct from all other assemblages. Apparently, it represents a drainage that was distinct from that of the San Francisco Bay trough throughout much or most of Pliocene times.

Southern Continental Divide

Evidence of drainage connections across the present continental divide in the southwestern United States and northwestern Mexico comes from two areas: the southeastern Colorado River drainage, and drainage of the Rio Conchos.

Two species of bivalves in the Gila River drainage of Arizona represent northwestern occurrences. *Sphaerium triangulare* (Fig. 23) is otherwise widespread on the Plateau of Mexico, *Musculium transversum* (Fig. 24) not only on that plateau but in eastern North America. The mussel *Anodonta californiensis* is widespread in the west, with a few occurrences known in interior basins of northwestern Chihuahua. These data support those of fish distribution (Miller 1981) implying faunal transfer between the upper Gila River drainage and interior drainage to the southeast.

A puzzling occurrence is that of *Valvata tricarinata* in Pleistocene lacustrine deposits near Concho, Arizona, in the upper Little Colorado River drainage. The nearest localities are in eastern New Mexico (also Pleistocene), and the nearest modern occurrences are much further east. Was the species once widespread in the Colorado River drainage?

The Rio Conchos of northern Mexico is now tributary to the Rio Grande, and has a fauna much like that of the lower Rio Grande save for being sparser. In particular the most significant Conchos mussel is *Anodonta imbecillis*, widespread in the eastern United States, but unknown in

Mexico outside of the Rio Grande drainage. South of the Rio Conchos on the Mexican Plateau, the only freshwater mussel is *Anodonta impura*, most closely related to *A. californiensis* of the western United States. In between these two species the east American fauna of the Conchos is interposed. The implication is that the present course of the Rio Conchos, and at least some of its fauna, is relatively young. The Rio Grande elements of its fauna are thus thought to have been added to the river after it was diverted from interior drainage (Fig. 25) to that of the Gulf of Mexico.

Former Gulf of California

In Miocene times the present Isthmus of Panama did not exist, and a wonderfully rich marine molluscan fauna was found through the former seaway (Woodring 1965). Rise of the land divided the fauna, eliminating the former community of reproduction. Descendants of the Miocene fauna have diverged to varying degrees on either side of the isthmus, commonly forming pairs of sister-species. Extinction of lineages has predominated in the Gulf of Mexico–Caribbean Sea region, so that the modern Panamic-Pacific tropical fauna is richer than the Panamic-Caribbean. Lineages surviving in the Caribbean region but extinct on the Pacific side have been termed Caribphiles, those of the opposite survivorship, Paciphiles (Woodring 1966).

Two species of brackish-water molluscs found as fossils at the head of the former Gulf of California in California and Arizona represent extinct sister-species of living Caribphiles. These are the Pleistocene *Rangia lecontei* (Fig. 30), most closely related to the living *Rangia cuneata* of the Gulf of Mexico; and Miocene *Batillaria californica* most like the living *B. minima* of the Caribbean Sea region (Taylor 1983). Just as the fossil *Rangia* and *Batillaria* have never been found together, so their modern close relatives do not occur together.

Rangia cuneata is found in brackish estuaries along the northern Gulf of Mexico, from northeastern Mexico to Florida, and on the Atlantic Coast of the southeastern United States. It may be enormously abundant, making up the dominant species in its special salinity belt. Occurrence of the extinct *R. lecontei* in the Colorado Desert is similar. It may be the only fossil present locally in the Borrego Formation, suggesting the presence of a brackish embayment in the desert. Peculiarly, the only other occurrence of *Rangia lecontei* is in late Pleistocene deposits near the former mouth of the Santa Ana River in southern California (Fig. 30). Could there have been a transient brackish-water connection between the Colorado Desert and the Los Angeles basin?

Batillaria minima is found widely in the Caribbean and western Atlantic, from Brazil to Bermuda and southern Florida, westward through the Antilles and as far north as southern Mexico. In comparison to *Rangia*, it is thus more tropical, and occurs widely on islands as well as on the mainland, with tolerance for higher salinities. The fossil *Batillaria* in the lower Colorado River valley has been found only in the Miocene Bouse Formation, and in association with a depauperate marine fauna. Again, close similarity of the habitat with that of the modern relative is implied. *Batillaria* in the Colorado River valley is part of the evidence for the estuary that extended well north of the present head of the Gulf of California.

Molluscan evidence is opposed to the occurrence of a brackish water estuary as far north as Needles. Fossils from probable Bouse Formation in the Lake Havasu area are freshwater species (Taylor 1983). Those from still farther north, in the calcareous facies of the Muddy Creek Formation, are also strictly freshwater, but have been misinterpreted (Blair and Armstrong 1979). The most abundant gastropod is *Fluminicola avernalis*, now living in mildly thermal springs at the head of Muddy River, Clark County, Nevada.

In the modern Mojave Desert fauna, only one genus is probably derived from the fauna of the Miocene Bouse estuary. Inland occurrences of the predominantly seashore snail *Assiminea* in the Death Valley area are thought to indicate the evolution of species living in saline or freshwater springs from a former ancestor that lived along the Miocene embayment (Taylor, in prep.).

Lower Colorado River

Tantalizing hints in the fossil record raise the possibility of drainage in later Miocene times from southeastern Idaho across what is now the Bonneville basin into the lower Colorado River. *Lutrilimnea dineana* in the late Miocene Bidahochi formation, Arizona (4, Fig. 13) is closely similar to *L. gentilis* of southeastern Idaho (2, Fig. 13). Specimens of *Valvata* in the Muddy Creek and Bouse formations of the Colorado River valley are much like *Valvata idahoensis* of southeastern Idaho (Fig. 12), and perhaps conspecific (Taylor and Smith 1981, Taylor 1983). Duration and course of such drainage are uncertain; the diagrammatic arrow (Fig. 15) indicates southward drainage of the Bonneville basin. In the modern fauna the only species providing possible evidence for such stream connections is *Tryonia protea* (Fig. 35).

While the Miocene Bouse embayment (the former head of the Gulf of California) extended up the lower Colorado River valley and its major tributaries, the strictly freshwater faunas of the tributaries were isolated from one another. The present-day faunas of various streams reflect this ancient separation: molluscs of the Gila River system in Arizona are distinct from those to the west; molluscs of the White River and Amargosa drainages of southern Nevada are related but distinct from those both east and west; and those of the Owens River drainage, southeastern California, are further distinct. The sharp distinction between the modern faunas east and west of Death Valley is more impressive than that of those east and west of the Colorado (Taylor, in prep.); could an arm of the Bouse embayment have extended so far? A long late Cenozoic history of arid climate and saline waters with rare interconnection of freshwater streams could account for the marked endemism and localized occurrence of species as one or two populations. Yet why are the sparse freshwater species so distinct on either side of Death Valley? *Tryonia* to the west is *T. protea* (Fig. 35); to the east *T. clathrata* and relatives. *Fontelicella* in Death Valley and westward is of the *F. stearnsiana* series; in the upper Amargosa drainage it is *Fontelicella* (*Microamnicola*) *micrococcus*, a local monotypic subgenus. All of the small "*Fluminicola*" species related to *F. avernalis* are only east of Death Valley. Field and laboratory studies in progress are providing more details of distribution and more taxonomic refinements that will help resolve some of the questions in historical zoogeography of the area.

Throughout the entire Mojave Desert, only one living species is restricted to both the Death Valley area and to the Colorado River drainage. This is an inland species of the usually seashore genus *Assiminea*, that lives in the humid belt next to the outflow of springs in the region. This provides the only living parallel to the pupfishes, *Cyprinodon*, that show similar distribution. This *Assiminea*, distinct from the local endemic *A. infima* of Badwater, Death Valley, is found in Death Valley and up the Amargosa drainage to Ash Meadows, Nevada, as well as along the shores of Lake Mead near the mouth of the Virgin River. Latest Pleistocene/Holocene fossils from an archaeological site in northern Panamint Valley, where it is associated with *Tryonia protea* (Fig. 35), are the northwesternmost occurrence of the species. Evolution of this inland species is likely related to the Miocene Bouse embayment.

After formation of a freshwater drainage of the lower Colorado valley in later Pliocene times, exchange of freshwater fauna between the Gila River, Arizona, and streams in southeastern California was possible. So far as present data reveal, expansion of range was accomplished by two species from the Gila that spread westward, none in the reverse direction. The sphaeriid clams *Sphaerium triangulare* (Fig. 23) and *Musculium transversum* (Fig. 24) are both known from late Pleistocene lake deposits well northwest of their present living occurrences. *Tryonia protea* in the Owens River drainage, California (Fig. 35) presumably spread into that area from the lower Colorado River valley during this interval also.

The sphaeriid clams are species of strictly perennial fresh waters, and there is no reasonable doubt as to the former stream connections between the Owens drainage and lower Colorado.

Only euryhaline fishes indicate such relationships (Smith 1978:38), hence Pleistocene or Holocene episodes of aridity may have eliminated both the molluscs and lower Colorado River primary freshwater fishes.

Like the fishes, the molluscs of the Owens River drainage include species with relationships to both the lower Colorado River drainage, on the southeast, and to the Lahontan basin on the north. Species of presumed northern derivation are *Valvata utahensis* (Fig. 16), *Lymnaea kingi* (Fig. 18), *Helisoma newberryi* (Fig. 19), *Pisidium ultramontanum* (Fig. 21), and *Vorticifex effusus* (Fig. 22). These are thought to indicate late Tertiary drainage southward from what is now the Lahontan basin; no northward spread of species from the Owens River valley has been recognized. One species, an undescribed form of *Fontelicella*, is living in both Owens Valley and the East Fork of Walker River, and marks a likely site of former faunal spread. In between the Mono Basin is barren of perennial-water molluscs. Destruction during collapse of the volcano-tectonic depression of Mono Basin during the last few million years (Bateman and Wahrhaftig 1966) seems plausible.

Most of these species of northern derivation have left no direct evidence that they ever spread as far southeast as the Colorado River valley. *Valvata utahensis* (Fig. 16) is the only one known from Pleistocene deposits in the Colorado Desert, indicating it once occurred in much of the lower Colorado River.

Pleistocene deposits of Lake Manix, in the Mojave River drainage, have yielded three species thought to have spread southward from the Owens River: *Valvata utahensis* (Fig. 16), *Helisoma newberryi* (Fig. 19), and *Vorticifex effusus* (Fig. 22). They presumably spread upstream through a former course of the Mojave River, as indicated by Fig. 25.

A former Mojave River with connections to the Colorado River and Owens Valley could explain the occurrence of the mussel *Anodonta californiensis* in coastal southern California. The species does not occur in Pacific drainage to the south, and occurs only well to the north, but is living in Mojave River, the lower Colorado drainage, and formerly Owens Valley. Uplift of the San Bernardino Mountains in the Pleistocene, and drainage transfer of its broad valleys from Mojave to coastal streams, could account for this one species. The other mussel in southern California is *Gonidea angulata*, whose occurrence remains puzzling. It is found nowhere to the south, nowhere near the Colorado River drainage, and its nearest occurrences are in the Central Valley of California.

With the breakup of external drainage in the Mojave Desert, various basins became isolated to varying degrees, with Death Valley now receiving much of the stream systems. According to present interpretations, various species lasted for various lengths of time in separate basins such as the modern Owens River, Pleistocene Lake Searles, Lake Panamint, and Lake Manix. Spillover from Lake Panamint into Death Valley, and into former Lake Manly in Death Valley, had no recognizable effect on mollusc distribution.

Rio Grande

The Pleistocene occurrence of brackish-water or marine molluscs along the Pecos River in New Mexico and Texas is one of the surprises of recent years. One species is the brackish-water clam *Rangia cuneata*, living along the Gulf Coast (Fig. 30). Transport by waterfowl could explain this single form, as suggested by Metcalf (1980). A parallel case is that of the marine clams *Cardium* and *Scrobicularia*, evidently brought by waterfowl across more than 100 miles of desert into the lake Birket Quarûn of the Egyptian Faiyûm (Arkell 1956:594). The four species of gastropods (Andrews 1977), however, are unknown in the Gulf of Mexico. Their occurrence implies a history of evolution and isolation in an inland saline environment.

The original colonizing molluscs in the Rio Grande Valley are assumed to have been derived from marine species of the Gulf of Mexico that adapted to saline flowing waters. Their existence requires that the Rio Grande was a salty river like parts of the middle Pecos, not the modern freshwater Rio Grande. Previous studies have indicated that until mid-Pleistocene times the upper Rio Grande was tributary to interior basins in northern Chihuahua and westernmost Texas (Hawley and Kottlowski 1969 and references therein). It is hypothesized here that in late Tertiary times the Rio Conchos was also a stream of interior drainage (Fig. 25), excluding much of the present freshwater sources of the Rio Grande. Geological studies have further indicated that the upper Pecos was, until mid-Pleistocene time, part of an eastward-flowing drainage (Jelinek 1967 and references therein). Thus there is independent evidence that much of the present fresh water flowing to the lower Rio Grande valley was formerly unavailable. The source of saline waters is assumed to be Permian salt, formerly as now rising under artesian pressure in southern New Mexico, but then contributing a far more significant part to the lower Rio Grande. The gastropods of marine affinity found along the lower Pecos are minute and may have lived interstitially. Thus it is possible that they may have survived in saline waters among stream gravel beneath a seasonally or temporarily freshwater river. In any case, the inland occurrences are assumed to have been isolated after the Rio Conchos and upper Rio Grande were integrated into the lower Rio Grande, converting the latter into a freshwater river.

Inland occurrences of the generally seashore snail *Assiminea* in the Rio Grande drainage as far upstream as New Mexico (Taylor, in prep.) are a modern parallel to the fossils. The species are thought to have adapted to inland saline springs from an ancestral habitat along a brackish lower Rio Grande. The only other inland occurrences of *Assiminea* in North America are in the lower Colorado River valley and Death Valley. A similar origin is postulated for the species there, where a former marine to brackish estuary has been documented by fossils. Implications of this hypothesis are that traces of marine fauna are to be expected far more widely along the Pecos and lower Rio Grande. Further, the subsurface solution of salts in southeastern New Mexico has continued during the later Tertiary as well as the Pleistocene.

ACKNOWLEDGMENTS

My interest in unraveling the combined histories of molluscan faunas and past drainages was stimulated by the interpretations of fish distribution by C. L. Hubbs and R. R. Miller (1948), who later shared their time and knowledge. The critical emphasis on detailed data of distribution, as opposed to theoretical preconceptions, is derived largely from works by L. Croizat (1958, 1976, 1982), who has reiterated that "earth and life evolve together." Importance of the joint study of modern and fossil faunas was taught by C. W. Hibbard, with whom much of my paleoecological thinking developed. Over a number of years I have benefited from the biological and geological knowledge of R. C. Bright and G. R. Smith. Parts of the present paper were prepared with stimulus by E. B. Leopold, who invited a lecture on "The Snake and The Snail" at the Quaternary Research Center, University of Washington.

Research has been supported (in part) by National Science Foundation grant DEB-7822584; and the New Mexico Bureau of Mines and Mineral Resources; New Mexico Department of Game and Fish; U. S. Fish and Wildlife Service Office of Endangered Species (field offices in California and Idaho).

Most distribution maps in the figures in this paper are plotted on part of the United States Drainage Map, published by the University of Michigan Museum of Zoology, used by permission.

LITERATURE CITED

Allison, I. S. 1966. Fossil Lake, Oregon. Oregon State University, Corvallis, Ore. 48 pp.
Alvarez, W., et al. 1982. Iridium anomaly approximately synchronous with terminal Eocene extinctions. Science 216:886-888.
Andrews, J. 1977. Shells and shores of Texas. University of Texas Press, Austin, Tex. 365 pp.

Arkell, W. J. 1956. Jurassic geology of the world. Oliver and Boyd, Edinburgh. 806 pp.

Baker, F. C. 1928. The fresh water Mollusca of Wisconsin. Wisc. Geol. Nat. Hist. Surv. Bull. 70(1):1-507.

Bateman, P. C., and Wahrhaftig, C. 1966. Geology of the Sierra Nevada. Bull. Calif. Div. Mines Geol. 190:107-172.

Blair, W. N., and Armstrong, A. K. 1979. Hualapai Limestone Member of the Muddy Creek Formation: The youngest deposit predating the Grand Canyon, southeastern Nevada and northwestern Arizona. U. S. Geol. Surv. Prof. Pap. 1111:1-14, pl. 1-19.

Bright, R. C. 1967. Late-Pleistocene stratigraphy in Thatcher basin, southeastern Idaho. Tebiwa 10:1-7.

Call, R. E. 1884. On the Quaternary and Recent Mollusca of the Great Basin, with descriptions of new forms. U. S. Geol. Surv. Bull. 11:1-66.

Clarke, A. H. 1973. The freshwater molluscs of the Canadian Interior Basin. Malacologia 13:1-509.

Croizat, L. 1958. Panbiogeography. Wheldon and Wesley, Codicote. 2 vols.

Croizat, L. 1962. Space, time, form: The biological synthesis. Caracas (published by the author). 881 pp.

Croizat, L. 1976. **Biogeografía analitíca y sintética ("Panbiogeografía") de las Américas.** Bibliot. Acad. Ci. Fis., Mat. y Nat., Caracas 15:1-454, 16:455-890.

Croizat, L. 1982. Vicariance/vicariism, panbiogeography, "vicariance biogeography," etc.: A clarification. Syst. Zool. 31:291-304.

Dall, W. H. 1894. Monograph of the genus *Gnathodon*, Gray (*Rangia*, Desmoulins). Proc. U. S. Nat'l. Mus. 14:89-106.

Davis, P., et al. 1977. Paleomagnetic study at a nuclear power plant site near Bakersfield, California. Quatern. Res. 7:380-397.

Feth, J. H. 1963. Tertiary lake deposits in western coterminous United States. Science 139:107-110.

Ganapathy, R. 1982. Evidence for a major meteorite impact on the earth 34 million years ago: Implication for Eocene extinctions. Science 216:885-886.

Gehr, K. D. 1980. Late Pleistocene and recent archaeology and geomorphology of the south shore of Harney Lake, Oregon. M. A. Thesis. Portland State University, Portland, Ore.

Gehr, K. D., and T. M. Newman. 1978. Preliminary note on the late Pleistocene geomorphology and archaeology of the Harney basin, Oregon. Ore Bin 40:165-170.

Hawley, J. W., and F. E. Kottlowski. 1969. Quaternary geology of the southcentral New Mexico border region. New Mex. Bur. Mines Min. Res. Circ. 104:89-115.

Hsü, K. J., et al. 1982. Mass mortality and its environmental and evolutionary consequences. Science 216:249-256.

Hubbs, C. L., and R. R. Miller. 1948. The Great Basin, with emphasis on glacial and post-glacial times. II. The zoological evidence: Correlation between fish distribution and hydrographic history in the desert basins of western United States. Univ. Utah Bull. 38(20):17-166.

Jelinek, A. J. 1967. A prehistoric sequence in the middle Pecos Valley, New Mexico. Univ. Mich. Mus. Anthrop., Anthrop. Pap. 31:1-175, pl. 1-16.

Keep, J., and J. L. Baily, Jr. 1935. West coast shells. Stanford University Press, Stanford, Calif. 350 pp.

Lemke, R. W., W. M. Laird, M. J. Tipton, and R. M. Lindvall. 1965. Quaternary geology of northern Great Plains. Pages 15-27 *in* H. E. Wright, Jr. and D. G. Frey, eds. The Quaternary of the United States. Princeton University Press, Princeton, N. J..

Lillegraven, J. A. 1979. A biogeographical problem involving comparison of later Eocene terrestrial vertebrate faunas of western North America. Pages 333-347 *in* J. Gray and A. J. Boucot, eds. Historical Biogeography, Plate Tectonics, and the Changing Environment. Oregon State University Press, Corvallis, Ore.

MacGinitie, H. D. 1969. The Eocene Green River flora of northwestern Colorado and northeastern Utah. Univ. Calif. Pub. Geol. 83:1-140.

MacGinitie, H. D. 1974. An early middle Eocene flora from the Yellowstone-Absaroka volcanic province, northwestern Wind River basin, Wyoming. Univ. Calif. Pub. Geol. 108:1-103.

Metcalf, A. L. 1980. Fossil *Rangia cuneata* (Mactridae) in Eddy County, New Mexico. Nautilus 94:2-3.

Miller, R. R. 1958. Origin and affinities of the freshwater fish fauna of western North America. Pages 187-222 *in* C. L. Hubbs, ed. Zoogeography. Amer. Assoc. Adv. Sci. Pub. 51.

Miller, R. R. 1981. Coevolution of deserts and pupfishes (genus *Cyprinodon*) in the American Southwest. Pages 39-94 *in* Naiman, R. J., and D. L. Soltz, eds. Fishes in North American Deserts. John Wiley & Sons, New York, N. Y.

O'Keefe, J. A. 1980. The terminal Eocene event: Formation of a ring system around the Earth? Nature 285:309-311.

Robinson, G. D. 1963. Geology of the Three Forks quadrangle, Montana. U. S. Geol. Surv. Prof. Pap. 370:1-143, pl. 1-3.

Sarna-Wojcicki, A. M. 1976. Correlation of late Cenozoic tuffs in the central Coast Ranges of California by means of trace- and minor-element chemistry. U. S. Geol. Surv. Prof. Pap. 972:1-30.

Skinner, M. F., and C. W. Hibbard. 1972. Early Pleistocene pre-glacial and glacial rocks and faunas of north-central Nebraska. Amer. Mus. Nat. Hist. Bull. 148(1):1-148.

Smith, G. R. 1975. Fishes of the Pliocene Glenns Ferry Formation, southwest Idaho. Univ. Mich. Mus. Paleont. Pap. Paleont. 14:1-68.

Smith, G. R. 1978. Biogeography of intermountain fishes. Great Basin Nat. Mem. 2:17-42.

Smith. G. R. 1981a. Late Cenozoic freshwater fishes of North America. Ann. Rev. Ecol. Syst. 12:163-193.

Smith, G. R. 1981b. Effects of habitat size on species richness and adult body sizes of desert fishes. Pages 125-171 *in* R. J. Naiman and D. L. Soltz, eds. Fishes in North American Deserts. John Wiley & Sons, New York, N.Y.

Snyder, C. T., G. Hardman, and F. F. Zdenek. 1964. Pleistocene lakes in the Great Basin. U. S. Geol. Surv. Misc. Geol. Invest. Map. I-416.

Stearns, R. E. C. 1890. Descriptions of new West American land, freshwater, and marine shells, with notes and comments. Proc. U. S. Nat'l. Mus. 13:205-225.

Stearns, R. E. C. 1901. The fossil fresh-water shells of the Colorado Desert, their distribution, environment, and variation. Proc. U.S. Nat'l. Mus. 24:271-299.

Sterki, V. 1895. Two new Pisidia. Nautilus 8:97-100.

Taylor, D. W. 1957. Pliocene fresh-water mollusks from Navajo County, Arizona. J. Paleont. 31:654-661.

Taylor, D. W. 1960. Distribution of the freshwater clam *Pisidium ultramontanum*. Amer. J. Sci. 258A:325-334.

Taylor, D. W. 1965. The study of Pleistocene nonmarine mollusks in North America. Pages 597-611 *in*: H. E. Wright, Jr. and D. G. Frey, eds. The Quaternary of the United States. Princeton University Press, Princeton, N. J.

Taylor, D. W. 1966. Summary of North American Blancan nonmarine mollusks. Malacologia 4:1-172.

Taylor, D. W. 1981. Freshwater mollusks of California: A distributional checklist. Calif. Fish & Game 67:140-163.

Taylor, D. W. 1983. Late Tertiary mollusks from the lower Colorado River valley. Univ. Mich. Mus. Paleont. Contr. 26:289-298.

Taylor, D. W., and G. R. Smith. 1981. Pliocene molluscs and fishes from northeastern California and northwestern Nevada. Univ. Mich. Mus. Paleont. Contr. 25:339-413.

Wolfe, J. A. 1978. A paleobotanical interpretation of Tertiary climates in the northern hemisphere. Amer. Sci. 66:694-703.

Woodring, W. P. 1965. Endemism in middle Miocene Caribbean molluscan faunas. Science 148:961-963.

Woodring, W. P. 1966. The Panama land bridge as a sea barrier. Proc. Amer. Phil. Soc. 110:425-433.

Yen, T.-C. 1946. Late Tertiary fresh-water mollusks from southeastern Idaho. J. Paleont. 20:485-494, pl. 76.

POLLEN PROFILES OF THE PLIO-PLEISTOCENE TRANSITION IN THE SNAKE RIVER PLAIN, IDAHO

ESTELLA B. LEOPOLD AND V. CRANE WRIGHT
University of Washington, Seattle, WA 98195

The diversity of grazing ungulates, including horses, in the Miocene of the Pacific Northwest has led to the belief that the vegetation of that interval was largely savanna grassland and woodland. However, both fossil leaf and pollen data from the Snake River Plain of Idaho do not support this idea.

Leaf floras of the middle and late Miocene indicate that deciduous hardwood and montane conifer forests were widespread in Idaho. Scattered pollen data support that conclusion.

Pliocene records from both pollen and leaf data suggest the development of a simplified montane conifer forest with sporadic appearance of open grassland and an increasing importance of sagebrush/salt bush steppe. During mid-Pliocene time a highly diverse land mammal fauna with abundant grazing horses (*Plesippus*) is recorded on the Snake River Plain.

Early to middle Pleistocene pollen samples, few in number, suggest a continuation of steppic vegetation while forest elements are poorly represented. Diatoms indicate the development of alkaline lakes and high evaporation rates.

A complex sequence of fossil-bearing rocks along the floodplain of the Snake River in southwestern Idaho provides one of the best-documented records of the Plio-Pleistocene transition in continental North America. Its sediments of the late Pliocene and early Pleistocene (Blancan Provincial age; Berggren and Van Couvering 1974) contain fossils that yield information on many aspects of this ancient ecosystem. Even with major gaps in the thick depositional record, these fossils offer a basis for inferring environmental changes and major biotic shifts that occurred during Blancan time. In this paper we describe pollen spectra that record the nature of vegetation during the interval between 3.4 and 1.3 million years (my) B.P. Though Tertiary leaf floras make it clear that mesic forest vegetation existed regionally through the Miocene, the abundance of grazing ungulates has led many paleontologists to suppose that extensive grasslands existed in the Snake River Plain during Neogene time. Our question is whether and when such grasslands actually did develop, as shown by direct evidence through pollen analysis. We are interested in the antiquity of the modern sagebrush steppe vegetation in the region. In this paper we adhere to the western definitions of steppe as shrub-dominated and grassland as herbaceous-dominated vegetation (Daubenmire 1970).

Detailed mapping of the Snake River Plain by Malde and Powers (1962, 1972), Malde, Powers and Marshall (1963) and Malde (1972) and tephrachronology (Swirydezuk et al. 1979) provide a stratigraphic framework for the extensive paleontological studies carried out by other workers. The Hagerman Lake beds, now considered to be chiefly of early Blancan age (late Pliocene) and which are the main source of pollen data for this report, are famous for their contained fauna: birds (Brodkorb 1958; Ford and Murray 1967), carnivores (Bjork 1970; Fine 1963), abundant horses (Gazin 1936; Shotwell 1961), rodents (Zakrzewski 1969), fishes (Smith 1975), and snails (Taylor 1960, 1966, and unpublished). Early reports of fossil leaves, wood and pollen from the local Neogene units are mentioned by Malde and Powers (1962), and some of Malde's pollen collection localities for the Glenns Ferry Formation are shown on their geological map

Copyright ©1985, Pacific Division, AAAS.

(Malde and Powers 1972). Diatoms are described by Bradbury and Krebs (1982), and by K. E. Lohman (unpubl. data).

Fossil plant localities for Miocene deposits in Idaho are numerous. At least 13 Miocene leaf localities are known from the western part of Idaho (summarized by Axelrod 1964). Of special interest in this volume is the Clarkia flora of northwestern Idaho, which demonstrates the luxuriant nature of early to middle Miocene forest vegetation of the region and documents a surprisingly strong representation of warm-temperate trees of the Old and New World (Smiley, Gray and Huggins 1975; Smiley and Huggins 1981; Rember and Smiley 1979, and this volume). Pollen of the Clarkia beds studied by Jane Gray (this volume) corroborate diverse genera represented by leaves and include some taxa not yet found in the leaf flora. The Clarkia flora suggests a humid, equable summer-wet climate that clearly preceded the lifting of the high Cascades and the consequent rain shadow that developed about 12-10 my ago (Smiley 1963).

In southern Idaho, the Trapper Creek flora of late Miocene age resembles living conifer-hardwood forests in the humid, cool-temperate summer-wet parts of western and eastern North America and northeastern Asia (Axelrod 1964). The site lies 90 miles (145 km) southeast of the Hagerman Lake beds along the south flanks of the Snake River Plain (see Fig. 2).

LOCATION AND STRATIGRAPHIC SETTING

The study area is set in the western part of the wide, arid Snake River Plain west of Twin Falls and east of the Bruneau River in southwestern Idaho (see Figs. 1 and 2). The plateau surface lies between 3000 and 3500 feet (915 and 1068 m), and the canyon of the Snake River, up to several hundred feet deep, winds through the middle. The Snake River flows generally westward from western Wyoming, cuts through the Teton Range into southern Idaho, and then trends northwestward into eastern Oregon and southeastern Washington to join the Columbia River. The Snake River Plain is the late Tertiary floodplain of the river. Rocks of the Plio-Pleistocene transition in the Snake River Plain were deposited in a subsiding basin about 50 miles (80 km) wide below Twin Falls, Idaho, and extend westward across the state. The units pertaining to this report include the Idavada volcanics and the overlying units of the Idaho Group. Their thicknesses and character are described in Table 1. Beginning in the late Miocene, volcanic sediments including lava flows filled the basin, periodically clogging the flood plain. Repeated subsidence down-faulted these so that remnants of late Miocene and early Pliocene sediments are now found only on the edges of the basin (see Fig. 3). Younger units, specifically the Glenns Ferry Formation and the overlying Bruneau Formation that span the Plio-Pleistocene transition, are the main deposits in the middle of the basin.

Late Miocene rocks, including the Idavada Volcanics and lower units of the Idaho Group exposed along the southern margin of the Snake River Plain, are composed of volcanic ash and basalts.

The Glenns Ferry sediments, some 1700 feet (518 m) thick, represent a wide range of floodplain, lacustrine, and valley-border environments. Bruneau sediments, including lake and stream deposits and basaltic lava flows, unconformably lie on the Glenns Ferry or fill deep canyons cut into the Glenns Ferry (Fig. 3). Malde and Powers (1962) have established that basaltic flows of the Bruneau at times dammed canyons cut into the Glenns Ferry and that lakes so formed received great thicknesses of the highly silicic sediments also of the Bruneau Formation.

The Glenns Ferry Formation is dated by a number of potassium/argon dates ranging from about 3.5 near the base to 3.4-3.2 near the middle (Malde 1972). More recently, Armstrong et al. (1975) obtained a series of isotope dates that are far older (ca. 6 my). These older (whole rock) dates conflict with the host of evidence from fossil vertebrates and invertebrates (summarized by Malde and Powers 1962; Malde 1972) and are "anomalously older than those of Evernden et al.

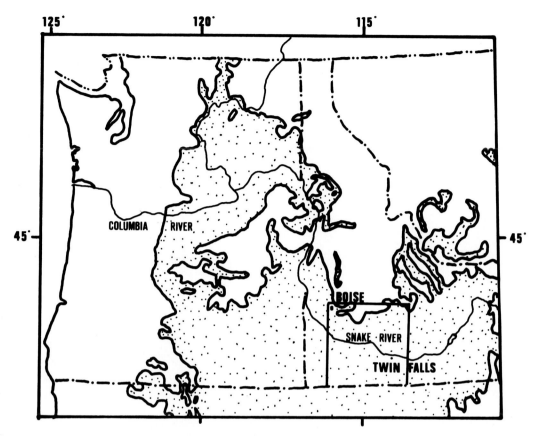

Figure 1. Map of the Pacific Northwest (Washington, Oregon and Idaho) showing the outlines of desert scrub, grassy steppe and saline basin vegetation in the stippled area (after Küchler 1970). The study area in the Snake River Plain of southwestern Idaho is enclosed by box.

(1964) and Obradovich" (Neville et al. 1979:517). Vertebrate evidence places all but the uppermost beds of the Glenns Ferry in the lower Blancan with certainty (middle and late Pliocene as calibrated by Berggren and Van Couvering 1974). The currently accepted age-range of the Blancan is about 4.0 to 1.7 my (Savage and Curtis 1970). Fossil snails clearly support the vertebrate age (Taylor 1966). Hence, for now, it seems best to accept the 3-my-old dates for this unit.

Recent magnetic stratigraphy studies by Neville et al. (1979) confirm conclusions by Hibbard and Zakrzewski (1967) and Zakrzewski (1969) that the uppermost Glenns Ferry sediments containing the Grand View fauna of Jackass Butte belong to the earliest part of the Matuyama epoch. The presence of mammoth (*Mammuthus*) indicates a Pleistocene age (White *in* Neville et al. 1979:519). Neville's excellent study establishes that the lower parts of the Glenns Ferry extend from the Cochiti event of the Gilbert reversed epoch through the Gauss normal epoch. The Hagerman fauna of Fossil Gulch "characterizes early Blancan and the Grand View fauna characterizes the late Blancan" (Neville et al. 1979:523).

In summary the Glenns Ferry Formation correlates with the late Ruscinian and Villafranchian faunas of Europe according to faunal, isotopic, and magnetic evidence.

The Bruneau Formation with its thick diatomite sediments documents the last of the large lakes that existed in the western Snake River Basin (Bradbury and Krebs 1982). The Bruneau is dated by potassium/argon ratios at about 1.8 and 1.36 my (Armstrong et al. 1975; Evernden and James 1974). Irvingtonian vertebrates and mid-Pleistocene molluscs are present. Some readings of

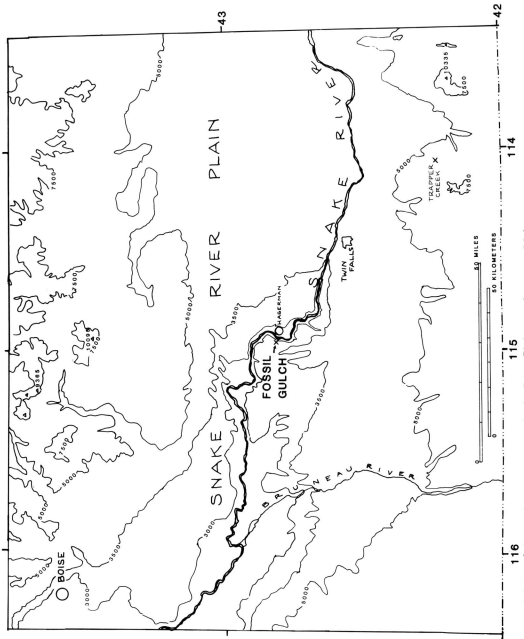

Figure 2. Map of the study area, western Snake River Plain, southwestern Idaho.

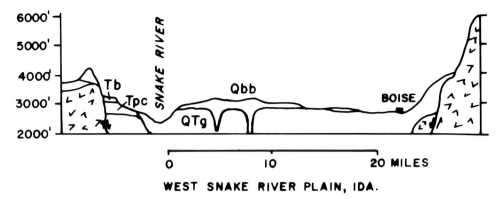

Figure 3. Diagrammatic geologic cross-section in western Snake River Plain from Boise to Reynolds Creek (after Malde and Powers 1962). Qbb = Bruneau Formation, basaltic lava flows; Tb = Banbury Basalt; Tpc = Poison Creek Formation. Tg = Glenns Ferry Formation. Granites on the edge of subsiding basin are of the Idaho Batholith.

reversed polarity from the formation may relate to the Matuyama Reversed Epoch. A rich mollusc assemblage is "strikingly more modern" than those of the Glenns Ferry and "indicates a major stratigraphic gap" (Malde and Powers 1962). In short, mammalian and molluscan fossils indicate a middle Pleistocene age, while a date of 1.8 my suggests that the Bruneau may also include earliest Pleistocene sediments.

Sediments of the Plio-Pleistocene from these two formations present a diversity of documented fossil types unrivaled in the Neogene of North America. The Glenns Ferry sediments are particularly rich in fossil remains, with 82 species of vertebrates, 2 species of crustaceans, more than 27 species of snails, 58 taxa of vascular plants, and 78 species of diatoms and other algae. The Bruneau biota is less diverse but contains remains for each of these categories.

In order to put the Plio-Pleistocene transition in perspective with the Neogene biota, we include the following phases in describing our composite sequence (Table 1):

1. Late Miocene or early Pliocene: Salt Lake Formation (equivalent to Idavada Volcanics);
2. Late Pliocene: Glenns Ferry Formation;
3. Early to middle Pleistocene: Bruneau Formation.

FIELD COLLECTIONS AND METHODS

The field collections for fossil pollen were made in large part by Harold Malde in the course of his field mapping in the area. We collected the suite of samples from USGS paleobotanical localities D5302 and D5299 at Fossil Gulch below the Hagerman Horse Quarry during the summer of 1975. Samples (5-30 grams each) were treated with HCl, HF and acetylation. Over 450 rock samples were processed in the pollen laboratories of the U. S. Geological Survey and the University of Washington, but only 35 yielded enough pollen to provide adequate counts. Because of the precisely defined ash and basalt stratigraphy, the field samples can be put in a stratigraphic order. Malde's collections represent largely isolated localities over a 30-mile (49 km) area between the Bruneau River on the west (Fig. 1) and Hagerman on the east (listed in Appendix 1), and the results are put together in a composite pollen sequence as a histogram (Fig. 6). Generally 200-350 pollen grains were counted at each level, though some sparse samples contained as few as 135 pollen and spores. Tallies for 12 of these are given in Appendix 2, according to their depositional environments.

Our collections from the famous Fossil Gulch where more than 100 horses have been excavated (sec. 16, T. 7 S. R. 13 E.), come from between 150 to 184 feet (46-56 m) below the Horse

TABLE 1. LATE NEOGENE AND EARLY TO MIDDLE PLEISTOCENE FORMATIONS
OF THE WESTERN SNAKE RIVER PLAIN (after Axelrod 1964; Malde and Powers 1962)

	Formation	Thickness (approx.)	General lithologic characters	Fossils	
IDAHO GROUP	BRUNEAU	800 ft. (244 m)	Volcanic ash and diatomites in lake and stream deposits; basaltic lava flows	Lower and middle Pleistocene mammals, mollusks, diatoms, and pollen	PLEIST.
			(unconformity)		
	GLENNS FERRY	2000 ft. (610 m)	Nonindurated, interfingering lake and stream deposits of light-colored siltst., ss., and sh., often ripple-marked, some oölitic beds, algal ls., and cgl. of more local occurrence. Local flows of olivine basalt, basaltic ash, and siliceous volc. ash	Leaves of aquatic plants, Blancan mammals, mollusks and pollen	PLIOCENE
			(unconformity)		
	CHALK HILLS	500 ft. (152 m)	Siliceous volc. ash, light-colored siltst. and ss., generally varicolored	Wood, Hemphillian mammals, mollusks and pollen	LATE MIOCENE
			(unconformity)		
	BANBURY BASALT	1,000 ft. (305 m)	Decomposed olivine basalt flows, with assoc. ss. and cgl., as well as diatomite, siltst.	Mammals, pollen	
			(unconformity)		
	POISON CREEK	100 to 400 ft. (30-122 m)	Silic. volc. ash and tuffaceous clayst., ss., some arkosic ss. and gravel, and thin beds of basaltic ash	Leaves, Clarendonian mammals, and pollen	
			(unconformity)		
	IDAVADA VOLCANICS	1,000 to 3,000 ft. (305-914 m)	Upper part consists chiefly of flows and ignimbrites of silicic latite (Cougar Point fm.), lower part of vitric ash, ashy sh., siliceous sh., ss., and local ignimbrites (Jenny Cr. fm. of Axelrod, 1964).	Leaves, mammals, mollusks, diatoms and pollen	

Quarry and are from fluviatile silts and sands. Results are shown as a pollen histogram (Fig. 7), and pollen counts are listed in Appendix 3.

In order to compare the fossil-pollen percentages with modern pollen-rain data, we collected an altitudinal transect of modern surface samples from the crest of the Teton Range (Teton Pass, State route 22) westward along the Snake River valley to the Hagerman area and the Bruneau River.

Our modern pollen transect was patterned after that of McAndrews and Wright (1969) in Wyoming and the northern Great Plains. Samples of dry organic detritus (duff) from the soil surface were obtained from beside shrubs or under trees where wind and rain splash may be minimal. A few samples represent pond mud. Sampling sites were on the average 15 miles (24 km) apart, except on the slopes of the Teton Range, where collections were about 2 miles (ca 3 km) apart. Laboratory preparation followed that described by McAndrews and Wright (1969) and pollen material was mounted in glycerine jelly. Results are shown in Figure 4.

MODERN VEGETATION, CLIMATE AND POLLEN RAIN IN SOUTHERN IDAHO

The present-day climate in the Snake River Plain is distinctly continental with the coldest month averaging below zero and the warmest month around 25° C. The forests above the Plain have colder winters but cooler summers. Continentality as measured by the difference between the coldest and warmest month is greater for the steppe (28°C) than for the low-altitude forest areas (ca. 25°C). Precipitation falls primarily in winter and is 2-3 times higher in the forest area than in the steppe; the growing season is longer on the steppe (5 months compared to 3-4 months). Evaporation is equal or slightly less than precipitation in the low-altitude forest areas, but is twice that of precipitation in the steppe. The vegetation biomass (standing weight per hectare) is 10 to 20 times greater in woodland and low-altitude forest areas than in the steppe, reflecting the fact that forest is a more productive type of ecosystem (see Table 2).

The native vegetation of the Snake River Plain (Weaver 1917) is steppe chiefly dominated by *Artemisia tridentata* and *Agopyron spicatuum* (sage brush/wheat grass). This vegetation is a part of more extensive shrub and grassy steppe extending eastward from the Columbia Basin to the Rocky Mountains (Fig. 1). The average annual rainfall of this steppe area is below 15 inches (38 cm), with desert areas in low basins of the steppe having as little as 8 inches (20 cm) of precipitation. The forest-steppe border (locally at 6300 ft, 1920 m), is closely associated with the 15 and 20 inch (38 and 50 cm) isohyets (Peter Dunwiddie, unpubl. data). Bunch-grass steppe with rare pinyon-pine trees on north-facing cliffs characterizes the higher steppe while shrub steppe predominates at lower elevations. Mack and Thompson (1982) point out that the carrying capacity of the original steppe of this region was low for ungulates because of the susceptibility of perennial bunch grasses to trampling. In historic time most of the native bunch grass below an altitude of about 5000 feet (762 m) has been eliminated by grazing and is largely replaced by the exotic annual cheat grass (*Bromus tectorum*; Mack 1981). Saline basins or flats in the Snake River Plain are characterized by species of *Atriplex* (salt bush), especially *A. confertifolia* (shadscale), *Sarcobatus vermiculatus* (greasewood) and, in some locales, *Grayia spinosa* (spiny hopsage). Riparian vegetation along the Snake River includes *Populus angustifolia* (narrow-leaved cottonwood), *Salix* (willow) and *Eleagnus commutata* (wolf willow), *Chrysothamnus* (rabbit brush) as well as *Typha* (cattail), and other aquatic macrophytes.

Each mountain range flanking the basin has different forest tree communities. The mountains to the north contain a variety of forest types: *Pinus ponderosa* (yellow pine) at low elevations, *Picea engelmannii-Abies lasiocarpa* (Engelmann spruce-subalpine fir) above 8000 feet (2438 m) and *Abies grandis-Pseudotsuga menziesii* (grand fir-Douglas fir) on middle-montane slopes. Two species of *Tsuga* (*T. mertensiana* and *T. heterophylla*, mountain and western hemlock)

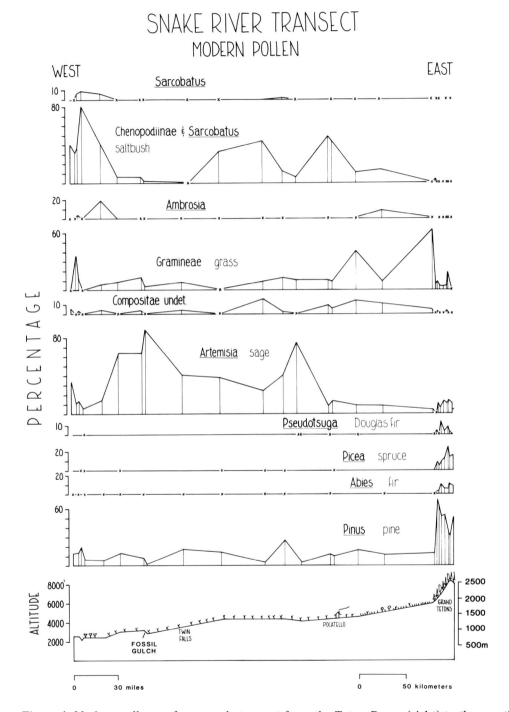

Figure 4. Modern pollen surface sample transect from the Teton Range (right) to the mouth of the Bruneau River (left) in southern Idaho. Topography and vegetation type is shown below; mixed conifer forest is indicated in mountainous terrain above 7000 feet, grassy steppe on slopes between 5000-7000 feet, and desert scrub occurs below 5000 feet. Data are percent total pollen.

TABLE 2. CLIMATE DATA FROM THE WESTERN SNAKE RIVER PLAIN (USDA 1941) AND BIOMASS FOR LOWER FOREST, WOODLAND AND STEPPE (estimated from Branson 1975)

CLIMATE AND RELATED DATA, WESTERN SNAKE RIVER PLAIN, IDAHO

		Woodland and lower forest	Steppe
Precipitation	Annual	18-23" (46-58 cm)	7-15" (18-38 cm)
	Winter	12-16" (30-46 cm)	4-8" (10-20 cm)
Temperature	Coldest month average	-7° C	-3° C
	Warmest month average	18° C	25° C
	Range	25° C	28° C
	Frost-free period	120-100 days	150 days
	Precipitation/Evaporation	1.0-1.3	0.5
Plant biomass (standing weight in kilograms/hectare)		4000-6000	400-2000

grow in the mountainous part of northern Idaho. The mountains to the east are largely *Pseudotsuga* forest, while those to the south contain extensive *Pinus monophylla-P. flexilis-Juniperus scopulorum* (pinyon-juniper) woodland, especially at lower elevations. Woodland southeast of the Snake River Plain in northwestern Utah contains extensive stands of *Quercus gambellii-Cercocarpus* shrublands (white oak-mountain mahogany). The lower edge of forest lies between 6250 and 7000 feet (1905 and 2133 m), but scattered pinyon-juniper woodland is found at 5000 feet on northern exposures.

The purpose of our modern pollen rain study (Fig. 4) is to demonstrate the main patterns of arboreal shrub and herb pollen types in an altitudinal transect in relation to the forest/steppe ecotone in Idaho. We wished to determine whether pollen rain can distinguish broadly different types of vegetation. Results from similar studies in the western U.S. (eastern Washington, Mack et al 1978; Wyoming and northern Great Plains, McAndrews and Wright 1969; Colorado, Kapp, 1965; southeastern Idaho, Bright 1966; Yellowstone Plateau, Baker, 1976; Snake River Plain, Idaho, Bright and Davis 1982) have shown that such a transect may be useful for this purpose.

Climate interpretations comparing modern with fossil pollen rain tend to imply that past habitats were like those of their living equivalents. Yet modern climates are probably atypical even in the Pleistocene. Our comparisons here are used chiefly as an index of vegetation type and as a guide to how various taxa are represented.

A transect of vegetation types and pollen rain from Glenns Ferry eastward along the Snake River Plain to the Wyoming border is shown along with topography in Figure 4. The modern pollen rain on the west side of the Teton Range reflects the local forest composition. Above 6250 feet (1905 m) conifer pollen is dominant. *Picea* and *Abies* together account for 30% of the pollen above 7000 feet (2135 m). *Pseudotsuga* pollen is underrepresented (5-10%) at middle altitudes, and *Pinus* dominates percentages at lower altitudes from 7000 feet (2135 m) to the forest-steppe border at 6300 feet (1920 m).

On the plains, shrub and herbs dominate in the modern pollen rain. In the higher steppe from near treeline (6300 ft, 1920 m) down to 4000 ft (1220 m), grass pollen (probably of bunch grass) is well represented (40-60%), while the lower steppe (2000-4000 ft; 609-1220 m) is dominated by members of the Chenopodiineae (salt bush), *Artemisia* and other Compositae. The low percentages (<15%) of grass pollen in the lower steppe are surprising and suggest that the exotic annual cheat grass (*Bromus tectorum*), which now makes consistent groundcover in the steppe below 4000 ft (1220 m), does not seem to contribute importantly. Apparently this annual is a low pollen producer, and according to Richard Mack (pers. commun. 1983), it is cleistogamous.

Conifer pollen, poorly represented on the Snake River Plain (pine 8-20%; spruce and fir <1%) drifts in from higher elevations. This pattern is similar to results of Bright and Davis (1982 and Fig. 5) in Idaho, and to our unpublished transect of surface samples along the Raft River in southern Idaho; however, for the northern Great Plains (McAndrews and Wright 1969) and eastern Washington, results suggest that conifer pollen drifts eastward from mountain fronts forming up to 50% of the pollen rain 50 miles (80 km) or more downwind from the conifer forests.

Modern pollen rain at Glenns Ferry (2200 ft, 670 m) near the Horse Quarry at Fossil Gulch

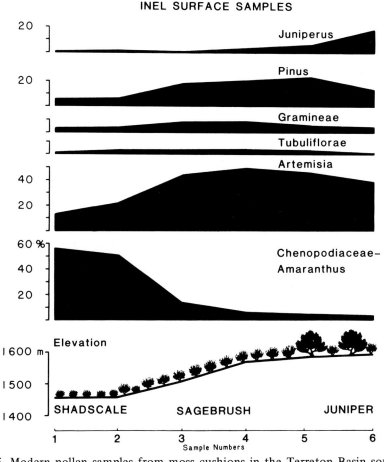

Figure 5. Modern pollen samples from moss cushions in the Terreton Basin southwestward to West Twin Butte, south-central Idaho. Data are percentages of all pollen and spores in each sample (from Bright and Davis 1982).

is dominated by *Artemisia* and other Compositae (75%), Chenopodiineae and *Sarcobatus*. Grass pollen is uncommon (<1%) and total trees (mostly pines) are only 4% of the count. This site is ca. 37 miles (60 km) from the nearest conifer forest.

DESCRIPTION OF THE FOSSIL SEQUENCE

A composite pollen histogram shown as Figure 6 serves as a framework for discussion of the Neogene paleontologic sequence.

Late Miocene and Early Pliocene

Poison Creek, Salt Lake, Banbury Basalt and Chalk Hills Formations. Because pollen from the Idavada volcanics was not found, 4 samples in the lower part of this diagram come from the correlative Salt Lake Formation (of Mapel and Hail 1959; equals Jenny Creek Formation of Axelrod 1964) from exposures near Trapper Creek (Fig. 2). These are included to demonstrate the importance of diverse broad-leaved trees in the late Miocene flora of the region. These pollen samples (USGS Paleobot. loc. D1208; Leopold in Mapel and Hail 1959) are dominated by *Pinus* and contain up to 10% *Ulmus-Zelkova*, and smaller amounts of *Quercus, Juniperus*-type, *Abies, Pterocarya, Carya, Juglans, Sarcobatus* and *Ephedra*. The hardwoods represented are either exotic to North America (*Pterocarya*, lingnut), or to the region (*Ulmus-Zelkova*, elm; *Juglans*, walnut; and *Carya*, hickory). *Quercus* does not now occur in Idaho but has its northern limit in Utah 70 miles to the south. These elements can be considered Tertiary relicts. Leaves of *Quercus brownii* have been reported from the Idavada volcanics (Malde and Powers 1962).

Remains of the horse *Neohipparion*, presumed by Shotwell (1961) to be a savanna type, have been collected from the Idavada volcanics; fossil snails from the unit are similar to those in the late Miocene Teewinot Formation of the Jackson Hole area (Malde and Powers 1962).

Fossil leaves from the late Miocene Poison Creek Formation include *Quercus prelobata* and cf. *Q. chrysolepis* (J. A. Wolfe, pers. commun., 3/21/61). *Quercus prelobata* is a white oak that could be related to *Q. alba* of the eastern U.S., or to the local *Q. gambellii*, a scrub oak of Utah and the southern Rockies. *Q. chrysolepis* type is particularly significant because it is a holly-leaved evergreen oak typical of the humid equable climate of the Great Valley and surrounding hills of California. One pollen sample (D1699) is dominated by *Pinus* with small amounts of *Abies* cf. *grandis, Abies* cf. *lasiocarpa, Carya, Ephedra, Juniperus* type and *Sarcobatus*. The green alga *Botryococcus* is abundant, reflecting a pond or lake environment.

The Banbury Basalt contains fossils of beaver (*Dipoides*, Hemphillian age), and a rodent (*Microtoscopites disjunctus*). No plant remains have yet been found in the Banbury Basalt, but one pollen sample from silt interbeds is reported here (D1696).

Mammalian evidence indicates a late Miocene—early Pliocene age. The abundance (50%) of an unidentified tricolpate pollen, which we took to be an herb, may not be correctly judged. The sample contains small amounts of conifer pollen (*Pinus, Picea*), rare hardwoods, chenopods, grass, and is the oldest sample in our Snake River Neogene suite to contain substantial amounts of Compositae pollen (26%) including *Artemisia*.

The Chalk Hills Formation has a diverse Hemphillian (transitional late Miocene—early Pliocene) fauna: camel, rhinoceros, rodents, beaver, fish, molluscs and snails (Malde and Powers 1962), but no pollen was found. Abundant fossil wood, mostly of hardwoods, includes at least *Carya, Ulmus, Fraxinus* (ash), *Alnus* (alder), *Abies*, and Salicaceae (R. W. Brown, pers. commun., 12/9/46; R. A. Scott now has the collections at the U.S.G.S., Denver, CO). Hickory and a fossil bracket fungus (*Fomes idahoensis*; Brown 1940) based on their present distribution suggest a climate more humid than now. The stratigraphic position of these fossil woods is shown at the unconformity at the base of the Glenns Ferry (Fig. 6).

The combined evidence from the units described above indicates a mixed conifer-hardwood

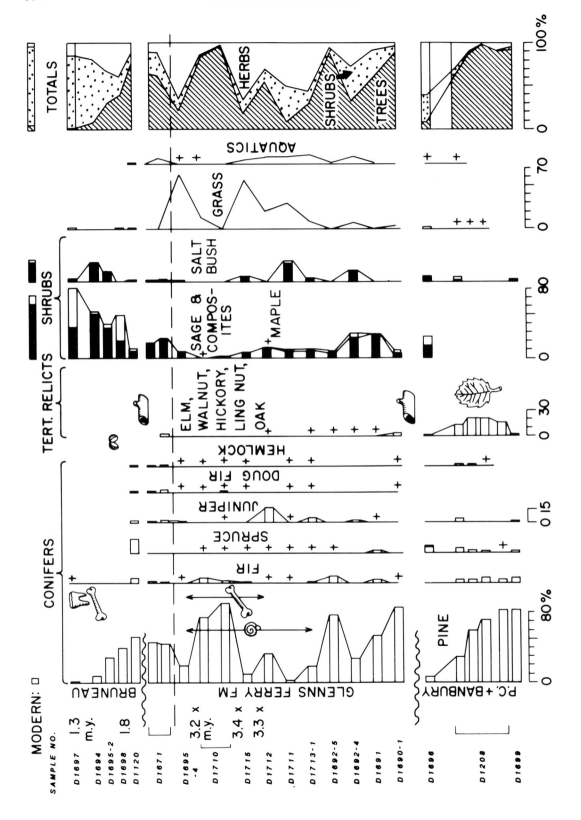

forest or woodland on the Snake River Plain during Mio-Pliocene time. The Mediterranean-type holly-leaved oak in the Poison Creek Formation could indicate woodland or forest but the high frequency of tree pollen in D1699 suggests forest vegetation. Holly-leaved oaks of the western U.S. are limited to humid areas with an oceanic climate, i.e. California oakwoods and savanna; Texas oak-hickory savanna and forest (of Küchler 1970).

Late Pliocene (early Blancan)

Glenns Ferry Formation. The Glenns Ferry pollen samples can be divided into two phases of vegetation: based on the composite pollen diagram in Fig. 6, the lower samples represent the early phase in which *Pinus* and mixed conifers seem to dominate over shrubs and herbs, and Tertiary elements are regularly present in trace amounts. In the middle and upper samples, grasses rise to prominence and *Juniperus* is sporadically well represented; a dominance of *Pinus* alternates with prominent peaks of grass pollen, and Tertiary elements are lacking or are poorly represented. A local sequence of pollen samples (Fig. 7) from just below the Horse Quarry at Fossil Gulch belongs in the upper part of the formation and will be discussed in sequence below.

Sedimentary facies of the Glenns Ferry Formation including valley border, lacustrine, floodplain environments have been mapped by Malde (1972). Pollen counts from each of these sediment types are arranged in environmental groups in Appendix 2; from these data one can see that there is little difference in pollen rain between the sedimentary environments cited by Malde.

A. Lower part of Glenns Ferry Formation. In the basal sediments of the Glenns Ferry Formation, pollen assemblages are dominated by pollen of *Pinus* associated with lesser amounts of *Artemisia* and Chenopodiineae. Other conifers (*Abies, Picea, Juniperus, Pseudotsuga, Tsuga*) are present, as are trace amounts of Tertiary relict hardwoods. The pollen spectra seem to resemble the present-day pollen rain from conifer forest above 8000 feet (2440 m) in all respects except for the trace amounts of hardwood trees (see surface pollen transect, Fig. 4). At this time a mixed conifer forest probably occupied the Snake River Plain and included small amounts of riparian temperate trees of the walnut and elm families.

B. Middle and upper parts of the Glenns Ferry Formation. In the middle beds pollen of *Pinus* is temporarily less prominent while Gramineae pollen reaches values of 50-70% of the count. *Artemisia* and Chenopodiineae are present, and *Juniperus* attains its highest values (15%). The spectra suggest an open landscape, woodland or perhaps savanna vegetation. The *Juniperus* along with the low amounts of *Pinus* (undifferentiated) pollen suggest that pinyon-juniper woodland was present in the valley bottom. Present-day pollen rain of grassy steppe and pinyon pine vegetation between 5000 and 7000 feet (1524 and 2135 m) closely resembles this assemblage

---

Fig. 6. Composite pollen histogram, Neogene and Pleistocene samples from the western Snake River Plain. (Percent total pollen, selected types). Late Miocene collections (bottom) are from Poison Creek Fm (P.C.), Salt Lake Fm from the Goose Creek area, and interbeds from the Banbury Basalt. The stratigraphic position of fossil wood from the Mio-Pliocene (Clarendonian) Chalk Hills Formation is indicated at the base of the Glenns Ferry Formation. The late Pliocene Glenns Ferry Formation collections include beds equivalent to the Horse Quarry (stratigraphic level indicated by the bone symbol and arrow). The horizontal dashed line is drawn beneath beds equivalent to the Grandview mammal locality of Hibbard and Zakrewski (1967). Early and middle Pleistocene samples of the Bruneau Formation are shown above the unconformity at the top of the diagram. Fossil fruits of *Celtis* (symbol) are associated with lower Bruneau sediments. The Tuana gravel that lies between the Glenns Ferry and Bruneau Formations contains fossil wood of *Quercus*. Data from modern pollen rain samples from Glenns Ferry, Idaho, are averaged and plotted at the top of the chart as a point of reference. K/Ar dates are shown on left.

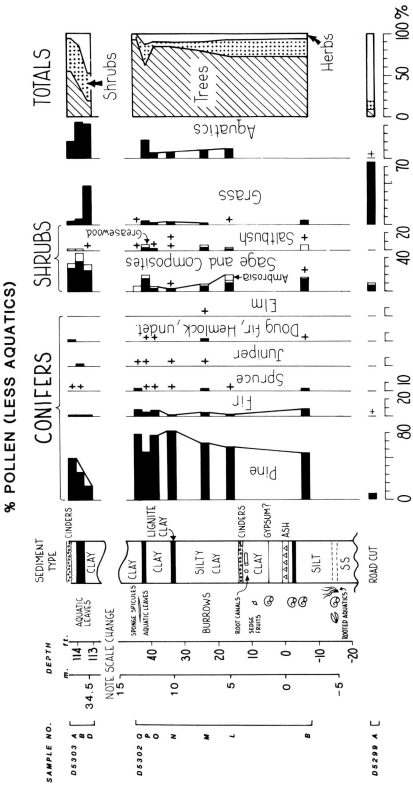

Figure 7. Percentage pollen diagram, below Horse Quarry at Fossil Gulch, Hagerman, Idaho. Vertical distance above and below datum level ash shown at left. Silts below the datum horizon are cross-bedded. Compositae pollen data show sagebrush in black and undetermined types in white unless otherwise indicated.

except that we found little or no *Juniperus* pollen in our modern-pollen transect; however, in modern-pollen surface samples from the Swan Valley area in the southeast corner of Idaho, Bright (1966) found significant amounts of *Juniperus* (10%), but only from samples collected in juniper stands; he reports smaller amounts (3%) up to 2 miles from the nearest juniper. These are similar to results in Bright and Davis (1982) diagram in Figure 5. The combined data suggest that juniper was growing locally on the Snake River Plain in mid-Glenns Ferry time.

The upper middle portion of the Glenns Ferry Formation includes beds equivalent to the measured section at Fossil Gulch, Hagerman, Idaho; the stratigraphic position of the Horse Quarry is shown by the bone symbol on Figure 6. The pollen diagram from just below the Horse Quarry measured section is discussed here to provide added detail from a local sequence.

The pollen diagram shown as Figure 7 spans a 134-ft (41-m) section. About 50 samples were collected from the beds at and below the Horse Quarry, but in four-fifths of the samples, probably due to the coarse lithology, pollen was absent or very degraded and could not be identified.

The sequence contains pyroclastic material at the top of the pollen diagram (cinders, Fig. 7) probably 40 feet (12 m) below the Shoestring Road lava flow (Bed I of Malde 1972); another cinder bed and a silicic volcanic ash occur near the base of our measured section (Fig. 7). Locally the Shoestring Road lava flow lies 100 feet below the Horse Quarry at Fossil Gulch at an elevation of 3165 feet (965 m). As in the composite pollen sequence (Fig. 6), this diagram (Fig. 7) shows sporadic peaks of grass pollen. These peaks suggest perennial bunch grasses, because in local shrub steppe (Fig. 4) and at other sites (California, Leopold 1967; Idaho, Bright and Davis 1982, Fig. 5; Columbia Basin, Mack and Bryant 1974) annual grasses, however abundant, are poorly represented in the pollen rain. At Fossil Gulch *Pinus* is generally the dominant pollen type and is accompanied by moderate values of *Artemisia* with other Compositae. Small amounts (2-12%) of *Picea, Abies* and *Juniperus* pollen are consistently present. Desert scrub elements, Chenopodiineae, *Sarcobatus, Ephedra*, are represented consistently in small numbers (5-20%).

Though not strictly analogous (see p. 331), the pollen spectra from the Horse Quarry section can be compared with those from modern surface samples shown in Figure 4. The percentages of *Pinus* (50-80%) indicate that forest or woodland was growing on the Snake River Plain. Based on Bright's (1966) data, even the small values of *Juniperus* pollen may indicate that it too was growing locally. The low amounts of other conifer pollen suggest that these genera grew some distance from the site, though it is possible that in the Pliocene they were minor elements within the *Pinus* woodland. Desert scrub taxa and *Artemisia* were important parts of the understory or woodland openings. Perhaps the closest analogue from the modern landscape is found near the lower limit of present treeline between 6000-7000 feet (1828-2135 m) where pinyon-juniper and lodgepole pine occur in woodland and forest along with desert-scrub elements and bunch grass.

Details of the sediments and their macroscopic remains lend further information in discerning the local environments. At the base of the section (below the datum level) cross-bedded silts suggest flood-plain environments and contain fossils of rooted aquatics standing in place; snails and clams are abundant. At the datum level (marked zero on Figure 7) is a thin volcanic ash (Fossil Gulch ash layer of Malde, 1972, dated as 3.3 my). Five feet (1.5 m) above the datum ash, a layer of salt (gypsum?) occurs, suggesting high evaporation rates. Intermittent lignite layers underlain by clay throughout the section are apparently algal oozes, because they contain little pollen but very abundant algae, resting cysts of *Spirogyra* and occasional aquatic monocot leaves. Sedge fruits are seen 10 feet (3.1 m) above the datum and sponge spicules are abundant 45 feet (13.7 m) above the datum. These sediments were probably deposited in floodplain ponds. A surprising abundance of *Ambrosia* pollen from sediments overlying a cinder bed 12 feet above the datum and from scattered occurrences upsection may represent local succession in a volcanic landscape.

The fauna associated with this part of the Glenns Ferry was extremely rich and diversified. The floodplain sediments at the Horse Quarry contain fossil vertebrates including remains of shrew, gopher, vole, weasel, otter, rabbit, peccary, camel, antelope, horse, and mastodon, as well as those of fish, reptiles and birds (Malde and Powers 1962). While many of the vertebrates represent extinct species and several extinct genera, most of the birds represent living species and genera. The birds include many taxa of migratory waterfowl: stork, crane, grebes, swan, pelican, and cormorant (Brodkorb 1958).

In the uppermost part of the Glenns Ferry section pollen counts (shown on Figure 6 as the two samples above the dashed horizontal line) are similar to the assemblages just described, but grass pollen is absent. *Pinus* is associated with moderate levels of *Artemisia* pollen, suggesting a woodland with sagebrush.

Early to Middle Pleistocene

Bruneau Formation. The Bruneau unconformably overlies the Glenns Ferry Formation. Samples from the lower Bruneau record declining percentages of *Pinus* pollen and increasing percentages of desert-scrub taxa including: Chenopodiineae, Compositae, and *Artemisia*. Other conifers including *Picea* are present only in the basal sample. Tertiary-relict genera (hardwoods) are absent. Except for the basal sample containing spruce, analogues are found in modern-pollen surface samples from the present-day Snake River Plain below 4000 feet (1220 m) where the vegetation is shrub steppe. Fossil fruits of *Celtis* (hackberry) have been found associated with pollen assemblages characteristic of steppe vegetation, suggesting this tree was present along canyon walls or in riparian environments as it is today in Idaho (Daubenmire 1970). Diatoms in the Bruneau suggest evaporating lakes and alkaline waters (K. Lohman, unpubl. data).

DISCUSSION AND CLIMATIC INTERPRETATION

Vegetation of the Snake River Plain shows marked changes in character during the Neogene. In the late Miocene a mesic and diverse mixed hardwood-conifer forest grew on the floodplain of the Snake River. Xeric elements are present (Chenopodiineae, Gramineae, Compositae) in the pollen record but are not abundant ($<15\%$) in the available samples. Similarity in generic composition to living coniferous-hardwood forests in the humid temperate parts of western and eastern North America and northwestern Asia suggests a summer-wet climate where annual precipitation is somewhat above 35 inches (89 cm). Live oaks suggest an oceanic equable climate. The precipitation probably exceeded evaporation, and the frost-free season was probably in excess of seven months.

During the Pliocene (early Blancan), vegetation was woodland or open forest and at times savanna-like in character and hardwood trees were probably confined to riparian environments. Steppe (up to 45%) and grassland (up to 50%) elements were more abundant, suggesting that annual rainfall was reduced (ca. 20 inches; 51 cm) and largely restricted to the winter season. Precipitation equalled or was slightly less than evaporation. The climate was less equable than before and was probably of cool-temperate character. The growing season may have been somewhat shorter based on the loss of equable climate indicators such as holly-leaved oaks. During this time the vegetation on the nearby hills and mountains may have been woodland and/or forest based on the fact that steppe today is restricted to areas below the forested belt in Idaho.

The early to middle Pleistocene vegetation records a modernization of the flora and vegetation. Tertiary hardwoods were eliminated and extensive steppe at times replaced woodland on the Snake River Plain. A major faunal change was probably related to a development of a continental climate; evaporation exceeded precipitation. Summer-dry conditions were more pronounced than in the Pliocene and, based on analogues in the present landscape, the precipitation was at times below 15 inches (38 cm) annually. Whether the increased continentality was the result of global climate change or was chiefly the result of regional uplift that occurred at the end of Glenns Ferry time (Malde and Powers 1962) can only be conjectured. Doubtless the development of a more continental climate was not due solely to the uplift of the high Cascades and the formation of the consequent rain shadow, because tectonic development had already occurred by ca. 10 my BP.

Volcanism through disturbance of forest vegetation may have been a factor in promoting the intermittent increase of xeric communities: succession on volcanic landscapes and soils, such as those described by Taggart and Cross (1980), Cross and Taggart (in press), and Taggart, Cross and Satchell (1982) for the mid-Miocene Succor Creek flora in eastern Oregon, may account for some of the apparent oscillations between forest and xeric elements during the Neogene of the Snake River Plain. However, the progressive impoverishment of the flora attests to the hypothesis that a fundamental change from a summer-wet to a summer-dry climate was the main forcing function. The "temporary" increases of xeric elements do not seem to follow volcanic episodes at the Horse Quarry section, but in some cases precede them. We are therefore inclined to interpret the main trends of the pollen data as resulting from long and short-term changes in climate. Long-term shifts may also be the result of regional uplift at the end of Glenns Ferry time (Malde and Powers 1962).

RECONSTRUCTION OF SNAKE RIVER PLAIN ENVIRONMENTS IN GLENNS FERRY TIME

The pollen data presented give evidence that terrestrial habitats of the Snake River Plain during the late Pliocene (early Blancan) time were characterized by pine woodland or open forest vegetation with diverse herbs and steppe elements. Pinyon-juniper woodland was probably an important conifer community on high ground and dry sites above the floodplain and along the valley border. Riparian trees and shrubs included willow, alder, birch and elm. Judging from the riparian lithologies mapped by Malde and Powers (1962), floodplain vegetation with freshwater lakes and sloughs probably extended over the entire Snake River Plain. The shoreline habitats included abundant rooted aquatics, particularly sedges, and the standing water contained rich blooms of colonial algae and resting cysts of green algae just as the modern reservoirs on the Snake River do today. A host of other aquatic organisms, including fish and invertebrates, attest to the fact that a diverse aquatic ecosystem existed at the Hagerman lakes at this time. Clearly this vast swampy land and slough area was an ideal habitat for waterfowl and their predators, and the nine species of mustelids may have depended on the eggs of water birds for their food source.

Diverse upland game, especially megafauna and grazing ungulates, probably used the floodplain habitat and depended on it as a water supply. Carnivores undoubtedly found this terrain hospitable, because the ratio of carnivore to ungulate taxa is surprisingly large compared to that in the present Artemisian faunal province (Bjork 1970).

ORIGIN OF STEPPE IN THE PACIFIC NORTHWEST

An important question is: what was the upland vegetation on the hills and mountains above the Snake River Plain during Glenns Ferry time? The dominance of grazing over browsing ungulates during the late Tertiary in western United States has led paleontologists to assume that extensive grasslands or steppes had developed in parts of western U.S. by the Mio-Pliocene (i.e. Hibbard 1960; Axelrod 1979:21) or Pliocene (Shotwell 1961:213). As Shotwell correctly pointed out, however, direct evidence of extensive Tertiary grassland vegetation is so far lacking in the U.S. paleobotanical literature, though an ample record of xeric forest and woodland is documented (Axelrod 1979). Pollen of steppe elements are intermittently abundant at Jackson, Wyoming, during the late Miocene (Barnosky 1984).

The finding of over 100 horse skeletons in the Horse Quarry at Hagerman (Gazin 1936) led to the supposition that vast grasslands or grassy steppe (as portrayed in the mural at the Smithsonian Institution reconstructing the landscape of the Hagerman) covered the uplands around the

Snake River in the late Pliocene. However, in order to have upland steppe above woodland requires that the woodland be riparian in nature—that is, dependent on ground water as a moisture source. But the composition of the Snake River Plain vegetation was only in small part arboreal phreatophytes, and it is unlikely that pine was a phreatophyte. Pollen of mountain conifers other than pine was represented in trace amounts on the Snake River Plain then as it is now; this suggests that conifer forest grew above the Snake River Plain as it does today.

So far, there is no paleobotanical evidence of extensive steppe anywhere in the Pacific Northwest during the Tertiary. This report presents some of the first paleobotanical evidence to indicate that steppe and grassland associations were a significant part of the vegetation regionally. Our evidence suggests a mosaic of steppe and grassy steppe with woodland or forest in the late Pliocene and regional upland steppe in the early to middle Pleistocene.

We surmise that the portrayal of the Hagerman landscape in the mural at the Smithsonian Institution is incorrect in depicting the hills surrounding the Snake River Plain as covered by grassland or steppe. That scenario with steppe would be more appropriate for the early or middle Pleistocene during Bruneau time. Since it is more likely that the hills and floodplains surrounding the Snake River were a woodland or open forest environment where a rich steppic and grassy understory occupied the openings, it was probably this vegetation that supported the diverse fauna of browsing and grazing animals in the Pliocene. Conifer forest communities probably characterized the high elevations.

The association of abundant grazing horses (*Plesippus*) with the first Neogene records of abundant grass pollen at the Horse Quarry in Idaho may be significant. The several types of grazing horses noted by Malde and Powers (1962) in this sequence may have had different habitats as suggested by Shotwell (1961). *Neohipparion* of the Idavada Volcanics and *Hipparion* of the Poison Creek and Chalk Hills Formations may have been associated with a vegetation of forest and perhaps some woodland, while *Plesippus* (Hagerman Lake beds, upper Glenns Ferry Formation) may have been associated with more open vegetation such as woodland, local grassland and grassy steppe. Shotwell predicted this idea for the Northern Great Basin, but supposed that grasslands were extensive by Blancan time.

SUMMARY

This report documents the development of a moderately continental climate during the mid and late Pliocene (early Blancan) in Idaho. Summer-dry conditions helped eliminate humid forest elements and promoted a more drought-resistant vegetation during the Pliocene. The regional vegetation of the Snake River Plain and its surrounding hills was probably woodland with steppic elements as understory. Contrary to earlier conjecture, extensive steppe and a strongly continental climate did not develop in Idaho until some time in the early or middle Pleistocene. Grasslands were either not extensive or only developed during brief phases of the Pliocene in the Snake River Plain.

ACKNOWLEDGMENTS

We wish to thank Harold E. Malde for his excellent help over several seasons in the field and in discussion of the material. We are indebted to Cathy Barnosky for her comments on the manuscript, and to Ellen Daniels and Rudy Nickmann for lab preparation of materials and to Rudy who counted the modern pollen rain samples and the Horse Quarry samples. David Clemens assisted with the illustrations, and Jeanette Pederson did a remarkable job in typing the manuscript. Field work was supported by the U.S. Geological Survey.

LITERATURE CITED

Armstrong, R. L., W. P. Leeman, and H. E. Malde. 1975. K-Ar dating, Quaternary and Neogene volcanic rocks of the Snake River Plain, Idaho. Amer. J. Sci. 275:225-251.

Axelrod, D. I. 1964. The Miocene Trapper Creek Flora of Southern Idaho. Univ. Calif. Pub. Geol. Sci. 51:148 pp.

Axelrod, D. I. 1979. Desert vegetation, its age and origin. Pages 1-71 in J. R. Goodin and D. K. Northington, eds. Arid Land Plant Resources. Proc. Int'l. Arid Lands Conf. on Plant Resources, Texas Tech. University, Lubbock, Texas.

Baker, R. G. 1976. Late Quaternary vegetational history of the Yellowstone Lake area, Wyoming. U.S. Geol. Surv. Prof. Pap. 729E.

Barnosky, C. W. 1984. Late Miocene vegetational and climatic variations inferred from a pollen record in Northwest Wyoming. Science 223(4631):49-51.

Berggren, W., and J. Van Couvering. 1974. Late Neogene time scale. Paleogeogr. Paleoclimatol. Paleoecol. 16(2):216 pp.

Bjork, P. R. 1970. The Carnivora of the Hagerman local fauna (late Pliocene) of Southwestern Idaho. Trans. Amer. Philos. Soc., n.s. 60(7):3-54.

Bradbury, J. P., and W. Krebs. 1982. Neogene and Quaternary lacustrine diatoms of the Western Snake River Basin, Idaho and Oregon, USA. Acta Geol. Acad., Sci. Hungar. 25(1-2):97-122.

Bright, R. C. 1966. Pollen and seed stratigraphy of Swan Lake, southeastern Idaho. Tebiwa 9:1-47.

Bright, R. C., and O. K. Davis. 1982. Quaternary Paleoecology of Idaho Nat. Eng. Lab., Snake River Plain, Idaho. Amer. Midl. Nat. 108(1):21-33.

Brodkorb, P. 1958. Fossil birds of Idaho. Wilson Bull. 20(3):237-242.

Brown, R. W. 1940. A bracket fungus from the later Tertiary of southwestern Idaho. Wash. Acad. Sci. J. 30(10):422-424.

Cross, A. T., and R. E. Taggart. In press. Causes of short-term sequential changes in fossil plant assemblages: Some considerations based on a Miocene flora of the northwest United States. Proc. Sympos. Plant Geographical Results of Changing Cenozoic Barriers, XIII Int'l. Bot. Congress, Sydney, Austr., 1981. Ann. Missouri Bot. Garden.

Daubenmire, R. 1970. Steppe vegetation of Washington. Wash. Agric. Expt. Stat. Tech. Bull. 62:131 pp.

Evernden, J. F., and G. T. James. 1964. Potassium-Argon dates and the Tertiary floras of North America. Amer. J. Sci. 262:945-974.

Evernden, J. F., D. E. Savage, G. H. Curtis, and G. T. James. 1964. Potassium-argon dates and the Cenozoic Mammalian Chronology of North America. Amer. J. Sci. 262:145-198.

Fine, Mary D. 1963. An abnormal P_2 in *Canis* cf. *C. Latrans* from the Hagerman fauna of Idaho. J. Mammal. 43(3):483-485.

Ford, N. L., and B. G. Murray, Jr. 1967. Fossil owls from the Hagerman local fauna (upper Pliocene) of Idaho. The Auk 84(1):115-117.

Gazin, C. L. 1936. A study of the horse remains from the upper Pliocene of Idaho. Proc. U.S. Nat'l. Mus. 83(2986):281-320.

Hibbard, C. W. 1960. The President's Address: An interpretation of Pliocene and Pleistocene climates in North America. Michigan Academy, Report for 1959-60. 30 pp.

Hibbard, C. W., and Zakrewski, R. W. 1967. Phyletic trends in the late Cenozoic microtine *Ophiomys* gen. nov. from Idaho. Mus. Paleontol., Univ. Mich. 21(12):255-271.

Kapp, R. O. 1965. Illinoian and Sangamon vegetation in southwestern Kansas and adjacent Oklahoma. Contr. Mus. Paleont., Univ. Mich. 19(14):243-252.

Kühler, A. W. 1970. Potential natural vegetation. *In* National Atlas of the United States of America, U. S. Geological Survey.

Leopold, E. B. 1967. Summary of palynological data from Searles Lake. Pages 51-66 in G. I. Smith, ed. Pleistocene Geology and Palynology, Searles Lake, California. Guidebook for Friends of the Pleistocene, Pacific Coast Section.

Mack, R. N. 1981. Invasion of *Bromus tectorum* L. into Western North America: An ecological chronicle. Agro-Ecosystems 7:145-165.

Mack, R. N., and V. M. Bryant. 1974. Modern pollen spectra from the Columbia Basin, Washington. Northwest Sci. 48(3):183-194.

Mack, R. N., V. M. Bryant and W. Bell. 1978. Modern forest pollen spectra from eastern Washington, and northern Idaho. Bot. Gaz. 139(2):249-255.

Mack, R. N., and J. N. Thompson. 1982. Evolution in steppe with few large, hooved mammals. Amer. Nat. 119(6):757-773.

Malde, H. E. 1972. Stratigraphy of the Glenns Ferry Formation from Hammet to Hagerman, Idaho. U. S. Geol. Surv. Bull. 1131-D:19 pp.

Malde, H. E., and H. A. Powers. 1962. Upper Cenozoic stratigraphy of Western Snake River Plain, Idaho. Geol. Soc. Amer. Bull. 73:1197-1220.

Malde, H. E., H. E. Powers, and C. E. Marshall. 1963. Reconnaissance Geologic Map of West-Central Snake River Plain, Idaho. U. S. Geol. Surv., Misc. Geol. Investigations. Map I-373.

Malde, H. E., and H. A. Powers. 1972. Geologic map of the Glenns Ferry-Hagerman area, west-central Snake River Plain, Idaho. U. S. Geol. Surv. Map I-696.

Mapel, J., and J. Hail, Jr. 1959. Tertiary geology of the Goose Creek district Cassia County, Idaho, Box Elder County, Utah, and Elko County, Nevada. Geol. Surv. Bull. 1055-H:217-254.

McAndrews, J. H., and H. E. Wright, Jr. 1969. Modern pollen rain across the Wyoming Basins and the northern Great Plains (U.S.A.). Rev. Paleobot. Palynol. 9:17-43.

Miller, R. R., and G. R. Smith. 1967. New fossil fishes from Plio-Pleistocene Lake, Idaho. Occ. Pap. Mus. Zool. Univ. Michigan 654:1-24.

Neville, C., N. D. Opdyke, E. H. Lindsay, and N. M. Johnson. 1979. Magnetic stratigraphy of Pliocene deposits of the Glenns Ferry Formation, Idaho, and its implications for North American mammalian biostratigraphy. Amer. J. Sci. 279:503-526.

Rember, W. C., and C. S. Smiley. 1979. Dominant plants and seral analyses of the Clarkia Fossil beds. Pacific Division, AAAS Sixtieth Annual Mtg., Abstracts:23.

Savage, D. E., and G. H. Curtis. 1970. The Villafranchian Stage-Age and its radiometric dating. Geol. Soc. Amer. Spec. Pap. 124:207-231.

Shotwell, J. A. 1961. Late Tertiary Biogeography of horses in the Northern Great Basin. J. Paleont. 35(1):203-217.

Skinner, M. F., et al. 1972. Early Pleistocene pre-glacial and glacial rocks and faunas of north-central Nebraska. Bull. Amer. Mus. Nat. Hist. 148(1):77-148.

Smiley, C. J. 1963. The Ellensburg Flora of Washington. Univ. Calif. Pub. Geol. Sci. 35(3):159-276.

Smiley, C. J., J. Gray, and L. M. Huggins. 1975. Preservation of Miocene fossils in unoxidized lake sediments, Clarkia, Idaho. J. Paleont. 49(5):833-844.

Smiley, C. J., and L. M. Huggins. 1981. *Pseudofagus Idahoensis*, n. gen. et sp. (Fagaceae) from the Miocene Clarkia Flora of Idaho. Amer. J. Bot. 68(6):741-761.

Smith, G. R. 1975. Fishes of the Pliocene Glenns Ferry Formation, southwest Idaho. Mus. Paleont., Univ. Mich., Pap. Paleont. 14:68 pp.

Swirydczuk, K., G. P. Larson, and G. R. Smith. 1979. Tephra stratigraphy of the Glenns Ferry and Chalk Hills Formations, Western Snake River Plain. Idaho Bur. Mines Geol.

Taggart, R. E., A. T. Cross, and L. Satchell. 1982. Effects of periodic volcanism on Miocene vegetation distribution in eastern Oregon and western Idaho. Proc. 3rd N. Amer. Paleont. Convention:1-6.

Taggart, R. E., and A. T. Cross. 1980. Vegetation change in the Miocene Succor Creek flora of Oregon and Idaho: A case study in paleosuccession. Pages 185-210 *in* D. L. Dilcher and T. N. Taylor, eds. Biostratigraphy of fossil plants—successional and paleoecological analyses. Down, Hutchinson & Ross, Stroudburg, Pa.

Taylor, D. W. 1960. Distribution of the freshwater clam *Pisidium ultramontanum*: A zoogeographic inquiry. Amer. J. Sci., Bradley Volume 258-A:324-334.

Taylor, D.W. 1966. Summary of North American Blancan nonmarine mollusks. Malacologia 4 (1):1-172.

U. S. Dep. of Agriculture. 1941. Climate and Man, Yearbook of Agriculture. Government Printing Office. 1248 pp.

Weaver, J. E. 1917. A study of the vegetation of southeastern Washington and adjacent Idaho. Wilson Bull. 20(3):237-242.

Zakrzewski, R. W. 1969. Rodents from the Hagerman local fauna, upper Pliocene of Idaho. Contr. Mus. Paleont. Univ. Mich. 23:1-36.

APPENDIX 1. LIST OF FOSSIL POLLEN LOCALITIES FOR NEOGENE AND PLEISTOCENE POLLEN SAMPLES, WESTERN SNAKE RIVER PLAIN, IDAHO

Formation	USGS Paleobot. locality	Site Description
Bruneau	D1694	Measured section 55, 2400 ft W, 450 ft N of SE corner, Sec. 1, T. 6, R. 8 E., Sand Point at base of Hammett Basalt, alt. 2848 ft (868 m). Glenns Ferry quad., Idaho. Malde coll. 60M372.
Bruneau and Glenns Ferry	D1695-2, 4	Measured section 57, 1300 ft W, 1500 ft N of SE corner, Sec. 6, T. 8 S, R. 10 E, Dove Springs, alt. 2946 ft (ca. 898 m). Glenns Ferry quad., Idaho. Malde coll. 60M383.
Bruneau	D1698	SW 1/4 NW 1/4, Sec. 16, T. 7 S, R. 13 E, "Yahoo Soil," at Hagerman Cliffs, alt. 3180 ft (969 m). Hagerman quad., Idaho. Malde coll.
Bruneau	D1120	Near center E side NE 1/4 Sec. 17, T. 7 S, R. 13 E, soil at base of "Dolman" Fm. and top of "Yahoo" Fm. 3180 ft. elev. (969.5 m). Bed 4, section 6. Hagerman quad., Idaho. Malde coll. 537.
Banbury Basalt	D1696	700 ft W, 1700 ft W of NE corner, Sec. 7, T. 8 S, R. 14 E, from 510 ft depth (2380 ft alt.) (725 m) in Thousand Springs well. Thousand Springs quad., Idaho. Malde coll. 55P132.
Salt Lake	D1208-1 to 4	Sec. 22, T. 16 S, R. 21 E. Barrett zone lignite from mine along Coal Banks Creek, Goose Creek District, Cassia Co., Idaho. Mapel coll. MM-212-3, 214-4 ca 850 ft above base of Salt Lake Fm.
Poison Creek	D1699	NW 1/4 SE Sec. 31, T. 3 S, R. 1 W, East bank Sinker Creek, 6 ft above stream, 7 mi. SSE of Murphy, Idaho. Malde coll. 59P144.
Glenns Ferry	D5303A-D	N 1/2 Sec. 16, R. 13 E., T. 7 S. Twin Falls Co. cinder ledge ca 40 ft below basalt directly up slope and W of loc. 101 ca. 100 ft (30.5 m) below horse quarry. Fossil Gulch section. Hagerman quad., Idaho. Leopold and Wright coll. EL-74-102A-D.
Glenns Ferry	D5299A-B	Section 10, R. 13 E, T. 7 S. Twin Falls Co. Rd. cut hairpin turn N exposure. Hagerman quad., Idaho. Leopold and Wright coll. EL-74-99.
Glenns Ferry	D5302A-Q	N 1/2 Sec. 16, R. 13 E, T. 7 N, Twin Falls Co., base of Fossil Gulch. Hagerman quad., Idaho. Leopold and Wright coll. EL-74-101.

APPENDIX 1. Continued

Glenns Ferry	D1671	Measured section 52, carbonaceous shale. 1300 ft W, 250 ft N of SE corner, Sec. 32, T 5 S, R. 9 E. The Narrows, 100 ft above Snake River, alt. 2570 ft (78.4 m). Glenns Ferry quad., Idaho. Malde and Powers coll. 60M303.
Glenns Ferry	D1710-1, 2	Measured section 3, 1000 ft E, 2400 ft N of SW corner, sec. 22, T. 7 S, R. 11 E, Twenty Mile Butte, alt. 3110 and 3111 ft (ca 948 m). Pasadena Valley quad., Idaho. Malde and Powers coll. 60M3, 60M4.
Glenns Ferry	D1715	Sec. 16, T. 7 S, R. 13E, unit 31, carbonaceous paper shale; between Lower and Upper Salmon Falls, alt. 3115 ft (950 m), Idaho. Malde and Powers coll. 55P289.
Glenns Ferry	D1712	Measured section 18, 2300 ft W, 1500 ft N of SE corner, Sec. 16, T. 6 S, R. 11 E, Deer Gulch, 57 ft above Snake River, alt. 2599 ft (792.5 m). Pasadena Valley quad., Idaho. Powers and Malde coll. 60M54.
Glenns Ferry	D1711	Measured section 17, 1800 ft. W, 1100 ft N of SE corner, Sec. 16, T. 6 S, R. 11 E, Deer Gulch, 38 ft above Snake River, alt. 2580 ft (787 m). Pasadena Valley quad., Idaho. Powers and Malde coll. 60M50.
Glenns Ferry	D1713-1	Measured section 36, 900 ft E, 700 ft N of SW corner, Sec. 2, T. 5 S, R. 10 E, alt. 2616 ft (797.5 m). Kings Hill quad., Idaho. Powers and Malde coll. 60M156.
Glenns Ferry	D1692-4, 5	Measured section 35, 2100 ft W, 1600 ft N of SE corner, Sec. 36, T. 4 S, R. 10 E, Walker Ditch, alt. 2865 ft (873 m). Kings Hill quad., Idaho. Powers and Malde coll. 60M147, 60M148.
Glenns Ferry	D1691	Measured section 26, 2250 ft W, 1900 ft S of NE corner, Sec. 31, Buckbush Draw, 3 ft below Chevkinite Ash, 82 ft below cinder bed, alt. 3002 ft (915 m). Kings Hill quad., Idaho. Powers and Malde coll. 60M110.
Glenns Ferry	1690-1	Measured section 25, 2500 ft E, 1050 ft S of NW corner, Sec. 11, T. 5 S, R. 11 E, 68 ft above Clover Creek, alt. 2778 ft (847 m). Kings Hill quad., Idaho. Malde and Powers coll. 60M99.

APPENDIX 2. SELECTED POLLEN COUNTS FROM COMPOSITE POLLEN DIAGRAM (FIG. 6)

Glenns Ferry Fm.:	Transitional flood plain & valley border	Valley border facies North			Lacustrine facies				Flood Plain Upstream			Valley border south	Flood Plain Downstream
USGS paleobot. loc.:	D1690-1	D1691	D1692-4	D1692-5	D1713-1	D1711	D1712	D1715	D1710-1	D1710-2		D1695-4	D1671
Abies cf. grandis		2.3		6.9	1.1	0.4	0.7	1.7	1.9	4.9		0.4	1.5
Abies cf. lasiocarpa	0.5	0.8		0.3				0.4	1.2	1.5			0.5
Juniperus type		0.8	2.6	0.5	6.2	0.4	17.0	1.3	0.3			2.1	2.9
Ephedra cf. nevadensis			0.4	0.5									1.0
Picea sp.		0.8		0.8		0.6	0.7	0.4	0.3				1.0
P. cf. engelmanii		1.5			0.7				0.3	0.4			
Pinus undet. & total Strobi group	86.3	52.0	26.9	77.7	17.4	1.6	32.6 1.5	9.4	90.5	75.6		18.2	46.5
Pseudotsuga menziesii		0.8					0.7	0.4	2.2	0.4			
Tsuga cf. heterophylla	0.5			0.5	0.4		0.7	0.4	0.3	0.4		0.4	1.0
Ulmus-Zelkova					0.4			1.3				0.8?	
Juglans						0.4							2.0
Pterocarya						0.4							
Carya						0.4							
Sequoia?						0.7		0.4?					
Carpinus?													0.5
Platanus?													1.5
Quercus	0.5		0.4	0.5	0.7	0.7							0.5
Acer					0.4							0.4	
Alnus				0.3	0.4								1.0
Betula			0.7					0.4					0.5
Populus type			0.4	0.5				2.1					1.0
Celtis						3.3	0.7						1.5
Subtotal (trees)	87.8	59.0	31.4	88.5	27.7	8.9	54.6	18.2	97.0	83.2		22.3	62.9
Salix													1.0
Rosaceae?				0.3	1.1								
Chenopodiaceae			3.0		3.3	20.3	0.7	4.7				1.2	0.5
Sarcobatus			11.2	0.5		2.2		0.9				2.0	1.5
Compositae-low spine		3.1	0.4		0.7	7.8	3.0	3.0	0.3	0.4		0.5	2.5
high spine	0.5		4.5		0.4	1.1							
fenestrate	0.5		6.0										
Artemisia	1.1												
	5.2	27.1	23.5	5.8	8.4	7.4	10.4	6.4	1.9	0.7		8.4	22.0
Myrtaceae						0.4		0.4					
Malvaceae						0.4							
Subtotal (shrubs)	7.3	30.2	45.9	6.6	13.9	40.3	14.1	15.4	2.2	1.1		12.1	27.5

APPENDIX 2. (CONTINUED)

Glenns Ferry Fm.:	Transitional flood plain & valley border	Valley border facies North			Lacustrine facies				Flood Plain Upstream		Valley border south	Flood Plain Downstream
USGS paleobot. loc.:	D1690-1	D1691	D1692-4	D1692-5	D1713-1	D1711	D1712	D1715	D1710-1	D1710-2	D1695-4	D1671
cf. *Nerium*												
Aizoaceae type												0.5
Anemone				0.5	1.1	0.7		0.4				0.5
Cruciferae												
Yucca?										0.4		
Polemoniaceae				0.8	0.7	0.7						
Graminae	2.6	0.8	6.0	1.0	8.4	3.7		2.1		13.7	61.7	1.0
						27.3	19.2	55.5				
Aquatic plants												
Potamogeton					0.7		0.7	1.7		0.4		1.0
Sparganium		0.8	0.7	0.8	2.5	1.5	6.0	0.4			0.8	4.9
Lemna					0.4	5.2		0.4		0.4		
Nuphar			5.2		1.1	0.4						
Nymphaea					0.7							
Scirpus					3.6			0.9				0.5
Cyperaceae					0.4							
Myriophyllum								0.4				0.5
Typha latifolia				0.3								
Subtotal (aquatics)		0.8	5.9	1.1	9.4	7.1	6.7	3.4		0.8	0.8	6.9
Eriogonum	0.5											
Dicots, undet.	1.0	2.3	2.6	1.1	14.5	2.9	1.5	8.4				1.5
Monocots, undet.	0.5	7.0	4.5	0.3	24.0	7.7	4.5			1.5	2.9	
Lycopodium			1.1									
Trilete spores		0.8	0.4		0.4	0.4	0.7					
Monolete spores						0.7			0.6			
	99.7	100.8	100.5	99.9	99.4	99.7	99.8	99.6	99.8	100.7	99.4	100.8
Total tally:												
pollen, spores	(190)	(129)	(268)	(378)	(275)	(270)	(135)	(234)	(313)	(263)	(238)	(205)
algae, cysts	(1032)*	(101)	(212)*	(386)*	(8)	(137)*	(66)	(103)*	(71)	(334)*	(155)*	(44)
Grand total of tallies	1222	230	480	764	283	407	201	337	384	597	393	249
% pollen & spores in total tally	18.7	56.0	55.7	49.5	97.2	66.0	67.0	66.5	81.0	44.0	60.5	82.0

* extrapolated from block counts

APPENDIX 3. POLLEN TALLIES FROM FOSSIL GULCH SECTION, GLENNS FERRY FORMATION

Sample number:	D5299 A	D5302 B	L	M	N	O	P	Q	D5303 D	B	A
Pinus	7.6	56.9	66.8	68.8	82.1	79.2	57.4	79.5	17.5	32.8	50.6
Abies	0.8	9.5	2.4	4.2	1.4	3.8	2.6	8.8	1.3	1.5	1.3
Picea		3.0	0.5	1.4	0.2	0.6	0.4	1.2		0.5	0.6
Undet. conifer				1.4		0.6	0.9	1.5			
Tsuga undet.		0.7						0.3			
T. mertensiana				0.7							0.6
Pseudotsuga											0.6
Juniperus				0.7	0.9		0.4	0.3	0.4	1.0	
Juglans					0.2					0.5	
Ulmus				0.7							
(total tree pollen)	8.4	70.1	69.7	77.9	84.8	84.2	61.7	91.6	19.2	36.3	53.7
Alnus		1.0									
Betula			1.0	0.4							
Salix	0.3										
Artemisia	7.6	13.8	10.6	4.9	3.2	2.6	14.4	7.0	25.4	28.9	29.8
Other Compositae	2.0	0.3	2.4	2.1	0.9	1.9	4.8		6.1	11.3	5.8
Ambrosia			4.8	0.7	0.9	0.3	0.4			1.0	0.6
Liguliflorae						0.3					
Ephedra cf. nevadensis	0.3		2.4	1.0	0.9	1.0	0.4				
E. cf. torreyana				0.4							
Rosaceae				0.4							0.3
Chenopodiineae		0.7	1.4	2.4	0.2	1.3	2.6	0.3			
Sarcobatus		6.9	1.0	1.0	0.2	0.6	3.0		0.4	1.5	1.6
(total shrub pollen)	10.2	22.7	23.6	13.3	6.3	8.0	25.6	7.3	31.9	42.7	38.1
Cruciferae											0.6
Gramineae	74.8	4.3	0.5	2.1	3.9	3.8	5.2	0.6	46.5	7.4	3.9
Eriogonum								0.3		5.4	
Umbelliferae			0.5								
(total herb pollen)	74.8	4.3	1.0	2.1	3.9	3.8	5.2	0.9	46.5	12.8	4.5
Unknown	6.8	3.0	5.8	6.9	4.6	4.2	7.4	0.3	2.2	8.3	3.5
TOTAL	100.2	100.1	100.1	100.2	99.6	100.2	99.9	100.1	99.8	100.1	99.8
TOTAL POLLEN TALLY	397	304	208	288	431	313	230	331	228	204	312
% TOTAL AQUATIC POLLEN	1.3	0.3		11.0	7.6	6.1	22.3		13.1	44.4	19.8

SPICULAR REMAINS OF FRESHWATER SPONGES FROM A MIOCENE LACUSTRINE DEPOSIT IN NORTHERN IDAHO

JOHN L. WILLIAMS
Cities Service Oil & Gas Corp., 1600 Broadway (Suite 900), Denver, CO 80202

Sponge spicules comprise a large part of the siliceous microfossil assemblage present at the Miocene Oviatt Creek (P-35) locality. Over a half-dozen spongillid taxa are represented from this northern Idaho site. The genera represented here include *Anheteromeyenia, Corvospongilla, Eunapius, Radiospongilla, Spongilla,* and *Trochospongilla.* The paleoecology of this lacustrine environment appears to have had a water temperature of approximately 26-30°C, moderate alkalinity with fairly high mineral content and color, low to moderate bicarbonate content, and a pH range of about 6.7-7.1.

Most freshwater sponges belong to the family Spongillidae and have a known geologic range of Jurassic to Recent (Moore 1955). This range is largely defined by the siliceous spicules that help support the flimsy tissues of the sponge. Even though rare occurrences of fully intact spicule arrangements are known in the fossil record (Racek and Harrison 1975), dispersed spicular remains comprise the major evidence for the presence of spongillids.

As Racek and Harrison (1975) pointed out, these spicules are abundant in lake sediments of Tertiary age; most of them are comparable with modern forms. This is true of the Miocene lake deposits in northern Idaho. Of these deposits, the Oviatt Creek site (P-35) was chosen for this spicular study.

METHODS

Several small samples were obtained from between 75 and 210 cm below the top of the exposure at the Oviatt Creek (P-35) locality (NE¼ Sec. 12, T. 39 N., R 1 E.). These were combined into one composite sample and chemically treated using the "dry sample" technique described by Schrader (1973). Slides were made of the siliceous residue using Permount as the mounting medium.

The preparation of this material for scanning electron microscopy was very similar to that employed for light microscopy. Following the chemical process, the residue was diluted with distilled water and a few drops of this solution were then placed on a cover glass and allowed to dry. When the preparation was completely dry, double-sided tape was used to mount the cover glass to an aluminum SEM stub. The cover glass was then coated with a thin layer of gold-palladium. An ISI-60 scanning electron microscope was used to examine this material, using an accelerating voltage of 30kV.

DISCUSSION

Sponge spicules are very significant tools for stratigraphic and paleoenvironmental studies of lentic sediments (Racek 1966). These siliceous structures are highly durable, further adding to their importance. An abundance of these skeletal components has been found at the Oviatt Creek locality. The spicules represented here are comparable with those of extant freshwater sponges. Racek (1966; pers. commun., 1979[1]) found a similar relationship between recent and subfossil spongillids from Australia and between modern and Cretaceous forms from South America.

There are three general types of spongillid sponge spicules; all are found at the Oviatt Creek

---

[1] Dr. A. A. Racek, c/o Post Office, WOOLI, NSW 2462, Australia.

Copyright ©1985, Pacific Division, AAAS.

site. These spicular types include the megascleres (skeletal spicules), microscleres (flesh spicules), and gemmoscleres (gemmule spicules). The microscleres (lacking in some species) and the gemmoscleres appear to be the most diagnostic of the three, while megascleres are less important, with respect to their taxonomic value.

According to Pennak (1978), variations in spicule form are common even within the same sponge or gemmule. Variations also occur in the shape of the axial canals of megascleres and are a common feature of the spicules from Oviatt Creek.

Figures 25 and 27 represent megascleres that have a narrow, uniform canal, which is typical of most megascleres. A magnified view of this regularity is shown in Fig. 22. Numerous spicules display irregular cavities in their axial canals (Figs. 23, 24, 28). These botryoid-looking irregularities have projections extending towards the outer surface of the spicule. Racek (1966) has noted similar axial canal abnormalities. He believes these irregular cavities may be caused by the bacterial reduction of gypsum. This reduction would yield $Ca(OH)_2$ which would dissolve the hydrated silica. This hypothesis may also be true for the irregular axial canals found in spicules from Oviatt Creek.

Figure 26 represents a less common type of spicular irregularity. Both the axial canal and the outer surface of the spicule possess bulbous enlargements, which suggest either an immature form (Potts 1887) or an abnormal environment (Harrison and Simpson 1976).

Sponge spicules are not only useful as taxonomic tools but also as paleoecological indicators. This is true only when the spicular components can be accurately identified with those of modern forms. With this, and knowledge of the ecological parameters of modern spongillid species, one is able to define accurately the paleoecological characteristics of freshwater habitats (Harrison et al. 1979).

Together, Harrison (1974), Jewell (1935, 1939), Moore (1953), Old (1932), Penney (1954), Poirrier (1969), and Wurtz (1950) noted various water conditions from among hundreds of freshwater sponge habitats. The average pH and water temperature were calculated for each species sampled. These averages were used in reconstructing the paleoecology of the Oviatt Creek locality. Of the spicules recovered at this site, four taxa have been found to possess skeletal components comparable with those of modern forms. These taxa include *Anheteromeyenia ryderi, Eunapius fragilis, Radiospongilla crateriformis,* and *Spongilla lacustris.* These species indicate a habitat with the following characteristics of pH and water temperature: *S. lacustris* (6.9, 27°C); *R. crateriformis* (6.9, 30°C); *E. fragilis* (7.1, 26°C); *A. ryderi* (6.7, 26°C).

According to Harrison (pers. commun. 1981[2]), there is also an indication that the Oviatt Creek locality contained water of moderate alkalinity with a fairly high mineral content and color and low to moderate bicarbonate content. He bases these conditions on the habitats of the modern species *R. crateriformis, E. fragilis,* and *S. lacustris. Anheteromeyenia ryderi*, however, favors more acidic water of lower bicarbonate content than do the other three associated species. Unlike *R. crateriformis, E. fragilis,* and *S. lacustris,* spicules of *A. ryderi* were recorded very infrequently from the Oviatt Creek locality. It seems likely that *A. ryderi* was present in low population density, living under less than optimal conditions (Harrison, pers. commun., 1981).

CONCLUSION

Even though sponge spicules commonly represent a significant part of the siliceous microfossils in lacustrine sediments, little attention has been given to their stratigraphic and paleoecological usefulness. A paleoecological interpretation of the Oviatt Creek locality during the Miocene is derived by applying ecological parameters of modern species of sponges with spicules comparable to the fossil forms. The interpretation is a lacustrine environment having water temperatures between 26 and 30°C, moderate alkalinity with fairly high mineral content and color, low to moderate bicarbonate content, and a pH range of approximately 6.7-7.1.

[2] Dr. F. W. Harrison, Department of Biology, Western Carolina University, Cullowhee, NC 28723.

The spicules at this site are morphologically diverse. The major differences are in the ornamentation, shape, and size of the spicules, as well as in the structure of their axial canals. These morphological characteristics are important in identification and comparison with modern forms.

SYSTEMATIC PALEONTOLOGY
(Refer to Penney and Racek [1968] for discussion of synonymy)

Class DEMOSPONGIA Sollas, 1875
Order HAPLOSCLERIDA Topsent, 1898
Family Spongillidae Gray, 1867

Genus *Anheteromeyenia* Schroder, 1927

Anheteromeyenia ryderi (Potts) Penney and Racek, 1968. (Figures 15 and 16)

These gemmoscleres are microbirotulates having cylindrical shafts bearing one to few spines. Terminally flat to slightly umbonate with equal sized rotules bearing numerous small dentitions at their margins. Typical modern forms range from approximately 20-46 μm in length.

Anheteromeyenia cf. *A. ryderi.* (Figure 17)

This gemmosclere is a microbirotulate having a very stout, spineless, cylindrical shaft (possibly ribbed). Slightly umbonate at the apices with rotules of equal size. These rotules have finely dentate margins, giving them a crenulate appearance.

This gemmosclere most closely resembles that of *A. ryderi*, but because of its unusual broadly "ribbed" shaft, this placement is tentative.

?*Anheteromeyenia ryderi.* (Figure 23)

An abnormal, fusiform amphioxea bearing numerous small spines except at the extremities. Similar to the forms described by Rioja (1953). Typical modern forms range from approximately 190-240 μm in length.

Genus *Corvospongilla* Annandale, 1911

Corvospongilla sp. (Figures 19 and 21)

These microscleres are straight to slightly curved microbirotulates with smooth shafts and 4-5 terminal recurved hooks.

?*Corvospongilla* sp. (Figure 20)

This microsclere is similar to Figs. 19 and 21, except that the terminal hooks are very small and stubby (this may be an ecomorphic variation or perhaps a species of *Corvomeyenia*). Typical modern forms (similar to Figs. 19-21) range from approximately 19-70 μm in length.

Genus *Eunapius* Gray, 1867

Eunapius fragilis (Leidy) Penney and Racek, 1968. (Figures 9 and 10)

Gemmoscleres are somewhat stout, slightly curved semi-amphistrongyles covered with numerous conspicuous spines. Modern forms range from approximately 75-140 μm in length.

Genus *Radiospongilla* Penney and Racek, 1968

Radiospongilla crateriformis (Potts) Penney and Racek, 1968. (Figures 12 and 13)

These gemmoscleres are slender, inconspicuously to strongly curved amphistrongyles with a variable number of conical spines on their shafts. Terminally they appear to have one to several rows of radiating, slightly recurved spines that are arranged in such a way that pseudorotules are produced. Typical modern forms range from approximately 50-75 μm in length.

Radiospongilla cf. *R. crateriformis*. (Figure 4)

This megasclere is a hastate, sharply-pointed amphioxea that is sparsely microspined except at its smooth tips. Typical modern forms range from approximately 210-300 μm in length.

?*Radiospongilla crateriformis*. (Figure 5)

Megasclere is a hastate, sharply-pointed amphioxea that is sparsely microspined except at the tips where the spines are even more sparse to absent.

?*Radiospongilla* sp. (Figure 14)

Short, very stout and inconspicuously curved gemmosclere with numerous irregular, conical spines seemingly aggregated near its unidentical ends. This may be an ecomorphic variation of a species of *Radiospongilla*. Typical modern forms range from approximately 47-110 μm in length.

?*Radiospongilla cantonensis* (Gee) Penney and Racek, 1968. (Figure 6)

This megasclere is a slender, fusiform and nearly straight amphioxea bearing numerous acute spines except at the very tips where they are sparse and less conspicuous to completely absent. Typical modern forms range from approximately 180-230 μm in length.

Genus *Spongilla* Lamarck, 1816

Spongilla lacustris (Linnaeus) Johnston, 1842. (Figures 7 and 8)

Gemmoscleres are semi-stout to stout amphioxea that are inconspicuously to strongly curved. Scleres are sparsely covered with stout, recurved spines. Typical modern forms range from approximately 50-150 μm in length.

Genus *Trochospongilla* Vejdovsky, 1883

?*Trochospongilla pennsylvanica* (Potts) Annandale, 1911. (Figure 11)

This megasclere is a rather slender, inconspicuously curved amphistrongyla-amphioxea covered with small, sharp spines. Typical modern forms range from approximately 140-210 μm in length.

SPICULE TYPE A
Figure 18

This gemmosclere is a straight amphistrongyla covered with stout, slightly recurved spines. These irregularly-shaped spines seem to be concentrated near the ends of their shaft.

This spicule type is very rare in the samples from Oviatt Creek. There does not appear to be a comparable form illustrated in the literature; perhaps it is an ecomorphic variation.

EXPLANATION OF FIGURES (Pages 353-354)

Figures 3, 7, 8, and 22 are scanning electron micrographs. All others are light micrographs using bright field illumination. The measurements correspond to the length of the specimens. (Me= Megasclere; Mi=Microsclere; Ge=Gemmosclere; b=birotulate; mb=microbirotulate.)

Fig. 1. ?*Spongilla* sp., Me, 203 μm.
Fig. 2. ?*Spongilla* sp., Me, 188 μm.
Fig. 3. ?*Spongilla* sp., Me, 263 μm.
Fig. 4. *Radiospongilla* cf.*R. crateriformis*, Me, 230 μm.
Fig. 5. ?*Radiospongilla crateriformis*, Me, 240 μm.
Fig. 6. ?*Radiospongilla cantonensis*, Me, 190 μm.
Fig. 7. *Spongilla lacustris*, Ge, 101 μm.
Fig. 8. *Spongilla lacustris*, Ge, 104 μm.
Fig. 9. *Eunapius fragilis*, Ge, 100 μm.
Fig. 10. *Eunapius fragilis*, Ge, 105 μm.
Fig. 11. ?*Trochospongilla pennsylvanica*, Me, 188 μm.
Fig. 12. *Radiospongilla crateriformis*, Ge, 81 μm.
Fig. 13. *Radiospongilla crateriformis*, Ge, 74 μm.
Fig. 14. ?*Radiospongilla* sp., Ge, 43 μm.
Fig. 15. *Anheteromeyenia ryderi*, Ge (b), 34 μm.
Fig. 16. *Anheteromeyenia ryderi*, Ge (b), 32 μm.
Fig. 17. *Anheteromeyenia* cf. *A. ryderi*, Ge (b), 24 μm.
Fig. 18. Spicule Type A, Ge, 40 μm.
Fig. 19. *Corvospongilla* sp., Mi (mb), 18 μm.
Fig. 20. ?*Corvospongilla* sp., Mi (mb), 20 μm.
Fig. 21. *Corvospongilla* sp., Mi (mb), 26 μm.
Fig. 22. Axial canal of a broken spicule, 5300X.
Fig. 23. ?*Anheteromeyenia ryderi*, Me, 194 μm.
Fig. 24. Part of a spicule, Me, 1200X.
Fig. 25. ?*Spongilla* sp., Me, 210 μm.
Fig. 26. Part of a spicule, Me, 64 μm.
Fig. 27. ?*Spongilla* sp., Me, 185 μm.
Fig. 28. Part of a spicule, Me, 700X.

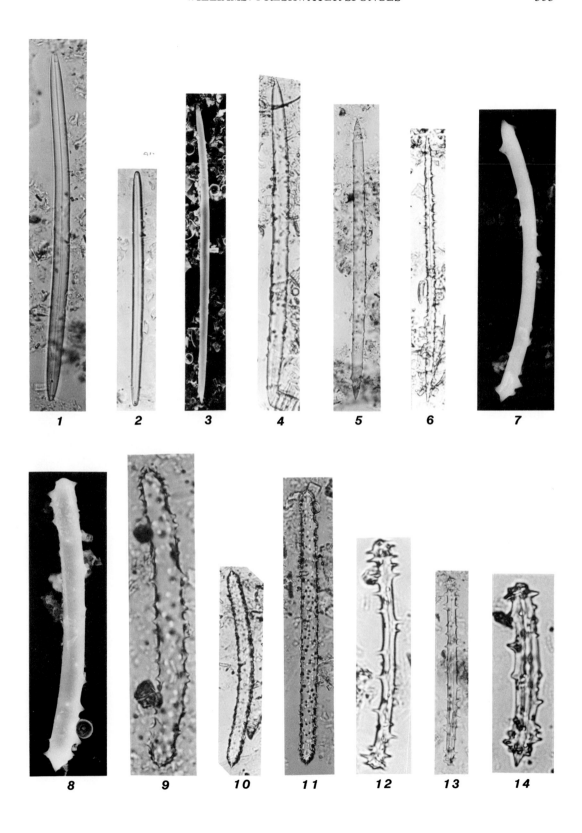

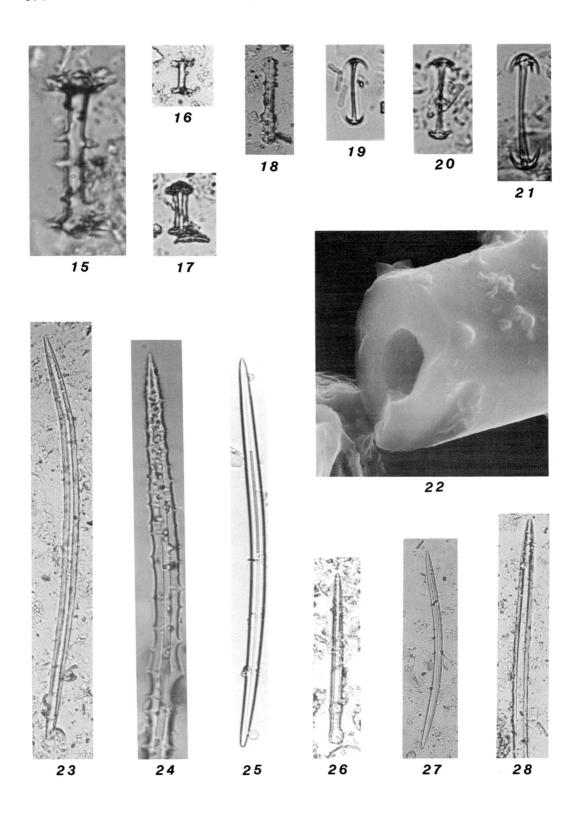

ACKNOWLEDGMENTS

Sincere thanks are due Dr. Frederick W. Harrison, Western Carolina University, and Dr. Frederick H. Wingate, Cities Service Oil & Gas Corporation, for reviewing the manuscript and making valuable suggestions.

LITERATURE CITED

Harrison, F. W. 1974. Sponges (Porifera: Spongillidae). Pages 29-66 *in* C. W. Hart and S. L. H. Fuller, eds. Pollution Ecology of Fresh-water Invertebrates. Academic Press, New York, N.Y.

Harrison, F. W., P. J. Gleason, and P. A. Stone. 1979. Paleolimnology of Lake Okeechobee, Florida: An analysis utilizing spicular components of freshwater sponges (Porifera: Spongillidae). Acad. Nat. Sci. Phila., Not. Nat. 454. 6 pp.

Harrison, F. W., and T. L. Simpson. 1976. Introduction: Principles and perspectives in sponge biology. Pages 1-46 *in* F. W. Harrison and R. R. Cowden, eds. Aspects of Sponge Biology. Academic Press, New York, N.Y.

Jewell, M. E. 1935. An ecological study of the fresh-water sponges of northern Wisconsin. Ecol. Monogr. 5:461-504.

Jewell, M. E. 1939. An ecological study of the fresh-water sponges of Wisconsin II. The influence of calcium. Ecology 2:11-28.

Moore, R. C., ed. 1955. Archaeocyatha, Porifera. Treatise on Invertebrate Paleontology. Vol. E. Geological Society of America and University of Kansas Press, Lawrence, Kan. 122 pp.

Moore, W. G. 1953. Louisiana fresh-water sponges, with ecological observations on certain sponges of the New Orleans area. Amer. Microsc. Soc. Trans. 32:24-32.

Old, M. C. 1932. Environmental selection of the fresh-water sponges (Spongillidae) of Michigan. Amer. Microsc. Soc. Trans. 51:129-137.

Pennak, R. W. 1978. Fresh-water invertebrates of the United States. John Wiley & Sons, New York, N.Y. 803 pp.

Penney, J. T. 1954. Ecological observations of the fresh-water sponges of the Savannah River Project area. Univ. So. Carolina Pub. Ser. 3. 1:156-192.

Penney, J. T., and A. A. Racek. 1968. Comprehensive revision of a world-wide collection of fresh-water sponges (Porifera: Spongillidae). U. S. Nat'l. Mus. Bull. 272. 184 pp.

Poirrier, M. A. 1969. Louisiana fresh-water sponges: Taxonomy, ecology, and distribution. Ph.D. Thesis. Louisiana State University. Univ. Microfilms Inc., Ann Arbor, Mich. No. 70-9083. 173 pp.

Potts, E. 1887. Contributions toward a synopsis of the American forms of fresh-water sponges. Acad. Nat. Sci. Phila. Proc. 39:158-279.

Racek, A. A. 1966. Spicular remains of freshwater sponges. Conn. Acad. Arts Sci. Mem. 17: 78-83.

Racek, A. A., and F. W. Harrison. 1975. The systematic and phylogenetic position of *Palaeospongilla chubutensis* (Porifera: Spongillidae). Linn. Soc. New South Wales Proc. 99:157-165.

Rioja, E. 1953. Estudios hidrobiologicos XI. Contribution al estudio de las esponjas de agua dulce de Mexico. Inst. Biol. Mex. Anal. 24(2):425-433.

Schrader, H. J. 1973. Proposal for a standardized method of cleaning diatom-bearing deep-sea and land-exposed marine sediments. *In* R. Simonsen, ed. Second Symposium on Recent and Fossil Marine Diatoms. Beihefte zur Nova Hedwigia 45:403-409.

Wurtz, C. B. 1950. Fresh-water sponges of Pennsylvania and adjacent states. Acad. Nat. Sci. Phila., Not. Nat. 228. 10 pp.

RECENT CLIMATIC VARIATIONS, THEIR CAUSES AND NEOGENE PERSPECTIVES

MAYNARD M. MILLER

College of Mines & Earth Resources, University of Idaho, Moscow, ID 83843
Foundation for Glacier & Environmental Research, Pacific Science Center, Seattle, WA 98109

The fugue of climatic variations is represented by Ice Ages through geologic time. Some factors in glacier behavior are considered, including causal relationships of global significance. Minor climatic fluctuations, exemplified by glacier variations in Alaska during late Neoglacial time, are emphasized, using the Juneau Icefield as a prototype.

Secular trends during the Little Ice Age and the Holocene suggest that if natural climatic controls prevail, both minor and major Ice Ages could be in the offing, the lesser one within a few centuries and a greater one in upwards of 10,000 years. Documented short-term glacial pulsations are shown to be paralleled by comparable time variations in solar energy, each independently revealing a superposition of periodicities of 5-17 years (averaging 11 years) and of 22-25 years. Another cycle, of about 90 years, is identified in the glacial record. This, too, closely compares with main oscillations in the solar cycle.

In an assessment of glacio-climatic conditions in the Southeastern Alaska glacier region, an apparent positive correlation is indicated between winter temperature trends and the solar cycle from the 1890s to the 1950s. Since then, to the 1980s, the comparison has been inverse in character. The implication is a south-to-southwesterly shift of the Arctic Front in this region, with the inferred solar relationship unclear or muted by other factors.

Multiple regression analysis of mean annual and mean January temperatures at coastal and inland weather stations near the Juneau Icefield substantiates the time position of high and low nodal points on the curve. In contrast to higher-than-normal sunspot numbers in the 1950s and 1960s, winter temperatures dropped over these years, accompanied by a downward shift of névé-lines and generally increased net accumulation on glaciers at intermediate and lower elevations. After 1969, a rapid reversal in trend took place, with strong increases in ambient temperature. Thus, over the past 15 years, low elevation glaciers have experienced accelerated down wastage and retreat, paralleled by notable increase in ice volume in some of the higher elevation cirques.

The coastal position of the Alaska Panhandle with its linear mountain ranges has accentuated the glacial variations. Based on mass balance trends since the 1930s at different elevations on the Juneau Icefield, a hypothetical model of surface atmospheric circulation is invoked to clarify the glacio-climatic sequence. Simplified, this involves periodic shifting of the Arctic Front across the Coast Ranges of the Alaska-Canada Boundary region and relates the rise and fall of freezing levels to elevations of maximum snow accumulation. Observed regimen trends consistent with the modular interpretation lend credence to the possibility of a solar mechanism affecting short-term glacier variations. Glaciological and geomorphological evidence also indicates longer glacial periodicities of about 180, 700-940, and 2400-2600 years.

Teleconnectional similarities with modern glacier behavior in Scandinavia, the southern Andes and New Zealand support global significance of the record. Comparative data on polar sea ice changes in historic time also reflect the general regime trends of terrestrial glacier ice. Because of the worldwide implication and relevance of the observations to past glaciations, an effort has been made to generate a logically consistent glacio-climatic model that fits the facts in the geological record and does not depart too far from known aerological characteristics and physical laws governing dynamic behavior of the atmosphere.

At time, stage and age intervals, British Columbia-Yukon-Alaska glacial stratigraphy and ocean core evidence have suggested longer-term intervals of glacial climate at approximately 10, 20, 40-50, 100 and possibly as much as 500 thousand years. In contrast to the minor variations dealt with in this study, there is a growing consensus that these larger intervals relate to geometric shifts in the orbital position and inclination of the earth with respect to the sun, through changes in its orbital eccentricity and in the tilt and precession of its rotational axis. Deductions from both long- and short-term paleoclimatic changes can be useful in developing models for biological and ecological interpretations of the Neogene.

Shifts in the magnitude and extent of recent surface water warming in the Central North Pacific and the Gulf of Alaska may play a role in short-term climate variations along the southern Alaska coast. Details of this effect and the long-term influences of oceanic factors, however, are uncertain due to lack of data. Closer integration of secular oceanographic and meteorological information is essential to understand the complexity of the mechanisms involved.

The problem of atmospheric pollution by the introduction of fine particulate matter in the stratosphere via recent volcanic eruptions and from the development of a carbon dioxide "Greenhouse Effect" through burning of fossil fuels introduces further elements in analysis of source energies driving the glacio-climatic model. These two processes have opposite effects, the first producing cooling and the second leading to warming trends. As a key to future climate warming, civilization's influence on the carbon process has the most significant implications for upsetting the short-term natural trends, whose characteristics have presumably persisted throughout geologic time. Recent evidence of natural carbon dioxide variations, even in the Pleistocene, is also cited, opening up a new realm of unanswered questions.

In the absence of a plausible explanation of the last 10-15 years of warming either from the solar cycle or from air-sea interactions, the concern is that a global carbon dioxide control on the general circulation may have begun during the 1960s. This could bear out current projections from numerical analyses that the man-made Greenhouse Effect is irreversible and may play a decisive role in worldwide climatic amelioration and the raising of sea level by the 1990s.

If this is true, and if it leads to obscuring other causal elements, to the extent that it upsets the natural climatic sequence, such as indicated by solar variations, it could make future exploitation of international coal resources undesirable and foster restrictions on the use of hydrocarbon fuels and the logging of pristine forest lands. To monitor the relative importance of these basic factors, as well as further to understand Neogene conditions, systematic glacier/climate studies and further critical tests of the sun-weather interaction should be continued throughout the remaining years of this century.

The problems and nature of climatic change through geologic time are topics of high interest to scientists in diverse disciplines, especially those concerned with geological and ecological changes, because environmental effects intertwine in the geologic record in both the inorganic and organic worlds. A key indicator of global-scale climatic variations is the glacial record, with the addition of comparative characteristics of glacial and interglacial ages (Brooks 1949). The Pleistocene has received the greatest attention, not only because it is the most recent and best documented glacial epoch, but because it represents a key interval in the evolution of man. This is also the epoch

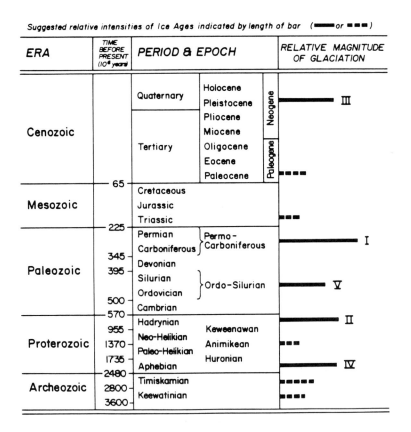

Figure 1. Epochs of glaciation throughout geologic time. This chart is based on a survey of current literature. Interpretations are very general, in view of new concepts in plate tectonics and the varied nature of Cordilleran (dashed lines) vs. Continental (solid lines) glaciations. Magnitudes I through V represent continental glaciation with allied global glacio-fluvial-pluvial conditions. Known details of pre-Cambrian glaciations are limited, with left column period terms applicable to the Canadian Shield and right column to parts of Upper Michigan and Wisconsin. The main uncertainty is interpretation of relative magnitudes, not the fact that glaciations occurred.

with the most widespread effects on the earth's surface, many of which are available for study.

In the geologic column there is scattered evidence of earlier Ice Ages, at least one of greater intensity and extent than the Pleistocene (Hambrey and Harland 1981; Crowell 1982; Fischer 1982). Five major and five lesser glacial epochs are shown in Figure 1, albeit incompletely. The lesser ones may not have been of true continental magnitude. Of pre-Pleistocene glaciations, the Permo-Carboniferous was more extensive than the Pleistocene, and was polar centered. Because of continental shifts since the Paleozoic, signs of this glaciation are found in mid-latitude belts from India to Africa to Brazil. These gave the German meteorologist Alfred Wegener (1924) and the American geologist Frank Taylor their first "proof" of continental migration. More recently, ocean cores have added to the evidence that geometric shifts in our planet's relation to the sun, through changes in the eccentricity of the earth's orbit and in the tilt and precession of its rotational axis, plus changing arrangements of the continents via plate tectonics, cause major climate changes such as those of the Neogene. Time correlation between ocean stratigraphy, glacial ages and geometric variation curves as outlined in a study by Hayes, Imbrie and Shackleton (1976) lends credence to this view. The chrono-sequence is remarkably in agreement with that first postulated quantitatively by Milankovitch (1920, 1930, 1938, 1941) and recently reviewed by Burger (1980).

Causal elements of lesser climatic perturbation during the Neoglacial, represented by the last 3000 years of the Holocene, are superimposed on the geometric basis of larger cycles (Miller 1973). With glaciers recognized as the most sensitive historians of climate change, attention is given in the following pages to historical records of secular climatic trends. Isotope dating, geochemical impurity alterations, and ice-melt core sample variations in 900-3000m boreholes in Greenland and Antarctica are paleoclimatically revealing (Purrett 1970; Langway et al. 1973; Herron et al. 1981; Dansgaard et al. 1982; Campbell 1984), but information on the fluctuation of existing glaciers in the middle and high mid-latitudes gives the best clue to recent climatic change. The pattern of glacial variation over the past five centuries is examined, an analysis embracing the Little Ice Age of late Neoglacial time. The mechanisms described for short-term climatic variation are inferred to have pertained in some measure to the evolution and distribution of biotic assemblages throughout the Neogene, with the larger geometric variations being the cause of some major biotic extinctions (Lewin 1984). Reference is made to main centers of modern glaciation in south coastal and southeastern Alaska as indicated in Figure 2 (Lats. 57°30' to 62°N). Significance of the pattern of glacier variations in this region is discussed and consideration given to the causal factors involved, and the worldwide teleconnectional implication.

LITTLE ICE AGE FLUCTUATIONS IN SOUTH COASTAL ALASKA

A review of overall Alaskan glacier fluctuations during and since the Little Ice Age has been presented previously by Field (1937), Lawrence (1950), Meier and Post (1962) and Miller (1955, 1963b, 1964). The pattern is shown in Figure 3 and Table 1; the latter, noting regimen patterns on 174 selected coastal glaciers since the 1920s, represents the variations of representative Alaskan coastal glaciers that do not have a history of "sudden slip" or catastrophic behavior. The aperiodic

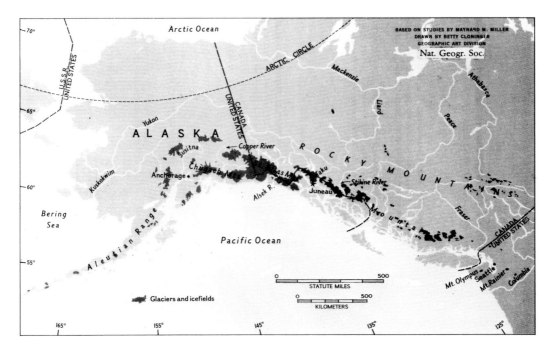

Figure 2. General map of Alaska and Western Canada showing principal locations and main centers of existing glaciers and icefields.

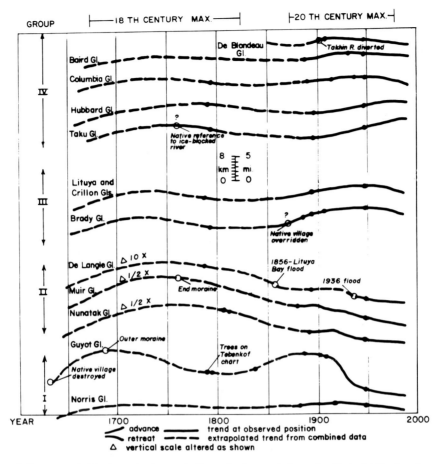

Figure 3. Pattern of Alaskan glacier variations during and since the Late Little Ice Age.

activity of surging glaciers (Post 1969) is eliminated from the analysis because in these cases the stress instability in ice probably does not relate directly to climatic cause. In such situations either earthquake avalanching or the build-up of impounded subglacial water, or both, are deemed more relevant than weather factors (Nye 1958; Bayrock 1967; Miller 1958b, 1967, 1973, 1974). Current research on surging glaciers, even in the Antarctic, is throwing light on this problem, but for now it is prudent to separate out from the Alaska regional pattern the behavior of any glaciers strongly characterized by surges. Thus we refer only to those with regimes producing normal discharge, so that we can have confidence that their fluctuations are climatically relevant.

Analysis of the record of several centuries reveals that regardless of geographic position, elevation, axial orientation or size of individual ice masses, all glaciers in the survey are correlative. Each out-of-phase fluctuation relates to a simultaneous advance or retreat of an adjacent terminus (Fig. 4). It also represents a long-term pattern with basic similarity to other glaciers in the region as well as elsewhere (Miller 1958b, 1971; Nichols and Miller 1952). Even the slowing down of a retreat in one glacier can be shown to correlate with the advance of another. Therefore, regardless of seemingly confusing dissimilarities at any one time, comparable patterns emerge in records of a century or more. Thus in the glacio-climatically sensitive coastal regions of Alaska, the retreat of any given glacier is followed by an eventual thickening and advance, and vice versa.

For documentation we now look to details of modern glacier fluctuations on a regional prototype in the Alaskan Panhandle. Conclusions will then be drawn regarding previous periodicities

TABLE 1

REGIME STATISTICS FROM 1946-58 RECONNAISSANCE SURVEYS OF MAJOR GLACIERS IN SOUTH AND SOUTHEASTERN ALASKA							
	NUMBER OF GLACIERS OBSERVED	REGIMEN PATTERN SINCE 1920'S					CLIMATIC CHARACTER OF NÉVÉ AREA **
		TERMINI DOMINANTLY SHRINKING	NEAR EQUILIBRIUM but GRADUALLY SHRINKING	IN EQUILIBRIUM	NEAR EQUILIBRIUM but GRADUALLY EXPANDING	STRONGLY OR RECENTLY ADVANCING AND NEAR POST-GLACIAL MAXIMUM	
A. STIKINE DISTRICT (a) Stikine Valley	10	5	3	2			COASTAL INTERIOR INT. & HIGH ELEV. SUB-CONTINENTAL
(b) Forest Fiords to Speel River	18	15	1		1	1*1	INT. ELEV. MARITIME TO HIGH ELEV. SUB-CONTINENTAL
B. TAKU DISTRICT (a) Taku-Llewellyn Glacier System and vicinity	16	14	1			1²	INT. ELEV. MARITIME AND HIGH ELEV. SUB-CONTINENTAL
(b) East side Lynn Canal	15	12	3				INT. ELEV. MARITIME
C. GLACIER BAY AND THE ALSEK RIVER	28	15	2	5	6		LOW & INT. ELEV. MARITIME
D. CHILKAT DISTRICT	22	7	5	6	4		INT. ELEV. SUB-CONTINENTAL
E. LITUYA BAY DISTRICT	7	1	1	1	1	3³,⁴,⁵	INT. & HIGH ELEV. MARITIME
F. ST. ELIAS DISTRICT (a) Yakutat Bay & Brabazon Range	29	24	3		1	1⁶	LOW, INT. & HIGH ELEV. MARITIME WITH INTERIOR AS HIGH ELEV. SUB-CONTINENTAL
(b) Icy Bay to Copper River Delta	12	10	2				LOW & INT. ELEV. MARITIME
G. PRINCE WILLIAM SOUND (a) Valdez Icefield	4				1	3⁷,⁸,⁹	INT. & HIGH ELEV. MARITIME
(b) West of College Fiord	13	3	2	4	3	1¹⁰	LOW, INT., & HIGH ELEV. MARITIME
TOTAL OBSERVED ON REGIONAL BASIS	174	106	23	18	17	10	

* 1. BAIRD GL. 2. TAKU 3. LA PEROUSE 4. CRILLON 5. LITUYA 6. HUBBARD 7. COLUMBIA 8. MEARES 9. HARVARD 10. HARRIMAN

** AS REFERRED TO IN THIS COLUMN:— LOW ELEVATION = 1-3000 FT.
INTERMEDIATE ELEV. = 3-5000 FT.
HIGH ELEVATION = 5000 FT. and above

and extrapolations rendered with respect to future behavior of key glaciers on this eastern shore of the far North Pacific.

Studies of the prototype glaciers were begun in 1946 and have been continued uninterrupted to the present under the auspices of the Juneau Icefield Research Program (JIRP). This program

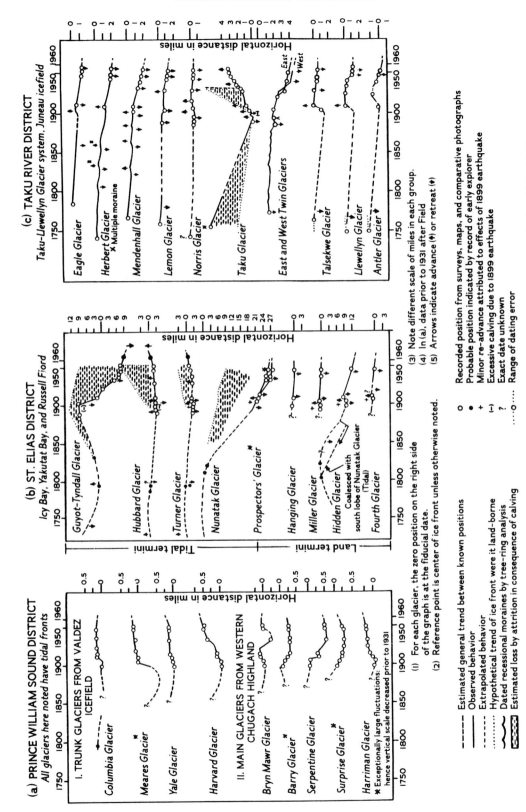

Figure 4. Fluctuations in lateral position of representative glacier termini in south and southeastern Alaska, 1750-1962.

received major support for its designed long-term field study and training activity from the U.S. Office of Naval Research through contract with the American Geographical Society (Miller and Field 1951). For the past 30 years JIRP has been carried forward under the aegis of the Foundation for Glacier and Environmental Research, based at the Pacific Science Center in Seattle, Washington. Support has continued through contracts with the National Science Foundation, the U.S. Army Research Office, the Federal Office of Water Resources Research (Department of Interior), NASA, the U.S. Forest Service, the State of Alaska, the National Geographic Society, and other agencies including Columbia and Michigan State Universities and the University of Idaho in the U.S., and Queens University, the University of New Brunswick, and the Nova Scotia Land Survey Institute in Canada.

Since 1959 the program has been allied with an annual Glaciological and Arctic Sciences Institute for training field scientists, supported by the National Science Foundation and currently under the academic aegis of the University of Idaho (Williams and Miller 1979-1982). The NSF programs supporting student training in recent years have reduced the original research aim of this activity, but still a number of monographs, published papers and open-file reports document JIRP scientific results. As a major emphasis since the early 1960s, the Glaciological Institute field training program has also been recorded in annual reports. The JIRP publication series began in the late 1940s and 1950s with a set of yearly monographs for the Office of Naval Research, the American Geographical Society and other supporting and cooperative agencies (Miller 1948-54; 1956, 1963, 1972, 1975; Giley 1954; LaChappelle 1954; Heusser 1952, 1953, 1960; Wilson 1959; Marcus 1964; and others to the present). Between 1949 and 1984 some 50 academic theses have further reported JIRP scientific results in many disciplines, including 12 Ph.D. dissertations. The investigations have also involved continuing liaison with scientists from over 40 universities in the United States, Canada and abroad. About a dozen of the studies have analyzed the glacio-climatic trends (e.g. Andress 1962; Egan 1965, 1971; Swanston 1967; Zenone 1971; Jones 1975, 1979; Miller 1956, 1963, 1972, 1975; Nelson 1978; Linder 1981; Price 1983; Lamb 1985).

A PROTOTYPE GLACIER/CLIMATE MODEL

The Alaska coast extends for 900 miles from Ketchikan to Anchorage (Fig. 2). Here glaciers of all morphogenetic types occur with a great range in size, length, elevation and geographic position. The fluctuational patterns in three key sectors are shown in Figure 4, representing examples from the Prince William Sound area of the Chugach Range, the Yakutat Bay area of the St. Elias Mountains, and the Taku River District in the Alaskan Panhandle. In the heart of this latter region is the Juneau Icefield, shown on the satellite imagery of Figure 5 to lie between Lat. 58° 30' and 60° N. The icefield shares half its area in Alaska and the other half in bordering areas of northern British Columbia and Canada's Yukon. The ultimate headwaters of the Yukon River lie at the crest of this icefield.

Positioned in the Northern Boundary Range of the Alaska-Canada Coast Mountains, the Juneau Icefield is the fifth largest on the North American continent. It has the distinction of being the most readily accessible icefield from a major center of expeditionary supply, in this case Alaska's capital, Juneau. For nearly 40 years systematic and consecutive annual records of the fluctuations of the main glaciers on the icefield have been obtained by JIRP, with earlier positions derived from aerial photographs, historical ground photos, early explorer charts, and from field studies in geomorphology, dendrochronology, lichenometry, soil science and radiocarbon dating of moraines. These data provide a unique record covering the Little Ice Age and extending back to the 1400s.

With respect to the timing of regime changes, the right-hand section of Figure 4 illustrates the out-of-phase advance and retreat of adjoining glacier systems emanating from the southern edge of the Juneau Icefield (Fig. 5). Figure 15 depicts the largest glacial system, the Taku, which embraces over 215 square miles of area. Although climatological details of the accumulation shift

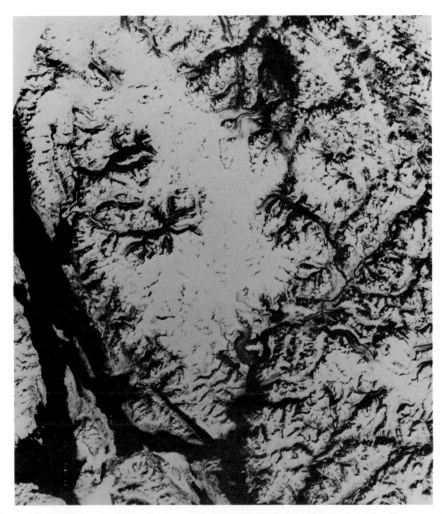

Figure 5. Landsat satellite photograph, Juneau Icefield, Alaska-Canada. From west to east (left to right) distance is 100 miles (170 km). Advancing Taku Glacier at lower center, showing portion of Taku Fiord and Taku River; receding Llewellyn Glacier at top center right. Transit date, August 18, 1979. (EROS Data Center, USGS, Sioux Falls, S.D.).

noted in Fig. 15 will be considered later, in general the lower névés of the Taku Glacier system experienced positive mass balance during the last half of the 19th and the first two decades of the 20th century, as have its upper névés since the 1920s. In contrast the Llewellyn Glacier (also Fig. 4), which is located in a comparably sized area to the north (Fig. 15) and flows into the interior continental sector from a névé adjacent to the uppermost névés of the Taku Glacier, received the heaviest accumulation on its lowest névés during the last two decades of the 19th and the first two decades of the 20th century. Since the 1920s negative regimes have prevailed throughout this large inland glacier system. The maritime Mendenhall and Norris glacier systems (adjoining to the west in Figs. 5 and 15; also note on Fig. 12) and also the Twin Glaciers to the east received their most substantial net accumulation during the first half of the 18th century, as did the lower névés of the Taku.

The seemingly disparate behavior of the Taku Glacier has been referred to as an anomaly by

Heusser et al. (1954). This term is somewhat misleading because we have not been dealing with a true anomaly but a consequence of natural differences in the area-elevation character of adjacent glacier systems. The situation is illustrated by the hypsometric curves in Figures 6a and 6b. On these graphs the main accumulation areas of the Mendenhall, Norris, Twin, and Llewellyn glaciers are seen to lie at elevations of 9-1500 m, whereas the main névé of the Taku Glacier is between 1200-2000 m. This means that the climatic amelioration at the end of the Little Ice Age (during the 1800s) caused the mean freezing level to rise some 600 m between the 18th and 20th centuries. The zone of maximum snowfall correspondingly also rose. This is because heaviest snowfall occurs at elevations close to the freezing level where ambient air temperatures are at or slightly below 0° C. On higher elevation surfaces well above the freezing level the available moisture drops off rapidly and snowfalls are lighter, while at lower elevations below the freezing level the snow turns to rain. Thus, during periods of maximum accumulation over many decades, accumulation on affected névés substantially thickens. The result is a given "flow lag," and only years later do the termini of the glaciers involved advance. On the Mendenhall Glacier, for example, the measured flow lag between the zone of maximum snowfall on its largest surface area—i.e. at its highest elevations of 1200-1500 m—is approximately 80 years. This is the number of years required for a significant accumulation increase to pass down valley and reach the terminus. For comparison, on the Taku Glacier, from its crestal névé to its terminal discharge, the flow lag is about 150 years, with some 80 years required for ice transfer from its large intermediate elevation nourishment zone. Thus the terminal variations of the Taku Glacier reflect retention of increased snowfall on its prime source areas many decades before and during quite different climatic conditions.

Flow lag numbers are drawn from surveyed surface velocity rates over the past 38 years, abetted by annual measurements in firn test-pits and from crevasse stratigraphy on the prime accumulation plateaus. To elaborate on modes and levels of maximum accumulation, comparative measurements have also been made during periods of individual storm (Miller 1956, Fig. 35) clearly showing shifting elevation zones of maximum accumulation. The cumulative effect has been well documented in annual test pit and crevasse stratigraphy reports, exemplified by the net accumulations and névé-line trend plots in Figure 7. The situation for the first half of this century is illustrated in the accumulation trend plots for the main Juneau Icefield névés given in Figure 8.

Observations on marginal scour zones and trimlines on the Juneau Icefield show that the delevelling process has proceeded on the lower névés, and reveal how later terminal recession has occurred because of a pronounced upward shift in freezing level and hence raised elevation of the zones of maximum accumulation in the late 1800s and again in the mid-1900s. This characterized the conditions at the end of the Alaskan Little Ice Age. The situation is best illustrated by the well-delineated trimlines and marginal scour zones since the 1700s on the East Twin Glacier at the icefield's southern margin (Fig. 9). The advance of the Taku Glacier over the past 90 years, and its distributary Hole-in-Wall Glacier, have also been documented by periodic photography and field mapping (Fig. 10). More recently, satellite imagery has helped to maintain a general record of these changes, although the photographic resolution is not yet adequate for detecting minor pulsations (e.g. Fig. 5).

Alaska Regional Glacier Patterns—1946-1983

The East and West Twin Glaciers also began to show slight resurgences in 1979, after some 80 years of continuing retreat. Also the east-central terminus of the Mendenhall Glacier has exhibited minor net forward thrusting since 1983. These events are paralleled by thickening of some highland cirque glaciers in the Alaska-Canada Boundary Ranges and could signal a return to more positive mass balances of other intermediate level névés on the Juneau Icefield. The continuing advance of the Taku and Hole-in-Wall Glaciers over the past 90 years typifies recent advances of normal discharge (non-surging) glaciers in the Alaskan Panhandle from the Stikine River to

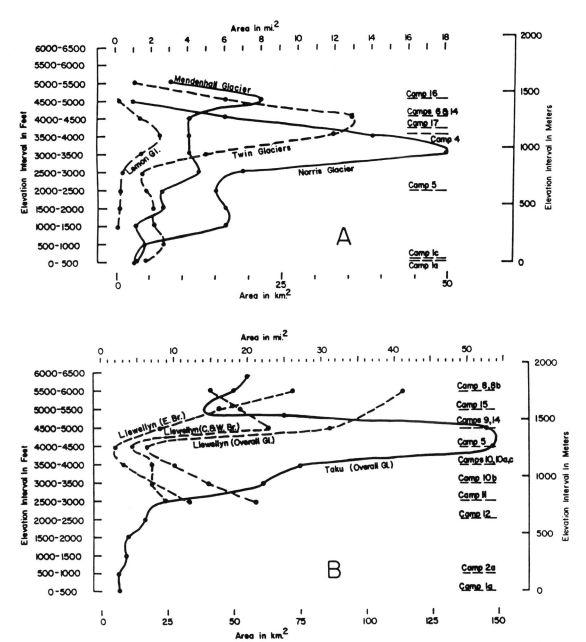

Figure 6. Hypsometric curves showing area-elevation character of (A) the Mendenhall, Lemon, Norris and Twin Glaciers; and (B) the Taku and Llewellyn Glacier Systems. Ice surface and névé areas represented between successive 500-ft (150 m) contours. Levels of key research camps on right.

Prince William Sound (Figs. 2, 10). In recent decades the other main advancing glaciers in this region (excluding surging ice fronts that are considered anomalous) are the Baird, LaPerouse, Crillon, Lituya, Hubbard, Meares, Harvard and Harriman Glaciers, and until recently the Columbia Glacier (Fig. 3 and Tables 1 and 2; also Miller 1964). Just four years ago the terminal regime of the Columbia Glacier shifted dramatically to a retreatal phase. However, in this century these advanced glaciers have been close to the maximum position they have attained in over 8,000 years.

The Columbia Glacier's spectacular present retreat behavior has been much accentuated by tidal action because the ice front shifted into deep water from a grounded maximum position it held for many years on an island in shallow water near the entrance to Prince William Sound (Miller 1964). Note is made of this because of the disproportionate attention given recently to this glacier's behavior and preoccupation with its iceberg production in view of its proximity to the Alaska pipeline oil export of Valdez. As for regime significance it may indicate that during the climatic cooling from the 1950s to the 1970s (Fig. 20) there has been a return to negative mass

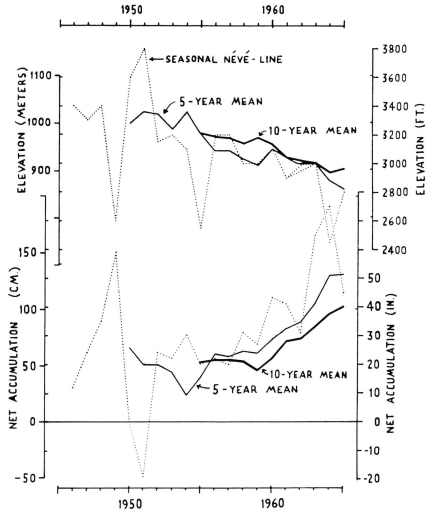

Figure 7. Comparative névé-line and net accumulation trends on the main nourishment plateau (1060-1365 m) of the Taku Glacier, 1946-65.

balances on some of its higher tributaries. Even though this is a complex multiple-névé glacier system, it is expected to herald the beginning of recession of other main trunk glacier systems in Alaska during the next several decades, including the prototype Taku Glacier. The Twin Glaciers noted above may be reflecting a lowering of freezing levels from this recent climatic change. The explanation again relates to the area-elevation and maximum accumulation factors noted in Figures 6, 7, and 8, and which will be considered later in more detail.

From the comparison in Table 2, it appears that about 8% of the main valley glaciers on the Juneau Icefield and elsewhere in coastal Alaska are currently experiencing thickening and terminal advance. On the other hand, some 63% are continuing terminal thinning and retreat. The positions

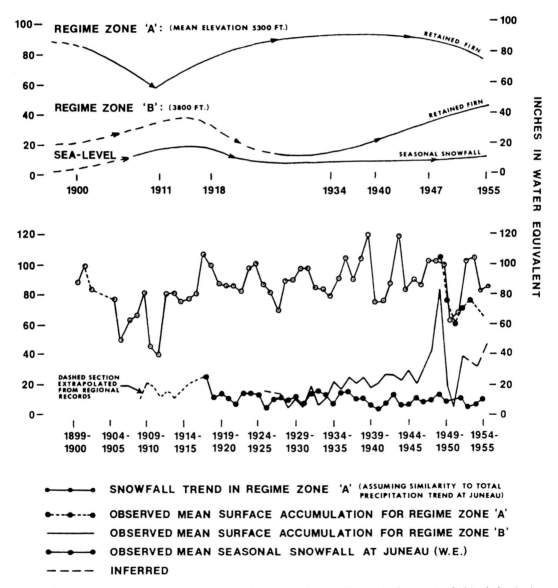

Figure 8. Net accumulation trends of retained firn on the main Juneau icefield névés, during the first half of the present century. Data from mean elevation of regime Zone A (5300 ft. or 1605 m) and of regime zone B (3800 ft. or 1150 m).

Figure 9. East Twin Glacier during retreatal phase, showing its well-delineated trimlines and marginal scour zones since the 1700s. (FGER Photo, July 1966).

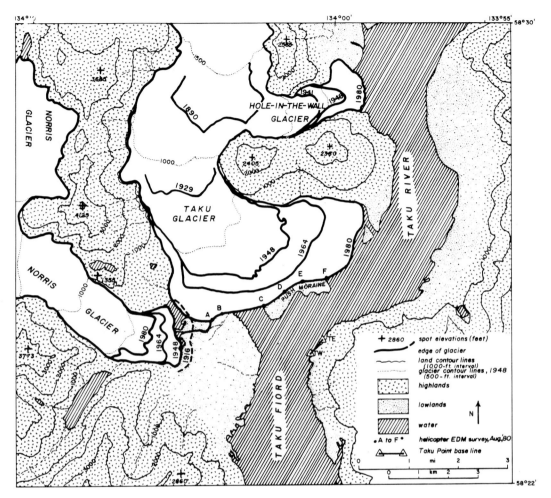

Figure 10. Map of advancing Taku and Hole-In-Wall Glaciers and receding Norris Glacier terminal positions between 1890 and 1980.

of about 30% of Alaska's coastal ice fronts, however, remain close to equilibrium and may be expected to shift one way or the other within a few years. Forty years ago, in the mid-1940s, some of the percentages were somewhat different, as was the statistical base for the comparisons (Table 2).

The tabulated increase in advancing glaciers is not a sacrosanct statistic, because the sample is small compared to all of the glaciers in Alaska. The fact is however, that this has been paralleled by minor resurgences of ice masses in the coast mountains of British Columbia and farther south in the Cascade Range of Washington State, where it is found that a larger proportion of high alpine cirque glaciers are experiencing positive mass balances, compared to only 3% forty years ago. Since the late 1960s similar reports of increased glacier volume and terminal advances in high cirque glaciers have been received even as far south as the Sierras in southern California (Trent 1983). A fuller record is available from the European Alps, Scandinavia and Iceland, where the percentage increases in positive mass balances on main valley glaciers since 1960 are comparable to Alaska. In Europe, the ratio is much greater on high-elevation cirque glaciers, as discussed in the teleconnectional section following (Slupetzky 1977-84, pers. commun.; Ostrem et al. 1980-82).

The regime changes in Austria have also been characterized by gentle "surges," considered to

TABLE 2. SHIFTS IN REGIME TRENDS OF REPRESENTATIVE
ALASKAN COASTAL GLACIERS FROM 1946 TO 1983

	No. of Trunk Glaciers Observed	Dominant Shrinkage	Near Equilibrium	Persistent Advance
1946	174	61%	33%	6%
1983	190[a]	63%	29%	8%

[a] Including most of the 174 surveyed in 1946.

be cyclic and connected to variations in climatic conditions, though not always strictly in phase with other glaciers in the Alps. A noteworthy 82-year periodicity was reported by Hoinkes (1969) based on studies on the Otztal Alps covering the years 1599 to 1969. The larger climatic significance of these trends and the similarity of records with respect to Alaska are considered next.

When viewed as a whole, the Alaskan glacier pattern in this century has appeared as one of dominating retreat. This could lead some to the inference that the world actually continues in climatic amelioration. This should also foster expectation of further thinning and down-wasting of most of the earth's middle and high-latitude glaciers. A closer scrutiny of the still relatively small percentage of thickening/advancing termini in Alaska and the size of the ice masses involved reveals quite an opposite situation. From the documented lowering of regional temperatures in Scandinavia and in the Alps, as well as coastal Alaska since the mid-1960s, it is to be expected that the 8% current volume increase of Alaskan coastal trunk glaciers also pertains to the prototype Juneau Icefield. Here the advancing Taku Glacier is the example, being the largest trunk glacier of the 30 under observation on that icefield. Planimetric comparison of the Taku Glacier's hypsometric characters and its geophysical depth characteristics confirms that this larger glacier system contains more area and volume of ice than any other system in the prototype district. It is probable, too, that in coastal Alaska as a whole the area and volume of ice involved in the total of all advancing valley glaciers represent considerably more mass of ice than all Alaska's receding glaciers put together. Comparatively, therefore, in this century, the higher areas of Alaska coastal mountain ranges have experienced a substantial build-up of glaciation, contrary to the general view of diminishment. Coincident with this, large quantities of melt-water are still debouching into the world's oceans from the ablation of low-elevation ice masses at temperate latitudes. Therefore, the question needs to be asked: how extensive will the increase in regional ice volume at intermediate and higher elevations become? And if we can assume that natural climatic factors will control, is another maximum phase of the Little Ice Age or even a greater Ice Age in the offing? These questions have also been discussed by others (Willett 1975).

Teleconnectional Implications

In a critical assessment of the climatic significance of glacier fluctuations, wherever possible, regime comparisons should be made between glaciers of similar size, form, gradient and elevation range, and also between those with névé areas of comparable size and elevation and termini at approximately the same levels. The extremely mountainous character of the Alaskan coastal region provides such diversity of morphological factors influencing glaciers that few simple or direct comparisons can be made. However, because our surveys have embraced about 90% of the main glaciers in south coastal Alaska, enough have been studied for the regional pattern to be recognized. Differences in district character and the general regional trend between the 1920s and the early 1980s are now explained with reference to the glaciological provinces cited in Figure 11.

An interesting consistency in pattern is found in glaciers in the Stikine and Juneau Icefields

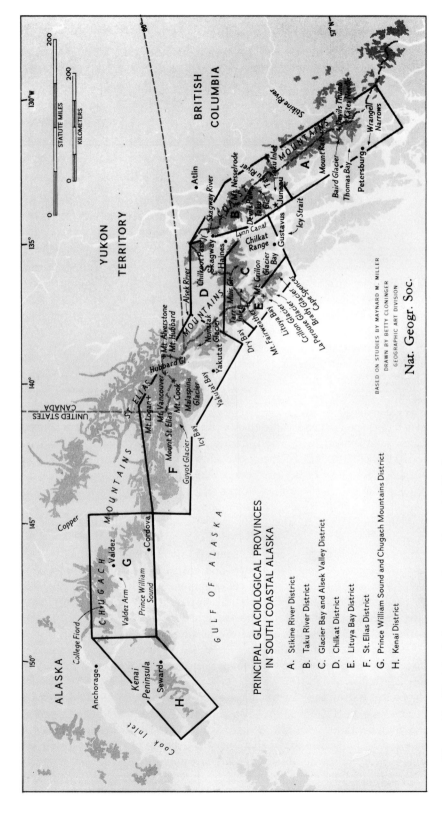

Figure 11. Glaciological provinces in south coastal Alaska.

in the Stikine and Taku districts, as well as in the coastal Alaskan sector of the St. Elias Mountains. A different pattern in glacial behavior is shared by the extreme maritime flank of the Fairweather and Chugach Ranges (Figs. 2, 11). This pertains to similarities in geographical position of the prime source névés in the high central sector of these ranges along the Alaska-British Columbia boundary and the inner St. Elias Mountains. Comparable conditions also are found on the continental flanks of these ranges in British Columbia and Canada's Yukon. The most heavily glacierized highlands of these regions are 160 km farther inland than the Fairweather and Chugach Ranges and are significantly more continental in climate than the icefields nourishing the tidal termini of Lituya Bay and Prince William Sound (Fig. 11; see also right column in Table 1).

Differences in regime between strongly maritime districts and those of continental climate character have also been taken into account. The role of geographical factors is invoked to explain the larger proportion of advancing glaciers in the maritime sectors. This is corroborated by the fact that the outer maritime provinces receive upwards of 100% more annual precipitation than the inner maritime and sub-continental sectors. The outer maritime provinces also receive as much as 800% more than the continental areas lying on the inland flanks of the Coast Ranges. Likewise sea-level stations at Cordova and Yakutat often receive between 200 and 300% more rainfall than recorded at the National Weather Service sea-level station at Juneau in the southeast Alaska control locality. Furthermore, the rainfall patterns at the mouth of Glacier Bay (Gustavus) and in the Chilkat District (Haines) are gradational between the continental and maritime extremes across the Juneau Icefield. On a northeasterly transect over the Boundary Range, precipitation changes from slightly less than 100 inches per year at Juneau to 200 inches on the Lemon and Mendenhall Glacier névés, dropping to only 80 inches on the upper Taku Névé (see Zenone 1972). In contrast, even farther inland the annual precipitation is only 15 inches of water equivalent per year at Atlin in northern British Columbia (Fig. 2; also see Thompson 1972). The transect involved in these comparisons is typical of coast-to-interior traverses elsewhere in the Alaskan Panhandle.

The conclusion is inescapable that in spite of marked differences in locational factors affecting glacier nourishment, within the maritime region of southeast Alaska the dominant characteristic since the 1910s has been terminal downwasting and shrinkage. The recessional rate on many glaciers, especially those with source névés at lower elevation, was much accelerated in the 1920s. On those with high névés, a slower retreat set in during the 1920s, though inland on the continental flanks of these ranges contemporaneously many strong advances took place culminating in the mid-1920s. By the 1930s and 40s, all of the inland glaciers experienced thinning and regression.

The only persistently strong departure from this trend was on the largest trunk glaciers, such as the advancing Columbia, Hubbard and Taku Glaciers. For each case of significant advance, however, there has been a marked and contemporaneous retreat of adjacent valley glaciers of comparable size and morphology. Invariably, this opposing behavior has occurred on glaciers stemming from the same or adjoining névés, but positioned in the interaction zone between maritime and continental sectors across which regional storm winds periodically shift.

In the global scenario, there has been teleconnectional significance in reports of simultaneous advance and retreat of adjacent glaciers in Iceland (Thorarinsson 1940) and Patagonia (Nichols and Miller 1951, 1952). Also there have been periodic reports on changing regimes of French, Swiss, Italian and Austrian glaciers (Hoinkes 1968, 1969; Kasser 1967, 1973; Jahresbericht der VAW-ETH Zurich 1983; also periodic IASH and UNESCO reports on fluctuations of glaciers). Summaries of these data show that from the late 1920s to the early 1960s, at least 90% of alpine and Scandinavian glaciers have been retreating, but since the early 1960s there has been a notable increase in the number of glaciers in equilibrium or showing pronounced expansion. As previously indicated, during the past decade a number of glaciers in the Alps comparable to those in Alaska have been reported to be advancing, with selected areas of Switzerland and Austria reporting much higher ratios (from 24 to 75% and 14 to 72% respectively between 1964 and 1980, with a reduction in these numbers down to about 40% in 1983). Over the past 20 years in Norway, thickening of glaciers has increased up to 44%, with a comparable pattern reported from Iceland.

Although these percentages may seem disturbingly high, it must be kept in mind that they mainly reference small, high-elevation cirque glaciers. It is of interest that they repeat growth patterns experienced by these glaciers during the first two decades of this century. Therefore, although significant with respect to climatic sensitivities, the statistics from these high-level cirque glaciers focus disproportionate attention on one end of the glacio-climatic cycle and should not be compared numerically with the integrated regime of large valley glaciers having multiple source areas, such as emphasized in our Alaskan surveys.

Yet the implied teleconnectional trends are significant. Even throughout the Archipelago of Arctic Canada and in west Greenland, noteworthy advances of small valley glaciers and some cirque glaciers have been reported since 1968, contrasting with simultaneous retreat of larger valley and icefield outlet glaciers in that region (Gordon 1980).

With this interpretive caution, Table 2 shows that in the broad regional sense the proportion of main valley equilibrium and advancing glaciers in Alaska is generally higher than in similar climate regions in the world. It should be kept in mind that this refers to trunk glaciers. Nowhere else has the contrast been on such a large scale. By teleconnection, the acute sensitivity of Alaskan coastal glaciers to global climatic change is clear. Also the Alaskan glacio-climatic pattern has been so consistently recognized over such a wide range of latitude (56° to 62°N; see Fig. 11) that a causal factor has to be invoked that is of not only regional but worldwide significance.

The Juneau Icefield—A Prototype of Normal Discharge Glaciers

Over the past 38 years of annual field observations in Alaska and elsewhere, the regional pattern of coastal glacier oscillations has been clarified and the relationship shown to earlier pulsations during the previous four centuries—i.e. over the interval of the Little Ice Age. The picture is complex and details continue to unfold. But one thing is certain: concern about changing termini without consideration of a glacier's total system and related orographical, meteorological, and geophysical parameters, will only result in partial answers. Therefore, to achieve a full understanding of the regime of Alaskan glaciers and their global significance during the Pleistocene, as well as their fluctuational behavior since the Thermal Maximum, long-term field study of selected ice masses is essential from termini to source area and in a three-dimensional context.

Because southeastern Alaska harbors the world's largest area of concentrated glacial ice outside of the polar regions, we chose to concentrate our investigations on the Juneau Icefield as a prototype glacier system, since it has all of the characteristics of the larger regional area (Fig. 11). This icefield (Figs. 5, 12) is uniquely suited for glacio-climatic research because it lies just outside of the geotectonic zone rimming the eastern shore of the Gulf of Alaska. This means that its glaciers have not been subjected to recent earthquake avalanching; their behaviors have been consistently of the climatically controlled or normal discharge type, and have not reflected aperiodic surges from any earthquake effects. In contrast, the Fairweather and Chugach Ranges (Lituya and Chugach districts in Table 1 and Fig. 11; also see Valdez Icefield, Fig. 4a) are located astride a narrow and tectonically sensitive belt, as is the coastal maritime sector of the St. Elias District (also Fig. 4b). It is not surprising, therefore, that in these provinces an unusually large number of catastrophically surging glaciers have been reported (Miller 1958, 1967, 1973; Post 1969).

This introduces the question of tectonic diastrophism and its influence on the irregular behavior of such glaciers. In our regional study earthquake effects are revealed by the presence of abnormally contorted or convoluted moraine patterns and other anomalous geomorphic

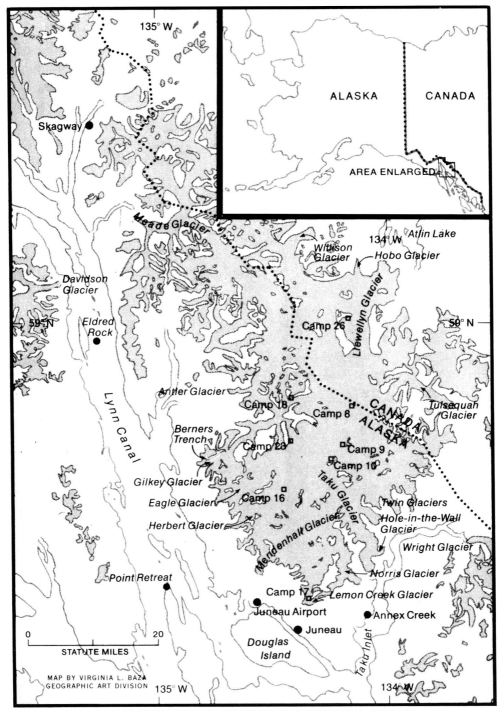

Figure 12. Key field research stations in the Juneau Icefield region, Alaska-Canada.

features. Where identified, such surging glaciers have been differentiated from the behavior of others not showing these characteristics and which presumably have variations of normal discharge type more directly controlled by climatic trends. For this reason the statistics in Tables 1 and 2 and Figure 4 concern only glaciers that have not been dominantly affected by earthquake avalanches. In fact Table 1 refers mainly to conditions observed in regional surveys between 1946 and 1958, conducted prior to the two most severe Alaska coastal earthquakes in this century—i.e. in 1958 and 1964, registering 8.2 and 8.4 on the Richter scale respectively.

Neoglacial Patterns and Secular Glacio-Climatic Trends

Dendrochronologic dating of a festoon of Neoglacial moraines at the Mendenhall glacier terminus near Juneau (Figs. 4, 13) has demonstrated that in this glacier system terminal fluctuations have been acutely sensitive to climate pulsations of the Little Ice Age. By adding known 1910-1913 and 1935-1939 moraines not documented by tree-rings, at least 15 well-differentiated

Figure 13. Mendenhall Glacier. Vertical air photo showing festoon of terminal moraines. Dendrochronology dates reveal this moraine complex was produced since the maximum outer moraine was established in the mid-1750s. Ongoing recession indicated by 1962 ice position. Carbon-14 dates of overridden forest remains portray a similar Neoglacial pattern attaining a comparable maximum near outermost Little Ice Age moraine about 2100 years B.P. (U.S. Forest Service Photo, July 4, 1962).

moraines developed between the 1760s and 1940s (Lawrence 1950). Periodicity in this sequence averages 11.6 years between readvances or successive still-stands (Fig. 14) with 22- and 26-year intervals relating the two latest major moraines. The moraine complex buries a sequence of earlier Neoglacial moraines, the most extensive of which involved ice-pushing and overriding of forest remains. The result was shearing off of tree tunks in the valley train area extending for three miles down valley from the present ice front. Radiocarbon dating of forest beds in association with the earliest of these moraines gives dates of 2000 to 2100 years B.P. This represents the first major readvance of ice in the Holocene since Valders time some 10,000 years B.P. As the pattern is repeated in tree-ring and radiocarbon dating of moraines on nearby Herbert and Twin Glaciers (Fig. 14), it is considered to reflect recent climate trends on the main Juneau Icefield.

From these and other field data and temperature and solar energy records through the 1950s we conclude that initiation of the Mendenhall moraine sequence related to a half century of cooling in the late 1600s and early 1700s, with significant changes in freezing levels and precipitation amounts. This corresponded with a coastal migration of the Arctic Front. A hypothetical and quite generalized model of the atmospheric circulation and freezing level conditions pertaining when low-pressure systems shift seaward is diagrammed in Figures 16 (left) and 17 (left). These figures illustrate how such circumstances produce increased accumulation on the lower névés on either flank of the Boundary Range. A number of years later the termini of glaciers so affected respond to the regime change in the nourishment zone by terminal thickening and advance. The timing of such terminal expansion is always dependent on the "flow lag" involved.

Similarly, in the late 19th and early 20th centuries increases in total glacier volume originated at high elevation and were paralleled by the dominance of low-pressure maritime air. This involved a rising of the freezing level and corresponding inland migration of the Arctic Front. A related decrease in volume took place on glaciers with low elevation névés. At such times the response of high glaciers is to thicken, with their termini advancing a given "flow lag" years later. The relevant atmospheric conditions are shown in Figures 16 (right) and 17 (right).

In the Juneau Icefield case the Taku, Mendenhall, Herbert and Twin Glaciers are continuing to exhibit this seesaw effect as supported by the plots in Figure 7 and as previously discussed with respect to Figure 5. The interpretation in Figure 15 is based on our records of firn stratigraphy, glacier fluctuation data, trimline observations and névé-line trends recorded up through the 1950s on this icefield. The nature of shifts in the zones of maximum and minimum accumulation is especially significant in Figure 15, which depicts changes in the Taku Glacier's prime areas of accumulation during a full regime cycle. In this figure the cycle represents essentially the Little Ice Age maximum. It is considered to have passed from an initial "cold" phase (P-1, e.g. in the 1400s and 1500s) through an intermediate cold phase (P-2, e.g. in 1600s and 1700s) to a "warm" phase (P-3, e.g. in the 19th and 20th centuries).

Since the 18th century, culminating advances of low-level and high-level glaciers over the two-phased period (Fig. 15) are highlighted by stippled and darkened arrows. Because the interpretation is based on field measurements, it is commensurate with the generalized explanation in the legends of Figures 16 and 17. The analysis mandates a total systems approach that considers all of the geographic and elevation factors as well as time variables. From this it should be clear that only observations over the whole of this glacier system, extending through a long period of years, can provide complete information leading to reliable interpretation. In our studies of the adjacent glaciers also, we have continued to emphasize systematic annual measurements in a program of continuing, sequential and long-term observations from their termini to highest source névés.

By extrapolating from these Juneau Icefield interpretations, which involve geographic and elevation boundary limits typical of the northeast Pacific region, conclusions are drawn with respect to ice fluctuations in the other glaciological provinces on Figure 11, where less detailed observations have been made. We have tried to remain aware of the limitations introduced by

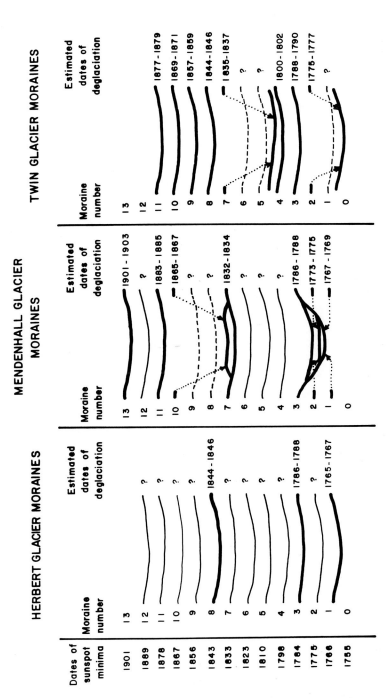

Figure 14. Little Ice Age moraine chronology of Mendenhall, Herbert and Twin Glaciers, based on tree-ring dates (after Lawrence 1950). From historical evidence and photographic records, the sequences for the Mendenhall and Herbert Glaciers also include bold moraines in 1910-1913 and 1935-1939.

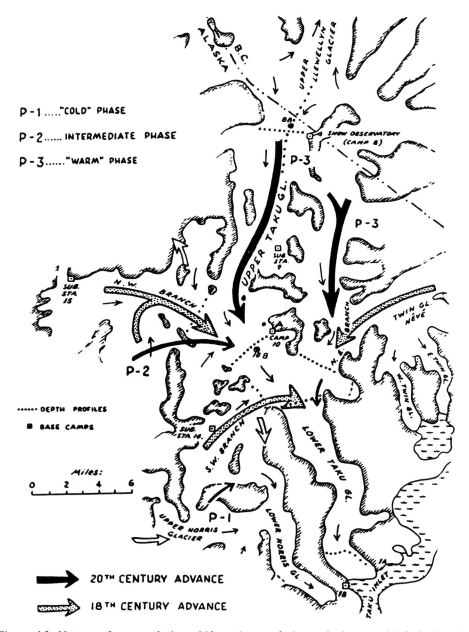

Figure 15. Nature of accumulation shift and out-of-phase glacier termini behavior in the Taku Glacier system, Juneau Icefield, during a full regime cycle.

incomplete information and by the oversimplification necessary in the hypothetical model, but are tantalized by the strong similarities revealed throughout the region. We conclude that simultaneous advance and retreat of major glaciers continues to be the pattern in the Chugach, St. Elias, Lituya, Glacier Bay, Chilkat, Taku, and Stikine districts (Fig. 10), in spite of the fact that not all of these districts are affected by earthquake avalanching.

Also significant are the high correlation coefficients calculated from the meteorological data in the records available since the 1880s at 64 near sea-level stations between Ketchikan and

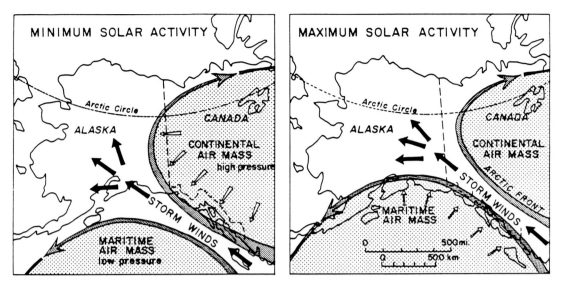

Figure 16. Hypothetical model depicting relationship of surface storm winds to cellular circulation during cool (left) and warm (right) climatic phases along the South Alaska-Canada coast. The diagram is very generalized, with shifted position of Arctic Front suggested for times of minimum and maximum solar activity (revised from Miller 1973).

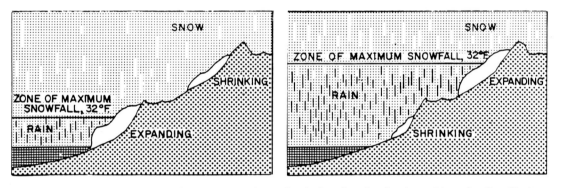

Figure 17. Diagrammatic representation of relative freezing-level positions in the Alaska-Canada Boundary ranges controlling elevation and maximum snowfall during cool (left) and warm (right) climatic phases. This simplified model shows comparative effects on glaciers with low versus high elevation névés (revised from Miller 1973).

Anchorage (Fig. 2). This includes those from Juneau and the Juneau airport on the perimeter of the prototype area, and from Sitka, 160 km to the southwest, where records have been maintained since the purchase of Alaska from Russia in 1867 (Fig. 24). Adding credibility is the apparent cross-correlation between solar energy variation, indicated by annual sunspot frequencies and changes in mean January temperature as graphed for the first half of this century in Figure 18. We may question the phase relationship, but the pulsations of these compared sets of data still seem to correspond. The solar-sunspot record is further elucidated by the plot of extended annual mean data since 1610 given in Figure 19. The similarities in these independent data have been interpreted as not being coincidental (Lawrence 1950; Miller 1956, 1958a). What distinguished this from other efforts to compare solar energy and climate is the introduction of information from

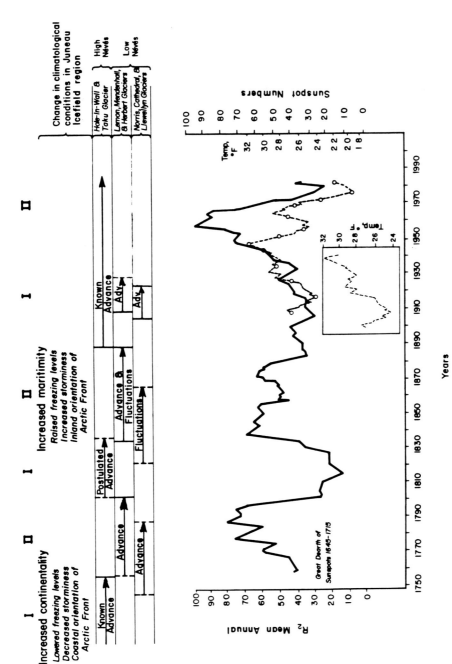

Figure 18. Comparison of solar energy (solid line), 1750-1982, and mean January temperatures at Juneau (dashed line), 1896-1982, with indication of periods of glacier growth and decay along Southeast Alaska coast. Plots represent 11-year running means of annual sunspot and winter temperature data. At top of figure, changes in glacial regimen and climatological conditions are noted, including inferred periods of increased continentality (I) and maritimity (II) during and since the 18th century Little Ice Age maximum. Inset is numerical plot of mean January temperatures displaced to left about 10 years to illustrate the similarity of solar and temperature curves during the first half of the present century. A generally reversed correspondence is indicated between these sets of data since about 1950.

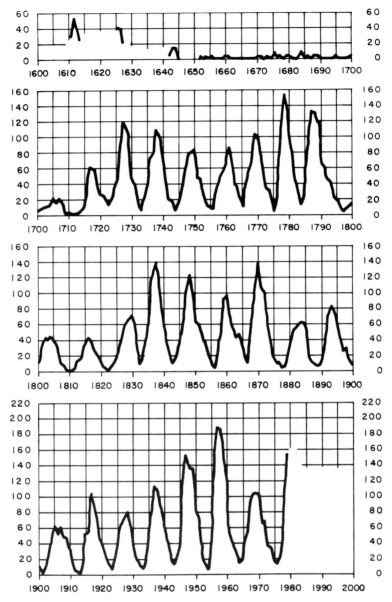

Figure 19. Annual mean sunspot numbers, 1610 to 1979, showing dearth of solar activity between 1545 and 1715. This is the 70-year Maunder minimum centered about 1670, as also noted in Figure 18 (after Eddy 1970).

the study of glacier variations, on the basis that the Juneau Icefield and others along this coast act as automatic weather stations in which annual ice regime fluctuations and longer-term secular trends have been sequentially recorded in natural detail. In this our aim has been to unravel and document that detail. This research caveat is noted with an awareness of the legitimate questions that scientists should have with respect to numerical data that attempt to support sun-weather relationships (e.g. cautions suggested by Pittock and Shapiro 1982; also Tucker 1964, and Gani 1975). In the present study, however, the glacier fluctuation record is not based on statistical analysis but on consecutive observations of ice front positions and documented records of related regime variations throughout the prototype glacier systems.

In Figure 18, plotted midwinter (January) mean temperatures for the NWS Juneau station and the solar curve 1760-1950 are compared, graphed as 11-year running means. Shown also are periods of glacier growth and decay along the Southeast Alaska coast, with notation of inferred changes in climatological conditions affecting the glacial regimes in periods of increased continentality (I) and maritimity (II) during and since the 18th-century Little Ice Age maximum. An apparent positive correlation is suggested between winter temperature trends and the solar cycle from the 1890s to the 1950s, with a separation of some 10 years indicated between these sets of data.

In the inset, similarity of the curves is shown by a numerical plot of mean January temperatures displaced left to a position that corresponds with the solar curve. It is recognized that the use of unweighted moving means may create a spurious cyclicity or even invert patterns via the Strutsky-Yule effect (Mosteller et al. 1973). Therefore, multiple regression analysis has been made of mean annual and mean January temperatures at coastal and inland weather stations (Juneau and Whitehorse) near the Juneau Icefield. This substantiates the time position of high and low nodal points on the curves (Fig. 20), and adds confidence that the trends and fluctuations shown in Figure 18 are valid. In turn, this reveals that in contrast to higher-than-normal sunspot numbers in the 1950s and 1960s, there was a noteworthy drop in winter temperatures over these years, accompanied by the previously noted downward shift of névé-lines and generally increased net accumulation on the glaciers at intermediate and lower elevations.

The sunspot record in itself is pertinent only as an indication of relative solar energy variations. If there is any real time lag in ambient temperature behind the highs and lows of solar radiation, it might reflect some amelioration along this coast produced by the heat sink character of the North Pacific Ocean. This concept is worth investigating and may be supported by changes in temperatures of oceanic surface water in the Gulf of Alaska, which have warmed over much of this century. In fact it has been reported by Scripps Institution in a paper by Namias (1979) that the 1961-67 period "averaged warmer than normal over this entire area of the central North Pacific...". He adds "we shall not comment now on this amazingly large and long-lasting abnormality except to say that it would be associated through complex lag effects with the climatic warming observed over many portions of the world from the turn of the century through the 1950s" (Willett 1950; Mitchell 1963). In this assessment, an increase of $+2°F$ ($+1°C$) above normal was reported in ocean surface water temperatures over a $10°$ square centered at $30°N$ and $165°W$, with upwards of a $1.5°F$ ($+0.8°C$) increase found in the northeastern Gulf of Alaska in a zone centered at $60-62°$ N and $140-145°W$. The winter atmospheric circulation generated by this warmer thermal regime in the upper layers of the sea during the 1960s "consisted of strong and southward displaced cyclogenesis ...resulting in climatic cooling over the eastern two-thirds of the United States" (Namias 1979).

Ocean surface water warming in the Gulf of Alaska during the 1950s is also supported by measurements obtained on other oceanic research vessels then working in the area (R. Fleming and Ian Fletchier, University of Washington; Ken Parker, International Pacific Halibut Commission, pers. commun.) and by studies of historic sea-surface temperatures around the globe (NOAA 1983). The information is too fragmentary, however, to delineate any short or long-term periodicities or to detail the extent of major shifts in these North Pacific warm water zones that could play an important role in the changing climate of Southeast Alaska. Such oceanographic considerations as they relate to the glacial conditions are somewhat beyond the scope of this present review, but the air-sea interface effect along this coast is recognized as a potentially key element that will have to be addressed in the future in more detail. This interpretation is somewhat simplistic in the light of research by Webster (1982), which considers contradictions and complexities in the atmospheric response to sea surface temperatures, including the importance of time scales involved and seasonal out-of-phase heating of the extra-tropical troposphere by sea surface waters.

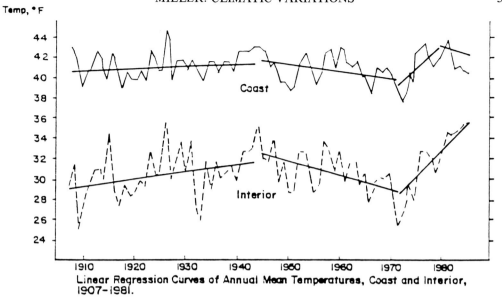
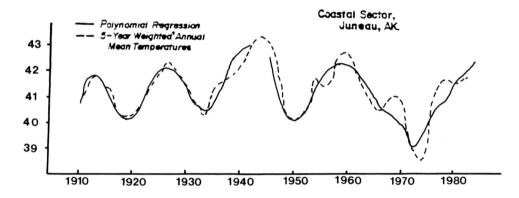
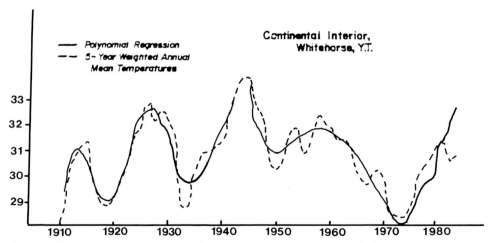

Figure 20. Polynomial regression (two-part) and five-year weighted moving average of mean annual temperatures since 1907 in the coastal sector and continental interior of the Northern Boundary Range, based on Juneau and Whitehorse records (after Jones 1975 and Lamb 1985).

Solar-Climate Considerations and Superposed Cycles

It is a precept of atmospheric physics that high sunspot numbers represent correlation with peaks of solar energy associated with maximum charged particle (corpuscular) radiation. Working backward along the chain of such curves shown in Figures 18 and 19, a series of lesser solar-energy cycles is revealed at roughly 11 and 22 years. The latter is the so-called Hale or Double Sunspot Cycle (Mitchell et al. 1979). These in turn are superposed on an approximately 90-year cycle involving high numbers which, in a generally regular way, trend to low and back again to high values.

Deviations that are not solar-induced can be analyzed by Fourier technique, although this reduces complexities in the curves and resolves fundamental components. With the regression analysis it is not wholly essential as the fact of cyclicity is well documented (Fig. 19) and the precise delineation of phasing of nodal points in these curves is not of major import in the analysis. Likewise details of positions of lesser deviations are of no immediate concern and the approximately 90-year periodicity is quite apparent (e.g. in Fig. 18, see low points in 1815 and 1905). A significant aspect, however, is that the shift from high to low values, and seemingly warm to cool conditions, and vice versa, takes place remarkably fast. This suggests a similarity of sharp changes in climatic stress. There is reason to believe that the 90-year cycle is superposed on sets of even larger cycles. In this respect other investigations reveal positive correlation between solar-energy variations and oxygen 18, measured in ice cores from Greenland (Langway, Dansgaard et al. 1973) and from Antarctica. Some of these suggest a 78-year periodicity, not far from the 82-year periodicity noted for glaciers in the eastern Alps (Hoinkes 1969), nor from the 80-90 year periodicity observed on some key glaciers in our Alaskan survey (Fig. 3 and Fig. 18, top). The assumption noted above is that these data have positive relationship with ambient temperature variations over the same years. That longer-range trends are also involved is suggested by inverse correlations calculated for the past 2,000 years between solar activity and carbon-14 variations (Bray 1967, 1968, 1971). Along with oxygen isotope dating, this further demonstrates validity of the solar activity index (Bray 1970) as a clue to solar-climate, solar carbon-14, and climate-glacier relationships over many centuries (Eddy 1977; Willett 1975).

On the question of secular climatic changes in the Neoglacial, it has been well documented that beginning about 2500 years B.P., there was a conspicuous climatic harshening, which culminated in the 1600s. This is defined as the low nodal point of the Little Ice Age. A period of minimum solar activity during this early part of the Little Ice Age is shown in Figure 18 by a great dearth of sunspots over the 70-year period between 1645 and 1715. This is well known as the Maunder minimum. The relationship is also demonstrated in the plot of annual mean sunspot numbers in Figure 19 covering the period 1610 to present (Eddy 1979). In Alaska, this cool period was followed by the multiple-phased oscillation of coastal glaciers indicated in Figure 3 (Miller 1946, 1963, 1971). The main ice advances represent successive cycles of about 90 years, but on the larger trunk glaciers this is expressed as a double cycle of 180-200 years. Appearing to corroborate this interpretation is an approximately 200-year periodicity between Little Ice Age culminations on the Taku, Hubbard and other large trunk coastal glaciers of Alaska (again refer to Figs. 3 and 5, the glacier behavior record noted in Fig. 18).

There is evidence from glacial stratigraphy in the Holocene (Denton and Karlen 1973; Miller 1973) including analyses of oxygen 18 in Greenland and Antarctic ice cores (Purrett 1970; Langway and Dansgaard 1973) that there have been climatic changes with comparable intervals and some larger, with the cooling spaced roughly at 180, 700-940, and 2400-2600 years. On such longer-term periodicities, the indicated 90-year solar pulsation is presumably superposed. Such secular trends during the Little Ice Age and the Holocene suggest that if natural climatic controls prevail, another lesser Ice Age may be in the offing. In round numbers, this could develop within 200 to 900 years. Such a portent is based on the aforementioned periodicities and is reinforced by our glacier research and by the solar-climatic interpretations of Willett (1975). In turn, the indicated

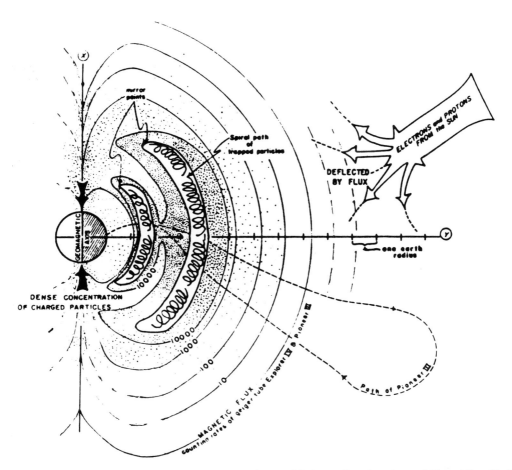

Figure 21. Schematic showing magnetosphere with intensity structure of Van Allen Radiation Belts and zones of charged particle concentration in the polar regions. In the equatorial zone, the magnetic field is compressed by the solar wind, with the flux distribution not uniform as depicted in this simplified sketch.

fluctuations both in the past and those expected to evolve in the future must be viewed as overlaying the larger variations of the Quaternary. This will be discussed in the later section that addresses additional geologic evidence and paleoclimatic interpretations.

Geomagnetically Channelled Corpuscular Radiation

One of the most significant geophysical discoveries of recent years is the identification of zones of geomagnetically trapped and ionized solar particles in the upper atmosphere. This was brilliantly theorized in 1912 by the Norwegian atmospheric physicist Carl Stormer. It was verified by the 1958 scientific satellites Explorer I and II, and further clarified by data from the trajectory of the Pioneer III rocket test, which first carried geiger tubes into space (Fig. 21). Russian scientists obtained confirmatory measurements in their Sputnik III test. Stormer's theory delineated what subsequently has been called the Van Allen Radiation Belts involving an immense region around the earth occupied by high-intensity zones of electrons and protons, some even beyond the ionosphere. Here charged particles are temporarily trapped in the earth's geomagnetic field, having been emitted from the sun's surface after transmutation of hydrogen to helium during solar storms and evidenced by prominences and flares immediately below the surface of the sun (Maran

1981). In simplistic terms, the four basic particles in the universe are neutrons, protons, electrons and positrons. In addition to the significant outflow of two of these as ionized solar particles (electrons and protons), important and variable amounts of energy are transmitted as wave radiation in the ultraviolet to extreme x-ray portions of the radiation spectrum. But it is the particle emissions that control the electrical character of the earth's upper atmosphere. It is probable that these may also influence atmospheric pressure nearer to the surface of the earth and from this global wind circulation, as well as changes associated with intense auroral storms in the ionosphere. The resulting energy input may also cause fluctuations in the wave-length of the circumpolar westerlies, changing the location of the tropospheric ridge along the coastal mountains of Alaska-Canada. The relative locations of the mean low-pressure systems in the Gulf of Alaska (Aleutian Low) and areas of anticyclogenesis in the interior (Canadian Polar High) would thus be a reflection of ridge location. In turn, temperature and accumulation changes in the Boundary Range would be affected.

Figure 21 also illustrates the intensity structure of the two main radiation zones and the two zones of charged particle concentration in the polar regions, producing the Aurora Borealis and the Aurora Australis. This figure is a two-dimensional schematic to be thought of as a three-dimensional figure of revolution around the earth's magnetic axis (Evans 1982). The periodic concentration of radiation corpuscles at each end of the earth's geomagnetic axis agrees with current concepts of an unsteady solar state (Mitchell 1965, 1976; Roberts 1976, 1976; Eddy 1978). Accentuation of variable solar energy in the polar regions through funneling of charged particles into the ionosphere above the earth's magnetic poles may exert a thermal influence on the general circulation of the atmosphere originating in the polar regions. Although the mechanism is unclear, this may affect expansion and contraction of the polar anticyclone and related shifts in the arctic front across the interaction zone between pressure cells on the Southeast Alaska coast (Fig. 16, and Miller 1956). This interpretation of pulsations in the larger-scale atmospheric circulation is not at variance with major paleoclimatic shifts suggested by a palynological study of Holocene bogs in the Taku and Atlin districts (Miller and Anderson 1974). This is discussed in the following section.

In Figure 18, an attempt is made to relate highs and lows of solar intensity to patterns of glacier advance and retreat. This is based on the previously cited modular concept that at times of high corpuscular radiation there appears to be increased temperature and precipitation along the northwest Pacific coast. The idea squares with the thickening of higher elevation névés and with corresponding advances of glacier termini a given "flow lag" years later. Again, timing of these events and duration of the lag depends entirely on geometric and geographic parameters of the glaciers involved. The model illustrates that, conversely, at times of low sunspot number with reduced corpuscular radiation, cooler conditions ensue, bringing less total precipitation and greater snowfall to low elevation sectors. This sequence is verified by our field measurements of increased accumulation on lower névés (e.g. as documented for Regime Zone B in Fig. 8), and subsequent advances of distributary termini a relevant number of "flow lag" years later. We have found not only the timing but also the intensity of these advances to be dependent on the combination of geographic constraints and the thermo-physical character of the ice (Miller 1974, 1976). Especially important here is the dissipator/accumulator ratio, which numerically relates the relative size of the areas of outflow below the mean névé line with the size of nourishment or input areas above the mean névé-line, which on temperate glaciers represents the equilibrium line.

Before proceeding to further consideration of the significance of these interpretations to paleoclimate, attention is again drawn to Figures 18, 20, and 24 and to the downward skewing of the running mean January temperature curve over the period of the last three decades. As for the five decades before this, in Figure 18 a shifting of the winter temperature curve to the left in the amount of the lag interval as discussed earlier reveals that the curves are tantalizingly alike. It should be noted here that the January temperature records are analyzed because they represent a period with the least cloud cover, due to the dominating spillout of clear air parcels associated with the polar anticyclone. In such circumstances any changes in character of the winter continental air mass will be revealed. Complicating the picture, however, is the fact that applying the same analysis to the curves since 1950 shows comparisons that are quite dissimilar and to some extent inverse in character, right up to the 1980s. Also characterizing the comparisons in these recent decades is a notably greater oscillation in the magnitude of each set of values, though this may further support basic solar control. Other implications of the striking changes in character of the comparisons over the past quarter century are considered at the end of this paper.

In spite of some seeming pattern inconsistencies suggested by short-term aerological conditions, the general character of the atmospheric circulation in this region is suggested through the simplified surface circulation model, discussed with respect to Figure 16. It should be clear that this model does not purport to represent climatological conditions at any one time, but rather graphically to depict the major parameters pertaining over long periods of geologic time.

The aim has been to fit the model to the geologic facts of glacial fluctuation and phasing and to be consistent with climatological conditions and physical laws that govern atmospheric behavior. Figure 22, derived from secular aerological information averaged on a monthly basis over a period of years, shows the major climatological storm tracks across North America, based on known passage of cyclonic centers and tropospheric troughs as they move from west to east across the coast of Alaska and British Columbia (Haurwitz and Austin 1944). Figure 23 shows comparisons between the observed sea-level pressure distribution of cyclonic and anticyclonic cells in the Gulf of Alaska and across western North America in the late spring and early winter (June and December), as well as the number of days of maximum development in these months (after Searby 1969).

As for glacier regimes, the descriptive model (Fig. 16) addresses this by showing lateral shifts of storm wind directions on the North Pacific Coast as they occur on the rotating edge of low pressure cells. The term storm winds refers to the direction of peripheral winds and the location of maximum storm energy in the frontal zone between opposing pressure cells. Thus defined, these moisture-laden surface winds are peripheral to the cyclonic cell, with the diagram generalizing their position along the Arctic Front.

According to the model, at times of high solar energy or warming conditions, the continental cell is postulated to receive increased solar energy from charged particle input in this region of the magnetosphere. In such circumstances, the maritime cells dominate. Without addressing the still-unanswered question of how concentrated corpuscular energy can reach the lower atmosphere, a simplifying assumption is that it does. This then means more energy received with some effect on the high-pressure area over the Arctic land mass, causing it to become warmer and less dense.

It is further hypothesized that in this circumstance the North Pacific low moves inland toward the continental margin. This would coincide with the Arctic Front shifting inland, also bringing the peripheral southeast storm winds inland over the Alaska-Canada Boundary Ranges (Fig. 16, right). In such a situation, low-level coastal glaciers, because of their more maritime location, would receive less accumulation than those at higher elevation on the continental flank of the range. The expected result would be a raising of freezing levels, with low-elevation névés receiving decreased nourishment on both flanks of the range. Correspondingly, it should be expected that high elevation névés receive more accumulation because of the encroachment of the maritime cell with its warmer and wetter condition (Fig. 17, right).

Conversely, at times of low solar energy, the continental pressure cell would become colder and its density increase. According to the model, this should be expected to expand the cell's periphery coastward (Fig. 16, left). Such a condition would drop the freezing level with consequent increases in accumulation on the lower névés. The greatest increase would be expected in the low elevation maritime sector (Fig. 17, left). Under this set of circumstances lower névés also receive low rainfall, thereby decreasing ablation and further tending to positive mass balances. With this a

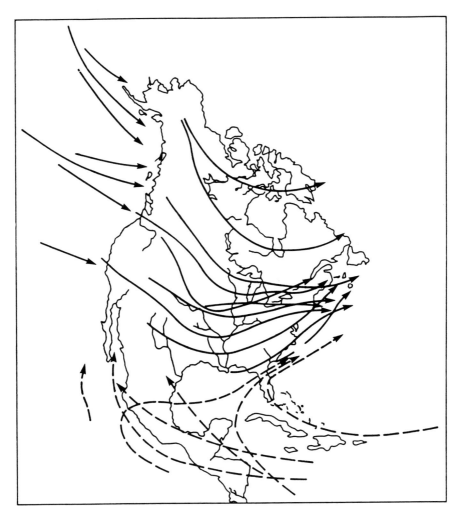

Figure 22. Major climatological storm tracks, based on passage of cyclonic centers over the North American continent (adapted from Haurwitz and Austin 1944).

lateral shift of the Arctic Front and its allied storm winds should take place. This generalized concept of the surface circulation provides a working model that is commensurate with our documented observations of out-of-phase advance and retreat on the high and low névé glaciers in the prototype locale, as well as throughout the Alaska coastal region (Figs. 2, 11, 12).

To illustrate the logical consistency of the model and its interpretive value, some meteorological relationships are reviewed during a recent high/low temperature transition at the Juneau station of the National Weather Service (Colman, pers. commun.).

Specifically, in November and December, 1983, the Alaskan Panhandle experienced extreme dry and cold conditions, as did most of the northwest as far south as Seattle and inland to north Idaho. In Alaska, this brought the Arctic Front substantially offshore to the southwest as depicted in Figure 16 (left). Our field records show that this produced unusually heavy accumulation at low elevations, as expected from the modular explanation in Figures 16 and 17 (left).

In contrast, through January and February 1984, the Panhandle came under continual maritime influence, with the Arctic Front remaining stationary far inland in Canada. This fits the model of Figure 16 (right). Thus, at Juneau, January temperatures were 10.3°F (5°C) above

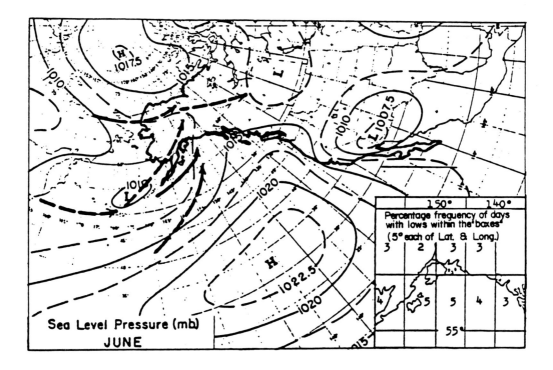

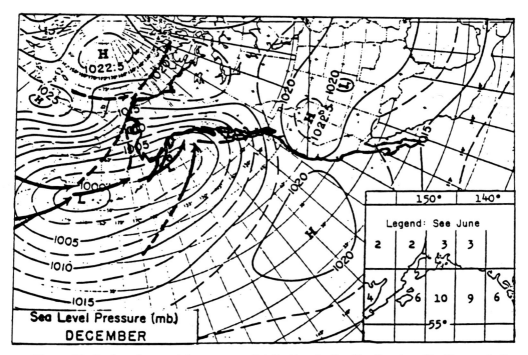

Figure 23. Surface barometric pressure distribution in the Northeastern Pacific area in June and December 1968, showing percentage frequency of days with lows (after Searby 1969).

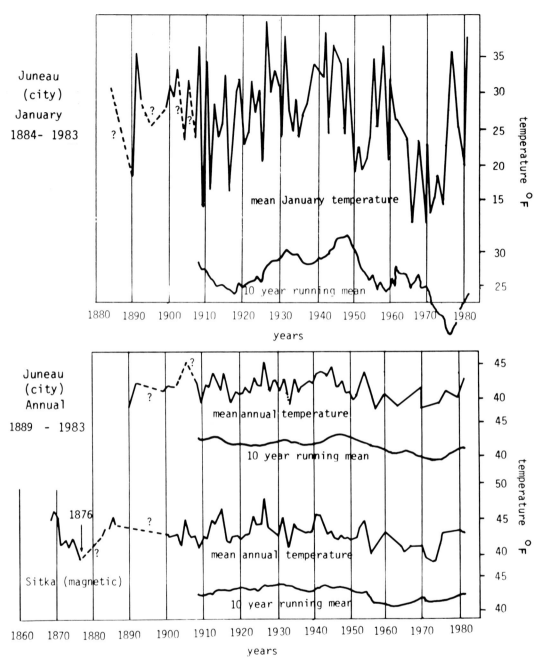

Figure 24. Annual and January mean temperature trends at Juneau City, 1884-1983, including 10-year running means.

normal, with precipitation 2.37″ (6 cm) above normal. February proved to be a full 8°F (3.8°C) above normal, with 1.8″ (4.6 cm) of precipitation above normal. With the warmer maritime air mass dominating, freezing levels were accordingly higher. This verified situation followed the model shown in Figure 17 (right). The integrated result of these shifting conditions during the 1983 autumn and winter accumulation season, abetted by 1984 summer ablation, was much lower than normal névé-lines for that budget year.

It may be coincidental that the 1983-86 interval represents the low energy years of predicted and discerned sunspot numbers in this decade. A view of longer-term cyclicities, however, is tantalizing. The fact that we are dealing with independently derived evidence of solar energy changes adds significance to the observation that these glaciers have experienced fluctuations approximating the 90-year cycle. Further support is given by a correlated 45-year half-cycle of ambient temperature variations shown in the Southeast Alaskan meteorological records in this century. Related here are the cool 1910s and the warm 1940s, as indicated on the temperature curve between the 1890s and the 1940s in Figure 18, and in the Juneau and Sitka winter temperature records shown in Figure 24, borne out also by the multiple regression plots in Figure 21. From this we can see that between the First and Second World Wars the city of Juneau's mean January temperature rose some 7½°F. (5°C). From lapse-rate analysis this would have caused a freezing level rise of some 615 m over that 30-year span. This brought the zone of maximum annual snow accumulation on the Juneau Icefield from the lower level névés to the higher névés. It also correspondingly raised the mean névé-line, as documented by our continuing glacier mass balance measurements. Of further significance is the documentation in Figures 20 and 24 that since the mid-1940s there was a double magnitude drop in mean annual and mean January temperatures at the Juneau station through the 1960s. This circumstance explains the consistent lowering of the mean and seasonal névé-lines on the Taku glacier over this more recent 30-year interval as shown in Figure 7. The corresponding increase in net accumulation that has been recorded on the lower Taku Glacier névés should also be noted. These névés lie at the same elevation as those nourishing the recently re-advancing Twin Glaciers immediately to the east.

This provides an explanation of the cause of periodicities in the glacier fluctuations record for the Alaskan Little Ice Age, and as depicted in Figures 3, 4, and 18. It also helps to explain teleconnectional similarities in glacial behavior in high middle and mid-latitude localities elsewhere in the world.

Additional Geologic Evidence and Paleoclimatic Interpretations

In broad terms comparative data on polar sea ice changes in historic time through the 1960s have mirrored the short-range glacio-climatic pulsations and trends on the Alaska coast (Miller 1956b; Wittmann and Miller 1972). In fact, by comparing satellite pictures to atlases and ship reports, it has been documented that summer pack ice around Antarctica has shrunk some 240 km since the 1930s, though a slight increase has been reported by the Royal Meteorological Society since 1979 (also see Chiu 1983), reflecting the fluctuational character of the marginal ice zone. Average spring and summer temperatures on the fringes of the Arctic Basin in northern Canada and Siberia are known to have risen some 2°F (1°C) between the mid-1930s and the mid-1970s (Kukla and Gavin 1981, 1982). There is some evidence, however, that a pattern of cooling that developed after 1940 "was more complex than the warming that affected the whole Arctic during the 1920's and 1930's" (Kelly et al. 1982). Even more dramatic are the comparative weather station statistics from Svalbard in the high Norwegian Arctic, where records reveal an 18°F (8.5°C) rise in mean winter temperature during the first half of this century (Willett 1950). This effect has global significance because of its relationship to changes in the thermal character and trajectory of the Atlantic Gulf stream. Trends in arctic temperatures have in general been similar to those for the northern hemisphere over the past century, but it is increasingly clear that certain regions differ greatly in the effect and are particularly sensitive to long-term variations. Short-term variations in temperature, and hence sea ice changes, in the arctic regions appear to be "most pronounced close to major regions of sea ice production and decay" (Kelly et al. 1982). On a broader scale annual means from 340 northern hemisphere reporting stations between Lat. 55° and 90° show a temperature increase of nearly 1°F (0.4°C) just in the 15 years since 1969.

With respect to earlier Holocene and Pleistocene climatic conditions in our prototype region, comparative field studies on Quaternary moraine sequences and palynology profiles in kettle and pit bogs have been made on the coastal and interior flanks of the Alaska-Canada Boundary Range. A transect of key sites extends inland from Juneau to Atlin Lake in northern British Columbia (Fig. 25). In the Atlin valley a brace of asymmetrical snow-drift trimlines on opposite side-walls of the Cathedral Massif Glacier corroborates significant changes in dominant storm-wind directions over the past century (Fig. 26). Such glacial geologic information lends credence to the surface circulation model depicted in Figure 16. Also out-of-phase glacio-climatic fluctuations during the Holocene (the past 10,000 years) are shown by comparing climatic characteristics and trends in the maritime versus continental sectors of this Cordilleran region. This gives insight into secular climatic changes that have occurred since Valders time (see following section).

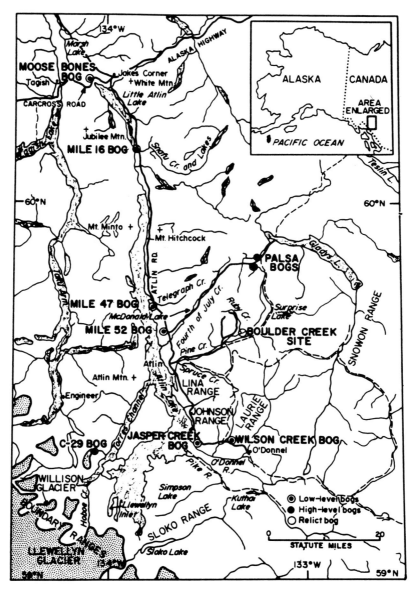

Figure 25. Atlin Lake region, showing location of Holocene bog sampling sites.

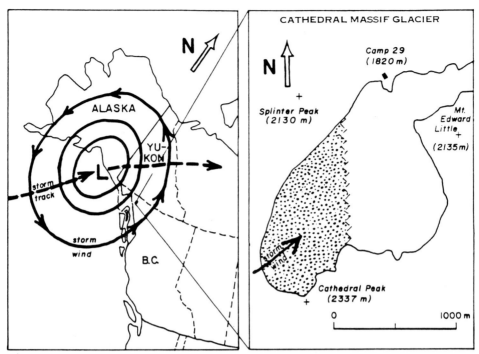

a) Storm track passes northward, with peripheral circulation bringing southwesterly storm winds.

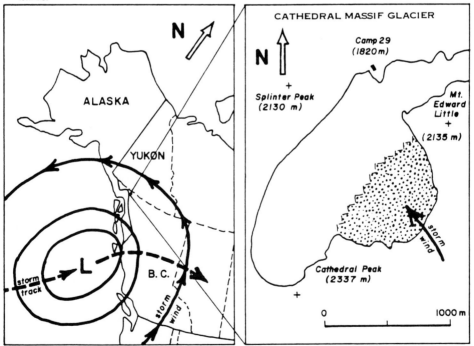

b) Storm track passes southward, with peripheral circulation bringing southeasterly storm winds.

Figure 26. Dominant autumn storm wind direction changes on Cathedral Massif Glacier over past century in Atlin sector of the Juneau Icefield. Southeasterly storm winds characterize present conditions.

Radiocarbon dating of key horizons in the bog stratigraphy of zones peripheral to the Juneau Icefield gives chronologic meaning to the tabulation in Figure 29 (Heusser 1960; Anderson 1970). This information stems from pollen and spore analyses such as from the Mile 16 bog illustrated in Figure 27. Sample palynology records are shown for this bog and also the Jasper Creek bog in Figures 28a and 28b. The table (Fig. 29) shows climatic trends for each region where today mean annual precipitation regimes are 228 cm at sea level in the temperate Sitka spruce and hemlock forest region of the Juneau sector (Taku District) compared to only 25 cm in the dry semi-arid white spruce and lodgepole pine forest region of the Atlin district. Also in this table references to "warmer and drier" and "warmer and wetter" designate precipitation regimes in each sector that characterize a period of increased storminess during the Thermal Maximum (Hypsithermal Interval). All other adjectives (e.g. "warm," "cool," etc.) are relative to the respective Thermal Maximum conditions in these geographically and climatologically different regions on opposite flanks of the Boundary Range. The Holocene climatic sequence interpreted from geobotanical evidence reveals pronounced secular shifts in the Arctic Front and related mean storm wind displacements along this North Pacific coast. The out-of-phase climatic character of the coastal vs. inland bog profiles reflects the same kind of peripheral storm wind shift described earlier with respect to recorded variations of glaciers in this region since A.D. 1500. It also coincides with analysis of meteorological trends in the coastal sector where weather records have been maintained sporadically since A.D. 1840, and consecutively so since 1889 (Fig. 24). A detailed report on the out-of-phase comparisons in this palynological record is given by Miller and Anderson (1974a, 1974b).

As for climatic changes in earlier geologic time, the literature is extensive (National Research

Figure 27. Mile 16 bog in glacial pit depression in forest fire burn area of the Atlin Valley (see Figure 25). (FGER photo, Sept. 1968).

Council 1982a). Stratigraphic and tectonic studies on the Tertiary (e.g. Wolfe and Poore 1982) and on the Paleozoic (e.g. Boucot and Gray 1982) have presented evidence of pronounced paleotemperature variations. In such studies, there are many differing conclusions about the intensity of warm intervals and the range of thermal fluctuations. Much of this confusion is the result of unidentified local and regional variations, even as those of today. These differences of opinion also stem from the use of imprecise sampling methods and the fact that records have been taken at widely spaced stratigraphic intervals, often leading to over-generalized correlation of samples. Inaccuracies have also been introduced by the varying techniques from which paleotemperature estimates are made, with further confusion arising from attempts to relate terrestrial and marine stratigraphic trends. In an effort to overcome such difficulties, the present study has emphasized working from the known of present glacio-climatic conditions into the unknown of the past.

Uniformitarianism is considered to apply, based on the premise that short-term solar-related climatic oscillations have prevailed throughout the Neogene. This assumption is also made regarding earlier geologic time and the larger climate variations caused by our planet's changing orbital geometry. This includes the approximately 10,000, 20,000 and 40-50,000-year intervals between intraglacial conditions, such as in the Wisconsinan Glacier Age, and indeed up to 100,000 and possibly 500,000-year variations between interglacial and glacial conditions as suggested by recent ocean core records as well as terrestrial stratigraphy (Hayes, Imbrie and Shackleton 1976; Fisher 1982; Miller 1975a). From this type of sequence we can justifiably return to the key question posed earlier in this paper, and state that it is not surprising to suggest that another major Ice Age may develop in the geologically near future of our planet. This would, however, probably not be before 10,000 years from now, and more likely not for another 30,000 years (Willett 1975), unless an entirely new set of climate controls comes to prevail. Regardless, there is no solid reason to conclude that the Quaternary Period has yet ended.

Reverting to even more ancient glacial times, a recent study of Australian Pre-Cambrian glacial varves relating sunspot periods to changing environmental conditions in a lithologic sequence apppears to corroborate the uniformitarianism principle. This study by G. E. Williams (1981) shows small-scale cycles remarkably similar to those we have found in the fluctuational history of modern Alaskan glaciers. Williams describes the Pre-Cambrian Elatina formation at Pichi Richi Pass (32°25'S, 137°59'E) and at other localities in South Australia's Flinders Range where he observed conspicuous cyclic laminations of alternating red and brown siltstones and fine sandstones over a thickness up to 60 m. The formation is interpreted as a deposit on a western shelf of the Adelaide Geosyncline during Marinoan (late Adelaidean) glaciation. Correlative proximal outwash sandstones have been found 150 km to the northwest, suggesting the provenance was an ancient ice-sheet in the low cratonic region west of the shelf. This glaciation was positioned at a low paleolatitude and reported at 680 my B.P. (see Fig. 1).

Cycles consequent to these southern hemisphere Pre-Cambrian varves are interpreted as strong climatic flucuations of about 11, 22, 145 and 290 years. A weaker yet distinct period of nearly 90 years is also noted. Williams' interpretation states that "the 11, 22, and 90-year cycles equate with sunspot maxima as indicated by tree-ring studies." The importance of his work in the present sense is that it closely yet independently parallels ancient ice-sheet glacio-climatic pulsations to the short-term periodicities we have found in existing Alaskan mountain and coastal glaciers. It supports our concept that such variations in the geological record have important implications deeper into the past, indeed far earlier than the late Neogene record that our studies document. This is also borne out by solar-terrestrial climatic patterns suggested in varved sediments of other geologic ages (Bradley 1929; R. Anderson 1961). With this information the synthesis of repeated glaciations noted in Figure 1 takes on further meaning. It gives support to the need for glaciologists and climatologists to exchange information with the stratigrapher, the sedimentologist and the stratigraphic paleontologist in order to improve the reliability of glacial interpretations and correlations.

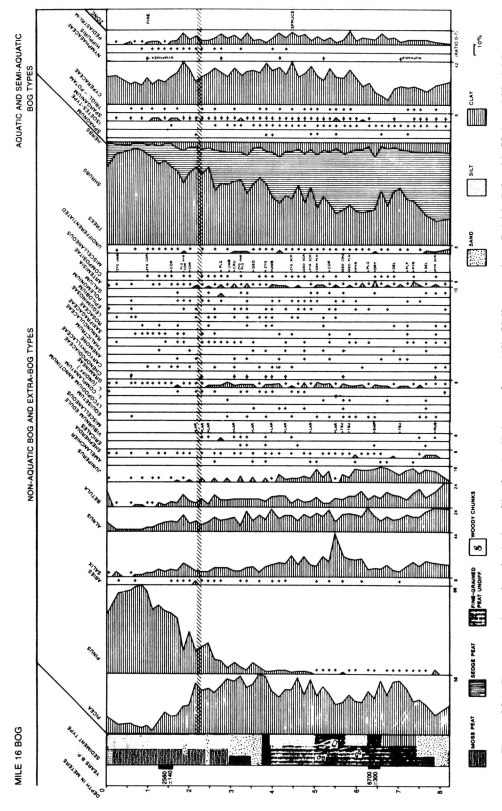

Figure 28A. Representative palynological profiles for the Holocene from Atlin Valley, northern British Columbia-Yukon. Mile 16 Bog; locations in Fig. 25. (After Miller and Anderson 1974).

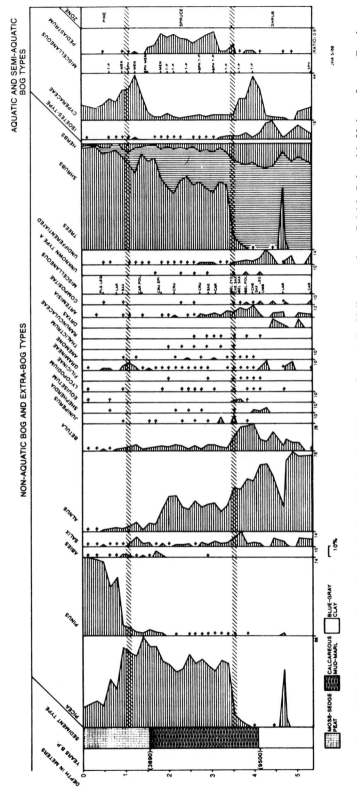

Figure 28B. Representative palynological profiles for the Holocene from Atlin Valley, northern British Columbia-Yukon. Jasper Creek Bog; locations in Fig. 25. (After Miller and Anderson 1974).

Trend Complications and Assessments

Rectilinear or plug flow is the usual type of discharge for most glaciers in the Alaskan inventory and also for those with teleconnectional similarity elsewhere in the world. The question of surging flow involves an apparently abnormal phenomenon, but it may still be indirectly important in a meteorological sense because glaciers do have unusual behavior in connection with climatic conditions (Hoinkes 1969). The nature of such behavior, however, needs to be clarified if the long record of Alaskan glacier terminal fluctuations is to be used with full confidence and the global climatic significance totally appreciated. Efforts should, therefore, continue to be made to study the possibility of climate-related periodicities in both large and small magnitude glacier surges and in kinematic waves, as well as in continuing surveillance of normal discharge glaciers of the climatically sensitive type. This need has stimulated emphasis on our continuing long-range total systems research.

Because we consider coastal glaciers of the Alaskan Panhandle to be prototypes of worldwide glacioclimatic trends, their investigation has included information from consecutive annual glacier surveys. These have been made every summer since the mid-1940s and have aided understanding of the nature of short-term glacier pulsations as well as longer-range secular change. As alluded to in the foregoing, the sequence of superposed solar "cycles" describes those of short duration as 11, 22, and 90 years. The latest 90-year ambient temperature crest was reached in the 1940s (see temperature curves in Figs. 18, 20, and 24). The analysis has been complicated by uncertainty as to whether solar radiation and ground temperature trends are in phase, which has led to the idea of lags relative to the thermal character of adjacent surface ocean waters affecting ambient air temperatures along this coast. Thus warmer annual and winter temperatures may not directly correspond to contemporaneous solar highs, which sometimes leads to a seeming contrast of warm winters on the coast being coincident with strong continental high-pressure systems and intensified winter cold inland. This latter effect could possibly be a consequence of expansion of the polar anticyclone during decades of lesser radiation, coincident with relatively warmer coastal temperatures associated with a lagging increase in maritimity in that sector. Clarification of this apparent paradox is attempted below, with further insights provided by incidences of unusual weather observed during recent decades. Here too an increasing concern is introduced about evidences of the effects of several types of atmospheric pollution.

By referring again to Figures 18, 20, and 24, and the trend of winter temperatures between the 1940s and the present, additional complications come to our attention. Noted first is the departure in positive correlation and parallelism in the 11-year running means of annual sunspot values and the 11-year running means of January temperatures at Juneau as observed since the last century. The pertinent years of parallelism embrace the interval of the 1890s to the 1940s. By the 1950s, however, this parallelism disappeared, almost reversing to an inverse correlation, with peaking of solar values still similar to thermal peaks but representing a much greater than usual dip in winter temperatures. The regional significance of this departure is underscored by unusually high correlation coefficients (0.90 to 0.94), found in the Juneau temperature trends when compared to the assemblage of data representing the mean of 64 coastal weather stations from Ketchikan to Anchorage. Even when adjusted for lag effects, an inverse correlation is displayed in the trends of solar energy and January temperatures from the mid-1950s to the 1970s. The strongest dips in mid-winter temperature have occured since the mid-1960s (Fig. 24). But regardless of this increased disparity the highs and lows in the solar energy and temperature curves have continued to have parallel trends. Something new in the process is suggested by the strong upward trend in winter and annual temperature means since 1969 shown in the Juneau and Whitehorse records, as well as at the 340 Arctic stations noted previously. The relationships are confusing and more sophisticated statistical analysis is needed to provide more reliable answers than suggested here. But nonetheless the apparent inverse correlation from the 1940s to the 1970s is of special interest

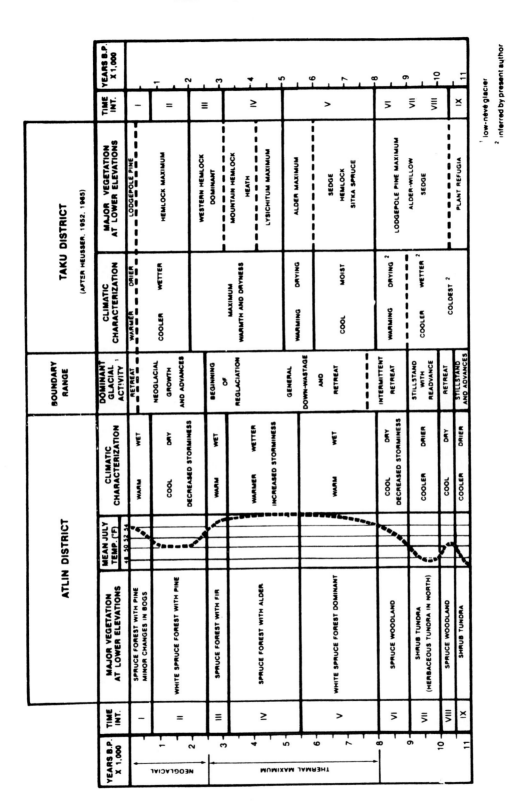

Figure 29. Chart of Holocene glacio-climatic chronology for the North Pacific continental margin, showing out-of-phase climatic characterization in the continental (Atlin) versus maritime (Taku) sectors of the Alaska-Canada boundary range.

because of recent estimates that by the end of this century global atmospheric pollution from industrial hydrocarbons should begin to exert a notable effect.

The results of an on-going multidisciplinary study at the Lamont Geological Observatory and the NASA Institute for Space Studies at Columbia University are pertinent in delineating the earth's probable climatological variation from atmospheric pollution. Answers have been sought in marine geology of the oceans where the record of the past has been locked in sediments. They relate to a greatly complex pattern of vertical and lateral currents which redistribute heat and anthropogenic but biologically mediated carbon dioxide (Plass 1959; Broecker 1984; Broecker et al. 1970; Hansen et al. 1981). The disciplines involved in this research go far beyond those discussed in this paper, for they include geochemistry, volcanology, micropaleontology, physical oceanography, and biology as well as geology, atmospheric sciences and glaciology.

Atmospheric Carbon Dioxide Trend Since the 1950s

Beginning to emerge from the Columbia study is a surprising picture of the timing, volume and aerial coverage of glaciers, and the ocean's temperature and biological productivity, plus new information on the driving mechanisms in these variations. One increasingly clear indication is that the global atmospheric system has been perturbed by an unusually large increase in carbon dioxide released in recent decades by the combustion of fossil fuels and the effects of rapid commercial deforestation of virgin timber lands, especially in the tropics. Recent reports by the Environmental Protection Agency and the National Research Council/National Academy of Science substantiate concern that global warming from the carbon dioxide Greenhouse Effect may be immediately in the offing and indeed may even already be underway (National Research Council 1982c, 1983). The scientific community has already become engaged in much dialogue on this issue as exemplified by a symposium held at the University of Alaska in April 1982 on the potential effects of carbon dioxide-induced climatic changes in Alaska (McBeath 1984).

In this connection relating to summer sea ice observations since the 1930s and especially since 1973 (Kukla and Gavin 1981), "observed departures in the two hemispheres qualitatively agree with the predicted impact of an increase in atmosphere CO_2," though unexplained details of the mechanism preclude a conclusion regarding a cause-and-effect relation. Also relevant is a recent general circulation model study of the atmosphere (Hunt 1984), which indicates that sea ice is readily produced at high latitudes regardless of the existence of land. It is suggested that although land may be necessary to promote glaciation, through geologic time polar ice sheets may be "the norm rather than the exception for the Earth." This would mean that no major perturbation in the global climate may be necessary for ice ages. In this case the problem would actually lie in accounting for interglacial warm periods. A most likely solution could involve large increases in atmospheric CO_2.

Dr. James Hansen and others at the Goddard Space Flight Center and its Institute for Space Studies, allied with Columbia University's Department of Geological Sciences and the Lamont Observatory, have calculated the expected carbon dioxide increase over the next century (Hansen et al. 1981). The forecast is a global warming of about 2.5° to 4°F (1½ - 3°C). Corroborating this, the EPA report predicts that levels of carbon dioxide in the air created by the burning of fossil fuels could result in an increase of 3.6°F (2°C) by the year 2040 and a 9°F (5°C) rise by 2100, representing "an unprecedented rate of atmospheric warming." The report further states that "temperature increases are likely to be accompanied by dramatic changes in precipitation and storm patterns and a rise in glacial average sea level." This could be as much as .6 m by the year 2025, due to the melting of polar ice (also see Etkins and Epstein 1982). This temperature prediction will be testable on a global scale by the 1990s, and if correct, by then the implications will be recognized relative to worldwide shifts in agricultural conditions leading in some areas to increased famine and flood plus potential disruption of environmental and economic systems and possibly severe stressing of political institutions. These portents beg the question whether the sensitivity of Alaska's coastal glaciers to climatic change as well as sea ice changes in the sub-polar and polar regions have indeed given notice of such a drastic global atmospheric influence.

Weather Anomalies and the Role of Volcanic Dust Clouds

Several significant short-term climatological events have been observed recently that have a bearing on the perspective of climatic influences during the Neogene. These are the unusual weather conditions experienced during the severe winter of 1977 in the eastern states, and later the exceptionally mild winter of 1982-83. The latter has been speculatively linked to a slight decrease in received solar radiation prompted by the explosive El Chichon volcanic eruption in Mexico on April 3, 1982 (Mitchell 1982) combined with that of Jalunggung Volcano in East Java just two days later on April 5, 1982. The 1977 situation is depicted in the diagram of Figure 30, which shows a strong departure of the early winter jet stream path over the western northern hemisphere. In strong contrast to the usual west-to-east circulation (zonal flow), which has been noted in Figure 22, there was a more meridional flow bringing moisture-laden air masses out of the Pacific and farther north than usual and in fact well into the high Arctic.

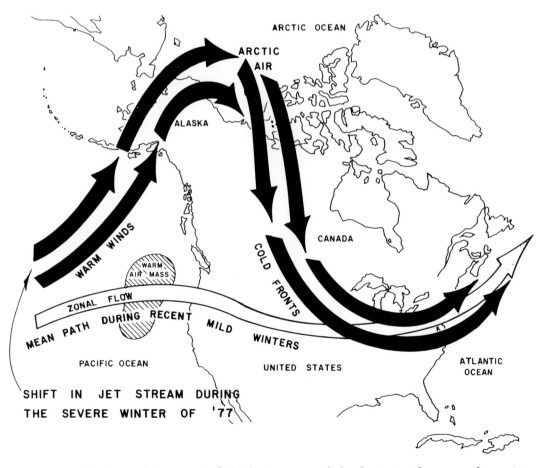

Figure 30. Unusual departure of the jet stream path in the troposphere over the western Northern Hemisphere in the early winter of 1977-78.

Related to this were abnormally elevated autumn temperature ranges in the Central Pacific tropics and a strong increase in sea surface temperatures in the Central Pacific and North Pacific, producing this notable shift from the more zonal mean flow path of recent mild winters (Horel and Wallace 1981; Kerr 1982). The result was a steering of unusually warm air into Alaska and abnormally cold air with excessively heavy snows into the eastern U.S. and Canada. This was influenced also by the Coriolus effect diverting the new path of the jet stream to the south out of the frigid Arctic Basin. This was the memorable winter of no snow in mountains from California to Utah and Colorado to Idaho with closure of nearly every ski resort in these states. Coincident, however, were abnormally heavy snows on the high névés of the Alaska-Canada Coast Mountains and also the best ski conditions in decades resulting from unusually heavy snows in the northeastern Appalachians and the Great Lakes region. On into the summer severe drought followed persistent floods and abnormally high water tables in the eastern, mid-Atlantic and New England states. This tropospheric condition is likened to that which may have pertained over many millennia of the maximum Pleistocene glacial ages resulting in growth of the vast Laurentide ice sheet of Quaternary time. It may also be significant that 1977 was a year of low sunspot numbers.

As for the El Chichon and Jalunggung effects, Murray Mitchell, senior research climatologist of the National Weather Service, has reported that in the ensuing year and a half, U-2 reconnaissance planes measured a cloud of solar energy-shielding volcanic debris 21 km up in the atmosphere carrying more eruptive dust than any similar cloud found in the northern hemisphere since the 1912 explosion of Alaska's Mount Katmai (Mitchell 1982). To give this perspective, El Chichon is credited with spewing some 25 cubic km of debris into the atmosphere, and Jalunggung with upwards of 40 cubic km. NOAA scientists at the Mauna Loa Observatory in Hawaii determined two weeks after these eruptions that this cloud of particulate matter was 140 times as dense as that measured over Hawaii following the 1980 eruption of Washington's Mt. St. Helens, which blew out less than 1 cubic km of its volcanic mass. The possibility for sudden and short-term climatic change from a cause of this dimension is large. It may also be underestimated in considerations of Pleistocene climate changes. The idea is supported by the Greenland ice-sheet core-drilling records, in which large pulses of volcanic activity are shown, going back as far as 10,000 years (Hammer et al. 1981; Langway et al. 1973).

Possibly related to the "monster" dust cloud dispersed into the upper atmosphere since 1982 and its effect on reducing radiant wave energy from the sun is the latest El Niño ocean current perturbation, to which so many disastrous global oceanic and atmospheric conditions have been ascribed (Cane 1983; Rasmussen and Wallace 1983; Canby 1984). Although we are still uncertain of its cause, El Niño involved so many unusual storms, floods, and droughts that new insights have been gained into the larger effects of abnormal ocean surface temperature changes, be they induced by direct solar energy causes or by indirect volcanic particulate cloud effects during the Neogene. Although particulate pollution can have a countering influence to global warming, and can also lead to grim consequences in the horrible event of a "nuclear winter," in the more long-term sense the effects of carbon dioxide contamination on global climate are probably more significant.

POTENTIAL FOR IRREVERSIBLE CLIMATIC CHANGE

What makes the potential of the carbon dioxide issue more alarming is the already documented 15-25% rise in carbon dioxide and other trace gases emitted into the atmosphere from refrigerators, plastics manufacturing, aerosol sprays, automotive exhaust, and oil and coal combustion over the past quarter century. The question is even more difficult because in spite of concentrated effects in the industrialized northern hemisphere where most pollutants are discharged, and regardless of the warming of some regions that has been described, the overall mean global temperatures noticeably decreased from the late 1940s to the 1970s. This is also the trend followed by the Alaska temperature plots in Figures 18, 20, and 24.

Recognizing the complicating nature of short-term climatic influences and the confusion introduced by disparate data from different regions of the world, the Columbia and NASA teams (Hansen et al. 1981) have taken into account geographic eccentricities in the earth's weather. In this they have included possible effects of natural solar activity and volcanic eruptions that could alter interpretations of the Greenhouse Effect through increases of carbon dioxide alone. From this approach they are convinced that looking at the whole century since 1900, the average global temperature has risen in these 80 some years 0.7°F (0.3°C). And now the EPA report cites a prospective further rise of 4°F (2.2°C) in the next 55 years. This is consistent with the 1%+ per year increase in atmospheric carbon dioxide measured at the Mauna Loa observatory since 1958 (Fig. 31), as well as with the noted increases in "smog" now being reported drifting across the Arctic Ocean to Alaska from industrial centers of the Soviet Union. From all of this evidence, the carbon dioxide issue has the potential of presenting to mankind an unparalleled dilemma. The problem is accentuated because the U.S. and U.S.S.R. have great supplies of coal and in this time of energy shortages are pushing for increased use of this carbon dioxide-producing commodity as an alternative to costly oil and controversial nuclear energy.

The problem is further exacerbated by the fact that not only has the Greenhouse Effect been given new credence by documentation in the Mauna Loa records and the realism presented in the 1983 NRC/NAS and EPA reports, but also by the fact that no technological breakthrough is yet in sight for removing carbon dioxide during the combustion of fossil fuels. A more fundamental difficulty is that carbon dioxide remains aloft for hundreds of years. This poses an even larger problem than simple global warmup because it introduces the issue of irreversible climatic

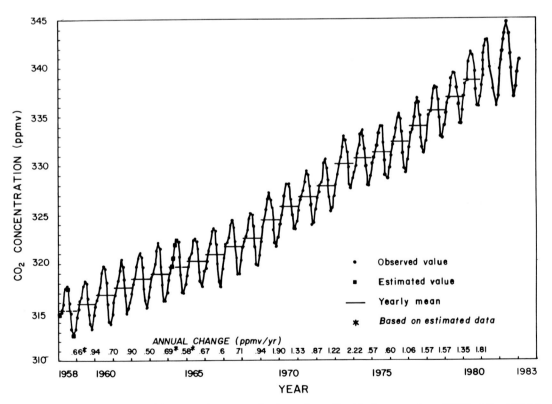

Figure 31. Measured increase in atmospheric carbon dioxide in ppm since 1957 at the NOAA Mauna Loa Observatory, Hawaii.

change. Of more immediate concern, however, is the probability that current political and economic factors affecting policies in energy use and fuel choice for civilization are so powerful that the carbon dioxide issue will not have a major impact on national or international energy decisions until there is pressing, unavoidable proof that global warming is at hand—and by then perhaps it will be too late. If this were realized by major world decisionmakers, such an effect could make highly inadvisable the larger exploitation of global coal reserves and indeed any further deforestation of the world's remaining pristine rain forests.

A General Systems Understanding of Paleoclimates

Adding to the complexity of global data correlation, the factors of temperature and precipitation often occur in out-of-phase variations in areas of comparable latitude both in the Northern and Southern Hemispheres (Willett 1950). In summary, however, by looking at the atmosphere as a general system the conclusion is apparent that atmospheric circulation is sensitive to variations in solar corpuscular radiation. This has been corroborated by teleconnectional evidence through the comparison of glacier behavior on icefields in Alaska, Iceland and Scandinavia, less so in parts of the European Alps, and strongly so in the southern hemisphere in Argentine and Chilean Patagonia and in South Island of New Zealand. These are all areas of existing glaciation at middle and high latitudes with most of them lying in interaction or mixing zones between continental and maritime surface air parcels and pressure cells. Also because these areas are positioned at opposite sides of the earth, the argument becomes convincing that the climatic cause of glacier variations is of global dimension.

The transportation of tropical warmth by the Kuroshio or Japanese current into the North Pacific and the Gulf of Alaska (Marr 1970) must also be taken into account. This is to the northeastern Pacific what the Gulf Stream is to the northeastern Atlantic. Not only have the surface water temperatures of the Gulf of Alaska notably increased in recent decades, but it is probable that these Pacific waters serve as that atmospheric heat sink that smooths out and possibly delays the effect of climatic change on the Alaskan coast. To understand the complicated character of worldwide natural climatic variations, such lag and associated temperature changes and cyclic relationships need to be considered in a general systems context.

Because today's environmental and energy issues are forcing consideration of the possible role of increasing atmospheric pollution, especially carbon dioxide and particulate concentrations, a number of interesting recent papers have addressed the problem. Some of these attempt to relate global temperature increases to carbon dioxide and also changes in global wind patterns to man-made pollution through the behavior of trace substances (Plass 1959; Newell 1971; National Research Council 1982b). During the 90-year interval between the 1870s and 1960, a documented increase of nearly 30 ppm by volume in atmospheric carbon dioxide represents a 9% rise, which is ascribed to the industrial revolution mainly through the burning of fossil fuels (Bolin 1970). This has been suggested to have caused the upward trend in temperatures experienced by our planet during that 90-year span. But Figure 18 suggests that natural solar radiation could also have produced much of this effect over the same interval. More alarming is the fact that there has been at least a 15% and more probably up to a 25% additional increase in atmospheric carbon dioxide since 1957 (Fig. 31). It is generally agreed that such can influence the Greenhouse Effect and lead to decreased albedo on exposed ground, but there has not been fully substantiated evidence that the carbon dioxide levels are yet of serious global significance. In this, we are reminded that the curve disparity and the more erratic and to some extent inverse correlation noted in the last 25 years of winter temperature trend shown in Figure 18 may indeed be significant.

The number of sunspots recorded at the 1958 solar peak was the highest on record over the 235 years since 1749, when the Zurich Observatory began continuous monitoring of solar phenomena. Sunspots manifest charged particle energy, which is not impeded by a carbon dioxide blanket. This could increase the relative magnitude of solar energy when applied to the over 500-year span of the Alaskan Little Ice Age covered by our glacier records. The conclusion is more compelling that these climatic trends were initiated by solar-energy causes well before the 1880s. But if carbon dioxide pollution continues to increase at the present rate, it could alter the situation by severely changing the atmosphere's capacity to absorb and retain solar energy. The resulting Greenhouse Effect could influence the heat sink capacity of the oceans. This could also mean an upset or a slowing down of the current cooling trend suggested by the data presented in this paper.

A Pandora's Box may be opening with the implication that carbon dioxide pollution is already changing climate. This is hinted by the January temperature curve in Figures 18 and 24 when extended from the 1950s to the present. This plot does appear to abort the similarities shown with the solar curve prior to 1950. And certainly since the 1960s on this graph, the peaks of solar energy correspond to much lower than usual winter temperatures, suggesting a more southerly migration of arctic conditions during the winter months. This may relate to the short-term warming of the arctic basin in recent decades and the fluctuating diminishment of polar sea ice determined by principal component analysis (Kelly et al. 1982). It may also ally to the 5°F (2.1°C) rise in mean annual and the 20°F (9.4°C) increase in mean January temperatures in the Juneau Icefield region since 1969, as well as the 1°F (0.4°C) Northern Hemisphere temperature rise average from some 340 arctic and high-latitude stations over the past 15 years.

Larger concerns in the scientific understanding of paleoclimates and appreciation of public environmental and energy issues mandate a continuing systematic study of global climate changes (Schneider and Londer 1984). For these reasons it is essential to monitor future effects resulting from natural or man-made causes or both. Evidence to the 1960s supports the fact that until recently natural causes of climatic change have been dominant (e.g. King 1975), but now it appears that we may be on the threshold of a new and overriding man-made causal factor of staggering impact. For the immediate future, however, it can be said that with the temporary El Niño-type influences subsided and climate allowed its natural cyclic course, a chilling trend should return accompanied by increasingly long winters to the turn of the century. If such is not borne out by climatic events into the 1990s and at the latest by the year 2000, then atmospheric pollution of the carbon dioxide type will indeed have proven itself to have exercised decisive global influence.

This issue is so critical and of such scientific importance to the future as well as to our interpretations of the past that it demands continuing systematic and teleconnectional study of the glacier regions on the North Pacific Coast.

Locational Relevance and an Epilogue for the Neogene

It is recognized that the complexity of the atmosphere-ocean-landmass system on earth makes the controls of climatic oscillation truly multi-faceted. With this caveat our study has relevance in its emphasis on independent lines of inquiry and, indeed, some tantalizing indications of agreement between three of them . . . i.e. observed glacier fluctuations since the mid-1700s and solar energy variations over the same time interval via records from the Zurich Observatory, plus the trend and variation of winter temperature means in Southeastern Alaska since the 1880s.

Added to the above is the unique location of the eastern rim of the Gulf of Alaska, which by teleconnection has been shown to be climatologically similar to the northeast Atlantic coast of Norway. This has lent itself to analysis of shifting storm winds and related snowfall trends in the surface mixing zone between blocking continental highs and oceanic lows in a prototype region that has experienced dynamic changes in glacial behavior of global significance. Abetting the climatic sensitivity of the region are major tectonic forces on the continental margin that have produced linear mountain ranges, which by their high elevation and adjacency to the sea are characterized by intensive glaciation. The glacier behavior in historic time has understandably reflected acute sensitivity to minor shifts in the Arctic Front. In this the similarities in solar energy pulsations and glacier regime beg the question of solar-control as a tripping mechanism unleashing a chain reaction of other elements in the process, beginning with the periodic expansion and contraction of the polar anticyclone.

Climatic sensitivity of the Juneau Icefield and its adjoining Stikine and St. Elias Icefields along the Southeastern Alaska coast is supported by the more than 15 significant advances and retreats of key coastal glaciers in this region since the early 1700s; some 480 km inland in the dry interior ranges of British Columbia and the Yukon, over the same interval only 2 to 3 glacier oscillations have left their record. Thus fine details of glacio-climatic variation are given acute expression along the Alaska coast, to an extent not seen elsewhere except in similar high-middle latitude maritime ranges of Scandinavia, Southern Chile and South Island of New Zealand. This gives us confidence in global relevance of this long-term study and the insights it provides regarding the fundamental role of solar energy on short-term climatic oscillation. Recent calculation of time-averaged insolation patterns at middle and high latitudes of the northern hemisphere based on combined effects of earth orbit perturbations (at times of equinox and solstice) and solar irradiance since 1874 adds credence to the view that solar activity has been the dominant factor in insolation changes over the past century (Borisenkov, Tsvetkov and Eddy 1985).

As this report attempts to relate natural trends of climate now and during the Neogene, it touches on the basic causal factors in short-term climate variations prior to any significant effects of man-made atmospheric pollution. As it similarly allies to the future, it is deemed essential to maintain continuity in the climate/glacier study record of this prototype region well into the next century. Also, on the possibility that the carbon cycle may not operate in the years ahead as it has in the past (Kerr 1983a, 1983b; Schneider and Londer 1984), further research should be conducted on pre-1958, on pre-industrial and on Pleistocene values of carbon dioxide buried deep within the ice. If not feasible on the high flow rate temperate glaciers of the sub-arctic we should look to the deeper ice of glaciers in the polar regions.

Some new research of this type seems already to be pointing the way. Measurements of atmospheric carbon dioxide variations in ice core samples from the Dye 3 drill site on the Greenland Ice Sheet are known to span a period of the late Neogene from 25-40,000 years B.P. At depths of 1800 to 1900 m on this profile four notably sharp peaks of carbon dioxide have been detected by Stauffer et al. (1984; see also Campbell 1984). These also parallel oxygen isotope changes and variations in other trace chemicals. The excursion rate on these peaks has been indicated to be of unexpectedly short duration, about a century. This is not unlike the rapid present-day buildup of man-produced carbon dioxide. Regardless of the limited extent of these data, the implication is that significant variations in carbon dioxide have occurred as a result of natural causes throughout the Pleistocene. This opens up a whole new realm of unanswered questions. Much more information, therefore, is needed on biogenic and geologic causal factors, but already the conclusion seems inescapable that the carbon cycle has played a significant role in global climatic change throughout geologic time. As for specifics on this issue, however, no one can yet afford to be dogmatic.

Against the backdrop of a documented century-long buildup of man-made carbon dioxide in the global atmosphere and the speculation of possible short-term effects from volcanic dust and other particulate matter, further research should yield useful results. As for the earlier Neogene, the principles elucidated by the total-systems glacier study and the generalized model derived in support of a fundamental solar-control mechanism for natural climatic change at high middle latitudes may be helpful to those concerned with the changing paleoclimatic character and processes of the past. This includes short-term effects that climate fluctuation has had on biotic assemblages and environmental conditions that have produced ancient rythmites and varved sediments. These principles should also have validity to others looking to the future of life on earth.

ACKNOWLEDGMENTS

Appreciation is extended to the following for careful review of the data presented and for helpful suggestions, which have broadened the conceptions and the conclusions in this paper. It is not implied, however, that these reviewers have necessarily adopted the conclusions presented in what has proved to be a most comprehensive interdisciplinary study. These are my own responsibility. Dr. J. H. Anderson, Institute of Arctic Biology, University of Alaska, Fairbanks, Alaska; Dr. Bradley Colman, Forecast Center, National Weather Service, Juneau, Alaska; Dr. Vernon K. Jones, Department of Atmospheric Sciences, University of Missouri, Columbia, Missouri; Dr. Scott Morris, Department of Geography and the Glaciological and Arctic Sciences Institute, University of Idaho, Moscow, Idaho; and Dr. A. H. Thompson, Department of Meteorology, Texas A&M University, College Station, Texas.

LITERATURE CITED

Anderson, J. H. 1970. A geobotanical study in the Atlin region in northwestern British Columbia and South-Central Yukon Territory. Ph.D. Thesis. Michigan State University, East Lansing, Mich.

Anderson, R. Y. 1961. Solar-terrestrial climatic patterns in varved sediments. Ann. N. Y. Acad. Sci. 95:424-439.

Andress, E. C. 1962. Névé studies on the Juneau Icefield, Alaska, with special reference to the glacio-hydrology of the Lemon Glacier. M.S. Thesis. Michigan State University, East Lansing, Mich.

Baker, H. G. 1984. The future of plants and vegetation under human influence. Spec. Pub. No. 2, Pacific Division, Amer. Assoc. Adv. Sci., San Francisco, Calif.

Bayrock, L. A. 1967. Catastrophic advance of the Steele Glacier, Yukon, Canada. Occ. Pub. No. 3, Boreal Inst., University of Alberta.

Berger, A. L. 1980. The Milankovitch astronomical theory of paleoclimates: A modern review. Vistas Astron. 24:103-122.

Bolin, B. 1970. The carbon cycle. Sci. Amer. 223(3):124-132.

Borisenkov, Y. P., F. V. Tsvetkov, and J. A. Eddy. 1985. Combined effects of earth orbit perturbations and solar activity on terrestrial insolation. J. Atmos. Sci. 42(9):933-940.

Boucot, A. J., and J. Gray. 1982. Paleozoic data of climatological significance and their use for interpreting Silurian-Devonian climate. Pages 189-198 *in* Climate in Earth History. National Academy Press, Washington, D.C.

Bradley, W. H. 1929. The varves and climate of the Green River Epoch. U. S. Geol. Surv. Prof. Pap. 158:87-110.

Bray, J. R. 1967. Variation in atmospheric Carbon 14 activity relative to sunspot-auroral solar index. Science 156:640-642.

Bray, J. R. 1968. Glaciation and solar activity since the fifth century B.C. and the solar cycle. Nature 220:672-674.

Bray, J. R. 1970. Solar activity index: Validity supported by oxgyen isotope dating. Science 168:571-572.

Bray, J. R. 1971. Solar-climatic relationships in the post-Pleistocene. Science 171:1242-1243.

Broecker, W. S. 1984. Carbon dioxide circulation through ocean and atmosphere. Nature 308:602.

Broecker, W. S., T. Takahashi, H. J. Simpson, and T. H. Peng. 1979. Fate of fossil fuel carbon dioxide and the global carbon budget. Science 206:409-418.

Brooks, C. E. P. 1949. Climate through the ages. McGraw-Hill, New York, N. Y.

Campbell, J. P. 1984. New data upset Ice Age theories. Nature 307:688-689.

Cane, M. A. 1983. Oceanographic events during El Niño. Science 222:1189-1195.

Chiu, L. S. 1983. Variation of Antarctic sea ice: An update. Mon. Weather Rev., Vol. 111 (Mar.):578-580.

Crowell, J. C. 1982. Continental glaciation through geologic times. Pages 77-82 *in* Climate in Earth History. National Academy Press, Washington, D.C.

Dansgaard, W., S. J. Johnson, J. Moller, and C. C. Langway, Jr. 1969. One thousand centuries of climatic record from Camp Century on the Greenland ice sheet. Science 166:377-381.

Denton, G. H., and W. Karlen. 1973. Holocene climatic changes, their pattern and possible cause. Quatern. Res. 3:155-205.

Eddy, J. A. 1977. Historical evidence for the existence of the solar cycle. Pages 51-72 *in* O. R. White, ed. The Solar Output and its Variations. University of Colorado Press, Boulder, Colo.

Eddy, J. A. 1978. Evidence for a changing sun. Pages 11-33 *in* J. A. Eddy, ed. The New Solar Physics. Westview Press, Boulder, Colo.

Eddy, J. A. 1979. A new sun. The solar results from Skylab, NASA SP-402. Nat'l. Aero. & Space Admin., Washington, D.C.

Egan, C. P. 1965. Firn stratigraphy and névé regime trends on the Juneau Icefield, Alaska. 1926-65. M.S. Thesis. Michigan State University, East Lansing, Mich.

Egan, C. P. 1971. Contribution to Late Neoglacial history of the Lynn Canal and Taku Valley sectors of the Alaskan Boundary Range. Ph.D. Thesis. Michigan State University, East Lansing, Mich.

England, J., L. Kerschaw, C. LaFarge-England, and J. Bednarski. 1981. Northern Ellsmere Island: A natural resource inventory. Spec. Rep. to Parks Canada. Dep. Geography, University of Alberta, Edmonton, Can. 207 pp. + App.

Etkins, R., and E. Epstein. 1982. The rise of global mean sea level as an indication of climate change. Science 215:287-289.

Evans, J. V. 1982. The sun's influence on the earth's atmosphere and interplanetary space. Science 216:467-474.

Field, W. O. 1937. Observations on Alaskan coastal glaciers. Geogr. Rev. 27(1):63-81.

Fischer, A. 1982. Long-term climatic oscillations recorded in stratigraphy. Pages 97-104 *in* Climate in Earth History. National Academy Press, Washington, D.C.

Gani, J. 1975. The use of statistics in climatological research. Search 6:504-508.

Gilkey, A. K. 1958. Geological structures on the Camp Ten Nunatak, Juneau Icefield, Alaska. Spec. Rep. No. 4, Foundation for Glacier and Environmental Research, Seattle, Wash.

Gordon, J. E. 1980. Recent climatic trends and local glacier margin fluctuations in West Greenland. Nature 284:157-158.

Hambrey, M. J., and W. B. Harland. 1981. Earth's pre-Pleistocene glacial record. Cambridge University Press, Cambridge.

Hammer, C. U., H. B. Clausen, and W. Dansgaard. 1981. Greenland ice sheet evidence of postglacial volcanic activity and its climatic impact. Nature 228:230-235.

Hansen, J., D. Johnson, A. Lacis, S. Lebedeff, P. Lee, D. Rind, and G. Russell. 1981. Climate impact of increasing atmospheric carbon dioxide. Science 213:957-966.

Hauritz, B., and J. M. Austin. 1944. Climatology. McGraw-Hill, New York, N. Y.

Hayes, J. D., J. Imbrie, and N. J. Shackleton. 1976. Variations in the earth's orbit: Pacemaker of the Ice Ages. Science 194:1121-1132.

Herron, M. M., S. L. Herron, and C. C. Langway, Jr. 1981. Climatic signal of ice melt features in southern Greenland. Nature 293:389-391.

Heusser, C. J. 1952. Pollen profiles for Southeastern Alaska. Ecol. Monogr. 22:331-352.

Heusser, C. J. 1953. Radiocarbon dating of the thermal maximum in S. E. Alaska. Ecology 34(3):331-352.

Heusser, C. J. 1960. Late Pleistocene environments of North Pacific North America. Spec. Pub. No. 35, Amer. Geogr. Soc., New York, N. Y.

Heusser, C. J., R. L. Schuster, and A. K. Gilery. 1954. Geobotanical studies on the Taku Glacier anomaly. Geogr. Rev. 44(2):224-239.

Hoinkes, H. C. 1968. Glacier variations and weather. J. Glaciol. 7(49):3-19.

Hoinkes, H. C. 1969. Surges of the Venagtferner in the Otztal Alps since 1599. Canad. J. Earth Sci. 6(4):853-861.

Horel, J. D., and J. M. Wallace. 1981. Planetary-scale atmospheric phenomena associated with the

Southern Oscillation. Mon. Weather Rev. 109(1):813-329.
Hunt, B. G. 1984. Polar glaciation and the genesis of Ice Ages. Nature 308, March 1, 1984.
Jahresbericht der VAW-ETH Zurich. 1983. Langenanderung der Gletscher in den Schweizer Alpen 1981/82 und 1982/83. Pages 48-49, Abb. 28.
Jones, V. K. 1975. Contributions to the geomorphology and neoglacial chronology of the Cathedral Glacier system, Atlin Wilderness Park, B. C. M. S. Thesis. Michigan State University, East Lansing, Mich.
Jones, V. K. 1979. Relationship between climate and crop yields. Ph.D. Thesis. Michigan State University, East Lansing, Mich.
Kasser, P. C. 1967. Report on fluctuation of glaciers, 1960-65. Int'l. Assoc. Sci. Hydrol., UNESCO.
Kasser, P. C. 1973. Report on fluctuation of glaciers, 1965-70. Int'l. Assoc. Sci. Hydrol., UNESCO.
Kelly, P. M., P. D. Jones, C. B. Sear, B. S. G. Cherry, and R. K. Tavakol. 1982. Variations in surface air temperatures: Part 2. Arctic Regions, 1881-1980. Mon. Weather Rev., Vol. 110 (Feb.):71-83.
Kerr, R. A. 1982. U. S. weather and the equatorial connection. Science 216:608-610.
Kerr, R. A. 1983a. Carbon dioxide and a changing climate. Science 222:49.
Kerr, R. A. 1983b. The carbon cycle and climate warming. Science 222:1107-1108.
King, J. W. 1975. Sun-weather relationships. Austronaut. Aeronaut. 13:10-19.
Kukla, G., and J. Gavin. 1981. Summer ice and carbon dioxide. Science 214:497-503.
Kukla, G., and J. Gavin. 1982. Icey signs of a warmer world. Science 82 (Jan.-Feb.):10.
LaChapelle, E. R. 1954. Snow studies on the Juneau icefield. J.I.R.P. Rep. No. 9, Amer. Geogr. Soc., New York, N. Y.
Lamb, J. 1985. Windrift glacier regimen and recent climatic change on the Juneau Icefield, Alaska-Canada. M.S. Thesis. University of Idaho, Moscow, Ida.
Langway, C. C. Jr., W. Dansgaard, C. J. Johnson, and H. Clausen. 1973. Climatic fluctuations during the late Pleistocene. Pages 317-321 *in* Mem. 136, Geol. Soc. Amer.
Lawrence, D. B. 1950. Glacier fluctuations for six centuries in southeastern Alaska and its relation to solar activity. Geogr. Rev. 40(2):191-222.
Lewin, R. 1984. A thermal filter to extinction. Science 223:383-385.
Linder, G. M. 1981. Reconnaissance glacial geology of Avalanche Canyon, Juneau Icefield, S. E. Alaska. M.S. Thesis. University of Idaho, Moscow, Ida.
Maran, S. P. 1981. Solving the mystery of the sun's violent flares—satellite report. Pop. Sci. (Jun.):66-69, 120.
Marcus, M. G. 1964. Climate glacier studies in the Juneau Icefield region, Alaska. Ph.D. Thesis. University of Chicago, Chicago, Ill.
Marr, J. C. 1970. The Kuroshio. East-West Center Press, University of Hawaii, Honolulu, Haw.
McBeath, J. H., ed. 1984. The potential effects of carbon dioxide-induced climatic changes in Alaska. Misc. Pub. 83-1. School of Agriculture & Land Resources Management, University of Alaska, Fairbanks, Alsk. 208 pp.
Meier, M. F., and A. S. Post. 1962. Recent variations in mass net budgets of glaciers in Western North America. UGGI, Assn. Int'l. Hydrol. Sci., Colloque d'Obergurgl, Oct.:63-77.
Milankovitch, M. 1920. Théorie mathématique des phénomènes thermiques produits par la radiation solaire. Gauthier Villars, Paris, France.
Milankovitch, M. 1930. Mathematische Klimalebre und Astronomische Theorie der Klimashwankungen. Pages 1-176 *in* W. Koppen and R. Geiger, eds. Handbuch der Klimatologie, ICA. Gebrder Borntraeger, Berlin.
Milankovitch, M. 1938. Stronomische Mittel zur erforschung der erdgeschichtlichen klimate. Handbuch der Geophysik 9:593-698.
Milankovitch, M. 1941. Kanon der Erdbestrahlung und Seine Andwendung auf des Eiszeiten problem. Ed. Spec. Acad. Roy. Serbe, Belgrade. (English Translation, U.S. Dep. Commerce.)
Miller, M. M. 1948-54. Juneau Icefield research project. Ann. Sci. Rept., Office of Naval Res., J.I.R.P. Reps. 1-7, Amer. Geogr. Soc., New York, N.Y.
Miller, M. M. 1955. Observations on the regimen of the glaciers of Icy Bay and Yakutat Bay, Alaska, 1946-1947. Spec. Rep. No. 1, Foundation for Glacier and Environmental Research, Seattle, Wash.

Miller, M. M. 1956a. Glacial geology and glaciology of the Juneau Icefield, S.E. Alaska. U.S. Off. Naval Res. Rep., Proj. ONR-83001, 2 vols.

Miller, M. M. 1956b. Floating islands of the polar sea. Nat. Hist. 65(5):233-239, 274, 276.

Miller, M. M. 1958a. Glaciers on the rampage. Sci. World 3(8):4-7.

Miller, M. M. 1958b. The role of diastrophism in the regimen of glaciers in the St. Elias District, Alaska. J. Glaciol. 3(24):79, 292-297.

Miller, M. M. 1963a. Taku Glacier evaluation study. Dep. Highways & Bur. Public Roads, State of Alaska, and U. S. Dep. Commerce.

Miller, M. M. 1963b. The regional pattern of Alaskan glacier fluctuations. Spec. Rep. No. 5, Foundation for Glacier and Environmental Research, Seattle, Wash.

Miller, M. M. 1964. Inventory of terminal position changes in Alaskan coastal glaciers since the 1750s. Proc. Amer. Philos. Soc. 108(3):257-273.

Miller, M. M. 1967. Alaska's mighty rivers of ice. Nat'l. Geogr. 131(2):194-217.

Miller, M. M. 1969. The Alaskan glacier commemorative project, Phase I. Pages 135-152 in Nat'l. Geogr. Soc. Res. Rep., 1964 Proj. Rep.; and pages 181-194, 1965 Proj. Rep.

Miller, M. M. 1971. Glaciers and glaciology. McGraw-Hill Encycl. Sci. Technol. 6:218-229.

Miller, M. M. 1972. A principles study of factors affecting the hydrological balance of the Lemon Glacier system and adjacent sectors of the Juneau Icefield, S. E. Alaska, 1965-69. Tech. Rep. No. 33, Inst. Water Res., Michigan State University.

Miller, M. M. 1973. Alaskan glacier commemorative project, Phase II, 1966: A total systems study of climate-glacier relationships and the stress instability of ice. Nat'l. Geogr. Soc. Res. Rep., 1966 Proj.

Miller, M. M. 1974. Entropy and the self-regulation of glaciers in arctic and alpine regions. Pages 136-158 in Research in Polar and Alpine Geomorphology. Proc. Guelph Univ. Symposium on Geomorphology, Ontario, Can.

Miller, M. M. 1975a. Pleistocene erosional and stratigraphic sequences in the Alaska-Canada Boundary Range. Pages 463-492 in W. C. Mahaney, ed. Quaternary Stratigraphy of North America. Dowden, Hutchison and Ross, Halstead Press, J. Wiley and Sons.

Miller, M. M. 1975b. Mountain and glacier terrain study and related investigations in the Juneau Icefield region, Alaska-Canada. Fin. Rep., U.S. Army Res.; Off.-Durham. Grants No. DA-ARO-D-31-124-71-G120; 72-G193; 73-G185. Foundation for Glacier and Environmental Research, Seattle, Wash. 250 pp.

Miller, M. M. 1976. Thermophysical characteristics of glaciers—toward a rational classification. J. Glaciol. 16(74):297-300.

Miller, M. M., and J. H. Anderson. 1974a. Alaskan glacier commemorative project, Phase IV: Pleistocene-Holocene sequences in the Alaska-Canada Boundary Range. Nat'l. Geogr. Soc. Res. Rep., 1967 Proj.

Miller, M. M., and J. H. Anderson. 1974b. Out-of-phase Holocene climatic trends in the maritime and continental sectors of the Alaska-Canada Boundary Range. Pages 33-58 in Quaternary Environments. Geogr. Monogr. No. 5, York University.

Miller, M. M., and W. O. Field. 1951. Exploring the Juneau Ice Cap. Res. Rev. Off. Naval Res. Dep. Navy NAVEXOS P-510, April.

Mitchell, J. M., Jr. 1963. On the world-wide pattern of secular temperature change. Pages 161-181 in UNESCO Arid Zone Research, Vol. 20.

Mitchell, J. M., Jr. 1965. The solar constant. Pages 155-174 in Proceedings, Seminar on Possible Responses of Weather Phenomena to Variable Extra-Terrestrial Influences. NCAR Tech. Note TN-8, Nat'l. Center Atmos. Res., Boulder, Colo.

Mitchell, J. M., Jr. 1976. An overview of climatic variability and its causal mechanisms. Quatern. Res. 6:481-493.

Mitchell, J. M., Jr. 1982. El Chichon—Weather-maker of the century. Weatherwise (Dec.):252-259.

Mitchell, J. M. Jr., C. W. Stockton, and D. M. Mekko. 1979. Evidence of a 22-year rhythm of drought in the western United States related to the Hale solar cycle since the 17th century. Pages 125-143 in Solar-Terrestrial Influences on Weather and Climate. B. C. McCormac and D. Reidel, Boston, Mass.

Mosteller, F. 1973. Periodicities and moving averages. Pages 113-120 *in* F. Mosteller, W. H. Kruskal, R. F. Link, R. S. Pieters, and G. R. Reising. Statistics by Examples: Weighting Chances. Addison-Wesley Pub. Co., Reading, Mass.

Namias, J. 1969. Seasonal interactions between the North Pacific ocean and the atmosphere during the 1960s. Mon. Weather Rev. 97(3):173-192.

National Oceanic Atmosphere Administration. 1983. Global sea temperature data. U. S. Dep. Commercial Fisheries.

National Research Council. 1982a. Climate in earth history. National Academy Press, Washington, D. C.

National Research Council. 1982b. Solar variability, weather and climate. National Academy Press, Washington, D.C.

National Research Council. 1982c. Carbon dioxide and climate—A second assessment. Rep. CO_2/Climate Rev. Panel, Nat'l. Res. Council. National Academy Press, Washington, D.C.

National Research Council. 1983. Changing climate. Rep. Carbon Dioxide Assessment Comm. Nat'l. Res. Council. National Academy Press, Washington, D.C.

Nelson, F. E. 1978. Patterned ground in the Juneau Icefield region, Alaska-British Columbia. M.S. Thesis. Michigan State University, East Lansing, Mich.

Newell, R. E. 1971. The global circulation of atmospheric pollutants. Sci. Amer. 224(1):32-42.

Nichols, R. L., and M. M. Miller. 1951. Glacial geology of the Ameghino Valley, Lago Argentino, Patagonia. Geogr. Rev. 41(2):274-294.

Nichols, Robert L., and M. M. Miller. 1952. The Moreno Glacier, Lago Argentino, Patagonia— Advancing glaciers and nearby simultaneous retreating glaciers. J. Glaciol. 2(11):41-50.

Nye, J. F. 1958. Surges in glaciers. Nature 181:1450-1451.

Ostrem, G., N. Haakensen, O. Liestol, S. Messel, and A. Tvede. 1980-82. Glasiologiske Under-Sokelser i. Norge. Rapports NRS 1-80, 1-81, 1-82. Vassdragsdirektoratet Hydrologisk Avdeling, Oslo.

Pittock, A. B., and R. Shapiro. 1982. Assessment of evidence of the effect of solar variations on weather and climate. Pages 64-75 *in* Solar Variability, Weather and Climate. National Academy Press, Washington, D.C.

Plass, G. N. 1959. Carbon dioxide and climate. Sci. Amer. 201(1):41-47.

Post, A. S. 1969. Distribution of surging glaciers in Western North America. J. Glaciol. 8(53): 229-240.

Price, J. 1983. Mass balance and hydrological trends, Cathedral Glacier, Atlin Provincial Park, Northern British Columbia, Canada. M.S. Thesis. University of Arizona, Tucson, Ariz.

Purrett, L. 1970. Ice cores: Clues to past climates. Science News 98(19):359-374.

Rasmussen, E. M., and J. M. Wallace. 1983. Meteorological aspects of the El Niño/Southern oscillation. Science 222:1195-1202.

Roberts, W. O. 1975. Relationship between solar activity and climate change. *In* W. R. Bandeen and S. P. Maran, eds. Possible Relationships between Solar Activity and Meteorological Phenomena. NASA SP-366, Nat'l. Aeron. Space Admin., Washington, D.C.

Roberts, W. O. 1976. Sun-earth relationships and the extended forecast problem. Pages 371-386 *in* Science, Technology and the Modern Navy. 30th Ann. Rep. (ONR-37), Off. Naval Res.

Schneider, S. H., and R. Londer. 1984. The coevolution of climate and life. Sierra Club Books, San Francisco, Calif.

Searby, H. W. 1969. Coastal weather and marine data summary for Gulf of Alaska, Cape Spencer westward to Kodiak Island. ESSA Tech. Mem. EDSTM8, pp. 18-23.

Siegenthaler, U., and T. Wenk. 1984. Rapid CO_2 variations and ocean circulation. Nature 308 (Apr.):624.

Slupetzky, H. 1980-83. Personal communication on International Hydrological Decade Glacier Research for Austria. Austrian National Committee.

Smiley, C. J. 1972. Plant megafossil sequences, North Slope Cretaceous. Geosci. & Man 4(Oct.): 91-99.

Stauffer, B., H. Hofer, H. Oeschger, J. Schwander, and U. Siegenthaler. 1984. Atmospheric CO_2 concentration during the last glaciation. Ann. Glaciol. 5:160-164.

Swanston, D. N. 1967. Geology and slope failure in Maybeso Valley, Prince of Wales Island, S. E. Alaska. Ph.D. Thesis. Michigan State University, East Lansing, Mich.

Thompson, A. H. 1972. Continentality across the Juneau Icefield in stormy and fair weather. Abstract in Arctic and Mountain Environments, Proc. Symposium at Michigan State University, April 1972. Foundation for Glacier and Environmental Research, Seattle, Wash.

Thorarinsson, S. 1940. Present glacier shrinkage and eustatic changes of sea-level. Geogr. Ann. 22:131-154.

Trent, D. D. 1983. The Palisade Glacier, Inyo County, California. Calif. Geol. (Dec.):264-269.

Tucker, G. B. 1964. Solar influences on the weather. Weather 16:302-311.

Webster, P. J. 1982. Seasonality in the local and remote atmosphere response to sea surface temperature anomalies. J. Atmos. Sci. 39(1):41-52.

Wegener, A. 1924. The origin of continents and oceans. Translated by J. G. A. Skerl. Methuen, London.

Willett, H. C. 1950. Temperature trends of the past century. Cent. Proc. Royal Meteorol. Soc.: 195-206.

Willett, H. C. 1975. Do recent climatic trends portend an imminent Ice Age? *In* Proc. Symposium on Atmospheric Quality and Climatic Change. University of North Carolina, Chapel Hill, N. C. 46 pp.

Williams, G. A., and M. M. Miller. 1979-82. NSF Undergraduate Research Participation Program, Juneau Icefield, Alaska: Annual Reports to National Science Foundation. Foundation for Glacier and Environmental Research, Seattle, Wash.

Williams, G. E. 1981. Sunspot periods in the late Precambrian glacial climate and solar-planetary relations. Nature 291:624-628.

Wilson, C. R. 1959. Surface movement and its relationship to the average annual hydrological budget of Lemon Creek Glacier, Alaska. J. Glaciol. 3(25):355-361.

Wittmann, W. I., and M. M. Miller. 1972. U.S. Navy Oceanographic Office sea ice surveillance in the Arctic Ocean and Greenland waters. Arctic and Mountain Environments Symposium, Michigan State University. Foundation for Glacier and Environmental Research, Seattle, Wash.

Wolfe, J. A., and R. Z. Poore. 1982. Tertiary marine and nonmarine climatic trends. Pages 154-158 *in* Climate and Earth History. National Academy Press, Washington, D.C.

Zenone, C. R. 1972. Glacio-hydrological parameters of the mass balance of Lemon Glacier, Juneau Icefield, Alaska, 1965-67. M.S. Thesis, Michigan State University, East Lansing, Mich.

OVERVIEW

CHARLES J. SMILEY
Department of Geology, University of Idaho, Moscow, ID 83843

The discovery of the Miocene Clarkia fossil beds at locality P-33, by Mr. Francis Kienbaum of Clarkia, Idaho, occurred during bulldozer construction of a snowmobile race track in 1972. Over the following years, various specialists were invited to work on different aspects of Clarkia research, both paleobotanic and paleozoologic, to initiate an interdisciplinary approach in the early investigation of a new Miocene fossil area. In 1979 these studies were organized as a symposium for the 60th Annual Meeting of the Pacific Division of the American Association for the Advancement of Science at the University of Idaho. The present volume is a compilation of reports on subjects discussed at these meetings.

The physical setting of Miocene Clarkia Lake is described in the first paper. This shows a long narrow lake in a confined drainage basin, surrounded by rugged hill-and-valley topography, that was developed on a Precambrian metamorphic (mainly mica schist) basement. The local drainage basin, apparently established in Paleogene time, was dammed by early Neogene lava, which backed up lake water for a distance of 35-40 kilometers in the valley of the main stream and the lower courses of its major tributaries on the west. The lake deposits of micaceous clays appear to have been derived from a residual soil that formed on the surface of the metamorphic basement, supplemented by periodic deposition of volcanic ash. A regional warm-temperate and humid climate, with short, mild winters, is inferred from rich fossil evidence of a taxonomically diverse and luxuriant vegetation on surrounding land and from the ecological requirements of limnetic organisms.

The second part of the book contains reports on various organisms of the limnetic realm, including diatoms, chrysophyte algae, thecamoebians, ?dinoflagellates, sponges, aquatic insect larvae, and fish. Most of the information on the fossiliferous Clarkia Lake deposits has been obtained from the 10-meter-thick exposed section at site P-33. At the base of this section is a 1-meter-thick churned turbidite unit, which may have resulted from the abrupt initiation of lacustrine conditions in a relatively stable area of diverse topography. In the middle of the section occurs 7.6 meters of unoxidized and finely laminated clays and discrete volcanic ash beds. At the top of the section is about 1.5 meters of oxidized and poorly laminated silty clays. The laminated portion of the section contains varve-like layers that are organic-rich, with common interbeds of thicker clay layers that are organic-poor. Virtually all layers, including the volcanic ash beds, show the same type of internal gradation (upward fining); a few of the thicker (organic-poor) clay layers contain internal churned structures of lacustrine turbidites. Volcanic ash beds 10 cm or more thick contain several sequential episodes of upward fining, probably resulting from initial airfall deposition followed by slope wash of ash that was deposited on surrounding land. Such sedimentary evidence, supplemented by taphonomy of fossil remains and by inferences on climatic conditions, suggests rapid rates of sedimentary infilling at the P-33 site, apparently resulting from frequent storms and slope wash in conjunction with accumulations of thick volcanic ash deposits. The duration of lacustrine conditions here has been estimated at less than 1000 years.

The unoxidized condition of the P-33 sediments and fossil remains, the presence of pyrite associated with fossil material, the absence of obligatory benthonic organisms (except for a few molluscs near the top), the lack of bioturbation of finely laminated and organic-rich lake deposits (except near the top), and the absence of benthonic trace fossils, indicate anoxic and probably

Copyright ©1985, Pacific Division, AAAS.

toxic conditions of the lake bottom. The burial of undisturbed fish carcasses, and the essentially unaltered condition of leaves derived from plants on surrounding land, suggest the absence of aerobic benthonic scavengers. Water depth at the P-33 depositional site is inferred to have been reduced by rapid sedimentation from an initial depth of 23-25 meters to 5-6 meters for the top of the laminated deposits. Evidence from study of fish taphonomy indicates depths ranging from about 8 to 12 meters for the two fish zones in the P-33 section, with bottom temperatures not higher than 15° C.

Apparent ecological requirements of the limnetic organisms suggest that the oxygenated surface and lake-border water was of near-neutral pH, was low in dissolved chemicals from the surrounding land, and was warm (perhaps in the range of 25-30°C) during much of the year. The forest association of land plants, associated fungi, and fish fauna indicate mild winters that would have precluded seasonal ice-closure of the lake; the high humidity and frequent storms throughout the year would have resulted in continual recharge of fresh water and low evaporation rates during the warm season.

From available evidence of sedimentology, taxonomy of floral and faunal remains, and ecological requirements of aerobic organisms in comparison to lake-bottom conditions required for their state of preservation in the P-33 section, one can infer that the small bay at site P-33 was stratified on the bases of (1) oxygen, (2) temperature, (3) toxicity, and (4) chemistry. This suggests a lake that was more or less permanently stratified and experienced little or no appreciable overturning—that is, a lake of the meromictic type.

The third part of the volume involves the terrestrial realm surrounding the Miocene lake. The allochthonous fossils include a diversity of plants and insects, commonly preserved as intact structures retaining original pigmentations and showing no evidence of abrasion by surface transport from land to depositional site. Plant megafossils and palynomorphs were derived from a diversity of habitats within the confined drainage basin. A diverse fungal flora indicates perpetually humid conditions of a climate not cooler than warm-temperate, which is consistent with climatic inferences from other paleontologic data.

Detailed stratigraphic analyses of plant megafossils and palynomorphs through the 7.6 meters of the P-33 lacustrine section show a stability of forest vegetation (and climate) during the short life span of Miocene Clarkia Lake. This was a taxonomically diversified mixed mesophytic forest similar to those now occupying the hill-and-valley topography of the southern foothills of the Appalachian Mountains, admixed with taxa now restricted to forests of similar climatic/topographic setting in eastern Asia. In the Miocene Clarkia basin, the moist and swampy bottomland forests were dominated by a cypress-tupelo (*Taxodium-Nyssa*) forest of eastern United States affinity; slope forests were dominated by taxa which now characterize the east Asian vegetation in and around the *Metasequoia* refugia; and the cooler uplands bordering the Miocene Clarkia basin were occupied by conifer forests of western American affinities. A temporal analysis of plant megafossils upward through the P-33 section shows a progressive change in the influx of plant organs derived from vegetation of different ecologic requirements: (1) an early dominance of riparian and floodplain taxa at the initiation of lacustrine conditions, (2) a dominance in the middle part of the section of taxa indicative of mesic but well-drained slopes, and (3) a dominance of taxa of the *Taxodium-Nyssa* swamp association during late lacustrine stages at the top of the column. The combined floral evidence is inferred to reflect a regional vegetation that remained essentially stable during the short time span of Miocene Clarkia Lake, but a lake-margin vegetation that changed markedly as the lake silted in and swampy borderlands expanded.

Insects are preserved as somatic parts, as aquatic larval cases constructed by caddisflies, as egg galls on various plant structures, and as mining of leaf tissue by larvae and adults. Many of the insects occur today as inhabitants of the ground litter on forest floors, or they require the leaves of living plants as a basic food source for their larvae. The insect fauna seems closely allied to modern faunas of southeastern United States, as does the Miocene forest association with which they occur in the Clarkia basin. The common presence of forest floor insects in the P-33 lacustrine deposits suggests that the humid and stormy climate inferred from other data resulted in sheet-flooding that rapidly dislodged and transported such forest-floor denizens from bordering slopes to the lacustrine depositional site.

The fourth part of the volume includes papers dealing with broader aspects of the late Cenozoic history of northwestern North America. It includes an analysis of Tertiary drainage systems based on the paleogeographic distribution of freshwater molluscs; the analysis of freshwater sponges from the Miocene Oviatt Creek basin of northern Idaho, representing a basin of somewhat younger Neogene deposits than those of the Clarkia basin just 30 kilometers to the north, and with sponge taxa common to both; a later Neogene palynoflora from southern Idaho, showing the late Cenozoic development of steppe vegetation in the interior as the Cascade Range became an effective climatic barrier between the western interior and the Pacific Ocean; and an analysis of Holocene climates and climatic fluctuations based on studies of modern glaciers in the southwestern panhandle of Alaska. This last part of the book is designed to set a geographic and historical perspective of the early Neogene Clarkia Lake of northern Idaho, in relation to (1) regional drainage patterns, (2) regional topographic development, (3) regional climatic changes, and (4) changes in regional biota, especially vegetation.

Clarkia investigations that have been conducted to date, as exemplified by the diverse reports in this volume, are significant not only for their own merits but also as indications of the future scientific potential of this new fossil area. The scope and precision of many of these studies can be accomplished at no other known medial Tertiary site in North America. The editor assumes full responsibility for this summary of new information, and for the abbreviated integration of various inferences which can be found in each of the separate chapters.

Charles J. Smiley